Series Editor: Rhodes W. Fairbridge
Columbia University

Published Volumes and Volumes in Preparation

ENVIRONMENTAL GEOMORPHOLOGY AND LANDSCAPE CONSERVATION, VOLUME I: Prior to 1900 / Donald R. Coates
RIVER MORPHOLOGY / Stanley A. Schumm
SPITS AND BARS / Maurice L. Schwartz
TEKTITES / Virgil E. and Mildred A. Barnes
GEOCHRONOLOGY: Radiometric Dating of Rocks and Minerals / C. T. Harper
SLOPE MORPHOLOGY / Stanley A. Schumm and M. Paul Mosley
MARINE EVAPORITES: Origin, Diagenesis, and Geochemistry / Douglas W. Kirkland and Robert Evans
ENVIRONMENTAL GEOMORPHOLOGY AND LANDSCAPE CONSERVATION, VOLUME III: Non-Urban / Donald R. Coates
BARRIER ISLANDS / Maurice L. Schwartz
GLACIAL ISOSTASY / John T. Andrews
GEOCHEMISTRY OF GERMANIUM / Jon N. Weber
WEATHERING AND SAPROLITE / Dan Yaalon
EARTHQUAKES: Geological Effects / Clarence R. Allen
ROCK CLEAVAGE / Dennis S. Wood
ENVIRONMENTAL GEOMORPHOLOGY AND LANDSCAPE CONSERVATION, VOLUME II: Urban / Donald R. Coates
VERTEBRATE PALEONTOLOGY / J. T. Gregory
PALEOMAGNETISM / Kenneth Creer
PALEOECOLOGY / John Chronic, Jr.
SOLID STATE GEOPHYSICS / Thomas J. Ahrens
OCEANOGRAPHY / Takashi Ichiye
PERMAFROST / J. D. Ives
PLAYAS AND CLAY-PANS / James L. Neal
GEOCHEMISTRY OF SULFUR / Isaac Kaplan
MODERN CARBONATES / Gerald Friedman and Roy D. Nurmi
PHILOSOPHY OF GEOLOGY / Claude C. Albritton

A *BENCHMARK* TM Books Series

BARRIER ISLANDS

Edited by
MAURICE L. SCHWARTZ
Western Washington State College

Stroudsburg, Pennsylvania

Benchmark Papers in Geology, Volume 9
Library of Congress Catalog Card Number: 73–12838
ISBN: 0–87933–050–3

Library of Congress Cataloging in Publication Data

Schwartz, Maurice L comp.
Barrier islands.

(Benchmark papers in geology, v. 9)
Includes bibliographies.
1. Barrier islands--Addresses, essays, lectures.
2. Coasts--Addresses, essays, lectures. I. Title.
GB471.S38 551.4'2 73-12838
ISBN 0-87933-050-3

Manufactured in the United States of America.

Exclusive distributor outside the United States and Canada: John Wiley & Sons, Inc.

Acknowledgments and Permissions

ACKNOWLEDGMENTS

American Association for the Advancement of Science—*Science*
"Barrier Island, Not 'Offshore' Bar"

American Geophysical Union—*Oceanology*
"Reasons for the World-Wide Occurrence of Barrier Beaches"

Department of Geology, Florida State University—*Coastal Research Notes*
"On the Cause of Present-Day Erosion of Barrier Bars"

National Academy of Science—*Quaternary Geology and Climate*
"Flandrean Transgression and the Genesis of Barrier Bars"

PERMISSIONS

The following papers have been reprinted with the permission of the authors and the copyright owners.

American Association for the Advancement of Science—*Science*
"Barrier Dune System Along the Outer Banks of North Carolina: a Reappraisal"

American Association of Petroleum Geologists
American Association of Petroleum Geologists Bulletin
"Criteria for Recognizing Ancient Barrier Coastlines"
"Chernier Versus Barrier, Genetic and Stratigraphic Distinction"

Recent Sediments, Northwest Gulf of Mexico
"Gulf Coast Barriers"

American Geographical Society—*The Geographical Review*
"Origin of the Sea Islands of Southeastern United States"

Koninklijk Instituut van Ingenieurs—*De Ingenieur*
"Some New Exploration Results About Sand Shores Development During the Sea Transgression"

Elsevier Publishing Company—*Sedimentary Geology*
"Sediment Budget Along a Barrier Island Chain"

Field Naturalists Club of Victoria—*Victorian Naturalist*
"Evolution of Shoreline Barriers"

Geological Society of America
Geological Society of America Annual Meeting, Abstracts
"Field and Laboratory Observations on the Genesis of Barrier Islands"

Geological Society of America Bulletin
"Influence of Island Migration on Barrier-Island Sedimentation"
"Barrier Island Formation"
"Barrier Island Formation: Discussion"

"Barrier Island Formation: Reply"
"Barrier Island Formation: Discussion"
"Barrier Island Formation: Reply"
"Development and Migration of Barrier Islands, Northern Gulf of Mexico"
"Holocene Evolution of a Portion of the North Carolina Coast"
"Development and Migration of Barrier Islands, Northern Gulf of Mexico: Discussion"
"Development and Migration of Barrier Islands, Northern Gulf of Mexico: Reply"
"Holocene Evolution of a Portion of the North Carolina Coast: Discussion"
"Holocene Evolution of a Portion of the North Carolina Coast: Reply"
"Holocene Evolution of a Portion of the North Carolina Coast: Discussion"
"Holocene Evolution of a Portion of the North Carolina Coast: Reply"

Gulf Coast Association of Geological Societies—*Gulf Coast Association of Geological Societies Transactions*
"Origin and Development of the Texas Shoreline"
"Bar and Barrier Island Sands"

Rijks Geologische Dienst—*Mededelingen van de Geologische Stichting*
"Coastal Barrier Deposits in South- and North-Holland"

Department of Geology, Tulane University—*Tulane Studies in Geology*
"Environments of Deposition on an Offshore Barrier Sand Bar, Moriches Inlet, Long Island, New York"

University of Chicago Press—*Journal of Geology*
"Tidal Inlets and Washover Fans"
"Submergence Effects on a Rhode Island Barrier and Lagoon and Inferences on Migration of Barriers"
"The Multiple Causality of Barrier Islands"

Series Editor's Preface

The philosophy behind the "Benchmark Papers in Geology" is one of collection, sifting, and rediffusion. Scientific literature today is so vast, so dispersed, and, in the case of old papers, so inaccessible for readers not in the immediate neighborhood of major libraries that much valuable information has been ignored by default. It has become just so difficult, or so time consuming, to search out the key papers in any basic area of research that one can hardly blame a busy man for skimping on some of his "homework."

This series of volumes has been devised, therefore, to make a practical contribution to this critical problem. The geologist, perhaps even more than any other scientist, often suffers from twin difficulties—isolation from central library resources and immensely diffused sources of material. New colleges and industrial libraries simply cannot afford to purchase complete runs of all the world's earth science literature. Specialists simply cannot locate reprints or copies of all their principal reference materials. So it is that we are now making a concerted effort to gather into single volumes the critical material needed to reconstruct the background of any and every major topic of our discipline.

We are interpreting "Geology" in its broadest sense: the fundamental science of the Planet Earth, its materials, its history, and its dynamics. Because of training and experience in "earthy" materials, we also take in astrogeology, the corresponding aspect of the planetary sciences. Besides the classical core disciplines such as mineralogy, petrology, structure, geomorphology, paleontology, and stratigraphy, we embrace the newer fields of geophysics and geochemistry, applied also to oceanography, geochronology, and paleoecology. We recognize the work of the mining geologists, the petroleum geologists, the hydrologists, the engineering and environmental geologists. Each specialist needs his working library. We are endeavoring to make his task a little easier.

Each volume in the series contains an Introduction prepared by a specialist (the volume editor)—a "state of the art" opening or a summary of the objects and content of the volume. The articles, usually some thirty to fifty reproduced either in their entirety or in significant extracts, are selected in an attempt to cover the field, from the key papers of the last century to fairly recent work. Where the original works are in foreign languages, we have endeavored to locate or commission translations. Geologists, because of their global subject, are often acutely aware of the oneness of our world. The selections cannot, therefore, be restricted to any one country, and whenever possible an attempt is made to scan the world literature.

To each article, or group of kindred articles, some sort of "Highlight Commentary" is usually supplied by the volume editor. This should serve to bring that article into historical perspective and to emphasize its particular role in the growth of the field. References, or citations, wherever possible, will be reproduced in their entirety—for by this means the observant reader can assess the background material available to that particular author, or, if he wishes, he too can double check the earlier sources.

A "benchmark," in surveyor's terminology, is an established point on the ground, recorded on our maps. It is usually anything that is a vantage point, from a modest hill to a mountain peak. From the historical viewpoint, these benchmarks are the bricks of our scientific edifice.

Rhodes W. Fairbridge

Contents

219. 249.

This volume is dedicated to
John Hoyt
whose tragic and untimely death is mourned
by all in the coastal-studies fraternity

Contents by Author

Introduction

Barrier-lagoon coasts comprise from 10 to 13 percent of the world's continental coastline. Ever-growing pressure upon, and concern for, the preservation of this valuable coastal environment has led to shoreline protection legislation in the United States and abroad. Conservation, however, demands not only knowledge of what needs to be done, but also requires that basic processes be fully understood first. Considering their fragile nature and intense occupancy by man, it is no wonder that these coastal features have long been the object of study by geomorphologists around the world. The following collection of papers traces these investigations over a period of more than 125 years.

Broadly defined, barrier-lagoon coasts are linear, detrital, present-day coastal features, rising less than 10 meters above sea level, backed at some time by a lagoon (see Paper 38 by Cromwell). Although this volume is devoted primarily to barrier islands, papers on barriers are included since the two are genetically related and the terms are often used ambiguously. Furthermore, such features as inlets, washover fans, and cheniers are considered as they relate to barrier islands. For the serious coastal student, two recent collected works are highly recommended as supplemental reading to this volume. The first is *Coastal Geomorphology* (D. R. Coates, ed., Binghamton, N.Y., Publications in Geomorphology, 404 pp., 1973). This fine proceedings volume of the Third Annual Geomorphology Symposium, State University of New York at Binghamton, contains several very good papers on various aspects of barrier island evolution published too late for inclusion here. The other collection is *Spits and Bars* (M. L. Schwartz, ed., published in this Benchmark Papers in Geology series). The papers to be read there cover the development of the spit and bar literature, an obvious precursor to the study of barrier islands.

Historically, the debate over barrier island development has run full circle. In 1845 Elie de Beaumont proposed bar emergence as the mechanism responsible for these coastal features. G. K. Gilbert, on the other hand, advocated shore drift building spitlike structures. Opposed to these ideas, W. J. McGee believed that ridge engulfment was the answer. After the turn of the century, D. W. Johnson reviewed these three hypotheses and, based on comparison of profiles, found bar development in place the most tenable. It was not until the late 1960s that J. H. Hoyt's ridge submergence paper rekindled the debate. In quick succession, J. J. Fisher advocated spit breaching and E. G. Otvos championed bar emergence. These were followed shortly by a paper reflecting my own opinion that there was a multiple causality of barrier islands. The reader, however, should follow the debates from early times to the present and decide for himself where the preponderance of evidence lies.

Apropos of an editor having a publication of his own in a collected volume, I vowed early in the present undertaking never to let personal prejudice influence objective commentary, or as the captain of the H.M.S. Pinafore hesitatingly put it, "well, hardly ever."

It is hoped that the editorial commentary before each paper will lead the reader through the maze of claims and counterclaims by extracting the essence of each of the writings. However, the introductions are no substitute for a careful, thorough, and rewarding reading of the papers. The biographical sketches of the authors are based, for the most part, on material furnished by the authors themselves. If you are disturbed by the various ways in which the names Leontiev and Zenkovich are spelled (Leontyev, Leont'yev, Zenkovitch), please be advised that the spellings used in the commentary are those which the authors employ in their correspondence with the editor.

Particular thanks for assistance in the preparation of this volume is extended to Joan Roley, Linda Wilcox, and the helpful staffs of the Wilson Library, Western Washington State College in Bellingham, and the several libraries at the University of Washington in Seattle. My personal gratitude is offered to Eric Bird for providing the original paper included.

Editor's Comments on Paper 1

1 de Beaumont: *Septième leçon*

Hardly a scholarly work on barrier islands has been written without mentioning this pioneering effort by Elie de Beaumont. The seventh lesson of his Leçons de géologie practique, reproduced here, contains one of the earliest qualified statements on the mechanism of barrier island construction.

In this section de Beaumont is concerned with the banks of sand or gravel that waves accumulate at the edge of the sea. First he discusses sediment transport by waves and the accompanying size sorting and reach that is dependent on wave energy. Later in the discussion, he establishes his hypothesis of barrier construction, which may be paraphrased as follows: wave action on a shallow bottom removes sediment and piles it up to form a bank or barrier that parallels the original shoreline, thus establishing a balanced profile above and below mean sea level. The reader will recognize this as a local, or emergent, bar origin, trending toward a profile of equilibrium.

De Beaumont then proceeds to give a series of examples of barrier chains around the world. In both words and diagrams, he cites coastal barriers bordering locales in and around Europe, North Africa, and North and Central America, with special emphasis on the North and Baltic Seas and the Gulf of Mexico. Quite an erudite treatment for his time! In his closing paragraph, de Beaumont states that the next lesson, or chapter, will be concerned with the protected area behind the barriers, but that he had not wanted to enter upon that discussion without first giving the reader an idea as to the mechanism by which the sea fashions those banks.

Jean Baptiste Armand Louis Léonce Elie de Beaumont was born in Canon, Caen, France in 1798. He studied at L'Ecole Polytechnique in Paris, became Chief Engineer of Mines in 1824, was named Professor of Geology at the L'Ecole des Mines in 1827, and Inspector General in 1847. In 1835 he became a member of the French Academy of Science; in 1844, Vice President; in 1845, President; and in 1853, Permanent Secretary. De Beaumont was a pioneer in the field of geology, authored a number of books on the subject, and aided in the preparation of a geologic map of France that was the basis for all later geologic work in that country. He died in 1874 after a prolific life in the service of his profession.

Reprinted from *Leçons de géologie practique,* P. Bertrand, Paris, 221–252 (1845)

1

SEPTIÈME LEÇON.

Le 6 janvier 1844.

Levées de sable et de galet.

Messieurs,

Je continuerai dans cette séance à vous parler des bords de la mer et des matières que les vagues accumulent sur les côtes. Ces matières sont généralement des sables, des coquilles et des galets ou cailloux roulés. L'action prolongée de la mer en produit des amoncellements qui ont une très-grande influence sur les formes des rivages et sur les phénomènes qui se passent le long des côtes.

Lorsqu'en approchant du rivage, la profondeur devient assez petite pour que le mouvement des vagues commence à être gêné, les molécules d'eau ne pouvant continuer à transmettre leur vitesse dans l'intérieur du fluide, l'agitation se concentre près de la surface; les vagues s'élèvent davantage, et finissent même par se déchirer, par se *briser,* en s'élançant plus haut sur le rivage qu'elles ne le font au large.

Mouvement des vagues sur la plage.

Il résulte de ce phénomène un mouvement assez compliqué des eaux sur la plage; le résultat de ce mouvement est, que la mer rejette de son sein une

Elles entassent les corps détachés en forme de bourrelet.

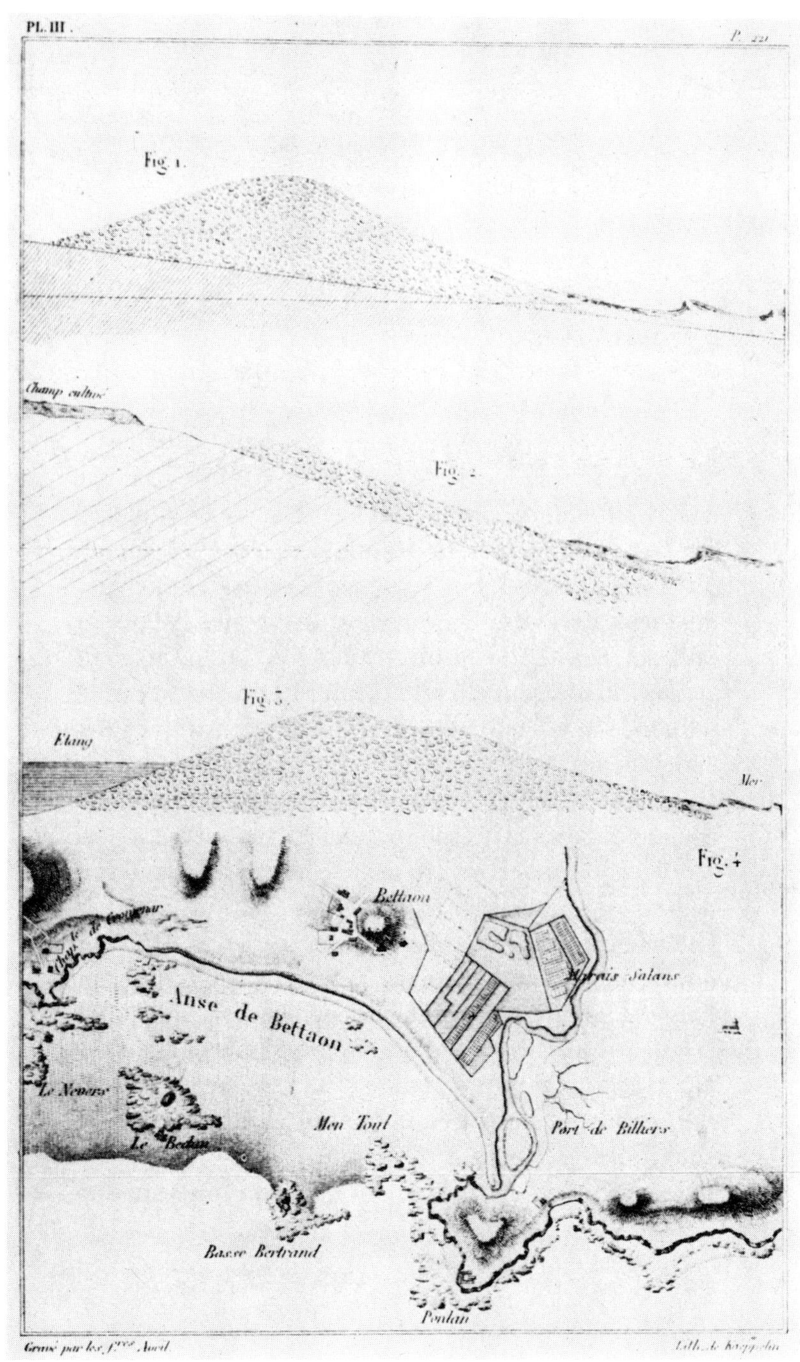
PL. III.
Fig. 1.
Champ cultivé
Fig. 2
Fig. 3.
Etang
Mer
Fig. 4
Bettaon
Marais Salans
Anse de Bettaon
Le Nevers
Le Bedan
Men Toul
Port de Billiers
Basse Bertrand
Poulan

certaine quantité de matières qui forment une espèce de talus, de levée présentant le profil, qui convient mieux au mouvement des eaux.

Par la manière même dont les vagues brisent, la mer aurait, sur un rivage presque plat, plus de tendance à rejeter vers l'extérieur les objets qu'elle agite, qu'à les ramener vers l'intérieur; mais elle donne à son fond, près du bord, une inclinaison qui ne devient fixe que lorsque les efforts opposés s'y compensent, et sur laquelle, par conséquent, elle n'a pas plus de tendance à rejeter vers l'extérieur qu'à ramener vers l'intérieur. L'inclinaison du talus, ainsi formé, va en aûgmentant jusqu'à la partie supérieure, ainsi que le montre la fig. 1, pl. III. La mer a plus de tendance à rejeter les grosses particules que les petites; elle repousse donc d'abord les gros galets, puis les petits, et enfin le sable.

Quand il n'y a pas de galets, la mer entasse des levées de sable, et lorsque ce sable est fin, il donne naissance à des dunes.

Plages de sable. Les plages de sable qui se trouvent au-dessous de la partie inférieure des digues de galets, sont quelquefois très-étendues, d'autres fois elles le sont très-peu : cela dépend de la quantité de sable que la mer charrie. La mer apporte et remporte ce sable; elle en remanie sans cesse la surface, et la disposition qu'elle lui donne dépend en partie de son abondance. Quelquefois on trouve des rochers plats à nu qui se montrent jusqu'à une hauteur assez grande au-dessus de la basse mer, de sorte qu'il n'y a qu'une plage de sable extrêmement étroite; d'autres fois il existe au pied de la levée de galets une plage de sable qui va jusqu'à la limite de la basse mer, ou qui s'étend même au delà. Quelquefois on n'en connaît pas les bornes, et la sonde rapporte du sable jusqu'à une très-grande distance de la côte. Lorsque le sable s'étend au loin, son inclinaison devient très-faible : elle n'est souvent que d'une fraction de degré, et dans certains cas elle finit même par devenir presque insensible.

Cordon littoral.

Le bourrelet de matières meubles que la mer élève sur ses bords, comme pour clore son domaine, pourrait être désigné assez convenablement sous le nom de *cordon littoral*. En y joignant les dunes, auxquelles le cordon littoral donne naissance lorsqu'il est formé de sable fin, non argileux, on pourrait appeler le tout *l'appareil littoral*.

Appareil littoral.

Une portion très-notable de la configuration des côtes est due à l'accumulation et au transport des galets et des sables, opéré par le mécanisme que je viens d'indiquer.

Quelquefois le cordon littoral s'applique sur les roches mêmes qui forment la côte. Dans l'île de Molène, située à l'extrémité de la Bretagne, entre la terre ferme et l'île d'Ouessant, j'ai vu, comme l'indique la figure 2, planche III, l'entassement des galets, appliqué simplement contre une portion d'un talus composé de rochers, contre lesquels la mer brise. Il y avait dans la partie supérieure du banc des galets très-gros, et au-dessous, des galets plus petits : le talus, dont l'inclinaison était seulement de 18 degrés, se prolongeait sous la mer. Le plus souvent de pareils talus se terminent vers le bas par des sables.

A Quemenès, petite île inhabitée, située de même entre la terre ferme et l'île d'Ouessant, la section de la côte m'a offert la forme représentée dans la fig. 3, planche III. L'inclinaison moyenne de la levée de galets est de 20 degrés.

Le profil du cordon littoral frappe généralement les yeux par ses formes géométriques, comme l'indiquent suffisamment les figures des planches III et IV; mais la régularité des formes produites par le phénomène qui nous occupe, est encore plus remarquable lorsqu'on les considère en projection horizontale. L'entassement se dispose naturellement de manière à couper la surface de la mer suivant une courbe très-simple, sur laquelle les vagues viennent se déployer.

La mer ne prolonge pas son mouvement jusqu'au fond des anfractuosités naturelles de la côte; elle forme devant chacune d'elles une digue de galets ou de sable, sur laquelle ses vagues viennent mourir.

Courbes régulières que forment les levées de sable et de galet.

Soit *a b c* (pl. III, fig. 4) la ligne irrégulière produite par l'intersection de la surface de la mer avec la terre ferme. Lorsque les vagues pénètrent dans une baie telle que celle qui s'étend du point *a* au point *c*, leur mouvement est altéré; l'altération qui se produit réagit d'un point sur l'autre, et il en résulte un mouvement d'ensemble par l'effet duquel la mer n'entasse pas les matières, qu'elle rejette dans toutes les anfractuosités de la baie; elle tend, au contraire, à les entasser suivant des courbes très-simples, telles que celle représentée par la ligne ponctuée *a d c*.

Il suffit qu'il y ait une très-petite baie d'une certaine forme pour que la mer produise ces phéno-

PL. IV.

Les Etangs d'Albion Estate à la Jamaïque

Fig. 7.

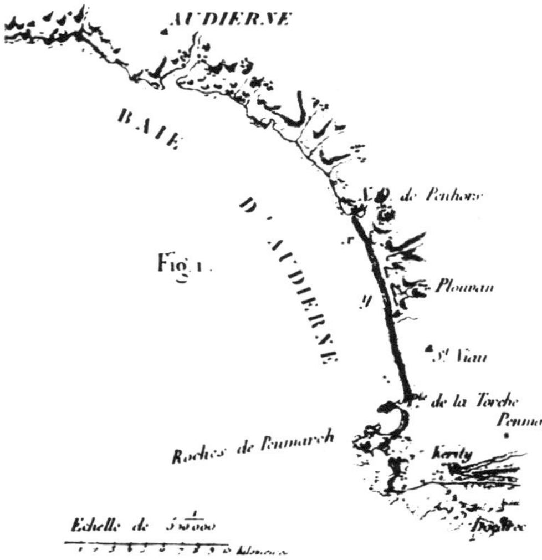

Banc de Galets

Mer

AUDIERNE

BAIE D'AUDIERNE

Fig. 1.

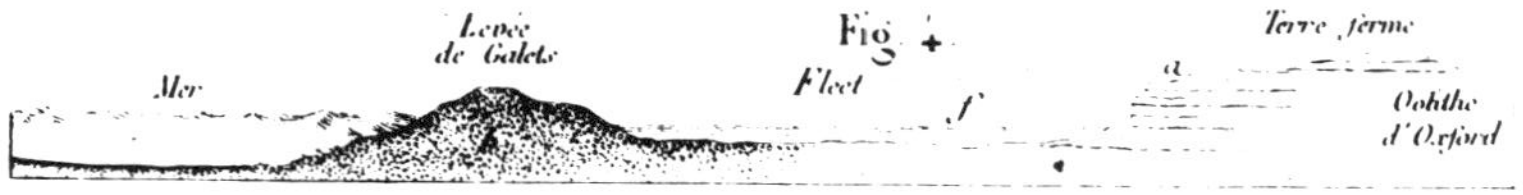

N.D. de Penhors

Plovan

St Vian

Pt de la Torche

Penmarch

Roches de Penmarch

Kerity

Echelle de 1/500000

Levée de Galets

Mer

Fig. 4.

Fleet

Terre ferme

Oolithe d'Oxford

Levée de Galets de la Baie d'Audierne

Fig. 2.

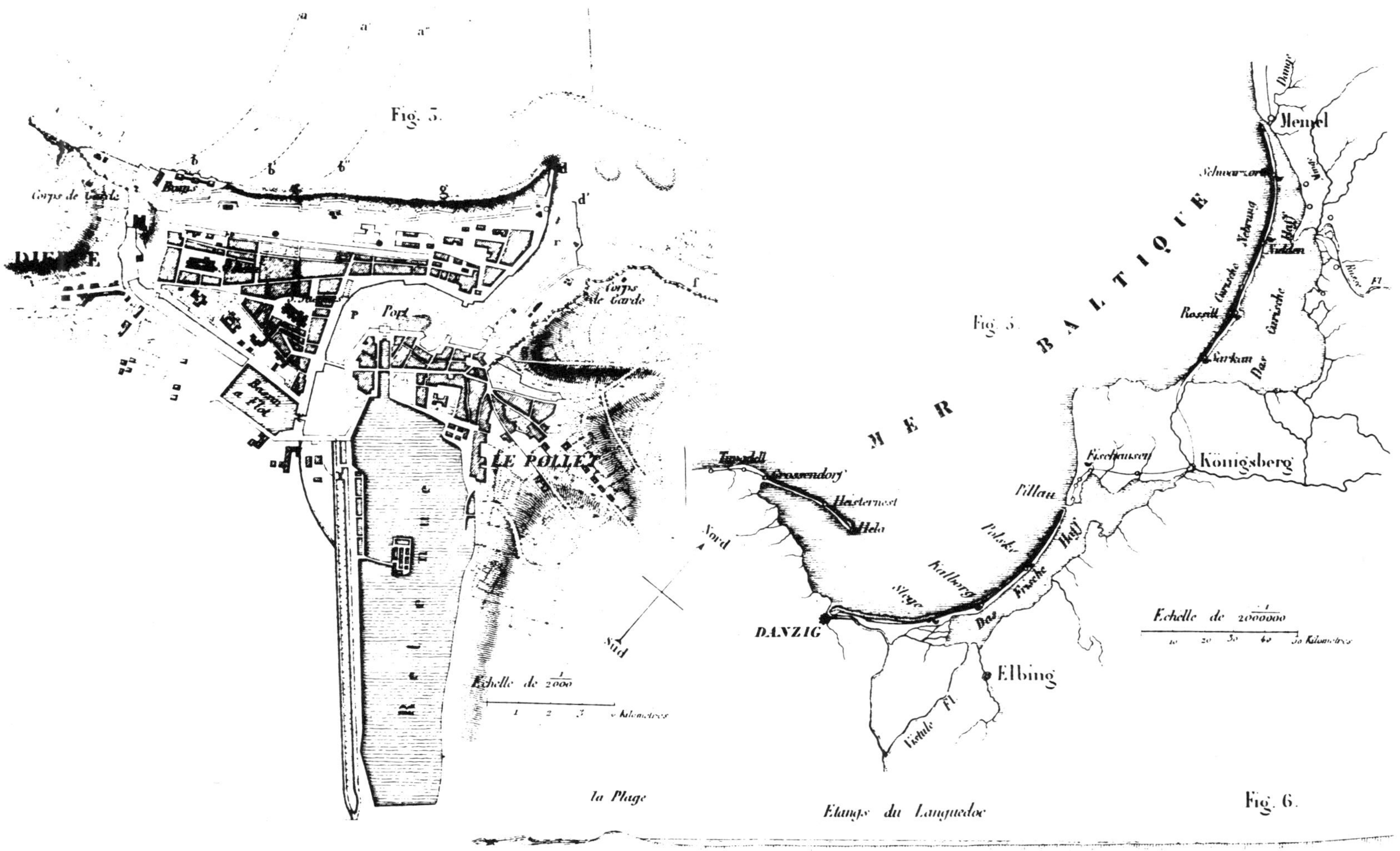

Fig. 5.

DIEPPE

Corps de Garde

Bains

Port

Corps de Garde

Bassin a Flot

LE POLLET

Nord

Sud

Echelle de $\frac{1}{2000}$

1 2 3 4 Kilometres

la Plage

Fig. 6.

MER BALTIQUE

Memel

Dange

Schwarzort

Curische Nehrung

Nidden

Das Curische Haff

Rossitt

Sarkau

Königsberg

Fischhausen

Pillau

Das Frische Haff

Polske

Kalberg

Steege

DANZIG

Elbing

Vistule Fl.

Hela

Heisternest

Grossendorf

Tupadell

Echelle de $\frac{1}{2000000}$

10 20 30 40 50 Kilometres

Etangs du Languedoc

Gravé par les Frères Avril, Rue des Noyers N° 33.

Lith. de Kaeppelin, Quai Voltaire N° 15

mènes. Elle ne s'arrange pas des enfoncements; son mouvement ne s'y développe pas à l'aise, et pour peu qu'il ne s'y trouve pas une grande profondeur, elle établit des digues régulières qui les interrompent. Il reste derrière ces digues des espaces bas occupés par des lagunes ou étangs, qui souvent ne communiquent plus du tout avec la mer.

Ce phénomène, quand on n'en voit qu'une partie, ne paraît pas mériter une grande attention; mais lorsqu'on l'examine dans son ensemble, on reconnaît qu'il a une grande influence sur une foule de faits importants.

Divers exemples sur les côtes de Bretagne.

Une multitude de petites baies de la Bretagne présentent vers leur fond un cordon littoral formant une courbe arrondie appuyée sur les deux caps de l'entrée, et derrière lequel s'étendent des étangs ou des marais. Presque toutes les baies où coulent de petites rivières, sont barrées par des levées de cette nature. La baie des Trépassés, près de la pointe du Raz; la baie de Pen-hir, près de la pointe du Toulinguet, en offrent des exemples remarquables. Je pourrais les multiplier presqu'à l'infini, mais je préfère vous décrire en détail quelques exemples où le phénomène se manifeste en grand, en vous rappelant que ce qui existe en grand, se reproduit en petit dans une infinité de localités.

Plus le *cordon littoral* est développé, mieux on voit la marche du phénomène qui le produit et le conserve.

Baie d'Audierne.

La baie d'*Audierne*, dans le département du Finistère, dont la planche IV, figure 1.re, indique la forme générale, est bordée en partie par des rochers qui la divisent en petites anses partielles, dont chacune présente en général le caractère indiqué ci-dessus; mais vers son extrémité sud-ouest se développe une courbe régulière et continue *xyz*, formée par une levée de galets qui a environ 12 kilom. de longueur : elle s'appuie aux deux extrémités sur des rochers, au nord sur ceux de Notre-Dame de Penhors, et au sud sur les roches de Pennmarck.

Derrière la levée s'étendent des terrains plats, plus ou moins complétement inondés: quelques parties sont occupées par des marécages. Il y a même un étang considérable.

Une section dans cette levée de galets présente la forme indiquée par la planche IV, figure 2. La ligne horizontale *m m'* indique des marais situés derrière la levée; la ligne courbe *mgdcba* figure le banc de galet.

Du côté de la plage, la levée a une forme variable, dont la mer remanie sans cesse le contour. Ordinairement on lui trouve le profil représenté par la ligne *abcd*.

Quand la mer est calme, ou qu'elle n'a que le mouvement causé par la marée, elle n'entasse les galets qu'à la hauteur *b*; quand elle est un peu plus forte, elle les entasse à la hauteur *c*; quand elle est très-forte, elle efface entièrement les deux premières lignes, et fait naître une forme représentée par la ligne ponctuée *ad*.

L'inclinaison en *d* est le plus souvent de 33 degrés, quelquefois de 34 ou de 35 degrés; plus bas on trouve des inclinaisons de 12 à 15 degrés. L'inclinaison de la surface du sable qui forme la plage, découverte à basse mer au pied de la levée de galets, est très-peu considérable; elle varie de 1 à 2 degrés.

La hauteur de cette levée est d'environ 5 mètres au-dessus de la plage de sable, qui ne dépasse pas le niveau des hautes marées ordinaires.

La levée du côté de l'étang ou des marécages situés derrière est généralement moins inclinée que du côté de la mer. La pente est là, en moyenne, d'environ 25°, et dans quelques parties elle se réduit à 3 ou 4°; elle a une forme arrondie et beaucoup plus stable, parce que l'inclinaison résulte ici d'actions très-prolongées : il s'y trouve même fréquemment un peu d'herbe qui pousse entre les galets.

Quelquefois la mer rompt cette digue et fait irruption dans les marécages par la brèche qu'elle s'est ouverte; mais comme la mer promène continuellement des galets le long de la côte, s'il se passe plusieurs mois sans grande tempête, la brèche se referme par l'action de la mer elle-même.

Les galets sont formés des roches des côtes voisines, de granite, de gneiss, de micachiste et de quartz blanc, qui constitue des veines dans cette dernière roche. On y trouve aussi du porphyre.

Quelques-uns des galets qui la composent sont très-gros; à l'origine septententrionale de la levée, près de Notre-Dame de Penhors, il y en a d'environ 30 centimètres de longueur; mais, après s'être entre-choqués pendant longtemps les uns contre les autres, ils finissent par s'amoindrir. On voit leur grosseur diminuer en suivant la levée dans la direction du nord au sud, ce qui montre que le transport des galets le long de la levée s'effectue dans cette direction. C'est une remarque importante que celle du transport du galet par la mer, le long de la plage, dans une direction déterminée. Je reviendrai encore ailleurs sur ce genre de mouvement. Il ne faut pas le confondre avec celui par lequel la mer entasse simplement le galet.

A son extrémité, près de la pointe de la torche, la levée n'est plus formée que de sable, d'abord très-gros, puis assez fin pour s'élever en dunes.

Les vents prédominants étant ceux du sud-ouest, l'action de la mer s'exerce dans cette même direction ; mais le courant violent qui débouche du *Raz de sein* lorsque la marée descend, a aussi une grande influence sur le phénomène.

Kérity. A Kérity, au sud des roches de Pennmark, le cordon littoral est uniquement formé de sable qui donne naissance à des dunes.

Loctudy. Plus à l'est, à Loctudy (Finistère), une grande plage que la haute mer couvre, est fermée et convertie en une sorte de lagune par une levée de sable que plusieurs ouvertures partagent en différents tronçons.

Port-Louis. A Port-Louis (Morbihan) on voit aussi une grande plage que la mer couvre, convertie en lagune à niveau variable, par une levée de sable qui la borde sans la fermer complétement.

Cordons rattachés à des îlots. Ce n'est pas seulement au fond des anses et des baies que la mer relève en forme de cordon les matières meubles qu'elle remue sur son fond dans les endroits où elle est à la fois agitée et peu profonde. Souvent elle les dispose avec une régularité frappante dans le voisinage des rochers isolés qui peuvent leur servir d'abri et de point d'appui. On en voit des exemples remarquables sur les rochers plats qui bordent en beaucoup de points les côtes de Bretagne et sur lesquels s'élèvent des pitons isolés ou de petits îlots rocheux. On voit fréquemment un cordon de galet s'étendre à une certaine distance sous le vent d'une de ces saillies, ou bien relier entre elles deux saillies situées l'une par rapport à l'autre dans la direction des vents dominants ou dans quelque autre condition favorable. Ainsi, d'après les belles cartes publiées au dépôt de la marine, sous la direction de M. Beautemps-Beaupré, les petites îles qui s'élèvent près de l'entrée de la rivière de Tréguier (Côtes du Nord), sur des plateaux de roches que la mer basse découvre en partie, sont souvent reliées entre elles par des levées de galet. La plus remarquable de ces levées porte un nom dans le pays : elle s'appelle *sillon de Talber*. Sillon de Talber.

La ville de Saint-Malo est bâtie sur un rocher que la mer haute entourerait complétement, s'il n'était lié à la terre ferme par une levée naturelle sur laquelle on a établi une route. Cette levée forme la seule limite qui sépare le port de l'anse du fort royal. Saint-Malo.

La presqu'île granitique de Quiberon serait de même une île, si la mer ne l'avait liée au continent d'une manière analogue. Mais ici le phénomène est plus étendu. Au lieu d'un simple cordon, il y a deux levées de sable qui partent du fort Penthièvre et vont en divergeant s'appuyer sur la terre ferme. Elles forment chacune de son côté le contour d'une anse arrondie, et elles tournent l'une vers l'autre les convexités de leurs courbures. Dans l'espace qu'elles laissent entre elles s'étendent des dunes et des étangs. Quiberon.

La Manche présente aussi beaucoup de levées de sable et surtout de galet. Je citerai d'abord comme exemple l'entrée de la vallée de la Béthune à Dieppe. Dans la figure IV, planche 3, *ff* sont des falaises de craie assez élevées; *r* est l'entrée de la Béthune; *p* le port. La ville se trouve située sur un terrain plat, protégé par une levée de galet *g*, qui vient jusqu'a l'entrée du port. La mer agite ce galet et tend à le pousser dans une certaine direction, qui est le plus souvent celle de l'ouest à l'est, parce que les vents prédominent dans cette direction, et que les vagues, qui viennent rencontrer le rivage, ont le plus souvent, dans les gros temps, la disposition indiquée par les lignes courbes ab, $a'b'$, $a''b''$, de manière qu'une même vague vient briser successivement en b, b', b''. De là la nécessité de construire des jetées d, d', très-solides, qui débordent le banc de galet, afin que ce galet ne remplisse pas l'entrée du port. Malgré cette précaution, le galet dépasse quelquefois la jetée et vient obstruer le port. On n'a pas trouvé de meilleur moyen pour déblayer le chenal, que de retenir, par une écluse, les eaux de la rivière gonflée par la marée. On ouvre cette écluse quand la mer est basse, et il en résulte un courant très-rapide, *une chasse*, qui emporte une partie du galet dans la mer. Ce moyen s'emploie dans presque tous les ports sujets à s'obstruer : cela ne se fait pas sans une dépense considérable, qui est un témoignage de la puissance des phénomènes que les ingénieurs ont à vaincre pour maintenir l'entrée des ports dans un état constamment praticable.

Les côtes de l'Angleterre présentent une foule

d'exemples analogues à ceux que je viens de décrire. M. de La Bèche cite le suivant, sur la côte méridionale du Devonshire. La mer a entassé au fond de la baie de Start, sur une longueur de 8 à 10 kilomètres, une levée de petits galets quartzeux; en arrière de la levée, qui barre les entrées de cinq vallons, s'étend un espace d'une certaine largeur situé au-dessous des hautes marées; dans cet espace il y a deux étangs qui communiquent l'un avec l'autre. Ils sont ordinairement remplis d'eau douce, qu'ils reçoivent de plusieurs petites rivières dont les dépôts ont comblé presque entièrement l'étang supérieur; elle filtre dans la mer à travers la levée de galet. Ces étangs renferment des poissons d'eau douce: truites, perches, brochets; il y a aussi quelques poissons de mer (carelets), qui sont habitués à l'eau douce. Quelquefois la mer fait irruption dans les étangs en s'ouvrant un passage à travers les levées de sable. Dans le mois de novembre 1824, pendant une violente tempête, la mer a rompu la digue et a fait invasion dans les étangs : tous les poissons d'eau douce sont morts, excepté ceux qui ont pu remonter dans la rivière. La mer a réparé elle-même ce ravage; la brèche s'est refermée, les eaux des étangs sont redevenues douces et les poissons d'eau douce les ont repeuplées.

Start-bay.

Ile de Portland
Chesil-bank.

Il y a dans la Manche une espèce d'île qu'on appelle l'*île de Portland :* elle présente des côtes escarpées. Derrière cette île se trouve la *baie de Weymouth;* puis un bras de mer très-prolongé, appelé *Fleet.* Ce dernier est séparé de la mer par une levée de galet de 16 milles anglais, ou 26 kilomètres de longueur, appelée *chesil-bank*. J'ai dit, une espèce d'île, parce que tant que la levée conserve sa continuité dans toute sa longueur, l'île de Portland n'est réellement qu'une presqu'île; mais la violence de la mer y détermine souvent des ruptures, et alors la masse rocheuse, qui n'est liée à la côte voisine par aucune crête de terrain invariable, devient une île véritable. La côte est formée par des couches de roches peu solides, et qui s'arrondissent par la seule action des agents atmosphériques, mais qui ne se dégradent pas et ne forment pas de falaises, parce que le *chesil-bank* les protége contre les vagues du large. Il est remarquable que les galets dont se compose le *chesil-bank* vont généralement en augmentant de grosseur de l'ouest à l'est. La figure 4, planche IV, représente une section qui coupe la côte *a*, le *chesil-bank* *b*, et entre les deux l'étang salé *f*, qui les sépare et qui est désigné sous le nom de *Fleet*.[1]

Les marées ne sont pas aussi nécessaires pour la production des levées de sable ou de galet que pour celle des dunes. L'absence de marées sensibles dans certaines mers est même cause que ces levées et les lagunes qu'elles protègent se dessinent avec une netteté particulière.

Sur les côtes de la mer Baltique il y a plusieurs exemples très-remarquables de ce genre de phénomène.

Curische Nehrung.

Memel, dernier port de la Prusse, en allant du côté de la Russie, se trouve à l'entrée d'un très-grand

1. De la Bèche, Manuel géologique, traduit de l'anglais par M. Brochant de Villiers, page 91. Paris. Levrault. 1833.

étang d'eau salée ou lagune, appelée *Curische Haff*, dans lequel se jette le Niémen. Cet étang (voyez fig. 5, pl. IV) est séparé de la mer Baltique par une digue très-étroite appelée *Curische Nehrung* : il a 110 kilomètres de longueur, et ne communique avec la mer que par un passage assez étroit et peu profond, où il existe une espèce de *barre*. Le port de Memel, situé en dedans de la barre, est vaste, sûr et profond.

Dans la mer Baltique il n'y a pas de marée sensible; mais les vents, suivant la direction dans laquelle ils soufflent, peuvent accumuler l'eau sur l'une ou l'autre côte de cette mer. Par suite de cette circonstance, la profondeur varie sur la barre de Memel de 4 à 6 mètres : les grands vaisseaux sont obligés de décharger en partie en dehors de la barre.

La langue de terre qui sépare l'étang de la mer, n'est qu'un entassement de sables rejetés par la mer et en partie couverts de végétation. Elle a assez de largeur et de solidité pour qu'on ait pu y établir une route de poste.

Les embouchures de la Vistule vont nous offrir un autre exemple du même phénomène.

Une langue de terre d'une courbure assez régulière, le *Frische Nehrung* (fig. 5, pl. IV), formée de sables accumulés et en partie couverts de végétation, et sur laquelle existent plusieurs lieux habités, sert de clôture à une grande lagune de 100 kilom. de longueur, le *Frische Haff*, qui reçoit les eaux du Pregel et de deux bras de la Vistule. Le port de *Pillau* est situé à l'entrée de cette lagune dans une position *Frische Nehrung.*

analogue à celle de Memel, en dedans d'une barre sur laquelle il n'y a que 3 ou 4 mètres d'eau. Les vaisseaux d'un fort tonnage sont obligés de jeter l'ancre en dehors de cette barre.

L'entrée du *Frische Haff* a 3,500 mètres de largeur. Elle est assez étroite pour que les Dantzikois aient eu une fois l'idée de l'obstruer; entreprise à laquelle les Prussiens s'opposèrent à main armée. Elle a subi, dit-on, de nombreux changements par l'action de diverses tempêtes arrivées de 1311 à 1510. Quant aux légendes, d'après lesquelles le *Frische Nehrung* tout entier aurait été produit par une tempête, M. de Hoff les range lui-même parmi les fables, et il cite des descriptions du neuvième siècle qui sont conformes à l'état actuel des choses.[1]

Le cordon littoral du *Frische Nehrung* se prolonge beaucoup plus à l'ouest que le *Frische Haff* dans son état actuel, et dépasse même la rade de Dantzig. L'une des routes de poste de Dantzig à Kœnigsberg le suit dans toute sa longueur, et c'est ainsi en longeant le rivage qu'elle traverse toutes les embouchures de la Vistule et du Pregel. La Vistule éprouve tout près de la côte une bifurcation qui la partage en deux bras, dont l'un tombe dans le *Frische Haff*, tandis que l'autre va passer sous les murs de Dantzig. Il est très-probable que cette bifurcation est due à l'existence du cordon littoral. On pourrait supposer que le *Frische Haff* s'étendait dans l'origine jusqu'à Dantzig et peut-être au delà, mais que les dépôts de la Vistule en ont comblé

1. Von Hoff. *Veränderungen der Erdoberfläche*: tome I.er, page 70.

une partie. Les terrains bas situés en arrière de ce cordon littoral, où la Vistule se ramifie, ont plus d'un rapport avec la Hollande : ce sont, pour ainsi dire, les *Pays-Bas baltiques*. Pays-Bas baltiques.

Au nord-ouest de Dantzig la côte de Prusse est encore bordée par un cordon littoral derrière lequel se trouvent plusieurs lagunes plus petites que les précédentes, et dont chacune n'a qu'une ouverture étroite. Ce cordon littoral, en se prolongeant à l'est, forme la longue pointe de Héla, qui s'étend de manière à fermer en partie la baie de Dantzig, qu'elle semble tendre à convertir elle-même en lagune. Pointe de Héla.

L'embouchure de la Duna, près de Riga, est barrée, comme celles du Niémen et de la Vistule, par un cordon littoral sablonneux duquel partent des dunes. Embouchure de la Duna.

L'Oder, comme la Vistule, se répand dans une lagune fermée par un groupe d'îles dont la continuité est due à une langue de terre très-étroite. Il verse ses eaux dans la mer Baltique à travers plusieurs ouvertures d'un long cordon littoral qui sert de clôture à cette lagune qui est divisée en deux parties, qu'on appelle *grosse Haff* et *kleine Haff*. Cette lagune aboutit à la mer en trois endroits principaux, entre lesquels s'étendent deux îles basses, très-étroites dans quelques parties, larges dans d'autres : l'île *d'Usedom* et l'île de *Wollin*. Les vaisseaux d'un fort tonnage ne peuvent pénétrer dans la lagune. Il n'y a que les vaisseaux tirant moins de 2 à 3 mètres qui puissent remonter jusqu'à Stettin. Les autres sont obligés de décharger en dehors. On a construit à l'une des ou- Embouchure de l'Oder. Usedom et Wollin.

vertures deux jetées, qui forment le port de Swinemünde. Il n'y avait sur la barre que 2 mètres d'eau; mais on a dragué, et aujourd'hui on peut faire arriver des vaisseaux tirant de 19 à 21 pieds d'eau (6 mètres à 6 mètres et demi) jusqu'à Swinemünde, qui est situé un peu en dedans de la barre.

La mer Baltique, par des causes que nous étudierons plus tard, ne présente de faits bien marqués de ce genre que dans sa partie méridionale; mais toutes les côtes de la Méditerranée fourmillent d'exemples de cordons littoraux, dont un grand nombre servent de clôture à des lagunes plus ou moins étendues.

Cordons littoraux des côtes de la Corse et de la Sardaigne.

En Corse, l'étang de Diana et d'Urbino, entre le Golo et le Fium-Orbo, et l'étang de Palo, au sud du Fium-Orbo, doivent leur existence à des cordons littoraux.

Il en est de même de l'étang de Biguglia, qui au sud de Bastia reçoit la rivière Bevinco.

Le golfe de Calvi est terminé par une plage demi-circulaire, sur laquelle s'élèvent des dunes et derrière laquelle s'étendent des marais que traversent le Ficarello et d'autres torrents.

L'embouchure du Liamone se replie derrière une levée de sable; il en est de même de celle du Tavaria et de plusieurs autres rivières de la même île et de celle de Sardaigne.[1]

Rade de Toulon; *les sablettes.*

A Toulon, la rade est protégée par le massif élevé de la presqu'île de Saint-Mandrier; c'est une montagne rocheuse terminée par le cap Cepet. De loin

1 Cartes marines des côtes de Corse, par M. Hell, publiées au dépôt de la marine.

on croit voir une île, parce qu'on n'aperçoit pas sa liaison avec la terre ferme, formée par une levée de sable appelée *les sablettes*.

Presqu'île d'Hyères.

Non loin de Toulon sont les îles d'Hyères et la presqu'île d'Hyères, qui termine la rade de ce nom, Cette presqu'île est élevée comme les îles Hyères elles-mêmes, auxquelles elle ressemble en tous points. Elle formait originairement une île; mais elle est maintenant réunie à la terre ferme par un terrain plat, bordé par deux longs cordons littoraux entre lesquels existe un étang, l'*étang du Pesquier,* qui ne communique avec la mer que par une seule ouverture.

Plus près de la ville d'Hyères se trouve un autre étang du même genre, appellé l'*étang de Fabrègues,* à côté duquel on voit des salines très-étendues, nommées les *salins d'Hyères*.

Sur la plage de Bormes on voit plusieurs ruisseaux se terminer dans de petits étangs derrière une levée de sable. Il en est de même dans la rade d'Agay.

Monte Argentaro

Le littoral de l'Italie offre un exemple dont sont frappés tous ceux qui parcourent les côtes de Toscane en bateau à vapeur. Sur ces côtes s'élève une montagne assez proéminente, appelée le *mont Argentaro:* elle est très-dentelée et composée de roches cristallines. Elle se présente de loin comme une île, et ne tient, en effet, à la terre ferme que par une langue de terre basse, formée par deux cordons littoraux. Dans l'intervalle de ces deux levées il y a un étang, joint à la mer par un canal. La ville d'Orbitello a son port dans l'intérieur de cet étang.

Monte Circeo, marais Pontins.

Les marais Pontins paraissent résulter du comblement d'une lagune du même genre, circonscrite par deux levées de sable, qui se dirigent de la terre ferme vers le *monte Circeo,* primitivement isolé au milieu de la mer.

Magnisi

Sur les côtes de Sicile on remarque la presqu'île de Magnisi, jointe à la terre ferme par un banc de galets, qu'on appelle la Chaussée. Ce banc s'élargit, et, dans la partie la plus large il y a un étang et des salines.

Scherschell.

En Algérie, la ville de Scherschell se trouve dans le même cas. Il y a une masse rocheuse réunie à la terre ferme par un grand banc de galets.

Gibraltar.

Gibraltar est une montagne très-élevée, formée de rochers escarpés, et formerait une île, s'il n'y avait une plage de sable entre Gibraltar et la terre ferme.

Examinons maintenant les côtes d'Espagne. Au nord du cap Palos il y a une petite île, puis des rochers qui font face à l'île, et dans l'intervalle un banc de galets et de sable; en arrière s'étend un grand étang qu'on appelle la Petite mer.

L'Albuféra de Valence.

Plus au nord se trouve Valence. Depuis l'embouchure de la rivière qui se jette dans la mer à Valence, jusqu'au cap *Cullera,* formé de rochers, il existe une ligne de rivage uniforme, qui résulte encore d'un entassement de sable ou de galets opéré par la mer. En arrière de ce cordon littoral se trouve un marécage très-étendu, qui reçoit deux rivières, et qu'on appelle l'*Albuféra:* il est très-poissonneux et peuplé d'un grand nombre d'oiseaux d'eau. Ce marécage est une propriété considérable; il constitue la plus grande partie du duché d'Albuféra, qui en a pris son nom.

Étangs du Languedoc et de Roussillon.

L'embouchure du Rhône présente au point de vue qui nous occupe des phénomènes d'une nature particulière, sur lesquels je reviendrai plus d'une fois. A partir des bouches du Rhône, la côte affecte jusqu'à Cette une courbure très-régulière : cette ville est adossée à un rocher, et celle d'Agde est située à la base de petites montagnes volcaniques, qui forment également un point d'appui pour la côte. Ensuite vient une côte régulière, qui s'étend, avec de très-légères inflexions, jusqu'à Collioure. La côte, telle qu'elle résulterait de l'intersection de la surface de la mer avec les roches en place, aurait une configuration beaucoup plus compliquée. L'intervalle entre les deux lignes est rempli par une foule d'étangs. A partir de Cette commence un grand étang, composé de trois parties, appelées l'étang de Maguelone, l'étang de Pérols et l'étang de Mauguio. Ce dernier s'étend presque jusqu'à l'embouchure du Rhône : dans l'origine il a reçu probablement les eaux du Rhône, et il reçoit encore maintenant les eaux de plusieurs rivières. Cet étang n'est séparé de la mer que par une digue de sable; de l'autre côté des étangs se déploie un rivage très-plat, qui est le bord de la terre ferme proprement dite.

De l'autre côté de la montagne de Cette, entre Cette et Agde, s'étend l'étang de Thau; puis, près de Narbonne, et plus au sud, se trouvent encore divers étangs, dont le plus considérable est celui de Leucate.

Leur origine indiquée par Astruc.

Tous ces étangs doivent leur origine à la même circonstance. Si vous faites une coupe, vous la trouverez toujours analogue à celle représentée par la figure 6, planche IV. « Il est visible, dit Astruc, que les étangs qui s'étendent le long de la côte du bas Languedoc, depuis Aigues-Mortes jusqu'à Agde, ont fait partie autrefois de la mer même, dont ils n'ont été séparés que par un long banc de sable qui s'est formé entre eux, connu sous le nom de *la plage*. Leur situation, leur niveau avec la mer, la salure de leurs eaux, ne permettent pas de douter de ce fait. »[1] L'analogie des *plages* du Languedoc avec les *Nehrungen* de la Baltique a frappé plusieurs auteurs, et particulièrement M. de Hoff.[2]

Ils ont été séparés de la mer par la formation de la plage.

Ils existaient déjà du temps de Pline.

Les étangs existaient déjà du temps de Pline, qui, en décrivant la province narbonnaise, remarque qu'il y avait peu de villes à cause des étangs qui s'y trouvaient. *Oppida de cetero rara, præjacentibus stagnis* (*Hist. nat., lib. 3, cap. 4*).[3]

La même chose a lieu en Égypte. Il y a près de l'embouchure du Nil des étangs analogues à ceux qui existent près de l'embouchure du Rhône. Tels sont le lac *Menzaleh*, le lac *Bourlos* et la lagune d'*Edkou*, séparés de la mer par des digues très-étroites.

Ces phénomènes ne sont pas particuliers à nos climats. Les côtes des mers tropicales y sont sujettes comme les nôtres. M. de la Bèche en a observé des exemples à la Jamaïque. Il en décrit un dans son

1. Astruc, Mémoires pour l'hist. nat. du Languedoc, p. 372.
2. Von Hoff, *Veränderungen der Erdoberfläche*; tome I.er, p. 369.
3. Astruc, *loc. cit.*, page 371.

Pl V.

LE GOLFE DU MEXIQUE

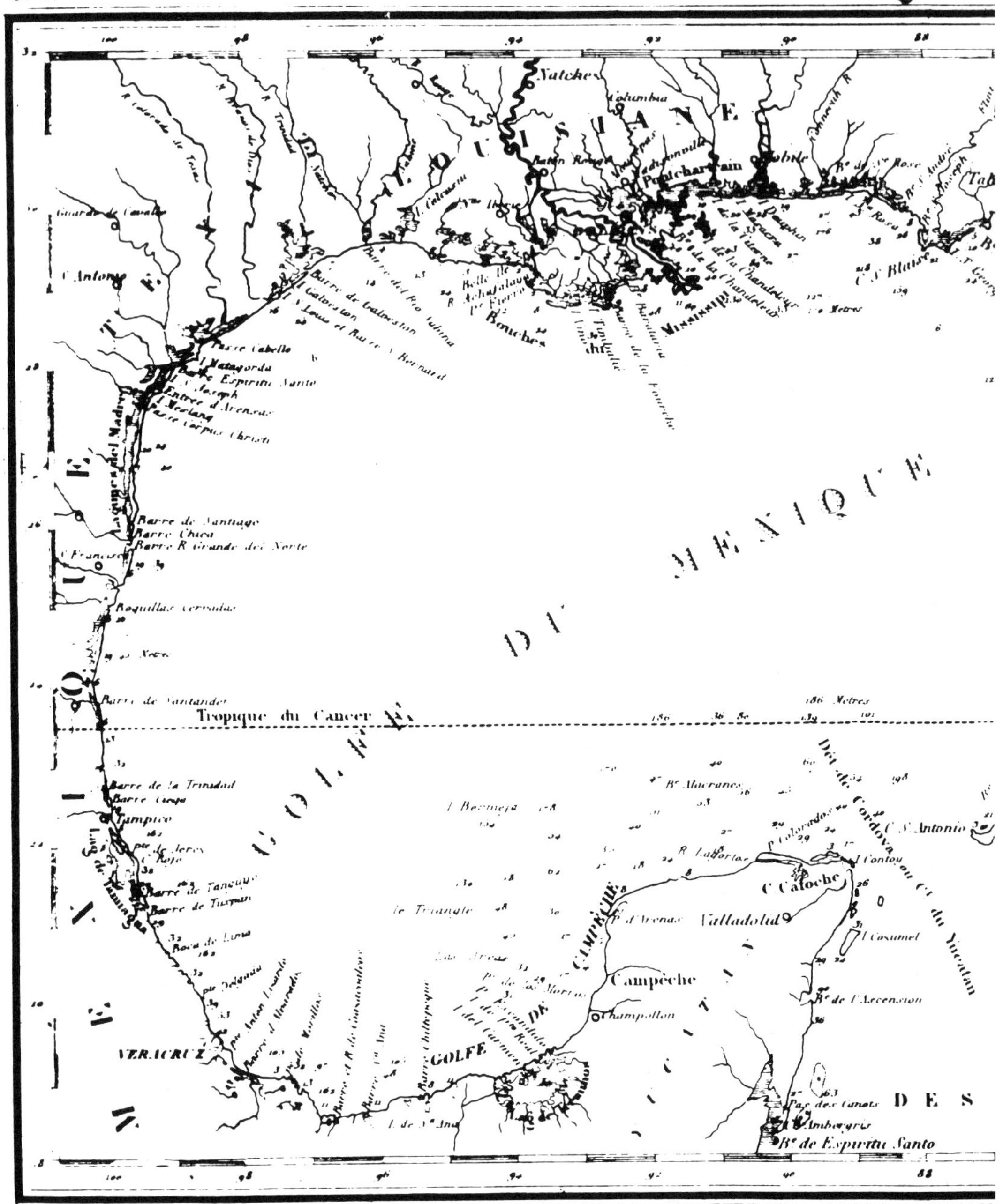

…ES GRANDES ANTILLES.

ouvrage intitulé *Sections and views illustrative of geological phænomena*. Le lac, dont la figure 7, planche IV, représente la coupe, « est défendu contre la mer par un banc de galets qui offre une petite ouverture au-dessus du niveau ordinaire de la mer; c'est par cette ouverture que s'écoulent les eaux du lac, qui est formé par l'écoulement des pluies, et par ce qu'il peut y arriver d'eau de la mer par-dessus le banc de galets dans les grandes tempêtes. Cette digue naturelle a été formée évidemment par les vagues. Au reste, ce lac n'est point une exception : il en existe, vers l'est, d'autres qui sont plus considérables.

« Il est habité par des crocodiles (*crocodilus acutus*, Cuv.) et par des poissons de mer : ces derniers y ont probablement été rejetés par les vagues lors des tempêtes, et ils se sont habitués peu à peu à l'eau saumâtre du lac. Il serait essentiel, pour les géologues, que l'on étudiât à fond les lacs de ce genre sous le rapport zoologique, afin de voir jusqu'à quel point des animaux d'eau douce et marine peuvent vivre et se multiplier dans une même nappe d'eau. Le fond de ces lacs est une vase molle, dans laquelle les crocodiles s'enfoncent lorsqu'ils sont poursuivis; cette vase est formée probablement par les détritus charriés par les cours d'eau, et par la décomposition de matières animales et végétales. C'est dans cette vase que les mangliers prennent racine. »[1]

Le golfe du Mexique présente de nombreux exem-

1. Coupes et vues pour servir à l'explication des phénomènes géologiques; par M. H. T. de la Bèche; édition française, publiée par M. H. de Collegno; Paris, Pitois-Levrault, 1839.

ples de lagunes littorales du même genre. Elles commencent presque à la pointe nord-est de la presqu'île de Yucatan. Les plus considérables sur les côtes méridionales du golfe, d'après la nouvelle carte publiée en 1843 par le dépôt de la marine, sont celles de Terminos, de Santa-Ana et d'Alvarado. Sur la côte occidentale il en existe un grand nombre, dont les plus étendues sont celles de Tamiagua et del Madre. Sur la côte nord se trouvent les lagunes de Galveston, de Sabine et l'anse de Hostiones ou anse aux huîtres : le Delta du Mississipi en embrasse plusieurs autres. Enfin, les côtes de la Floride occidentale en présentent une longue série, qui s'étend jusqu'à la baie d'Apalache. Ces lagunes rappellent complétement, par leurs formes, celles de la mer Baltique, celles de Venise, les étangs du Languedoc, etc.

Lagunes ou étangs qui bordent les côtes du golfe de Mexique.

Analogie avec les côtes de Gascogne.

La plus longue de toutes, la lagune del Madre, ressemble tout à fait à celles qui bordent les landes de la Gascogne, ou du moins à ce que deviendraient celles-ci, si le bassin d'Arcachon et les étangs de Bicarosse et autres tenaient les unes aux autres d'une manière continue depuis la Gironde jusqu'à l'Adour; les dunes de la côte de Gascogne formeraient alors une longue île tronçonnée, dont l'île del Padre et les autres îles qui bordent la lagune del Madre, offrent tout à fait l'image. Les renseignements imparfaits que nous possédons sur ces côtes, permettent de supposer que telle est, en effet, leur constitution.

Environs de la Vera-Cruz.

« Aux environs de la Vera-Cruz, dit M. Saint-Clair Duport, la côte du golfe du Mexique offre peu d'attraits au géognoste. A quelques lieues du bord de la mer le sol se compose de sable qui, sur plusieurs points, forme des dunes mouvantes presque dénuées de végétation, et qui augmentent par rayonnement la température de l'air au point de la rendre insupportable. Les roches ne se montrent nulle part à découvert, et les pierres de construction sont même tellement rares, à l'exception des roches madréporiques (*piedra de mucara*), avec lesquelles on a bâti le château d'Ulloa et une partie de la ville de la Vera-Cruz, que l'on trouve en ce moment avantage à faire venir, toutes taillées de New-York, les pierres employées à la construction d'un édifice pour la douane et aux réparations du môle de la Vera-Cruz.»[1]

«Sur la côte orientale du Mexique, celle qui borde le golfe de ce nom, il n'y a pas un seul bon port. On y trouve, à la vérité, les rivières de *Tula*, de *Tampico* et de *Tabasco*, mais elles ont des *barres* à leurs embouchures qui empêchent que de grands vaisseaux puissent y entrer.»[2] Mauvais ports.

Ces côtes présentent même une longue série de ports qui portent le nom de *barres*, telles que la *barre d'Alvarado*, la *barre de Tuxpan*, la *barre de Tanguijo*, la *barre de Tampico*, la *barre de Ciega*, la *barre de la Trinidad*, la *barre de Tardo*, la *barre de Santander*, la *barre de Santiago*, la *barre d'Espiritu-Santo*, la *barre de Saint-Bernard*, la *barre del rio Sabina*. M. Burkart, conseiller des Barres.

1. Saint-Clair Duport, De la production des métaux précieux au Mexique.

2. Mac-Culloch, Dictionnaire géographique: tome II. p. 314.

mines de Prusse, qui a résidé plusieurs années au Mexique, donne sur ces barres, et sur celle de Tampico en particulier, les détails suivants :

Barre et lagunes de Tampico.

« Les rivières du Mexique, bien que leur cours ne soit pas très-long (excepté toutefois le Rio-Bravo del Norte), entraînent une grande quantité de sable, d'argile et de gravier, qu'elles jettent dans la mer à leurs embouchures. Ces dépôts, qui portent le nom de *barres*, empêchent les grands vaisseaux de pénétrer de la mer dans les rivières. Ils sont très-préjudiciables au commerce, parce qu'ils s'élèvent ordinairement jusqu'à quelques pieds seulement de la surface, et changent souvent de place par l'effet des crues et des tempêtes.... Le *rio Tampico* est fermé par une barre de cette espèce, sur laquelle il n'y a que 2,^{m}20 à 3 mètres d'eau.... Près de l'embouchure du *rio Tampico* il n'y a que quelques huttes de pêcheurs et de mariniers, appelées *barre de Tampico*... La ville de *Pueblo-viejo-de-Tampico* se trouve à cinq lieues (2 myriamètres) plus haut. Dans tout l'intervalle les bords de la rivière sont plats, non cultivés, couverts de broussailles peu élevées, dans lesquelles vivent un grand nombre d'oiseaux de terre et d'eau. La rivière, aussi bien que les lagunes qui se trouvent plus haut en communication avec elle, sont peuplées d'alligators. »

Plus loin, M. Burkart mentionne des dispositions locales, dans lesquelles on reconnaît aisément l'action exercée par les rivières pour se tracer un lit à travers les lagunes littorales, en formant des digues sur leurs bords par l'effet de leurs propres alluvions.

Près de Tampico s'élèvent de petites collines composées de diverses roches : de l'une d'elles, appelée la *mira*, on a une belle vue de la mer, des *lagunes de Tampico* et de la contrée adjacente.

« *Pueblo-viejo-de-Tampico* se trouve sur la rive occidentale de la grande lagune de Tampico, qui n'est séparée que par une langue de terre étroite du bras principal de la rivière, avec laquelle elle communique par un canal naturel situé en face de la ville.

« *Tampico-de-Tamaulipas,* qui est devenue la ville principale, est située au sud-ouest de la *barre,* sur la rive gauche de la rivière, entre celle-ci et la lagune de *Carpintero.* »[1]

Cette dernière confine à une autre lagune très-grande, celle de Tamiagua, qui est séparée de la mer par une digue extrêmement longue et étroite, laquelle se termime à la barre de Tanguijo. Depuis cette barre jusqu'à celle de Tampico, sans interruption, règne un long cordon de littoral étroit, derrière lequel s'étendent ces différentes lagunes.

La grande *lagune del Madre,* située au nord du rio del Norte, présente plusieurs passes qui la font communiquer avec la mer; l'une d'elles est appelée *barre de Santiago;* une autre, *barre d'Espiritu-Santo,* etc. Lagune *del Madre.*

On trouve ensuite le Texas, qui nous présente une longue île étroite, appelée l'île Saint-Louis ou de Galveston, et derrière une série de lagunes.

« Toutes les rivières du Texas tombent dans le Barres et lagunes du Texas.

1. J. Burkart, *Aufenthalt und Reise in Mexico;* tome I.er, pages 23 à 36.

golfe du Mexique, ou plutôt (excepté le Brazos de Dios) dans les *baies* ou *lagunes*. Ces dernières ont une grande ressemblance avec les *haffs* qui se trouvent sur la côte méridionale de la mer Baltique, excepté qu'elles sont sur une beaucoup plus grande échelle. La côte, comme l'a indiqué M. de Humboldt, présente des obstacles formidables à la navigation dans les longues, basses et étroites langues de terre par lesquelles elle est défendue et qui bordent les lagunes, par le manque de ports pour des vaisseaux tirant plus de 12 pieds et ½ d'eau (4 mèt.), et par les *barres* aux embouchures des rivières.»[1]

Analogie avec les *haffs* de la mer Baltique.

De Tuxpan à Galveston le cordon littoral du golfe du Mexique, bordé intérieurement de lagunes presque continues, présente un développement de 180 lieues marines (100 myriamètres); développement qui surpasse d'un tiers environ celui de tout le cordon littoral de la partie méridionale de la mer d'Allemagne, depuis Calais jusqu'à l'Elbe.

La baie de Galveston forme le principal port du Texas. En continuant à suivre la côte, on rencontre une grande lagune dans laquelle tombe la rivière Sabine, et qui a pour entrée la barre del rio Sabina; puis quelques autres lagunes plus petites, jusqu'à ce qu'on arrive au Delta du Mississipi.

Barres et lagunes entre le Mississipi et la Floride.

De l'autre côté du Mississipi la même disposition se présente de nouveau. Depuis le Delta du Mississipi jusqu'à la baie d'Apalache, à partir de laquelle la Floride se détache de la masse du continent, la côte est bordée de baies et de lagunes, en avant desquelles on trouve des îles plates et sablonneuses, allongées

1. Mac-Culloch. Dictionnaire géographique; tome II. p. 314.

dans le sens du littoral ou des péninsules de même forme et de même origine, rattachées à la côte ferme par des atterrissements modernes. En commençant à la baie d'Apalache pour revenir vers l'ouest au Mississipi, on trouve d'abord l'île Saint-George, puis la baie de San-José, qui est une grande lagune enveloppée par une espèce de crochet, et dont la courbe saillante est appelée cap San-Blas; ensuite la baie de Saint-André, qui est également une grande lagune avec un cordon du même genre; la baie de Santa-Rosa et la longue île de Santa-Rosa, derrière laquelle se trouve la baie de Pensacola offrant plusieurs ramifications; puis la pointe de Mobile, formant la baie dans laquelle se jettent les rivières d'Alibamous et de Mobile; et enfin une série de barres et de petites îles qui vont jusqu'à l'embouchure du Mississipi.

A l'entrée de la baie de Mobile il y a une barre sur laquelle on trouve 15 pieds (4,^{m}57) d'eau à mer basse; mais il existe des bancs en dedans de l'entrée qui font qu'il n'y a que les vaisseaux tirant seulement 8 à 9 pieds (2,^{m} ½) d'eau qui puissent arriver jusqu'à la ville.

« Le meilleur port de tout le golfe du Mexique est celui de Pensacola, à l'extrémité d'un petit bassin fluvial du même nom, qui est situé entre ceux de la Mobile et de l'Alphalachicola. Il y a sur la barre 7 mètres d'eau. Ce littoral, bordé de lagunes, offre par là même aux petits navires du cabotage une ligne de navigation intérieure presque continue, dont on pourrait tirer un parti fort avantageux au moyen de quelques travaux. »[1]

1. Michel Chevalier. Des voies de communication aux États-Unis; tome I.er, page 81.

Barres et lagunes sur les côtes atlantiques des États-Unis.

« Les côtes des Carolines et de la Géorgie sont de même bordées de lagunes, qui, se développant parallèlement au littoral presque sans interruption depuis le cap Hatteras jusqu'au cap Féar, et de Charleston jusqu'en Floride, y établissent une ligne de navigation intérieure qu'il serait aisé de compléter, et dont le commerce tire parti depuis longtemps, quoiqu'on se soit peu occupé de la perfectionner. »[1]

Dans cette partie méridionale du littoral atlantique des États-Unis, les rivières ont généralement des barres. La rivière Saint-Jean a 5 mètres d'eau à la barre, puis 5 à 6 mètres et au moins 4 sur une longueur de 300 kilomètres. L'Alatamaha a 4,^{m}50 à la barre, puis un peu plus jusqu'au delà de Darien. La barre de Savannah est recouverte de 5,^{m}50 d'eau, et les bateaux à vapeur la remontent jusqu'à Augusta.[2]

Barres sur les côtes des mers intérieures et des lacs.

Les côtes des mers intérieures et des grands lacs offrent des phénomènes analogues. On en verra de très-beaux exemples sur la carte de la mer Noire et de la mer d'Azof, par M. Hommaire de Hell. Les bords de la mer Caspienne sont dépourvus de ports profonds; les fleuves qu'elle reçoit offrent à leur embouchure des barres formidables ou des rochers.... Les lacs américains, l'Ontario excepté, étaient pareillement privés de ports naturels. Les navires ne peuvent y aborder qu'à l'embouchure des rivières; et jusqu'à ce que les ingénieurs américains fussent intervenus, ces embouchures, toutes barrées par

1. Michel Chevalier, *loc. cit.*, page 29.
3. *Idem*, *ibid.*, page 29.

des sables, étaient inabordables pendant la presque-totalité de l'année : aucune n'avait à la barre plus de 2 mètres d'eau, et très-peu avaient cette profondeur.[1]

Un tiers des côtes du globe doivent leur configuration à ces phénomènes.

En général, cette forme de côtes est très-répandue, et je crois que je n'exagère pas, en disant qu'il y a un tiers des côtes du globe qui doivent leur configuration à des phénomènes de ce genre : elles sont plus arrondies qu'elles n'étaient destinées à l'être, si la mer eût été parfaitement calme.

Le fait de l'existence d'une *barre* à l'entrée d'un grand nombre de rivières est un des plus notoires parmi ceux auxquels la navigation est subordonnée. C'est un grand fait naturel, qui témoigne du changement de régime que les eaux éprouvent en s'éloignant des larges espaces ouverts et profonds qui sont les centres d'agitation de la surface de la mer. Les localités privilégiées, où ce changement de régime, cet amoindrissement de l'agitation, n'a pas amené d'ensablement, sont fort rares. Ce sont les *ports* par excellence : Rochefort, Brest, Plymouth, Portsmouth, Londres, Anvers en sont des exemples.

Remarques générales sur les cordons littoraux et les barres.

De tout cela il résulte que la mer, dans les endroits où elle n'a pas une grande profondeur, modifie la forme de son lit, en entassant les matières qu'elle met en mouvement, et en donnant au fond une certaine inclinaison qui est plus en harmonie avec ses mouvements. Elle agite les matières qui le couvrent, et elle tend à en élever une partie sur ses bords sous la forme d'un cordon qui marque les limites de son domaine. Les barres sont le prolonge-

1. Michel Chevalier, *loc. cit.*, page 39.

ment sous-marin de ces levées de galet, de ces accumulations de sable qui forment les dunes, qui seulement sont tracées un peu au-dessus du niveau des hautes mers. Au moyen de ce mécanisme la mer se renferme pour ainsi dire chez elle.

Cette ligne extrême de la terre, sur laquelle la mer vient exercer toutes ses fureurs, et qu'on peut appeler *la ligne des barres,* a frappé l'attention dans tous les temps. C'est à elle que s'applique cette phrase si souvent répétée « *tu viendras jusque là et tu n'iras pas plus loin; c'est là que tu briseras l'orgueil de tes vagues!* »

C'est à l'entrée des rivières qu'on a le plus d'occasions d'observer les barres.

En général, la mer obstrue les entrées des rivières, et celles-ci ont une profondeur assez considérable à une certaine distance de leur embouchure. En se rapprochant de la mer, il y a un endroit moins profond; c'est cet endroit qu'on appelle *la barre.* En dedans de la barre on est en rivière, en dehors on est en mer. La rade est en dehors, le port est en dedans. La question difficile pour entrer en rivière, n'est pas de franchir un endroit plus étroit, mais de passer l'endroit où les matières s'entassent et où la mer brise avec plus de force. Les matières ainsi entassées ne laissent que le vide nécessaire pour donner passage aux eaux de la rivière. Si elles s'entassaient plus haut, les eaux seraient arrêtées, et il se produirait une écluse de chasse naturelle : c'est là ce qui limite la hauteur de la barre.

D'après les exemples que j'ai cités et que j'aurais pu multiplier beaucoup, que je serai même conduit à multiplier dans les leçons subséquentes, il est rare que sur une barre il y ait plus de 6 mètres d'eau à basse mer; il y en a généralement beaucoup moins.

Cas particulier où il ne se forme pas de barres.

La ligne de démarcation du domaine de la mer n'est cependant pas toujours aussi tranchée que dans les exemples nombreux que je viens d'indiquer: il y a des exceptions importantes à noter. Quelquefois, par suite de circonstances locales, le cordon ne se forme pas, les choses s'arrangent autrement. Suivant ces circonstances, cette barre peut devenir un ensablement très-long, comme l'Elbe et la Seine en offrent des exemples; mais le plus souvent il est très-court.

Quand le phénomène, tel que je l'ai décrit, ne se produit pas, il se produit quelque chose d'équivalent.

Ainsi le Rhin, la Meuse et l'Escaut viennent se jeter dans un terrain bas, qui présente beaucoup d'analogie avec les lagunes qui nous occupent. Le Zuyderzée peut être considéré comme une lagune de cette espèce.

L'Ems se jette à 50 kilomètres de la mer dans une lagune. A l'extrémité orientale de la lagune, qui s'appelle le Dollart, est le port d'Emden : la lagune est séparée de la mer par une ligne d'îles et de bancs de sable.

Exemple de l'Elbe.

Mais l'Elbe présente des dispositions différentes : il y a seulement à son embouchure des bancs de sable très-nombreux et très-irréguliers.

La Seine n'a pas de barre; elle présente, comme vous le savez, une large embouchure; mais il s'y élève une digue de galets qui vient jusqu'à l'entrée de la rivière. Le Hâvre est situé, comme Dieppe, sur un terrain plat, protégé par une levée de galet, et son port étant continuellement obstrué par ce galet, il a fallu établir une *écluse de chasse.* Cependant le phénomène n'a pas eu assez d'énergie pour former une digue continue en travers de l'ouverture de la Seine et produire une lagune. La Seine est obstruée à son entrée; mais cette obstruction est étendue sur une longueur de plus de 60 kilomètres : elle va à peu près jusqu'à la Meilleraie; audessus de ce point la Seine est très-profonde.

Estuaires

Ces larges embouchures, où la mer entre sans obstacles et dont elle façonne le fond à son gré, sont souvent nommées *estuaires* (d'après le mot *æstuairy,* adopté en anglais); nous nous en occuperons plus tard; ils rentrent dans le domaine des phénomènes marins.

Nous examinerons dans la prochaine séance ce qui se passe à l'abri du cordon littoral, en dedans de la ligne des barres, dans ce que nous avons appelé le domaine de la terre; nous y verrons un grand nombre de phénomènes remarquables : mais je ne pouvais entrer dans leur examen sans vous avoir donné une idée générale du mécanisme par lequel la mer façonne ses rivages.

Pl. 11

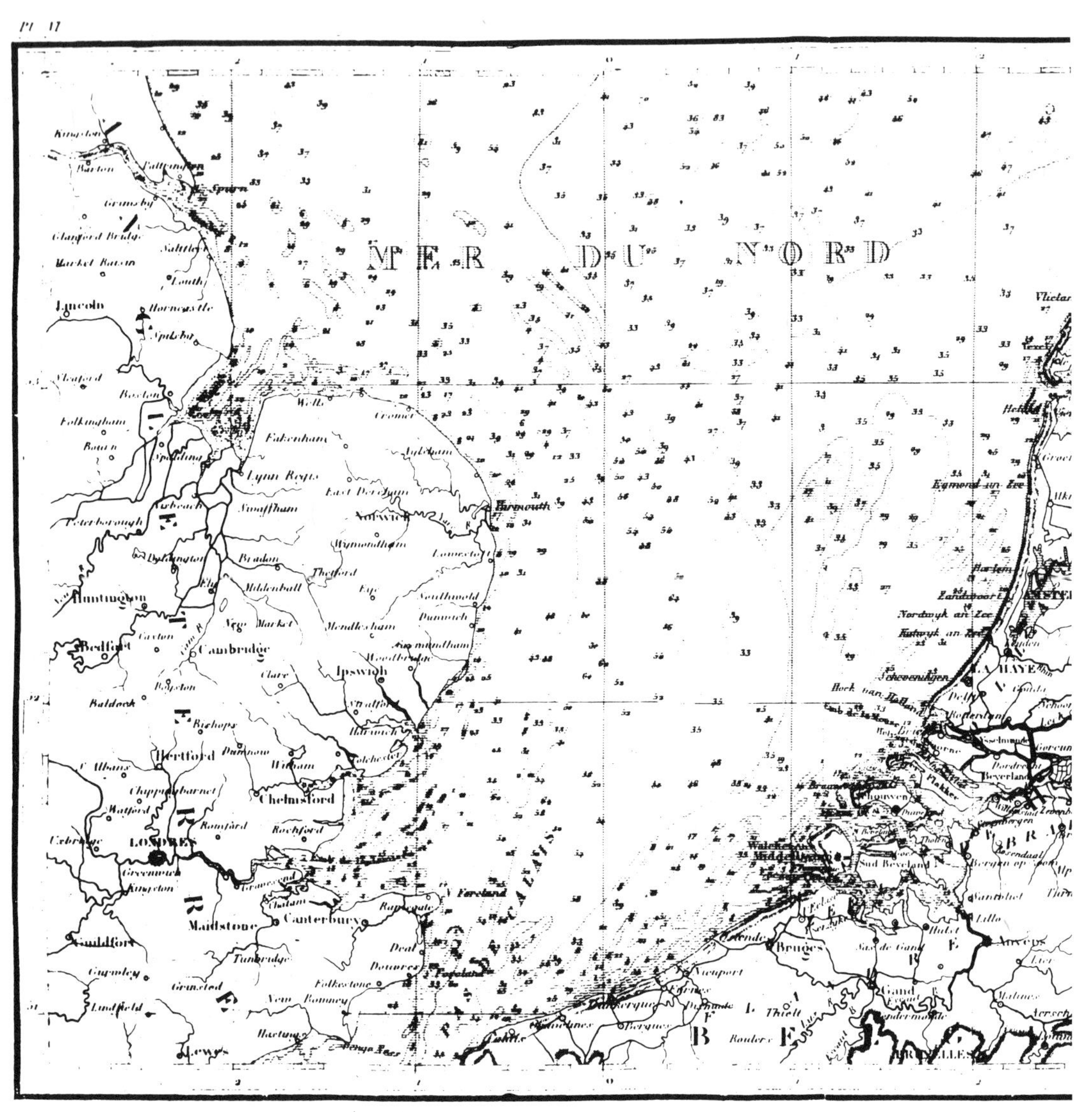
MER DU NORD
Lincoln
Boston
Lynn Regis
Norwich
Yarmouth
Huntingdon
Bedford
Cambridge
Ipswich
Harwich
Colchester
Hertford
Chelmsford
LONDRES
Maidstone
Canterbury
Dunkerque
Bruges
Gand
Anvers
LA HAYE
Rotterdam
Scheveningen
Nordwyk aan Zee
Katwyk aan Zee
Egmond aan Zee
Middelburg
Sud Beveland

PAYS-BAS
NEERLANDAIS
Echelles 1/2 000 000
Kilomètres
Milles Marins
Milles Anglais
Lubeck
Hambourg
Harburg
Bremen
Oldenbourg
Emden
Groningue
Leeuwarden
Assen
Zwolle
Kampen
Deventer
Meppen
Osnabrück
Minden
Celle
Hanovre
Brunswick
Lunebourg
Verden
Diepholtz
Bentheim
Munster
Dusseldorf
Elberfeld
Maestricht
Wesel
Dortmund
Gladbach
GRAND DUCHÉ
BAS RHIN

Editor's Comments on Paper 2

2 Gilbert: *The Topographic Features of Lake Shores*

During the span of his lifetime, from 1843 to 1918, Grove Karl Gilbert devoted a half-century to geology and geological organizations. He started his unique career as a museum worker under Henry A. Ward, founder of the famous natural history supply house. Gilbert then became an assistant on the Ohio Geological Survey, where, among other duties, he reported on elevated beach terraces bordering Lake Erie. Following this, he moved west with the Wheeling Survey and visited the Lake Bonneville region. Thus began a major study which led to the pioneering report reproduced here in part. Gilbert later joined the Powell Survey, and with the establishment of the U.S. Geological Survey in 1879, he became a Senior Geologist on its staff—an active and prolific association that continued until his death.

Although Gilbert was interested in a great many aspects of geologic science, his Lake Bonneville papers are landmarks in the study of coastal processes. Named by Gilbert to honor a military explorer of the early West, Lake Bonneville had once covered 20,000 square miles in northwestern Utah and attained a depth of 900 feet. As the surface level fluctuated (not necessarily in phase) with Pleistocene glacial cycles, wave action along its shores developed the stepped terraces seen today on the mountain slopes bordering the present Great Salt Lake. It was on these terraces that Gilbert found a subaerial inventory of coastal forms ideally preserved; from them he was among the first to report on the development of shoreline features.

In contrast to the work by de Beaumont, Gilbert saw shore drift as the origin of coastal barriers. He succinctly states that shore drift follows the agitation of the breaker zone, building a continuous outlying ridge on a gently sloping bottom. The barrier's crest stands a few feet above sea level and its seaward face has a typical beach profile.

Reprinted from *U.S. Geological Survey 5th Ann. Rept.*, 87–88 (1885)

2

THE BEACH.

The zone occupied by the shore drift in transit is called the *beach*. Its lower margin is beneath the water, a little beyond the line where the great storm waves break. Its upper margin is usually a few feet above the level of still water. Its profile is steeper upon some shores than others, but has a general facies consonant with its wave-wrought origin. At each point in the profile the slope represents an equilibrium in transporting power between the inrushing breaker and the outflowing undertow. Where the undertow is relatively potent its efficiency is diminished by a low declivity. Where the inward dash is relatively potent the undertow is favored by a high declivity. The result is a sigmoid profile of gentle flexure, upwardly convex for a short space near its landward end, and concave beyond. (See Fig. 3, Plate IV.)

In horizontal contour the beach follows the original boundary between land and lake, but does not conform to its irregularities. Small indentations are filled by shore drift, and a smooth, sweeping outline is given to the water margin.

THE BARRIER.

Where the sublittoral bottom of the lake has an exceedingly gentle inclination the waves break at a considerable distance from the water margin. The most violent agitation of the water is along the line of breakers; and the shore drift, depending upon agitation for its transportation, follows the line of the breakers instead of the water margin. It is thus built into a continuous outlying ridge at some distance from the water's edge. It will be convenient to speak of this ridge as a *barrier*. (See Fig. 4, Plate IV.)

The barrier is the functional equivalent of the beach. It is the road along which shore drift travels, and it is itself composed of shore drift. Its lakeward face has the typical beach profile, and its crest lies a few feet above the normal level of the water.

Between the barrier and the land a strip of water is inclosed, constituting a lagoon. This is frequently converted into a marsh by the accumulation of silt and vegetable matter, and eventually becomes completely filled, so as to bridge over the interval between land and barrier, and convert the latter into a normal beach.

The beach and the barrier are absolutely dependent on shore drift for their existence. If the essential continuous supply of moving detritus is cut off, not only is the structure demolished by the waves which formed it, but the work of excavation is carried landward, creating a wave-cut terrace and a cliff.

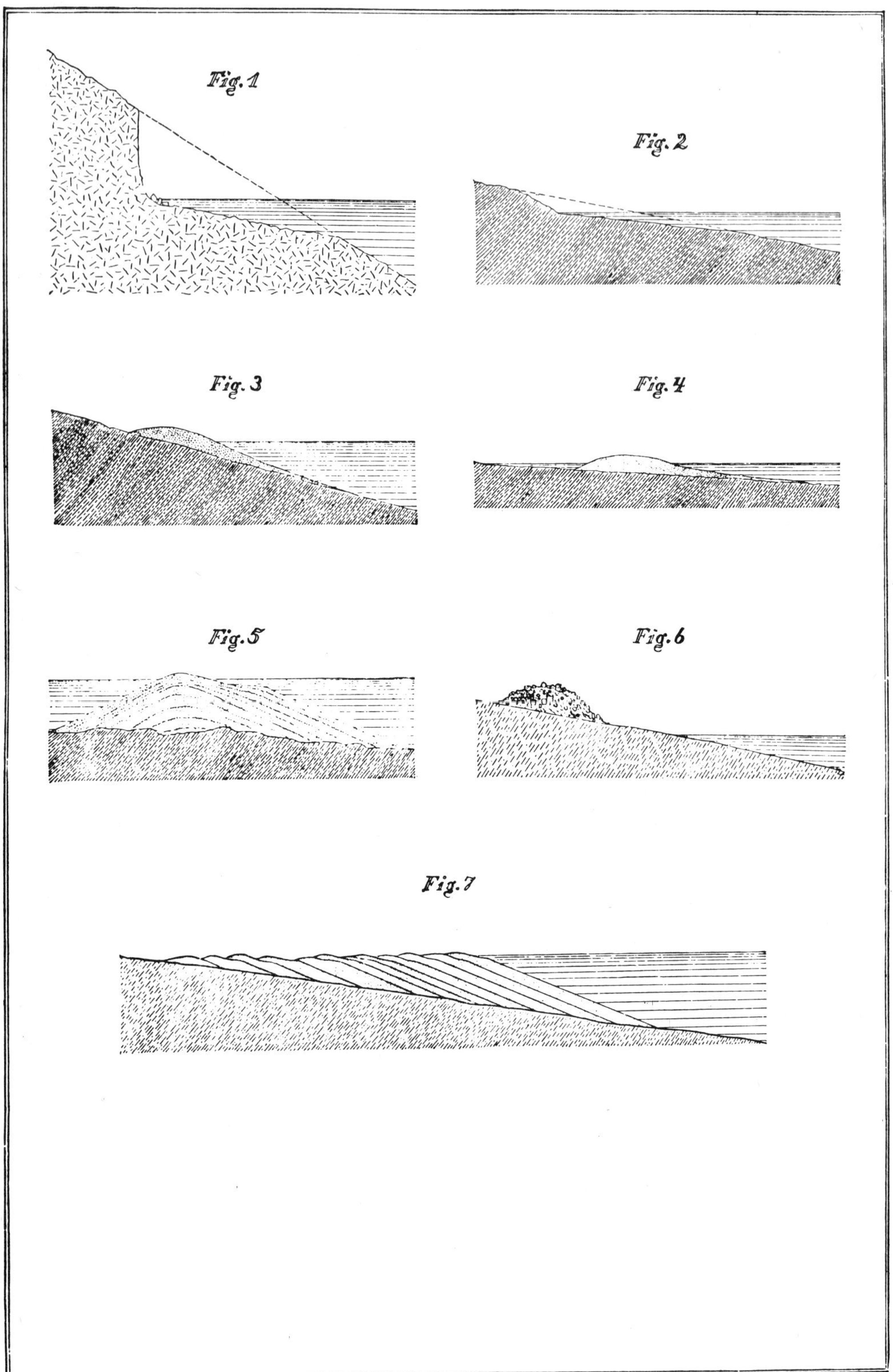

IDEAL CROSS-SECTIONS OF SHORES.

Editor's Comments on Paper 3

3 **McGee:** *Encroachments of the Sea*

Anticipating present-day environmentalists, W. D. McGee wrote in 1890 about the disastrous effects of sea encroachment upon coastal areas. In six lines of reasoning, he deals with the evidence and causes of such an inundation.

Of particular interest to us are McGee's references to "half-drowned keys." (See his pages 441, 442, and 443). Based on an assumption of submergence, McGee reasoned that strings of islands paralleling the coast, which he calls keys, were formed by engulfment of former shore ridges. While wind and wave action were, admittedly, moving the islands shoreward, the rapid rise of the sea along a gently sloping coastal plain quickly flooded broad areas behind them. In this manner, lagoons were formed over former terrestrial surfaces. As opposed to de Beaumont and Gilbert, McGee thought partial submergence of former beach ridges to be the cause of barrier island development.

We now have three hypotheses by which to explain barrier island genesis: local derivation, shore drift, and ridge submergence. Much discussion of these alternative views occurred at the turn of the century until the classic work by D. W. Johnson (*Shore Processes and Shore Line Development,* New York, John Wiley & Sons, 584 pp., 1919) temporarily put an end to the controversy.

Copyright of the Johnson work elsewhere precludes the inclusion of any portion of it here. The interested reader is strongly advised, however, to peruse Chapter 7 of this fine text. Here Johnson finds McGee's hypothesis untenable and turns mainly to the proposals put forth by de Beaumont and Gilbert. In this connection, he takes up the pros and cons of the nearshore bottom as a source of material for bars building up in place as compared to their development by shore drift. To test the validity of these views, Johnson compared actual profiles across barrier islands with the hypothetical profiles of (1) sediment deposited upon a preexisting nearshore slope (Gilbert's shore drift) and (2) erosion of, and deposition upon, the nearshore slope (de Beaumont's local derivation). He concluded that the evidence favored local derivation and threw in his lot with de Beaumont. Johnson held further that while the gently sloping nearshore bottom conducive to wave readjustment might be found along a neutral shore, it would more likely be produced during a period of emergence. Although, as you will see in later readings, more recent workers in this field have found fault with Johnson's conclusions, his views on the subject prevailed for many years.

William McGee was born in 1853, near Farley, Iowa. Largely self-educated, he tried blacksmithing and later studied surveying. As his interest turned to geology, he published reports on the geology of Iowa in the *American Journal of Science* from 1878 to 1882. While carrying out this project, he also studied the archeology of the region. Impressed by his achievements, Major J. W. Powell, in 1883, invited McGee to join the United States Geological Survey. From 1883 to 1894, McGee was in charge of a study of the Atlantic Coastal Plain, and he published extensively on his findings. He was also editor of the *Geological Society of America Bulletin* from 1888 to 1891.

In 1893, McGee transferred to the post of Ethnologist in Charge in the Bureau of American Ethnology, where Major Powell was director. McGee profoundly influenced the Bureau, and by the time of Powell's death in 1902, he was serving as acting director. In 1903, McGee left the Bureau to supervise the anthropological and historical exhibit at the Louisiana Purchase Exposition in St. Louis. President Roosevelt later appointed him to the Inland Waterways Commission, and it was a report to the commission that was McGee's last publication prior to his death in 1912.

Reprinted from *The Forum*, L. S. Metcalf, ed., The Forum Publishing Co., New York, **9**, 437–449 (1890)

3

ENCROACHMENTS OF THE SEA.

WITHIN the last two years a naval vessel was swamped in a dry dock in the lower Chesapeake that was supposed to be well beyond the reach of the highest tides; portions of the great chain of metropolitan dependencies stretching from Sandy Hook to Barnegat Bay, if not to Cape May, have been partially inundated; the mechanical world has been delighted and diverted by the spectacle of a team of locomotives hitched abreast and tandem to a grand caravansary—a miniature city in itself—to move it beyond the reach of the waves; on the shores of the Gulf, between Mobile Bay and the mouth of the Mississippi, villas embowered in fragrant orange groves and moss-festooned live oaks, are menaced by like perils and overtaken often by like catastrophes; broad lagoons and marshes stretching along hundreds of miles of coast have been submerged within a generation; thirty-four years ago, on August 10th, Last Island, a health and pleasure resort of New Orleans, was swallowed by the storm waves, with most of its transient population—"the wealth and beauty of the Creole parishes"—and naught but a tide-washed bank remains to mark its site; and more than once during later years villages and settlements upon the Gulf shores and upon the delta islands of the Mississippi, have been swept from the face of the land and made the prey of the insatiable waters. All of these occurrences, and scores of kindred events not mentioned, have a common feature: in all alike the sea encroached upon the land.

An immediate explanation of each disaster is readily offered: a high, perhaps unprecedented, tide; an unusual storm; a long-continued in-shore wind, by which the waves were driven upon the coast; great floods in the rivers discharging into bays or sounds. These "theories," severally or in conjunction, are promptly seized by the reportorial imagination and set forth in the daily press. But these immediate explanations, ready and rational though they be, fail to tell us why each great flux of the tides is unprecedented; why the waves of this decade wash hights beyond the reach of those of the last; why the land, fortified by every device which human ingenuity has been able to invent and human skill to apply, cannot hold its own against the ocean; they merely state, and do not explain, attendant conditions. The explanation that the ocean is overflowing the land by a secular Spring tide not yet fully in, or by reason of an ebb of the continent not yet fully out, is adequate; but it is a harsh, heartless, pessimistic explanation, opposed by the instinctive notions of the stability of the earth and the buoyant optimism with which vigorous-minded man is inspired. Such an explanation will not and should not be adopted without the most conclusive evidence. Yet the disasters are so many that it would seem wise to scan the evidence. Its aggregate volume is indeed far too great for record here, but the ascertained facts with respect to one or two typical coast stretches may be summarized.

1. Evidence of the encroachment of the sea upon the land is given by history. The older shore lines recorded in maps, deeds, and other documents do not coincide with the newer; and while in some cases the newer shore is locally pushed seaward across an estuary or in the line of a bar, the general change is one of expansion of the ocean at the expense of the land. Comparison of Long Beach, from Barnegat Inlet twelves miles southward, upon the United States Coast Survey maps of 1839 and 1871 respectively, shows that during the intervening period of thirty-two years the land line retreated from 0 to 930 feet, or an average of 545 feet—more than one tenth of a mile. Thence southward to Little Egg Harbor the land line indeed advanced a little way seaward; but here as elsewhere the landward shifting far exceeded the seaward. Comparison of maps of a part of the shore of Cape May County, shows that in the century ending in 1866 Dennis Creek was shortened at its mouth 2,310 feet, and East and West Creeks each 1,880 feet; and that the shore in general receded three quarters of a mile for the surveyed stretch of three miles. In the annual report for 1885 of the State Geologist of New Jersey (the late Dr. Geroge H. Cook, an eminently discreet investigator), many pages are devoted to accounts of coast changes

within the historical period, nearly all of which indicate considerable advance of the sea upon the land, either by the wearing away of shores or by the flooding of lowlands formerly beyond the reach of tide; and the conclusion is stated in these words:

"There is a general wear on the east shore, of the beaches along the Atlantic coast of New Jersey. As a result of this, and the action of wind and wave in carrying sand westward over the beaches, there is a change of position whereby most of what were formerly 'sand reefs,' are now mere accumulations of blown sand on the surface of a former tide marsh. This lateral movement has, in many cases, amounted to more than the breadth of the beach since the settlement of the State, and it is at present going on with undiminished activity. Although in places there has been a certain amount of eastward growth, this has in all cases been dependent on the action of currents which are governed by such local conditions as the position of sand bars, and may at any time be converted into agents of destruction. We must therefore accept it as a rule, on the east shore, that loss is absolute and gain but relative." *

Further southward there are parallel, though perhaps less decisive, indications. Along the Carolina coast the advance of the ocean upon the insular rice plantations has been noted and discussed by observant residents during three generations. About Florida the indications are less decisive, but on the central Gulf coast they again become evident even to casual observation. The general facts, gathered by hundreds of observers and garnered in dozens of printed records, are thus graphically summarized by Lafcadio Hearn:

"The sea is devouring the land. Many and many a mile of ground has yielded to the tireless charging of Ocean's cavalry. Far out you can see through a good glass the porpoises at play where of old the sugar cane shook out its million bannerets, and shark fins now seam deep water above a site where pigeons used to coo. . . . Grande Terre is going; the sea mines her fort, and will before many years carry the ramparts by storm. Grande Isle is going, slowly but surely; the Gulf has eaten three miles into her meadowed land. Last Island has gone!" †

2. Equally significant evidence that the tides now run higher than of old is given by submerged forests and meadows. In the official report for 1882 of the geologist mentioned above, there are accounts of ancient meadows and forests considerably below ordinary high-water mark, recently exposed by the wear of the

* Page 93. † "Harper's Magazine," August, 1888, p. 735.

waves upon the coast. None of the numerous stumps are of brine-loving trees, but only of such as grow well above the reach of tide; and some of them had been cut with the ax. The meadows bear abundant impressions of the hoofs of unshod horses and cattle, though the waters have now so far flooded the beach that it is practically abandoned by men and animals. Dr. Cook adds:

"This fact furnishes another proof of the slow advance of the sea upon the land which is going on along the entire eastern sea coast of the United States, and a reason for the increased effect of the waves in wearing away the shore." *

The tide-flooded forests and swamps of the Carolinas have been described and discussed by Toumey, Holmes, and others; and although the flooding has been by one student ascribed to the breaking down of wave-built barriers during great storms, the evidence is none the less decisive, for the alternate building and breaking down of barriers is the play of the advancing ocean. To-day the segment of the Gulf called Mississippi Sound has partly undermined Pascagoula City, at the mouth of the river of the same name; lines of piling, sometimes two and three deep, supported by fascines and protected by jetties, vainly strive to shield other portions; the waves are rippling and anon thundering over aboriginal habitations marked by shell heaps; charming villas flanking the sun-lit coast are threatened; a mile east of the village, the tap roots of upland pines are bound together by a younger mat of salt-swamp sod; and all, with a long-forgotten bit of "corduroy" road, are revealed by the retreat of a shifting sand levee a foot or more below mean tide.

Even more significant are the buried cedar swamps, which have given rise to a singular industry—the literal mining of timber. At several points in eastern New Jersey enormous quantities of white cedar, liquidambar, and magnolia logs, sound and fit for use, are found submerged in the salt marshes, sometimes so near the surface that roots and branches protrude, and again deeply covered with smooth meadow sod. Many of the trees overthrown and buried were forest giants. In the Great Cedar Swamp, on the creek of the same name, the logs reach a diameter

* Page 83.

of four, five, and even seven feet, and average between two and three feet in thickness. Sometimes the ancient forests are above tide level, and indicate only the slackening flow of the streams; but this is one of the ways in which the rise of the waters is felt upon the land.

But only a tithe of the forest kings sacrificed to implacable Neptune are honored even by unmarked tombs. All along the New Jersey coast from Cape May to Raritan River, along the Virginia and Carolina shores, and on the mainland and the half-drowned keys confining Mississippi Sound, stumps of upland trees peep from beneath the tidal waters; or aged oaks, cedars, and liquidambars, still living, though stunted and gnarled by the poison of the brine, stand here and there to give rude measure of the rate of Ocean's advance. Reviewing these measures in 1868, Dr. Cook estimated the rate of subsidence of the land at about two feet per century, or a quarter of an inch per year.*

3. Evidence of the encroachment of the seas is found also in the geography of the coast. One of the most strongly-marked natural boundaries on the globe divides the middle Atlantic slope into an ocean-fringing lowland, known to geographers as the coastal plain, and an upland springing some hundred feet above tide at its margin, and rising in the interior. The same line marks the inland reach of tide water, the series of cascades and falls known as the "fall line," and the zone of deflection of drainage at which the Delaware, Schuylkill, Susquehanna, Potomac, and other rivers, after maintaining direct courses through Appalachian ranges and Piedmont highlands alike, are turned at right angles literally by a sand bank little higher than their depth. This geographic feature has materially affected the culture of the country. The pioneer settlers ascended the tidal canals to the falls of the rivers, where they found, sometimes within a mile, clear fresh water, the game of the hills and woodlands, the fish and fowl of the estuaries, and abundant water power and excellent mill sites, easy ferriage and practicable bridge sites. Here the pioneer settlements and towns were located; and across the necks of the inter-estuarine peninsulas the

* Ib., page 362.

30

pioneer routes of travel were extended from settlement to settlement, until the entire Atlantic slope was traversed by a grand social and commercial artery stretching from New England to the Gulf States. As the population grew and spread, the settlements, villages, and towns along this line of nature's selection waxed, and many of them yet retain their early prestige; for Trenton, Philadelphia, Wilmington, Baltimore, Washington, Fredericksburg, Richmond, and Petersburg are among the survivors of the pioneer settlements, and the early stage route has become a great railway and telegraph line connecting North and South, as they were connected of old in more primitive fashion.

Now the subaërial and subaqueous surface of the coastal plain set off by this trenchant boundary is cleft by a labyrinth of estuaries—Long Island Sound, Kill von Kull, Arthur Kill, Raritan Bay, Delaware Bay, Chesapeake Bay with its confluent estuaries, the tidal Potomac, etc.—which are recognized by all geographers as "drowned" rivers; and the Hudson and Delaware have narrow, clear-cut channels prolonging their present land-bound courses scores of miles beyond, and hundreds of feet below, the present coast line. Indeed, the lowland fringe stretching from Cape Cod to Cape Hatteras is but the higher part of a great terrace or bench, mostly submarine, skirting the continent in a zone 75 to 150 miles broad. It is known to have been now land and again sea bottom, in many alternations, ever since the middle of the Mesozoic time of geologists; its surface is veneered with tide-scattered sands and gravels; and its drowned rivers indicate that the ocean is rising upon it to-day so rapidly that their channels remain unfilled by the sediments of the sea.

Related evidence appears on other shores; and nowhere is it more curious or decisive than along the Gulf coast east of the mouth of the Mississippi. The coast of Florida is skirted by elongated peninsulas and islands called "keys," separated from the mainland by sounds a fraction of a mile to three or perhaps five miles in width. Thus, in western Florida and Alabama, the bays of Choctawhatchee, Pensacola, Perdido, and Bon Secours are separated from the Gulf by characteristic keys, and nearly or quite connected by narrow sounds; but west of Mobile Bay the keys quickly retreat to five, ten, even fifteen miles from the mainland, and form the sea islands, Dolphin, Petit-Bois, Massacre, Horn, Dog, Ship, and Cat; while the intervening Mississippi Sound is a great slice of the Gulf, the rapidly-encroaching sea having outstripped the slow-moving keys and left them far behind. And most of the water courses of the eastern Gulf water shed, except the detritus-laden Mississippi, are, like their fellows of New Jersey, drowned rivers, with their mouths transformed into estuaries or lagoons; and the mainland is trenched with tidal canals where recent rivers ran.

4. Evidence of similar import is found in structural geology. Rivers gather detritus from the mountain top, the hill side, even the lowland field, and transport it seaward, rapidly during freshets, slowly at low stages; but much of the material is always dropped by the way, a part to be again taken up during later freshets, and another part to lie long and form new lands in a belt or "bottom" of sand, silt, and clay, commonly called alluvium. In the lower course of the river, where the declivity is slight, the alluvium accumulates rapidly, and is pushed out into the sea, bay, or lake, as a delta, like that of the Mississippi, the Nile, or the Ganges. But along the Atlantic slope between Capes Cod and Hatteras, and to some extent further southward, the rivers are not flanked by alluvium in their lower courses, and are destitute of deltas. On the Hudson, Delaware, Susquehanna, Potomac, and James, there is indeed a surprising dearth of the ordinary fluvial deposits. Year after year the rivers drop into their estuaries silt, sand, pebbles, ice-borne bowlders, unquestionably by the hundred tons, yet the receptacles are never filled, the tidal trenches barely shoaled. These rivers are anomalous; but the anomaly presents its own explanation: the rivers fail to fill their estuaries because the valley bottoms sink at least as rapidly as the detritus is poured in. Sedimentation in some of the rivers falling into the southern Atlantic and the Gulf indeed keeps pace with the sinking of the channels, and the bays are short and shallow, or pygmy deltas appear. The vast volume of detritus poured from the Mississippi far exceeds the capacity of the slowly-deepening trough, and so the Mississippi builds a delta after the normal habit of rivers. By the aid of its

bayous and distributaries it has wandered from side to side, adding material here, removing a little there, planing and fashioning the new-made land, but ever building up, and ever pushing out; yet when, through its own caprice or the intervention of man, the great river long abandons any of its radial lines of delta-building, the Gulf waters gradually invade the neglected lowland in lines of swamps and lakes like the modern Borgne and Pontchartrain. Only for a little space can the land hold its own against the hungry waters.

There is another way in which structural geology gives similar evidence. The coast line about Barnegat Bay, about Pascagoula Bay, or on any other typical coast stretch, is an alternation of low-cliffed headlands and reedy lagoons, separated from the tidal waters by low sand banks. Now, when the land stands stationary, the feeble waves and sluggish currents of a shoal offing are unable to clear the bases of the cliffs, the talus is in time bound together and protected by sward and shrubbery, and sloping banks and shelving shores are formed; while if the waters are rising relatively to the land, the wash of the waves at the cliff base keeps pace with the weathering at the cliff summit, and the stimulated currents sweep into estuaries and deep waters the steady gains of waves and weather. So, too, the surf builds a barrier of sand against the entering streamlet. If the land is stationary, the streamlet alternately drops its burden of sand and silt during each high tide, and collects it again at the ebb, and the barrier remains low and the embouchure of the stream narrow; while if the waters are encroaching, the flux of the stream going with the ebb of the tide fails to remove all the detritus, and it gradually accumulates to form a tidal meadow, and the wave-built levee is pushed further and further inland until the wooded hill sides are choked by the swamp muck of the widening lagoon, or until both forest roots and swamp sod reappear on the seaward side at a depth below tide, affording a rude measure of the rate of the land's sinking.

5. Circumstantial evidence that certain parts of the Atlantic and Gulf coasts are sinking, is found in dynamic geology. During the first period in the development of geologic science,

the sculpturing of the land and the accumulation of sediment were ascribed to great *débâcles*, or waves of translation, sweeping the earth from equator to poles; and the *débâcles* were ascribed to, or at least connected with, unexplained uplifting of continents and mountains and downthrowing of sea bottoms. This was the period of catastrophism. Gradually students of earth science, with Lyell at their head, perceived that the valleys are carved by the streams which now occupy them, that the sculpture of the land surface tells of rainfall and storm work perhaps not more rapid than that of to-day, and that the vast bodies of sediment forming the continents are like, both in kind and degree, those now accumulating in lakes and bays and on Ocean's shores; and with this perception came the realization that the surface and sediments of the earth were formed by agencies little if at all more potent than those now at work. This was the Lyellian period, or period of uniformitarianism. When American geologists, with Powell at their head, began to decipher the records of mountain-building inscribed in the tilted, fractured, and contorted sediments of past ages, they gradually perceived that when a continent or mountain range is lightened by the denudation of rivers, it rises, and that when a sea bottom is weighted by the deposition of sediment, it sinks; and indeed that the entire earth crust is in a condition of hydrostatic equilibrium, and relatively as sensitive to changing pressure as the beam of the assayer's balance. As this perception came it was realized that mountain-making, like land sculpture and sediment-forming, depends upon agencies in daily operation upon land and sea. This is the Powellian period, or period of rationalism in dynamic geology. It leaves unexplained but a single category of earth-building processes—the initial uplifting of continents by which the transfer of detritus is inaugurated.

Now the coastal plain from Long Island to Cape Hatteras is the focal tract upon which the great rivers of a vast hemi-ellipse concentrate their mud-charged currents; and a yet smaller tract in the Gulf coast is the *centrum* upon which the rivers of a third of the country converge, and about which their untold millions of tons of detritus are dropped. If the rational principle developed by Powell and now accepted by every competent geologist in every land be true—if it be true that weighted areas sink beneath a never-ceasing accumulation of load—then these tracts must be undergoing depression.

6. Direct evidence of the sinking of the Atlantic coast is given by the configuration of the land. Three periods in the development of geologic science have been characterized; a fourth is dawning. In two intellectual centers at home and one abroad—Washington, Cambridge, Paris—it has come to be recognized that world history may be read from the configuration of the hills as well as from the sediments and fossils of ancient oceans. The volcanic cone displays characteristic features; the uplifted mountain range has a characteristic physiognomy; the ice-swept hills and valleys exhibit unmistakable lineaments; and the tireless stream and frequent rainstorm give unmistakable form and expression to the face of nature. To-day these features, lineaments, forms, physiognomy, and expressions are discriminated and interpreted by half a dozen geologists, and thereby the field of the science is broadened by the addition of a coördinate province—by the birth of a new geology, which is destined to rank with the old, but for which no better name has yet been found than "geomorphology."

Now the student of earth forms perceives at a glance that the topographic configuration of the coastal plain was developed, the waterways outlined, the valleys carved, and the uplands fashioned when the land stood higher than now; and that the stream-carved configuration—which is never imitated by any agency operating below tide-level—passes into the sea or under the alluvium lining the estuaries. He perceives at a glance, too, that the topographic configuration of the Piedmont zone on the further side of the "fall line," was developed when that part of the continent stood lower than now, only the shorter gorges below the cascades of the Schuylkill, the Susquehanna, the Potomac, and their neighbors post-dating the uplift of the land. So the New Geology not only corroborates other evidence in the strongest manner, but at the same time locates the line along which the sinking of the coast is last felt; and this line is the wonderfully trenchant natural and cultural boundary already described.

On reviewing the evidence, it appears that historical records, submerged forests and meadows, geographic configuration, the phenomena of structural geology, the principles of dynamic geology, and topographic forms, all attest that in portions of the Atlantic and Gulf slopes the sea is encroaching upon the land. This evidence is not indeed all equally clear and apposite. The historical evidence is weak in quality because of the inaccuracy of early surveys, early maps, early tide marks, and early records of all kinds; but its volume is vast. Even by itself the historical record shows that, albeit imperceptible in a single year, the advance of the sea is considerable when decades are compared, and enormous when comparison is made between centuries. The evidence of submerged forests and meadows has not always been interpreted alike; but the cases are legion, their significance often unmistakable, and in the best-observed regions the testimony is conclusive. The evidence of geographic configuration —of drowned rivers, half-flooded islands, and outlying keys— proves that the land is either recently submerged or now sinking. The evidence of structural geology, and particularly that of the dearth of alluvium in the absence of deltas at the mouths of mud-charged streams, is of like tenor and value. The value of the evidence of dynamic geology depends upon the validity of the Powellian principle, which all competent authorities accept, though some might question its quantitative sufficiency in the given case. The evidence of geomorphology—of the forms of hills and the features of plains—is eminently apposite, clear, and conclusive; it applies not only to the coast, but to the entire coastal plain; and it might be made to give rude measure of the rate and amount of the earth movement. But however the several lines be weighted, the evidence is consistent and cumulative, and permits no escape from the conclusion that certain portions of our coast are yielding before advancing seas.

On reviewing the sum of evidence by areas, it is found to prove oceanic encroachment along the Atlantic coast from Sandy Hook to Cape Henry, and along the Gulf coast between the mouth of the Mississippi and Mobile Bay; to suggest a like condition all the way from Cape Cod to Cape Fear; and to give little indication as to change in the relations of sea and land about the shores of the Florida peninsula. Concerning the northern New England coast, the western Gulf coast, and the rugged promontories and flat sand beaches of the Pacific coast, where the records are scanty or equivocal, "this deponent saith not."

Men who haunt the shores for pleasure or for profit, naturally inquire the rate at which the sea is encroaching upon their domain. The cautious estimate of the rate at which the New Jersey coast is sinking made by the official geologist of that State, is two feet per century. Now the mean seaward slope of the coastal plain, including its subaërial and submerged portions, is perhaps six feet per mile; so that each century's sinking would give a third of a mile, and each year a rod, of lowland to the ocean; and this would appear to be below the rate of encroachment indicated by comparison of maps. This is probably the maximum rate for this country. Pending further observation and the scanning of other records, little more can be said.

Men of maritime lands naturally inquire whether the continent settles easily and uniformly, or whether it descends by successive starts at intervals; for rapid mass movement in the earth is justly believed to beget the earthquake, and perchance the tidal wave; but upon this point the evidence and science (aside from a hypothetical presumption in favor of *per saltum* movement) are silent. Last Island was indeed overwhelmed when, after a ten days' northeaster which forced the Gulf waters offshore and allowed the water-heavy silts and sands to settle in the lighter air, the wind veered to the south, and the surf swept back over sunken shores and delta islands; yet there were no deep-seated earth tremors. The land went down in the lower Mississippi region, it is true, during the greatest of American earthquakes—that of New Madrid, in 1811–13; but it has not been learned whether the earthquake caused the sinking or the sinking the earthquake. Slight quivers of earth, too, appear to run along the "fall line," and are commonly recorded many times annually in its vicinity; but it is more probable that they represent gradual and easy relief of earth stresses than that they are premonitory of a catastrophe.

Prophecy of evil is an ungrateful and ungraceful task, before which Science justly quails, for Science is no longer content only to make two blades of grass grow where one grew before. She is ambitious also so wisely to use the productions of the earth that one blade of grass will fill the place filled by two before, and moreover to wring from barren rocks and desert sands artificial substitutes for natural blades of grass, and thus to multiply indefinitely the gifts vouchsafed by unaided nature. So her devotees are the most lightsome of optimists, the most sanguine of philanthropists. But optimism must not be confounded with short-sight, or philanthropy with foolhardiness.

There is a broad lowland stretching from Sandy Hook to Cape Henry, and running inland to the line of metropoles, and another washed by Mississippi Sound, upon which the sea is encroaching. They are wave-fashioned plains, but recently wrested from the ocean, and Ocean reclaims its own. Already its octopus arms have seized the lowlands in horrid embrace, and day by day, month by month, year by year, generation by generation, the grasp is tightening, the monster creeping further and further inland. Each average year the water mark advances a rod. The seaside cottage with a broad lawn before it has an "expectation of life" of a decade or a generation; but the cottage at the verge of the cliff may go in a year and must go in a lustrum, unless human devices outwit and overpower the waves for an exceptional period. On most other eastern and southern coasts the waves are also encroaching, but their progress is slower. And the ocean's power is too great for puny man to oppose successfully; he can only provide against, and slowly retreat before, the invasion.

W J McGEE.

Editor's Comments on Paper 4

4 **Price:** *Barrier Island, Not "Offshore Bar"*

W. Armstrong Price was born in Richmond, Virginia, in 1889, received the A.B. degree from Davidson College in 1909, and the Ph.D. degree from Johns Hopkins University in 1913. Price worked with the Maryland, West Virginia, and U.S. Geological Surveys between 1910 and 1918. Since 1919 he has been an oil and gas exploration consultant. He also taught at West Virginia University and Texas A.&M. Price is an active member and officer of many geologic organizations and has published extensively, half of his papers being on coastal problems. He is now a consulting geologist and geological oceanographer and lives in Corpus Christi, Texas.

In this short paper, Price takes exception to the ambiguity of terms being used in 1951 to designate barrier islands. He maintains that an offshore bar is essentially a littoral-zone submerged bar, whereas a barrier island is an island or spitlike peninsula of sand, gravel, or shingle, situated offshore on a gently sloping shallow bottom. It is interesting to note that spits are included in the latter definition, a view not shared by many coastal geomorphologists whose work follows in this volume. The reader is urged to note particularly the evidence of changing thought on the topic when reading the paper by John J. Fisher. Nevertheless, Price's paper stands as one of the earliest efforts to correct a growing misuse of barrier island terminology; it is included here in recognition of that position in the literature.

Reprinted from *Science*, **113**, 487–488 (1951)

4

Barrier Island, Not "Offshore Bar"[1]

W. Armstrong Price
Agricultural and Mechanical College of Texas, College Station

It is time for students of shorelines to standardize their terminology by removing ambiguous terms, discontinuing double uses of the same term for different features and multiple terms for a single feature. The terminology of bordering islands ("barrier sand reefs") and miscellaneous "offshore" bars needs such standardization. Some specific changes are suggested here that seem to be especially needed at this time in view of the widespread attention being given to marine geological problems and the emergence of a specialty of geological oceanography. Other revisions of the terminology of barlike structures, including the modern use of the ancient eolian term "dune" for a somewhat dunelike underwater bar, are not discussed.

Douglas W. Johnson in his shoreline treatise (*1*) gave an extensive analysis of the form and origin of the sandy barrier bar or island of gently sloping shoreline bottoms, such as those of coastal plains. He concluded that the structure originated largely as a bar or bars formed offshore—normally submerged—and that the driving into shallow water of a series of such bars and their up-building there led to the emergence above normal tidal range of a barrier island. Hence, he adopted the term "offshore bar" for the island.

Johnson's hypothesis of the origin of the barrier island and the use—though somewhat ambiguously (*2*) —of his "offshore bar" as a criterion of "emergent shorelines" attracted much attention and fixed the use of the term "offshore bar" in the minds of many geologists. The class of normally submerged bars formed offshore which, in Johnson's view, contributed largely to the origin of the barrier island was left without a specific name, to be called merely *bars*.

Students of modern sedimentation who have not also been closely concerned with the geomorphology or structure of the barrier island itself seem to have overlooked Johnson's appropriation of "offshore bar" and continue to use the term for true bars lying in various offshore positions, typically just beyond and below normal low tide levels. This situation has caused time-consuming confusion, specifically for this writer in some transoceanic correspondence.

Textbooks of geomorphology and geology have not, so far, arrived at a full standardization of coastal terminology. This deficiency is evident in the two current texts on marine geology or geological oceanography. Thus, F. P. Shepard's *Submarine Geology* (*3*) speaks on one page of the barrier island as an "offshore bar," but on another uses the same term for any one of a series of normally submerged bars found at increasing depths on and seaward from a gently sloping beach. Ph. H. Kuenen's *Marine Geology* (*4*) seems to use "offshore bar" only in the Johnsonian sense, but repeats Shepard's figure (*3*, Fig. 33; *4*, Fig. 121) in which a bar is shown in the offshore zone of the beach without providing a specific qualifying term for such a bar. Kuenen's contribution to the confusion in the use of "offshore bar" is to term some of them "sandbanks" (*4*, 270, and Fig. 124), thus attempting to appropriate for a specific technical meaning a term long in wide common use for any kind of sandy beach, dike, or levee. Kuenen's "sandbank" is the ridge in the "low and ball" topography. Such bars have lately been termed "longshore bars" by Shepard (*5*).

The following terminology is proposed to correct the ambiguity and double uses noted:

1. *Offshore bar*: any normally submerged bar formed offshore in marine or fresh waters, found chiefly in the littoral zone. It includes, among other bars, the ridges of the low and ball topography that have been called "sandbanks" by Kuenen and "longshore bars" by Shepard, the latter being here preferred. To term such a ridge a "sandbank" has no more distinctiveness than to call it a "ball."

2. *Barrier island*: the island, or chain of islands of sand, or sand and gravel or shingle, lying offshore on a gently sloping shallow bottom. It is separated from the shore by a coastal lagoon or "sound." Its beach is the main line of resistance of the land to the attack of the major waves of the offshore region, the beach slope forming a profile of equilibrium. The term is used technically and includes, as a single barrier island or *barrier island chain*, such segments in the same alignment as may be separated by tidal inlets or other tidal openings. This barrier is commonly tied to land at one or both ends, either to a stream delta or to a headland of some other origin, as an erosional projection of a drowned coast. The segments of the island or chain may include spitlike peninsulas, the difference between spit and spitlike segment of a barrier island being one of size and individual usage, as in the case of hill and mountain. Synonyms: Barrier beach, offshore bar (Johnsonian sense), and barrier sand island. The term "barrier beach" is not favored because many barrier islands are much more than beaches, being widened by washover fans, tidal deltas, and eolian plains and prograded to form extensive beach plains and cuspate forelands.

3. *Barrier reef*: an organic, commonly partly coralline reef lying offshore from a continent or island much in the position of the barrier island of sand. It includes barrier reefs of islands—commonly volcanic—and such great barrier reefs as that of the northeastern Australian coast. The terminology of coralline reefs is fairly well stabilized and has recently been summarized, with some additions, by Kuenen (*4*, 423–26).

It seems remarkable that our geological terminology is as well organized as it is in view of such obstacles as (1) the characteristic use in English of

[1] Contribution from the Department of Oceanography of the Agricultural and Mechanical College of Texas, No. 9.

one word in multiple meanings, (2) the incompleteness in scope of our scientific dictionaries, (3) the general brevity and sketchiness of book indices, and (4) the relative scarcity of detailed treatises of the scope of Grabau's *Principles of Stratigraphy* and Twenhofel's *Treatise on Sedimentation,* in contrast with the abundance of textbooks suitable for review in a one- or two-semester course of graduate or undergraduate instruction. Some of these shorter textbooks of geology present somewhat individualistic terminologies.

References

1. JOHNSON, D. W. *Shore Processes and Shoreline Development.* New York: Wiley (1919).
2. PRICE, W. A. *J. Geomorphol.*, **2**, 357 (1939).
3. SHEPARD, F. P. *Submarine Geology.* New York: Harper (1948).
4. KUENEN, PH. H. *Marine Geology.* New York: Wiley (1950).
5. SHEPARD, F. P. *Tech. Memo. 15.* Beach Erosion Bd., U. S. Army C. of E. (1950).

Editor's Comments on Paper 5

This paper was presented at the Second Coastal Geography Conference, Baton Rouge, Louisiana, 1959, and subsequently included in the Wyoming Geological Association's *Symposium on Late Cretaceous Rocks,* 1961. It is one of many reports in the literature, too numerous to be included here, that discuss the extensive barrier islands bordering the northwest section of the Gulf Coast.

Le Blanc and Hodgson maintain that the Texas shoreline barrier islands developed during a stage of standing sea level that started approximately 5000 years ago, following an early Recent rise from the late Pleistocene low. Stillstand is thus introduced in comparison with the asymptotic and fluctuating curves invoked by other coastal workers. For the method of barrier island development, Le Blanc and Hodgson hold that barriers were generated by littoral transport of abundant deltaic sediment. They also state that washover fans broadened the islands and cite the ridge and swale topography as evidence of seaward progradation.

Rufus Joseph Le Blanc, Sr., was born in southern Louisiana in 1917. He received his B.S. and M.S. degrees in 1939 and 1941 at Louisiana State University, and was then employed by the Mississippi River Commission as an assistant geologist. In 1948 he joined the Exploration and Production Research Division of Shell Oil Company in Houston, and in 1965 transferred to the firm's Exploration Training Department. Le Blanc is a member of the Society of Economic Paleontologists and Mineralogists, the Houston Geological Society, and a Fellow of the Geological Society of America. He has authored a number of papers on the Recent and Late Pleistocene geology of the Gulf Coast.

W. D. Hodgson was born in Boonsville, Texas, in 1933. He received his B.A. and M.A. degrees in geology at Texas Christian University in 1955 and 1957, respectively, and did postgraduate study at Oklahoma University from 1963 to 1965. Hodgson was employed by the Shell Development Company in 1957, moved, in 1961, to the Sinclair Oil and Gas Company, and has been with the Atlantic Richfield Company since 1969. He is a member of the American Association of Petroleum Geologists and the Corpus Christi Geological Society.

Reprinted from *Gulf Coast Assn. Geol. Soc. Trans.*, **9**, 197–220 (1959)

ORIGIN AND DEVELOPMENT OF THE TEXAS SHORELINE

Rufus J. LeBlanc and W. D. Hodgson*

5

ABSTRACT

The shoreline of Texas generally parallels the regional trend of the coastal plain and the outer edge of the continental shelf. The seaward slope of that portion of the coastal plain which lies below 300 feet elevation is about the same as the slope of the inner continental shelf which is less than 300 feet deep. This slope varies from about 6 feet per mile south of Corpus Christi to about 2½ feet per mile east of Galveston, Texas.

The coastal features of Texas consist of (1) alluvial valleys and deltaic plains, (2) estuaries, (3) barrier islands, and (4) lagoons.

The Gulf coastal plain in Texas is drained mainly by nine rivers and their tributaries. The size of the alluvial valley and deltaic plain for any one stream is proportional to the size and silt load of that stream. Large streams, such as the Rio Grande and the Brazos River, have wide alluvial valleys and broad deltaic plains and empty directly into the Gulf of Mexico. Smaller streams, such as the Nueces and San Jacinto rivers, flow in comparatively narrow valleys and empty into shallow bays or estuaries which lie behind barrier islands. These estuaries, such as Baffin, Corpus Christi, San Antonio, and Galveston bays, are elongated perpendicular to the coast. The lagoonal type of bay lies directly behind the barrier island and is elongated parallel to the coast, or normal to the estuary type of bay. Examples of this type are Laguna Madre, Aransas, Matagorda, West, and East bays. The barrier islands of Texas range in length from about 15 miles to over 110 miles and rise as much as 50 feet above sea level. One of the characteristic features of these barrier islands is a series of abandoned beach ridges which are aligned parallel to the present seaward edge of the islands.

The shoreline of Texas consists of a Gulf shoreline and a bay shoreline. The Gulf shoreline is the seaward edge of the barrier islands and deltaic plains and the bay shoreline lies at the edge of the mainland and behind the barrier islands. The Gulf shoreline is generally very regular, arcuate, and characterized by well-developed sand beaches. In contrast, the bay shoreline is quite irregular and generally devoid of sand beaches. In many places the bay shoreline is associated with low-lying Recent marshes; low bluffs characterize the bay shoreline wherever wave action is eroding Pleistocene terrace deposits.

Some features of the shoreline are of late Pleistocene age while others are definitely related to the Recent epoch. During the last Pleistocene glacial stage when sea level was lowered approximately 450 feet, the coastal Texas streams deeply entrenched their valleys and the Gulf shoreline was probably 50 to 140 miles seaward of the present shoreline. With melting of the late Pleistocene glaciers and the accompanying rise in sea level, Texas streams alluviated their entrenched valleys. Sedimentation in these valleys did not keep pace with rising sea level and consequently the lower portions of the entrenched valleys were drowned to form a series of estuaries. The bay shoreline of Texas originated during this stage. During the standing sea level stage, which began about 5,000 years ago, the large Texas rivers, the Rio Grande, Brazos, and Colorado, filled their former estuaries and constructed broad deltaic plains which protrude into the Gulf. The smaller Texas rivers which carry lesser quantities of sediments are still in the process of filling their estuaries. A series of barrier islands was formed along the coast between the Rio Grande and Colorado-Brazos deltaic plains and east of the Colorado-Brazos deltaic plain, giving rise to the Gulf shoreline along these segments of the coast. The abandoned beach ridges and intervening low swales and mud flats, which are well preserved on these barrier islands, clearly demonstrate the seaward growth of the islands during the standing sea level stage. Although most of the Gulf shoreline of Texas has regressed seaward during the past few thousand years, there are a few notable examples of local marine transgressions.

* Shell Development Company, Houston, Texas. This paper was presented first at the Second Coastal Geography Conference at Baton Rouge, Louisiana, in April 1959.

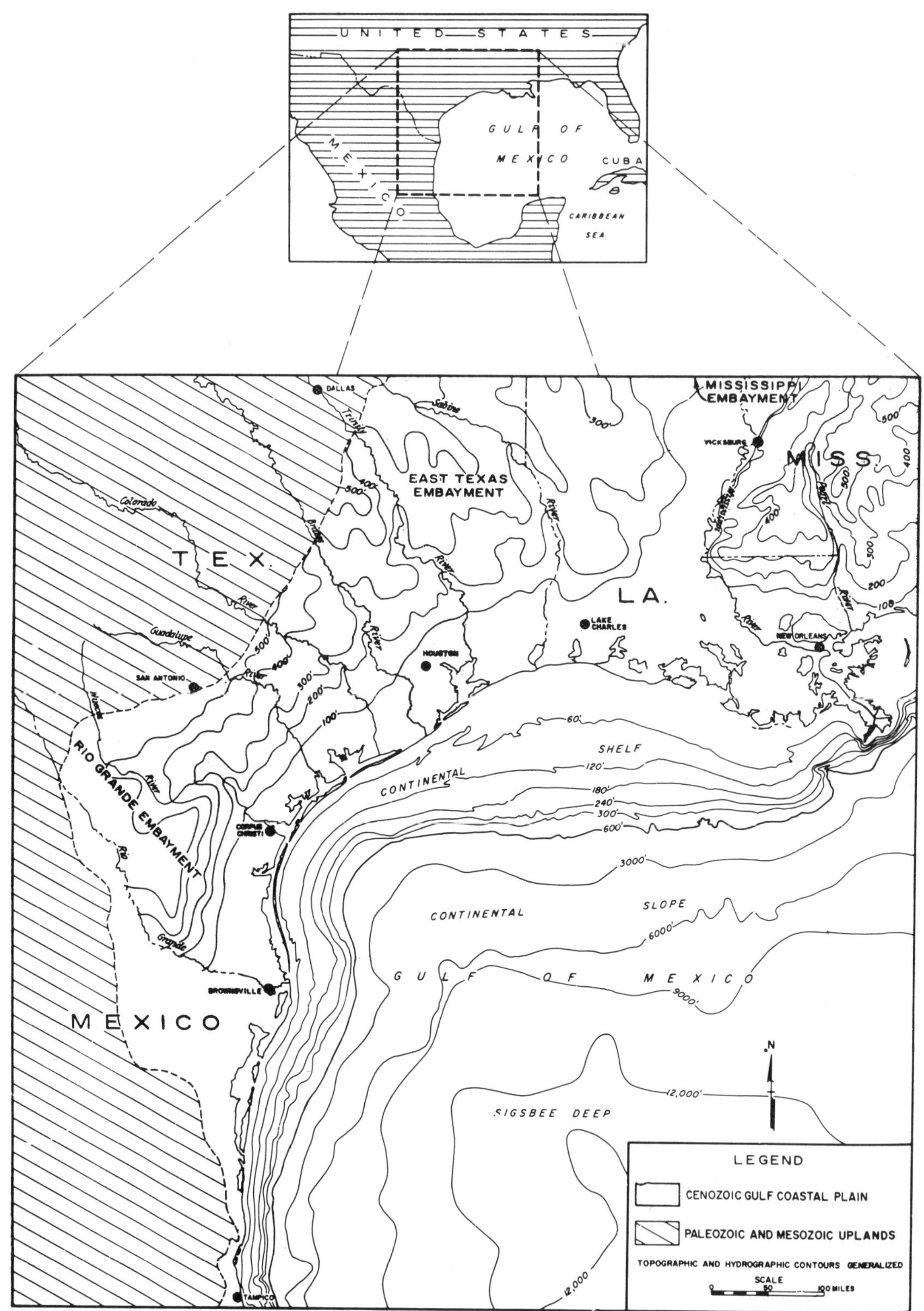

Figure 1. Relationship of the Texas shoreline to the Gulf coastal plain and Gulf of Mexico.

INTRODUCTION

The shoreline of Texas forms the boundary between the northwestern Gulf of Mexico and the western Gulf coastal plain for a distance of about 375 miles between the Rio Grande and the Sabine River. The shoreline trends in a northerly direction along the lower one-third of the Texas coast, then curves gently in an easterly direction along the middle and upper Texas coast. That this north-northeast trend generally parallels the regional trend of the western Cenozoic Gulf coastal plain and the outer edge of the continental shelf can be seen on figure 1.

Geological events associated with the late Quaternary have controlled the development of the Texas shoreline. Some features of the shoreline are of late Pleistocene age; others are definitely related to the Recent epoch.

Two general types of evidence indicate a late Quaternary age for the Texas shoreline. The most obvious consists of the various types of physiographic features which characterize the coastal region, such as alluvial valleys, abandoned river courses, and beach ridges. Geomorphological analyses based on integrated studies of aerial photographs and topographic maps, together with field observations, provide excellent tools for deciphering the orig'n and stages in the development of the Texas coast.

The second type of evidence consists of the various kinds of sediments which are associated with each physiographic feature. But because these sediments are seldom exposed on the coastal plain and in many cases occur under water, they are more difficult to study and interpret.

The Gulf Coastal Plain

The extent of the coastal plain in Texas and northeastern Mexico is shown on figure 1. In Texas the Cenozoic plain varies in width; it is about 280 miles in the East Texas embayment region, 150 miles between San Antonio and the Gulf, and 200 miles in the Rio Grande embayment. Southward in Mexico the Cenozoic coastal plain gradually decreases in width until it is only about 10 miles wide near Tampico.

The generalized topography of the Cenozoic coastal plain is also shown on figure 1. The inland half of the coastal plain consists of dissected Tertiary uplands ranging in elevation fom 150 to over 500 feet. Near the coast the plain is characterized by less-dissected, gently seaward-sloping Quaternary terraces. The average seaward slope of the youngest Pleistocene terrace and the Recent coastal plain is only about one foot per mile.

The western Gulf coastal plain is drained mainly by nine rivers and their tributaries. The course of each stream, with information regarding its drainage area, discharge, and average sediment load per year is shown on figure 2. Stevens (1951) has discussed the silt load of Texas streams in some detail. There is a direct relationship between the size and load of each stream and the configuration and characteristics of the shoreline where the stream reaches the Gulf. The larger streams, such as the Brazos, Colorado, and Rio Grande, have developed extensive deltaic plains which protrude into the Gulf. A number of smaller streams, which transport less sediment, have considerably smaller deltas which are being constructed at the head of shallow bays well removed from the shoreline of the Gulf (fig. 2).

Large variations in climate prevail along the Texas coast (Thornthwaite 1948, and Russell 1945). The upper Texas coast, between Galveston and the Sabine River, is characterized by high humidity (fig. 3), with an average annual precipitation of 45 to 50 inches. Southward from Galveston, the climate gets progressively drier. The lower Texas coast, between Corpus Christi and the Rio Grande, is semi-arid, with less than 25 inches of rainfall per year. Corresponding changes in vegetation and soil types occur in a southerly direction from the Sabine River region.

Subsidence contemporaneous with sedimentation has been a dominant process in the western Gulf coastal plain throughout the Cenozoic era. For example, the Miocene beds of Texas, which were initially deposited at or near sea level, have subsequently been uplifted to elevations of about 200 feet above sea level along the outcrop belt and tilted seaward south of the outcrop to depths of several thousand feet below sea level at the present shoreline. A similar seaward-tilting of all pre-Miocene and post-Miocene beds has occurred in the region, and the process of subsidence contemporaneous with sedimentation is active at the present time.

The Gulf of Mexico

The northwestern Gulf of Mexico can be divided on the basis of hydrographic data into the continental shelf (0-600 feet), the continental slope (600-9000 feet), and Sigsbee Deep (9,000-13,000 feet), as shown on figure 1. The continental shelf varies in width from about 40 miles east of Tampico to over 140 miles south of the Sabine River. The inner shelf (0-300 feet) is remarkably flat, and its seaward slope is about the same as that of the coastal plain which lies between the shoreline and the 300-feet-above-sea-level contour. South of Sabine River the inner shelf slopes seaward at the rate of about 2½ feet per mile, and east of Corpus Christi its average slope is about 6 feet per mile. The outer continental shelf (300-600 feet) has a greater seaward slope, which averages about 25 feet per mile. It is characterized by a few topographic highs, many of which are believed to be salt domes (Shepard 1937; Carsey 1950; Weaver 1950; Greenman and LeBlanc 1956).

The topography of the continental slope is considerably more rugged than that of the continental shelf. The average slope between the 600-foot contour and the 9,000-foot contour is about 70 feet per mile. Well-defined basins and ridges occur in this region with relief of as much as 3,000 feet (Gealy 1955).

The ocean currents, waves, and winds in the Gulf of Mexico, especially within the inner continental shelf and shoreline region, play a very important role in the configuration of the Texas shoreline (figs. 4 and 5). Although the exact nature of the surface and subsurface currents is not known, a considerable amount of general information is available (Leipper 1951 and 1954). The results of a survey of surface current directions for each month of the year are shown in figure 5. Hedgpeth (1953) made a study of these data and presented generalizations concerning current directions for each of the four seasons of the year.

The directions of prevailing and predominant winds in the northwestern Gulf region and their influence on coastal sedimentation were recently discussed by Lohse (1955) (see fig. 4).

The tides in the Gulf of Mexico are of two types: (1) daily tides with one high and one low for each 24-hour

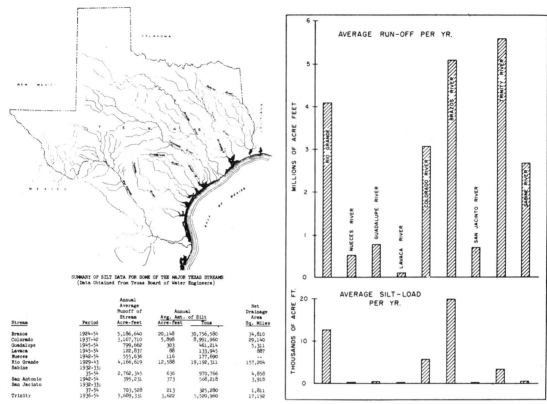

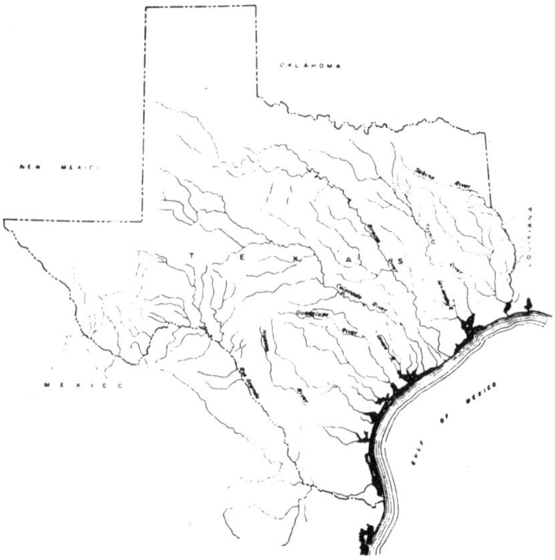

SUMMARY OF SILT DATA FOR SOME OF THE MAJOR TEXAS STREAMS
(Data Obtained from Texas Board of Water Engineers)

Stream	Period	Annual Average Runoff of Stream Acre-feet	Annual Avg. Amt. of Silt Acre-feet	Annual Avg. Amt. of Silt Tons	Net Drainage Area Sq. Miles
Brazos	1924-54	5,186,640	20,148	30,756,580	34,810
Colorado	1937-42	3,167,710	5,898	8,991,960	29,140
Guadalupe	1945-54	799,662	303	461,214	5,311
Lavaca	1945-54	122,837	88	133,945	887
Nueces	1942-54	555,636	116	177,690	--
Rio Grande	1929-43	4,166,619	12,588	19,192,311	157,204
Sabine	1932-33; 35-54	2,762,345	636	970,766	4,858
San Antonio	1942-54	395,231	373	568,218	3,918
San Jacinto	1932-33; 37-54	703,528	213	325,280	1,811
Trinity	1936-54	5,689,331	3,622	5,520,960	17,192

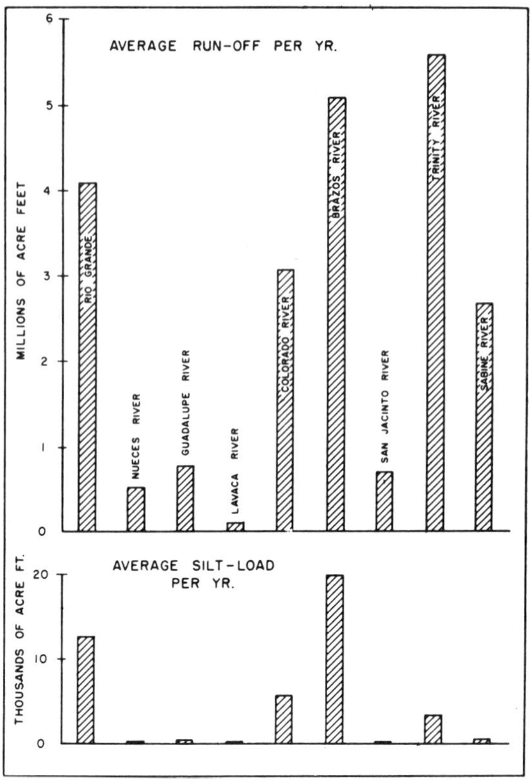

Figure 2. Silt load of Texas streams.

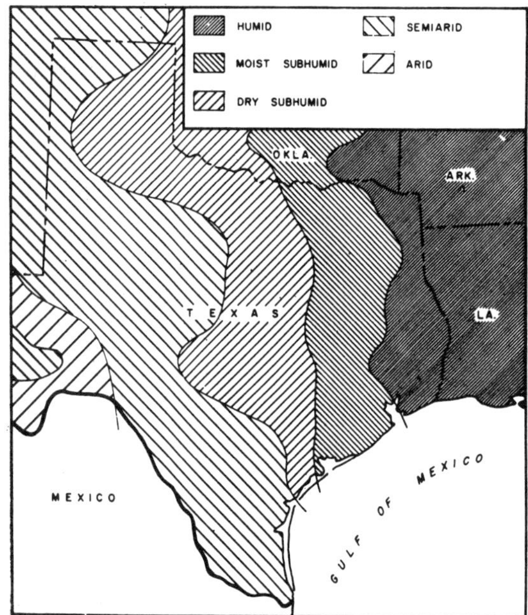

Figure 3. Climate of Texas (after Thornthwaite 1948).

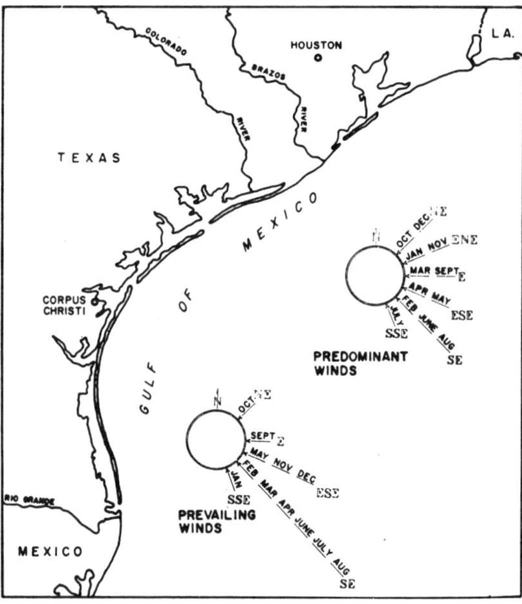

Figure 4. Prevailing and predominant winds of coastal Texas (after Lohse 1955).

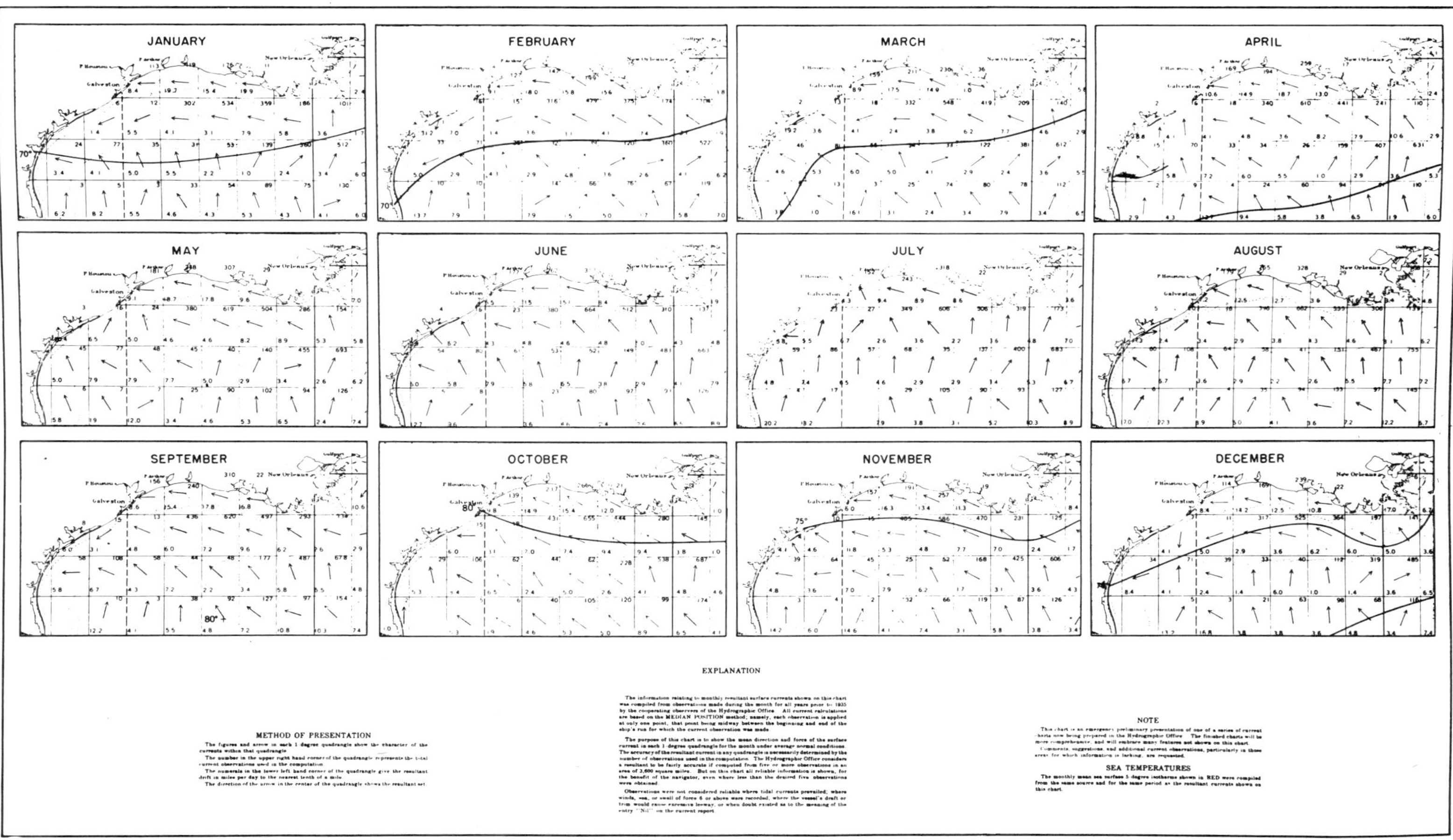

Figure 5. Surface currents in the Gulf of Mexico (data from U. S. Hydrographic Office).

PHYSIOGRAPHIC FEATURES AND WATER BODIES WHICH HAVE ORIGINATED AS A RESULT OF VALLEY ENTRENCHMENT DURING THE LATE PLEISTOCENE LOW SEA LEVEL STAGE AND SUBSEQUENT VALLEY ALLUVIATION DURING THE RECENT RISING AND STANDING SEA LEVEL STAGES. FEATURES LISTED BELOW ARE GENERALLY ALIGNED NORMAL TO THE TEXAS COAST.	
ALLUVIAL VALLEYS	ESTUARIES
RIO GRANDE	FILLED
SAN FERNANDO CREEK	GRULLO BAYOU - BAFFIN BAY
AQUA DULCE CREEK	ALAZAN BAY
NUECES RIVER	NUECES & CORPUS CHRISTI BAYS
ARANSAS RIVER	COPANO BAY
GUADALUPE RIVER	SAN ANTONIO BAY (GREEN LAKE & HYNES BAY)
LAVACA RIVER	LAVACA BAY
BRAZOS-COLORADO RIVERS	FILLED
SAN JACINTO RIVER	GALVESTON BAY
TRINITY RIVER	TRINITY BAY
SABINE RIVER NECHES RIVER	SABINE LAKE

PHYSIOGRAPHIC FEATURES AND WATER BODIES WHICH ORIGINATED AS A RESULT OF SHORELINE PROCESSES DURING THE LATTER PART OF THE RECENT RISING SEA LEVEL STAGE AND PARTICULARLY DURING THE PRESENT STANDING SEA LEVEL STAGE. FEATURES LISTED BELOW ARE ALIGNED PARALLEL TO THE TEXAS COAST AND NORMAL T[O] ESTUARIES AND ALLUVIAL VALLEYS.	
BARRIER ISLANDS & PENINSULAS	LAGOONS
PADRE ISLAND	LAGUNA MADRE
MUSTANG ISLAND	EASTERN 1/3 OF CORPUS CHRISTI B[AY]
ST. JOSEPH ISLAND	ARANSAS BAY
MATAGORDA ISLAND	MESQUITE BAY, SE 1/3 OF SAN ANTONIO BAY, ESPIRITU SANTO BAY
MATAGORDA PENINSULA	MATAGORDA BAY
GALVESTON ISLAND	WEST BAY
BOLIVAR PENINSULA	EAST BAY

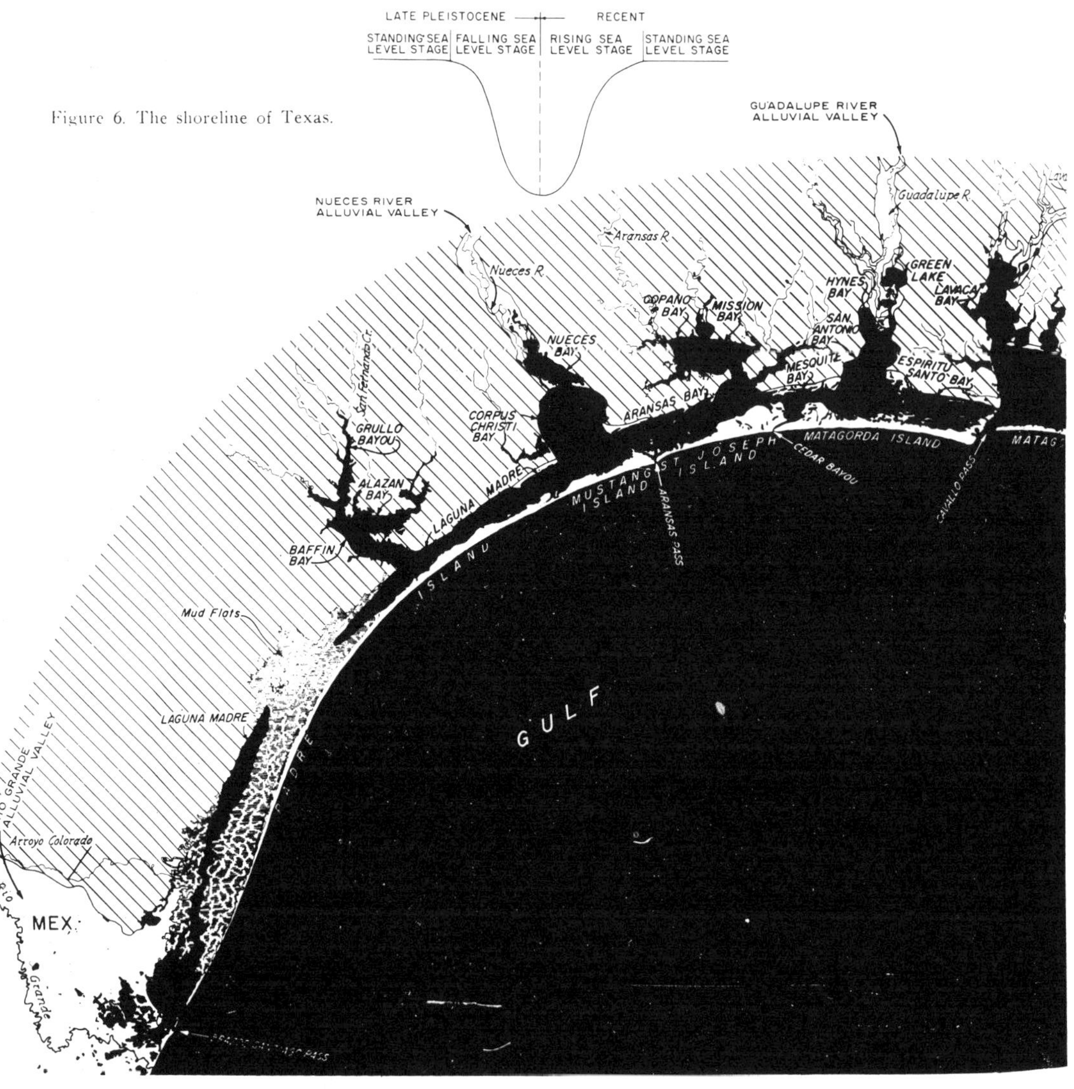

Figure 6. The shoreline of Texas.

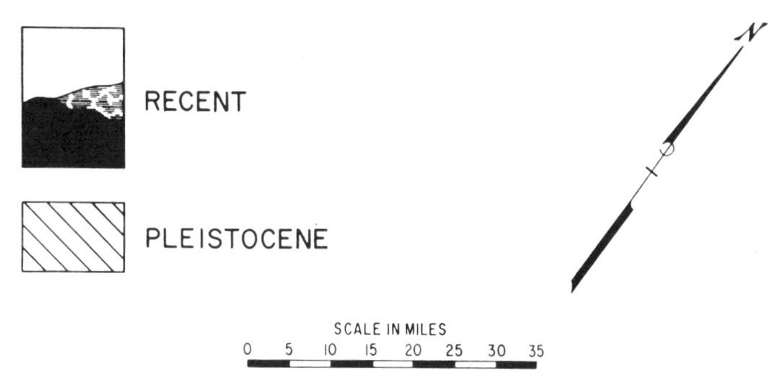

Figure 6. The shoreline of Texas.

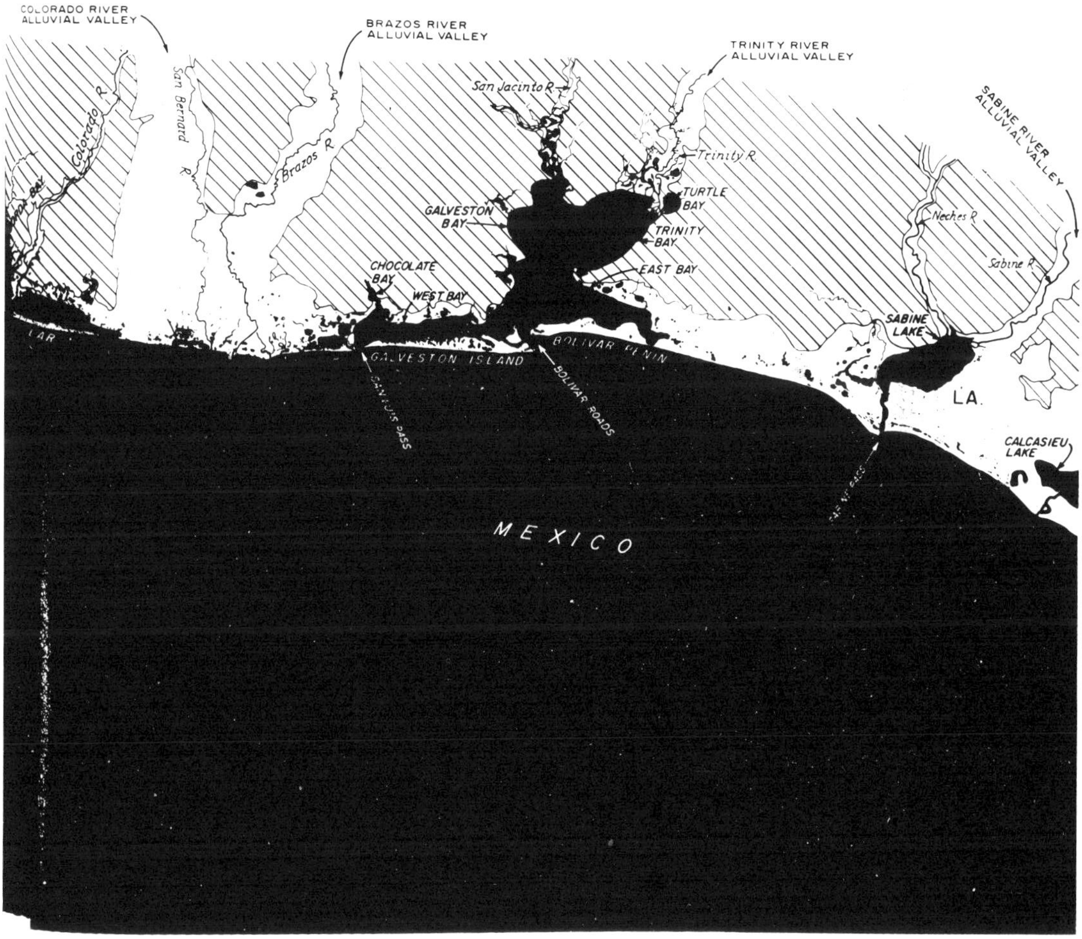

period, and (2) mixed, semi-daily tides with two highs and two lows for each 24-hour period (Hedgpeth 1953). The average range in tide is not more than a foot or two. From June through October many storms and hurricanes, accompanied by high tides up to 15 feet above sea level, move in a northwesterly direction over the Texas coast. The tides in the Gulf of Mexico are discussed more fully by Marmer (1954).

GENERAL CHARACTERISTICS OF THE TEXAS COAST

The coastal features of Texas can be divided into four main groups on the basis of their general characteristics and origin. These are (1) alluvial valleys and deltaic plains, (2) estuaries, (3) barrier islands, and (4) lagoons.

Each stream of coastal Texas flows within an alluvial valley of Recent age (fig. 6). These valleys generally lie a few feet below the flanking Pleistocene terrace surface, and their seaward slopes are generally less than one foot per mile. Each alluvial valley merges with a deltaic plain near the coast. The size of the alluvial valley and deltaic plain for any one stream is proportional to the size and load of that stream. Large streams, such as the Rio Grande and Brazos River, have wide alluvial valleys and broad deltaic plains. They empty directly into the Gulf of Mexico. The smaller streams, such as the Nueces and San Jacinto rivers, flow in comparatively very narrow valleys and empty into shallow bays rather than the Gulf of Mexico. The Colorado River, which flows in a very narrow alluvial valley just north of Matagorda Bay, occupied a much wider alluvial valley to the east a few thousand years ago (figs. 6 and 11).

The estuary type of bay, such as Corpus Christi, San Antonio, Lavaca, and Galveston bays, is elongated normal to the coast. Most of the coastal streams, with the exception of the Rio Grande, Colorado, and Brazos rivers, are presently constructing deltas in the upper parts of these bays or estuaries. The lagoonal type of bay is elongated parallel to the coast or normal to the estuary type of bay. Examples of this type are Laguna Madre, Aransas, Matagorda, West, and East bays. Both types of bays are very shallow, usually less than 10 to 12 feet deep, and generally have mud bottoms. A discussion of the forces and processes active in these bays has been presented by Price (1947).

The barrier islands of the Texas coast range in length from about 15 miles (St. Joseph Island) to 110 miles (Padre Island) and rise as much as 50 feet above sea level. In general the highest barriers occur along the lower Texas coast where well-developed sand dunes exist. One of the characteristic features of these barrier islands is a series of abandoned beach ridges (Price 1951a, and Shepard 1952 and 1956) which are aligned parallel to the present seaward edge of the island.

It can be seen from figure 6 that the shoreline of Texas is actually a compound feature. It consists of a Gulf shoreline and a bay shoreline. The Gulf shoreline is the seaward edge of the barrier islands and deltaic plains. The bay shoreline lies at the edge of the mainland and behind the barrier islands. The Gulf shoreline is generally very regular and arcuate and is characterized by well-developed sand beaches. In contrast, the bay shoreline is quite irregular and generally devoid of sand beaches. In many places the bay shoreline is associated with low-lying Recent marshes, and it is characterized by low bluffs wherever wave action is eroding Pleistocene terrace deposits.

LATE QUATERNARY GEOLOGIC HISTORY

The late Quaternary geologic history of the northwestern Gulf coastal plain is now well established as a result of physiographic and shallow subsurface studies conducted during the past two decades (Russell 1936 and 1940; Fisk 1944, 1947, 1952 and 1955; Price 1947; LeBlanc and Bernard 1954). The history of the Texas coast can be summarized briefly as follows:

1. Late Pleistocene falling sea level stage: During the last Pleistocene glacial stage sea level was lowered approximately 450 feet and the nine Texas streams referred to above deeply entrenched their valleys. These valleys extended seaward to a shoreline which was probably 50 to 140 miles seaward of the present shoreline (see fig. 7A).

2. The early Recent rising sea level stage: Melting of the Pleistocene glaciers was accompanied by a rise in sea level. During this stage the Texas streams alluviated their entrenched valleys. The lower portions of the entrenched valleys were drowned, thus forming a series of estuaries (see fig. 7B). The inner bay shoreline originated during this stage.

3. Recent standing sea level stage: Approximately 5,000 years ago sea level reached its present position and has remained constant since that time (LeBlanc and Bernard 1954; McFarlan 1955). The large Texas rivers, the Rio

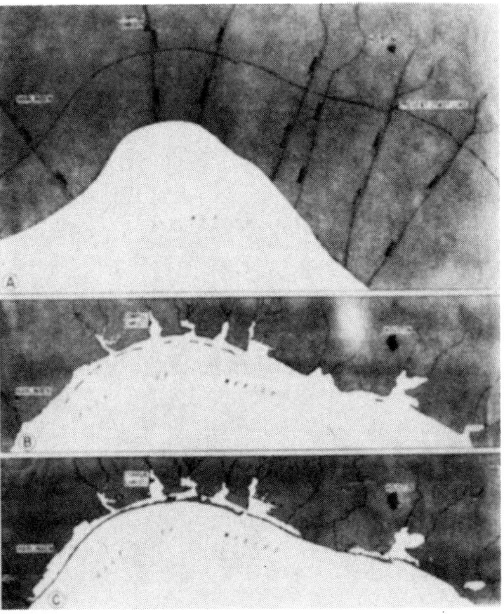

Figure 7. Origin and development of the Texas shoreline: (A) late Pleistocene falling sea level stage; (B) the early Recent rising sea level stage; (C) Recent standing sea level stage.

Grande, Brazos, and Colorado, have completely filled their estuaries which were formed during the early rising sea level stage and they have actually constructed broad deltaic plains which protrude into the Gulf. The smaller Texas rivers, which carry only small quantities of sediments, are still filling their drowned estuaries (see fig. 7C).

The Gulf shoreline, formed by the barrier islands and deltaic plains, originated during this stage (Hedgpeth 1953). The lagoonal bays behind these barriers are of the same age.

A combination of factors is responsible for the construction of the long arcuate Texas barrier islands. Most of the sands which accumulate on these islands are considered to be derived primarily from two sources (Bullard 1942). A large amount of the sand is introduced to the Gulf region by the Rio Grande and the Brazos and Colorado rivers and swept up and down the coast by longshore currents. Another source of sand is probably from the scouring of Recent and late Pleistocene sediments occurring on the Gulf bottom in the inner continental shelf region (Goldstein 1942).

Most of the Texas bays are presently receiving sediments from the small rivers feeding into the heads of these bays and also from the open Gulf during storm and hurricane periods. Shepard (1953a) has discussed the rate of filling which has occurred on these bay bottoms since 1852. Although sedimentation is generally taking place on the bottoms of most bays, a considerable amount of erosion of the bay shores has taken place during the last few thousand years.

THE RIO GRANDE SEGMENT OF THE SHORELINE

The principal physiographic features of the Rio Grande segment of the shoreline, including a small portion of Mexico, and the topography of the adjacent inner continental shelf are shown on figure 8. It is obvious that the large quantity of sediment introduced to this region by the Rio Grande has had a very pronounced influence on the configuration of the south Texas coast. It is also obvious that wind and current action in the inner continental shelf area have played an important role in the distribution of sediment along the coast. In order to understand better the origin and development of the shoreline of this region, it is necessary to analyze the physiography of the Rio Grande delta and flanking barrier islands.

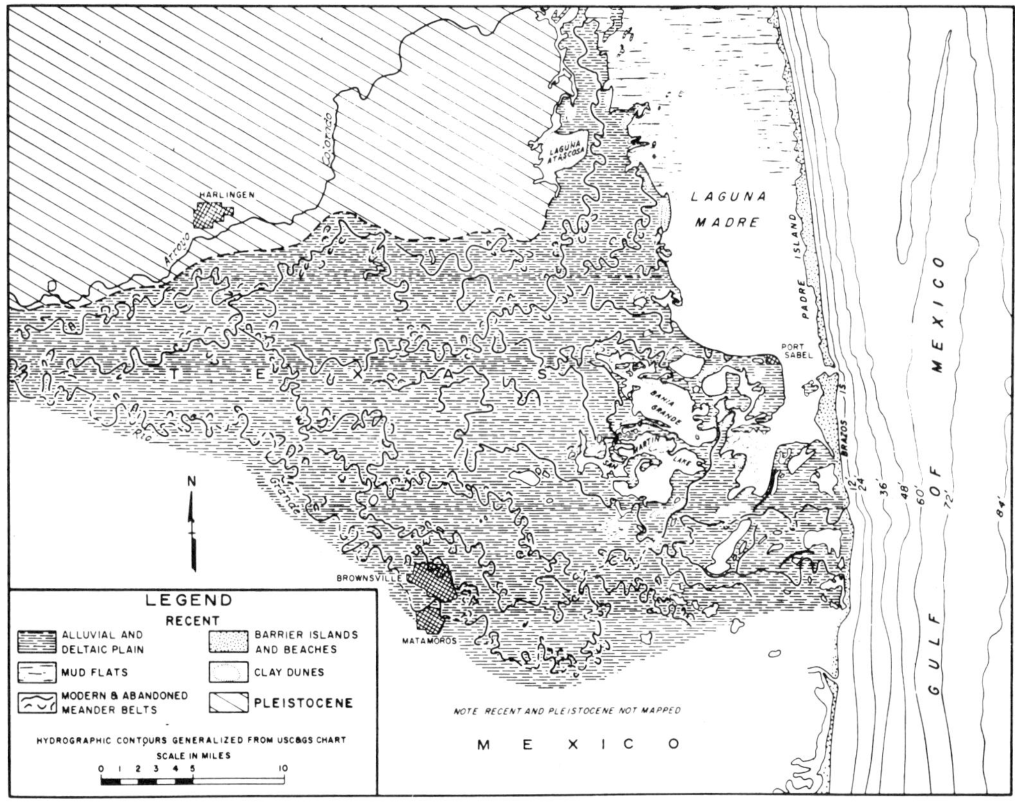

Figure 8. The Rio Grande segment of the Texas shoreline.

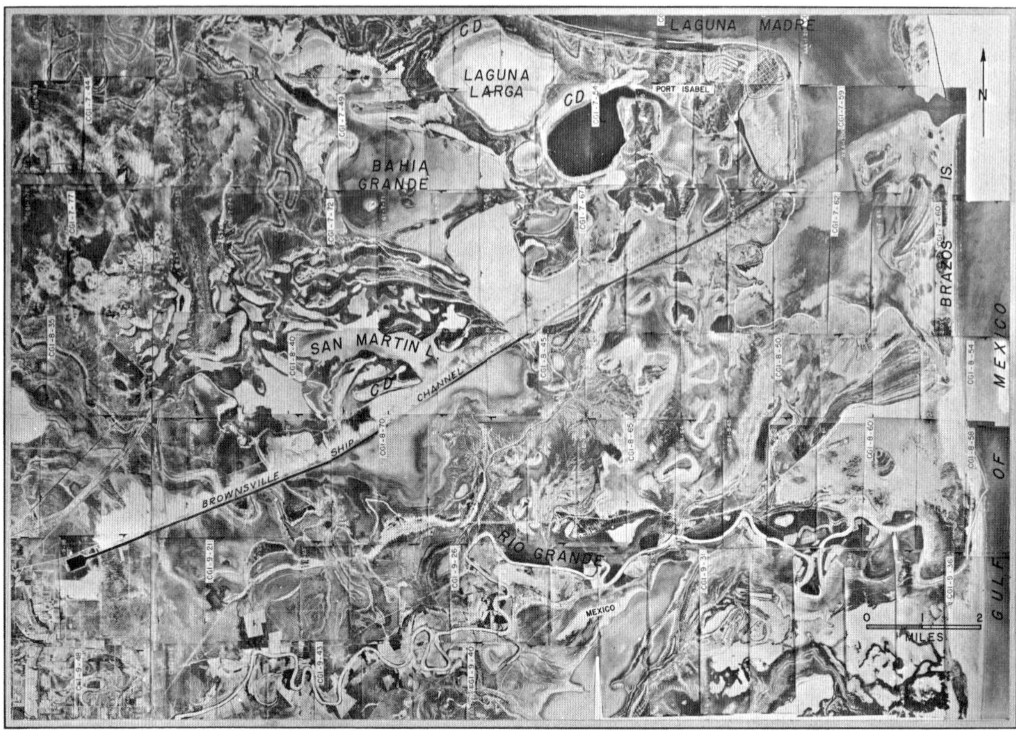

Figure 9. Aerial mosaic of a portion of the Rio Grande deltaic plain (mosaic by U. S. Dept. Agriculture).

The Recent alluvial valley of the Rio Grande is comparatively narrow as far downstream as the vicinity of Harlingen. Downstream from this point the alluvial valley merges imperceptibly with a much wider deltaic plain. An examination of aerial mosaics and topographic maps (figs. 8-10) reveals that this deltaic plain is characterized by more than a dozen well-defined abandoned meander belts, locally called resacas, mud flats, inter-meander belt lowlands, and clay dunes. Most of these Recent abandoned Rio Grande courses are shown on figure 9. The best developed meander belts, containing numerous oxbow lakes, average about 1.5 miles in width and rise about 5 to 15 feet above the inter-meander belt lowlands. Less well developed meander belts, which obviously were occupied for a shorter period of time by the Rio Grande, are somewhat narrower and lower in elevation. The seaward gradient along these abandoned courses is approximately the same as that of the modern Rio Grande—about 1.25 feet per mile.

Several of the abandoned Rio Grande courses can be traced seaward to the present shoreline of the Gulf, particularly in the region south of the present course. Many of the abandoned courses which occur north of the modern Rio Grande can be traced seaward only to the vicinity of Laguna Madre, which is some eight to ten miles from the present Gulf shoreline along Padre Island.

That portion of the Rio Grande deltaic plain which lies within 10 to 15 miles from the Gulf and south of Laguna Madre has a very distinctive type of physiography (figs. 8-10). The inter-meander belt areas consist mostly of extremely shallow water bodies (Bahia Grande and San Martin Lake) and also a number of large mud flats. Other distinctive topographic features of the area are the clay dunes, which reach elevations of 60 feet above sea level and vary in length from one-fourth mile to about eight miles. Huffman and Price (1949) and Price (1958) have shown that these clay dunes were formed by windblown fine-grained sediments derived from the adjacent mud flats occurring between the meander belts.

Three stages in the origin and development of the Rio Grande segment of the Texas shoreline are shown on figure 7. During the early rising sea level stage the Rio Grande entrenched valley was drowned and an estuary is believed to have existed in the area now covered by the deltaic plain. This estuary was gradually filled by the river sediments during the standing sea level stage. At the beginning of the standing sea level stage Laguna Madre probably extended as far south as the present course of the Rio Grande. Sediments contributed by the former courses of the river gradually filled the lower end of Laguna Madre. Bahia Grande, Laguna Larga, and San Martin Lake are remnants of Laguna Madre (figs. 8-10). Thus, it appears that the Rio Grande has constructed more than 600 square miles of deltaic plain in Texas and Mexico

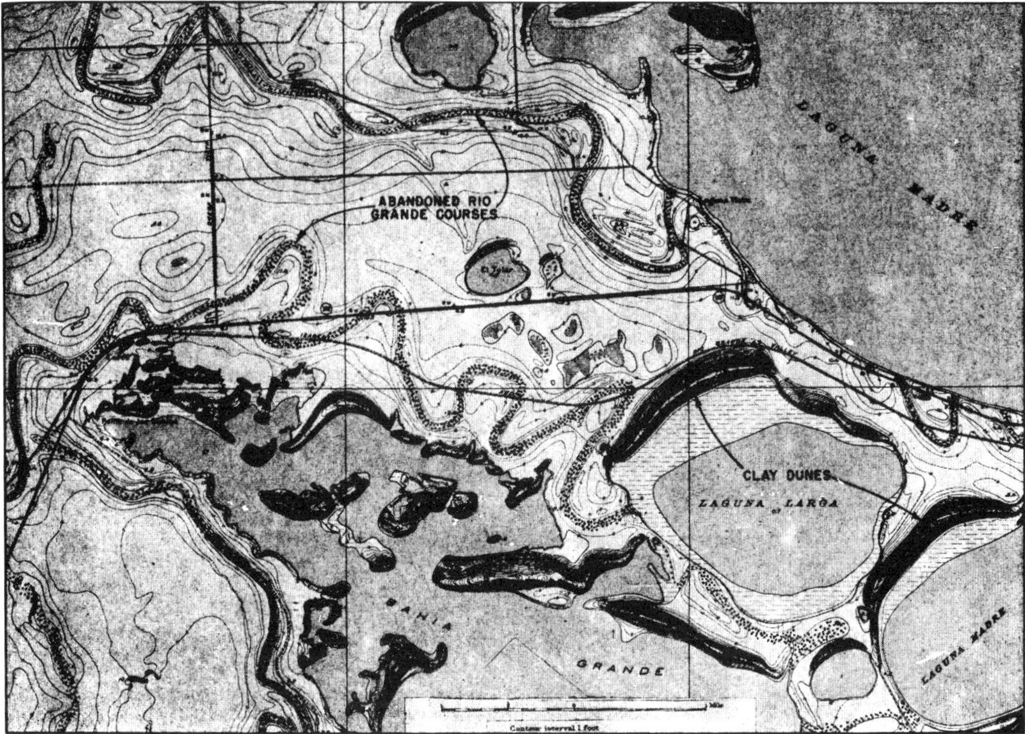

Figure 10. Topographic map of a portion of the Rio Grande deltaic plain (taken from Laguna Vista quadrangle).

over what was formerly shallow waters of Laguna Madre and the inner continental shelf region (Lohse 1958).

Although it is probable that sediments contributed to the Gulf by the Rio Grande have been and are currently being swept both in a notherly and southerly direction along the coast, a substantial portion of the material has been deposited in the immediate vicinity of the deltaic plain.

THE COLORADO-BRAZOS SEGMENT OF THE SHORELINE

Sedimentation associated with the development of the merging deltaic plains of the Colorado and Brazos rivers has exerted a marked influence on a 40-mile segment of the Texas coast between Galveston Island on the east and Matagorda Bay on the west (fig. 11). This portion of the coastal plain is similar in many respects to that of the Rio Grande region discussed above. The Colorado River south of Wharton, Texas, now flows in a very narrow alluvial valley to its junction with Matagorda Bay and the Gulf of Mexico. The development of the Colorado River delta as a result of dredging operations in the vicinity of Matagorda, Texas, has been discussed by Weeks (1945).

An analysis of the coastal plain in this region clearly shows that the Colorado River course between Wharton and Matagorda is quite a young feature. Prior to a few hundred years ago the Colorado River flowed within a broader alluvial valley southeasterly from Wharton. This alluvial valley is marked by several well-defined meander belts, the most striking of which is now occupied by Caney Creek, which flows through eastern Matagorda County (fig. 11).

The alluvial valley of the Brazos River between Richmond and Columbia, Texas, is also shown on figure 11. Although this valley is slightly narrower than that of the Colorado River, it exhibits typical Gulf coastal plain alluvial valley features, namely several abandoned meander belts and intervening backswamp areas. The meander belts of both the Brazos and Colorado rise several feet above the adjacent backswamps and their seaward gradient is less than a foot per mile. Near the vicinity of Columbia in Brazoria County, the alluvial valleys of the Brazos and Colorado merge to form a single alluvial plain which merges seaward with the deltaic plain of the two rivers. Within a 5- to 10-mile belt parallel to the coast, the deltaic plain is marked by marshes which rise only a few inches to a foot above sea level. Although small lakes and bays occur between the abandoned river courses of both the Brazos and Colorado, mud flats and clay dunes are absent in this region. The absence of broad mud flats and clay dunes is probably due to the fact that the Colorado-Brazos region lies within the more humid climate of the upper Texas coast which supports more abundant vegetation.

The Colorado-Brazos deltaic plain is flanked by barrier

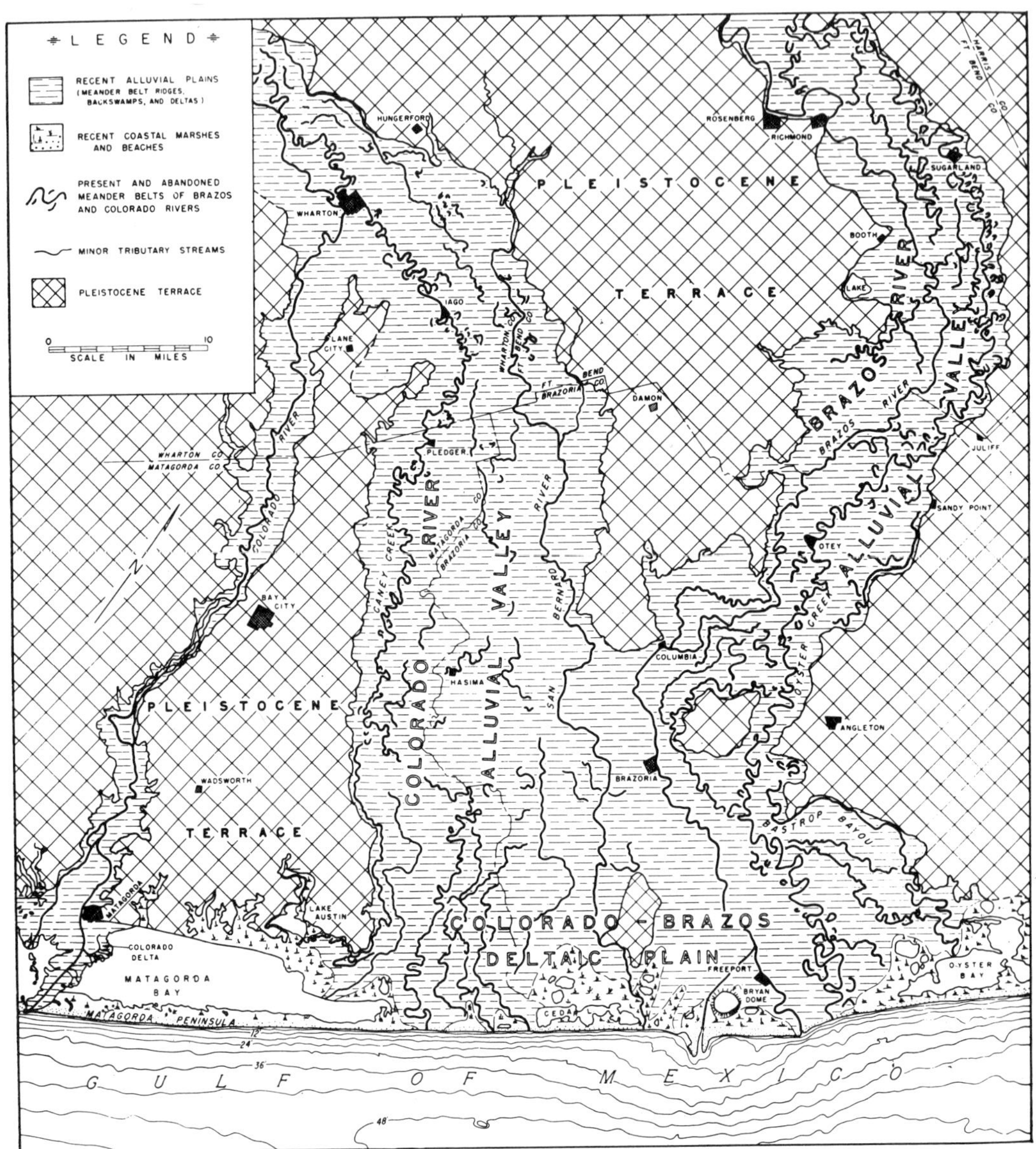

Figure 11. Colorado-Brazos segment of the Texas shoreline.

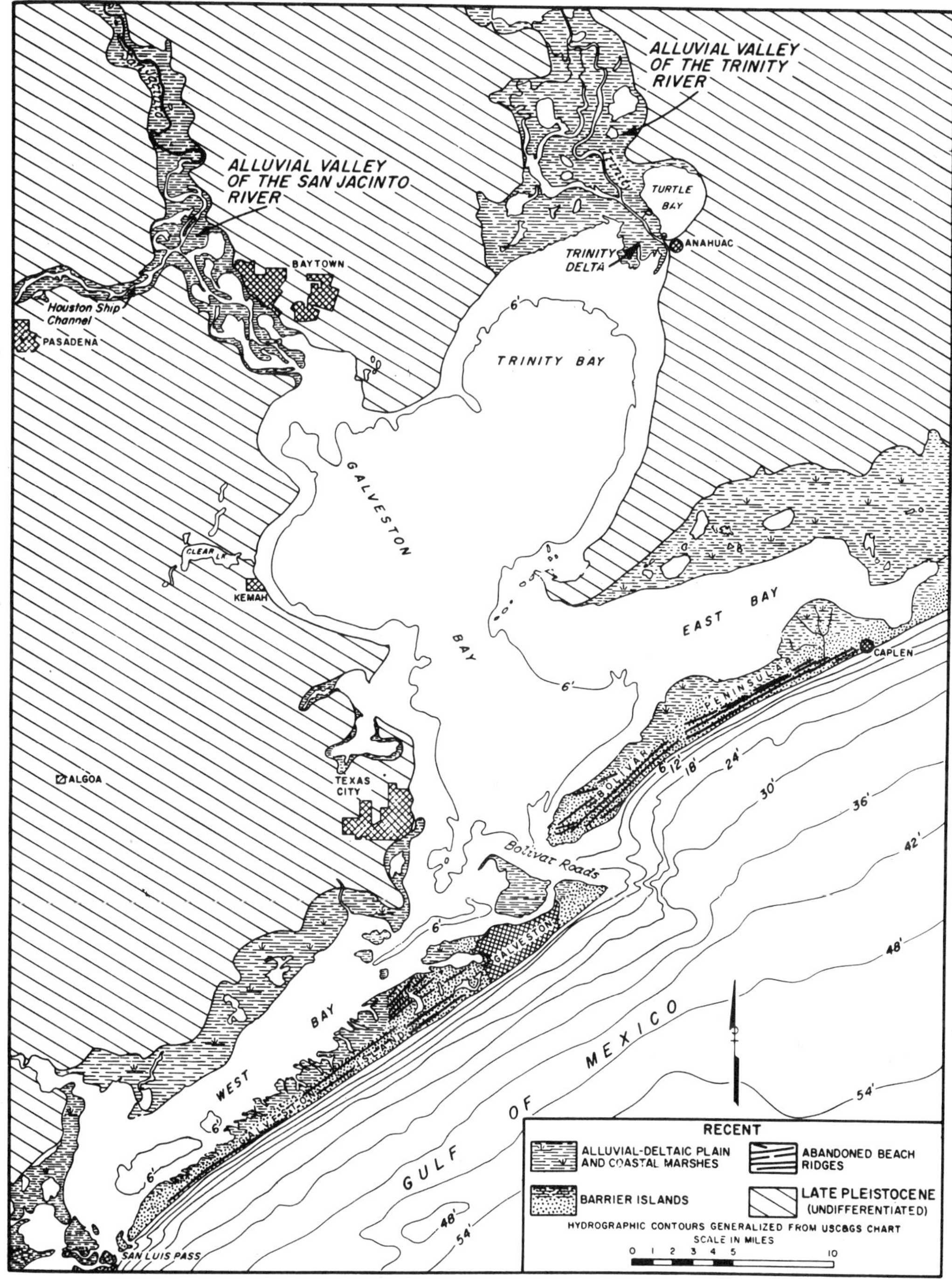

Figure 12. The Galveston Island-Bolivar Peninsula segment of the Texas shoreline.

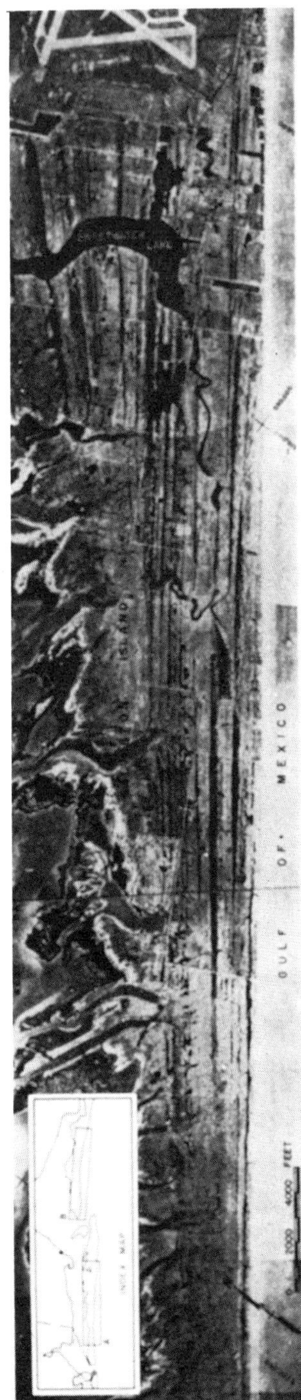

Figure 13. Aerial mosaics of: (A) a portion of Galveston Island; (B) the western portion of Bolivar Peninsula (Courtesy, Jack Ammann Photogrammetric Engineers, Inc., San Antonio, Texas).

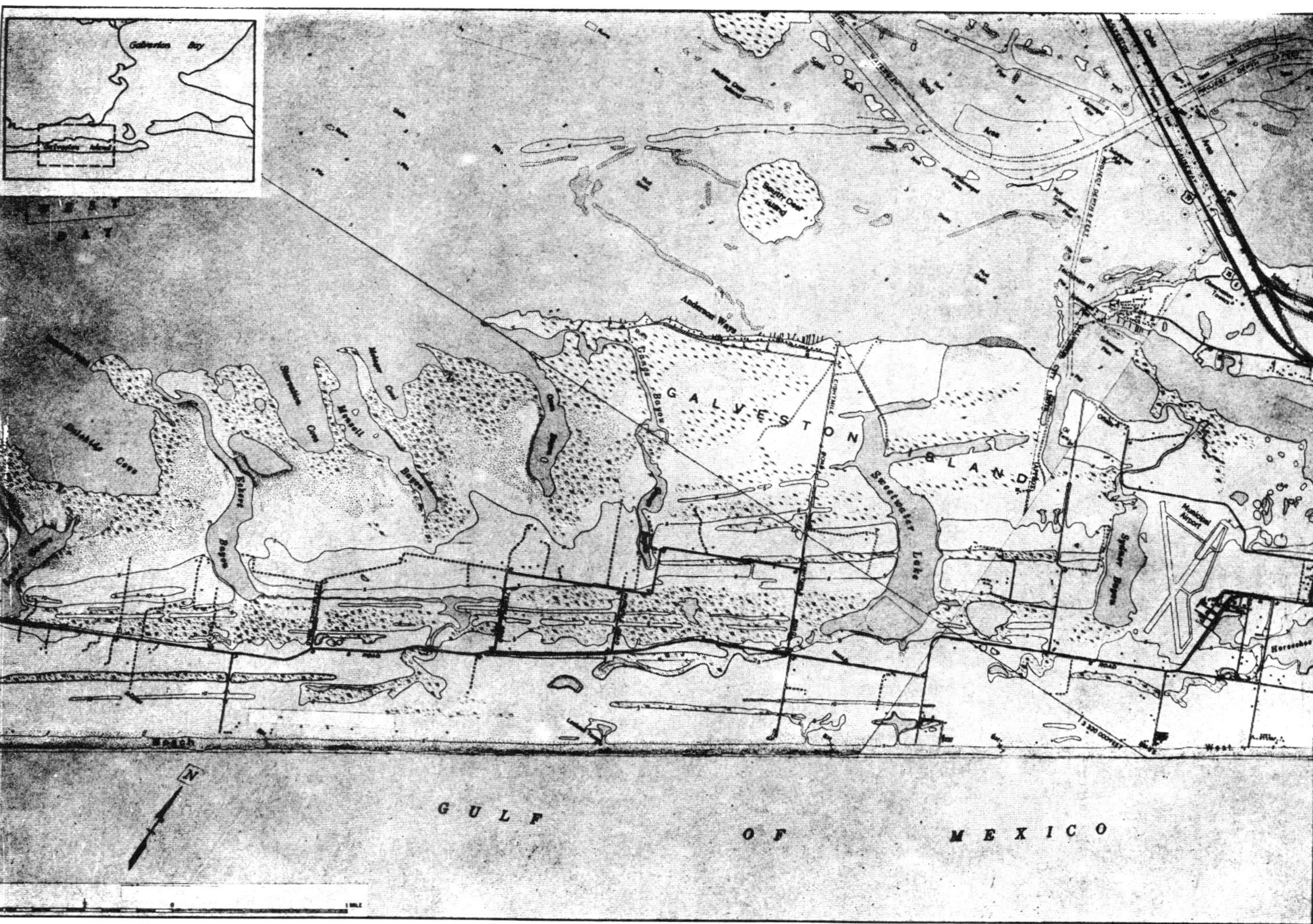

Figure 14. Topographic map of a portion of Galveston Island (taken from Galveston, Virginia Point, and Lake Como quadrangles).

islands and the lagoonal type of bays. There is much evidence to indicate that during the early rising sea level stage the lower Colorado-Brazos deltaic plain was occupied by an estuary similar to Galveston Bay and Lavaca Bay. The Brazos and Colorado rivers have carried a sufficient amount of sediment during the last 5000 years to fill this estuary and construct a deltaic plain which extends out into the Gulf.

THE GALVESTON ISLAND-BOLIVAR PENINSULA SEGMENT OF THE SHORELINE

The upper Gulf coast of Texas is markedly different in origin from the deltaic regions previously discussed. Here marine processes, rather than deltaic sedimentation, have been dominant in shaping the shoreline and in forming many of the physiographic features of the region. Some of the prominent features of this segment of the Texas coast are Galveston Island and West Bay, Bolivar Peninsula and East Bay, Trinity River alluvial valley and Trinity Bay, San Jacinto River alluvial valley and Galveston Bay, and the Pleistocene terrace uplands (fig. 12).

Figure 15. Oblique view of Galveston Island looking northeastward toward Galveston, Texas.

Galveston Island is perhaps one of the best known examples of the western Gulf barrier islands. It is about 30 miles long, 2.5 miles wide near its eastern end, and gradually tapers westward. Abandoned beach ridges and intervening low swales which are well preserved in the central portions of the island (figs. 13, 14, and 15) clearly demonstrate the seaward growth of the island by continued addition of sand brought in by longshore currents. These beach ridges trend in a north-northeast direction and rise 5 to 12 feet above sea level. The Gulf shoreline is very smooth and is characterized by a very broad beach. In contrast the shoreline along the back side of the island has a serrated outline with many small lakes aligned perpendicular to the axis of the abandoned beach ridges.

West Bay is only about 3 or 4 miles wide, generally less than 6 feet deep, and has a mud and sand bottom. It is separated from the Pleistocene uplands to the north by a narrow fringe of Recent coastal marshlands.

The surface of Bolivar Peninsula is also marked by numerous well-defined abandoned beach ridges (fig. 13). The trend of these abandoned beaches is generally parallel to the present Bolivar Peninsula shoreline; however, the western end of many of the beach ridges curves northwestward, indicating that the peninsula was developed in a southwesterly direction. Another prominent feature of the peninsula is the washover fan which has developed in the eastern part of East Bay just northwest of Caplen. Sand dunes are present on Bolivar Peninsula and Galveston Island; however, they are very small. The most prominent topographic features are the modern and abandoned beach ridges.

East Bay is very shallow and has the same general characteristics as West Bay. It is fringed by Recent marshlands along its northern and eastern shore.

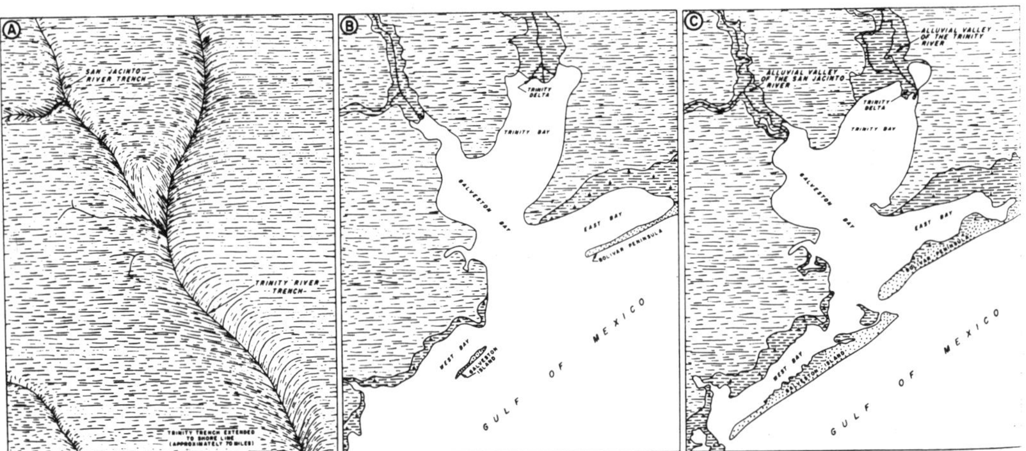

Figure 16. Origin and development of the Galveston Island-Bolivar Peninsula segment of the Texas shoreline: (A) late Pleistocene falling sea level stage; (B) the early Recent rising sea level stage; (C) Recent standing sea level stage.

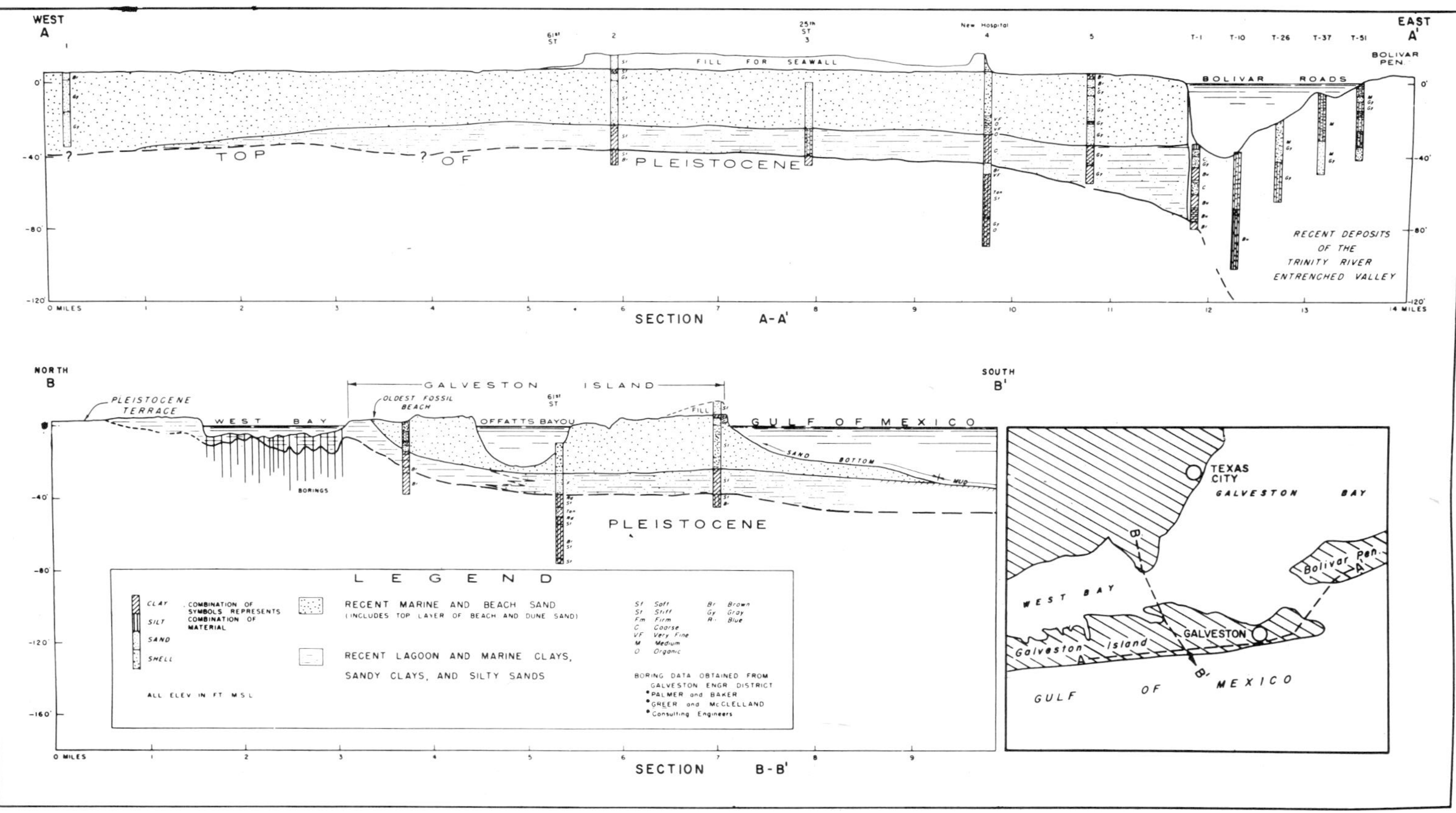

Figure 17. Cross sections showing the relationships of Galveston Island to West Bay and the late Pleistocene terrace.

Figure 18. Oblique view of the northern shore of Trinity Bay showing erosion (summer 1958) of the bay shoreline as a result of wave action (Courtesy, The Houston Chronicle).

The Trinity River flows largely through the humid and sub-humid belts of east Texas, and although it has a comparatively short course it actually has the largest discharge of any coastal Texas stream. The silt load, however, is rather small (fig. 2). It flows within a 4- to 10-mile-wide alluvial valley south of Liberty, Texas, and has its delta in the upper part of Trinity Bay just west of Anahuac, Texas (fig. 12). The San Jacinto River, one of the smaller Texas streams, empties into upper Galveston Bay. However, in contrast with the Trinity, it does not have a well-developed delta.

Galveston Bay and Trinity Bay represent the seaward continuation of the San Jacinto and Trinity alluvial valleys, respectively. These two bays merge some 15 miles southeast of the Trinity delta to form one of the largest estuaries of the Texas coast. The southern portion of Galveston Bay connects with the Gulf of Mexico through Bolivar Roads.

More than one-third of the area covered by Galveston and Trinity bays is less than six feet deep (fig. 12). The central portions of the bays have maximum depths of approximately ten feet and their bottoms consist largely of soft muds.

The topography of the inner continental shelf adjacent to Galveston Island and Bolivar Peninsula is shown on figure 12. With the exception of the local area just east of Bolivar Roads, the Gulf bottom is generally smooth. Its gulfward slope is about seven feet per mile within a four-mile belt parallel to the shore. Beyond this area the slope of the shelf decreases to about one or two feet per mile.

The late Quaternary geologic history of the Galveston Island region, as interpreted from physiography and subsurface data, is summarized in figure 16. The maximum depth and exact position of the Trinity River trench, which was developed during the last Pleistocene glacial stage when sea level was about 450 feet below the present level, is not well known. The position of the trench has been determined in the vicinity of Galveston by means of boring data obtained from Palmer and Baker, consulting engineers, and the U. S. Army Engineers (fig. 17). The trench is located beneath Bolivar Roads between the eastern tip of Galveston Island and the western end of Bolivar Peninsula. On the basis of information concerning the other entrenched valleys along the Louisiana coast, it is assumed that the Trinity trench extended approximately 100 miles across the continental shelf to the low sea level shoreline (fig. 16A).

The configuration of the shoreline during the early phase of the rising sea level stage is shown on fig. 16B. The rate of sedimentation in the lower San Jacinto and Trinity entrenched valleys was very low compared with the rate of rise in sea level; consequently, the alluvial valleys of these two streams were drowned for a distance of at least 60 miles up from the present coast. The mouths of the San Jacinto and Trinity rivers were probably several miles upstream from their present positions. The exact width of Trinity and Galveston bays during this stage is unknown; however, they were certainly narrower than they are today.

Galveston Island and Bolivar Peninsula originated at the end of the rising sea level stage and developed during the standing sea level stage (figs. 12 and 16B).* The oldest beach ridge on Galveston Island and Bolivar Peninsula marked the position of the Gulf shoreline during this stage. Galveston Island was probably not over two miles long, extremely narrow, and had a northeasterly trend (about N 43° E). Bolivar Peninsula probably originated just east of Caplen, and as a result of coastal deposition associated with longshore currents, it developed in a westerly direction.

Significant changes in the shoreline of Galveston and Trinity bays and along Galveston Island and Bolivar Peninsula have occurred during the past few thousand years (compare figs. 16B and 16C). These changes were associated with both erosional and depositional processes active along the bay shoreline and also along the Gulf shoreline. In the bay region, for example, shoreline changes related to sedimentation have been very limited in extent, being confined to the immediate vicinity of the Trinity and San Jacinto deltas. It is difficult to determine the growth of the San Jacinto River delta during the past few thousand years because of the numerous man-made changes in connection with the establishment and maintenance of the Houston ship channel. Few man-made changes have been made in northern Trinity Bay and it is obvious that the Trinity River has enlarged its deltaic plain over several square miles in what was formerly the northeasternmost segment of Trinity Bay.

Although sedimentation has taken place on the bottom of both Galveston and Trinity bays (Shepard 1953a), the predominant process along the bay shoreline has been erosion. Wave action along the shore zone, particularly during periods of storms and hurricanes, has resulted in recession of the bay shoreline. The process is still active today and much of the shoreline of Trinity Bay between Smith Point and Anahuac, and between the Trinity and San Jacinto deltas, has receded considerably in recent years (figs. 16 and 18). The shoreline along the western side of Galveston Bay between Kemah and Texas City, and north of Kemah, has also receded during recent years. The broad arcs in the shoreline at Kemah and just north of Kemah are believed to have developed since sea level reached its present stand. Recession is probably

* The origin and development of another Texas barrier island (Padre Island) is discussed by H. N. Fisk, 1959.

more rapid in local zones where the Pleistocene deposits fringing the bays consist of sandy material. Bank recession where the Pleistocene deltaic materials consist largely of backswamp clays occurs at a slower rate.

The most prominent changes in the shoreline of the Galveston Island-Bolivar Peninsula region have resulted from deposition associated with longshore currents in the littoral zone of the Gulf of Mexico and also with sedimentation in East and West bays. Some changes in the shoreline on the back side of Galveston Island and Bolivar Peninsula have occurred as a result of scouring of barrier island deposits by storm and hurricane forces.

Studies in the Gulf by the U. S. Army Engineers (see U. S. Congress 1953) have shown that the littoral drift along Galveston and Bolivar Peninsula varies considerably during the seasons of the year. Physiographic evidence on both Galveston Island and Bolivar Peninsula strongly indicates that the predominant drift has been in a southwesterly direction throughout the standing sea level stage. Beach ridges on Galveston Island clearly show that accretions to the island have progressed in a southerly and southwesterly direction, thus changing the general trend of the island from approximately N 43° E to N 55° E.

Beach erosion studies conducted by the U. S.. Engineers District Office, Galveston, have shown that sand is presently being accreted to Galveston Island and that this process has been active during the past 80 years. Although some segments of the Galveston beach have receded slightly during the past 50 years, the net result has been accretion of sand along the beach. Thus the process leading to seaward growth of the island, which began when sea level reached its present stand, is still active.

The source of the sand which has been deposited on Galveston Island and Bolivar Peninsula during the standing sea level stage is not definitely known. Bullard (1942) has shown that at least some of this material was probably derived from the Colorado-Brazos delta. Perhaps some sand material from the Louisiana coast is swept in a westerly direction and deposited on the beach in the Galveston region. The Trinity River contributes an unknown quantity of clays and silts to the Gulf, particularly during flood seasons; however, it is very doubtful if much sand is transported through Bolivar Roads.

Comparative studies by the U. S. Army Engineers of hydrographic charts constructed at intervals during the past 80 years offshore from Galveston Island demonstrate that the bottom of the Gulf of Mexico is presently being scoured in many places and that this condition has prevailed for at least 80 years. The engineers have calculated that on a line approximately six miles long normal to Galveston Island an average of 5 million cubic yards of material is being removed annually from the Gulf bottom. Although this material is not all sand, there is undoubtedly more than enough sands and silts derived from these deposits by winnowing processes to account for the coarse sediments now being added to Galveston Island.

Many changes in the shoreline on the back side of Galveston Island occurred contemporaneously with the seaward growth of the island. The lakes which occur in the northern half of the island (fig. 14) are aligned normal to the beach accretions and, as indicated above, are considered to have formed by scouring of sand deposits by current and wave forces associated with the major hurricanes.

Bolivar Peninsula has enlarged during the standing sea level stage in a manner similar to Galveston Island. A series of beach and spit accretions mark stages in the western and southwestern growth of the peninsula and a corresponding narrowing of Bolivar Roads. The north side of Bolivar Peninsula has been considerably modified by the development of large wash-over fans which have been built northward into East Bay.

Only comparatively minor changes have taken place in the bay shoreline on the northern side of West and East bays. Sedimentation at a very slow rate has resulted in a widening of the fringe of near sea level marshes. According to Shepard (1953a), there has been about one to two feet of filling in East Bay and practically no filling of West Bay during the past 80 years.

THE SAN ANTONIO BAY SEGMENT OF THE SHORELINE

Another excellent example of shoreline development associated mainly with marine shoreline processes is the region around San Antonio Bay, which lies approximately half-way between Corpus Christi, Texas, and the Colorado-Brazos deltaic plain (fig. 6). This part of the coastal plain is particularly interesting because it exhibits a well-defined late Pleistocene barrier island and lagoon in addition to the Recent barrier island and lagoon (figs. 19 and 20).

The San Antonio Bay region was extensively studied over a five-year period by the University of California, Institute of Marine Resources, in connection with the study of near-shore Recent sediments and their environments in the northern Gulf of Mexico. These studies were sponsored by the American Petroleum Institute, and numerous reports have been published by the project staff. A summary paper on the results was recently published by Shepard and Moore (1955).

Topographic features on Matagorda Island are quite similar to those of Galveston Island and Bolivar Peninsula. Well-defined beach ridges aligned parallel to the modern Gulf shoreline occur within a one-mile belt over most of the island (fig. 20). Locally these ridges are masked by younger sand dune deposits and in some places they have been truncated or masked by washover fans. The back side of Matagorda Island has a crenulated shoreline which is typical of most of the Texas barrier islands.

The topography of the inner continental shelf off Matagorda Island is essentially the same as that off Galveston Island. Sands occur on the Gulf bottom in a narrow zone which extends seaward to the 30-foot contour or about three miles from land.

Sedimentation within San Antonio, Mesquite, and Espiritu Santo bays has occurred at a rate of approximately 0.5 to 1.23 feet per century (Shepard 1953a).

The Guadalupe River has a comparatively small drainage area. It flows through carbonate rock terrain in its upper reaches; consequently, it transports only a small amount of sediment to the coastal region. It is presently developing its delta in the upper part of San Antonio Bay between Green Lake and Hynes Bay (fig. 19). The Guadalupe delta is very similar to the Trinity delta discussed above. It lies only a foot or two above sea level and is flanked by a slightly higher late Pleistocene deltaic plain.

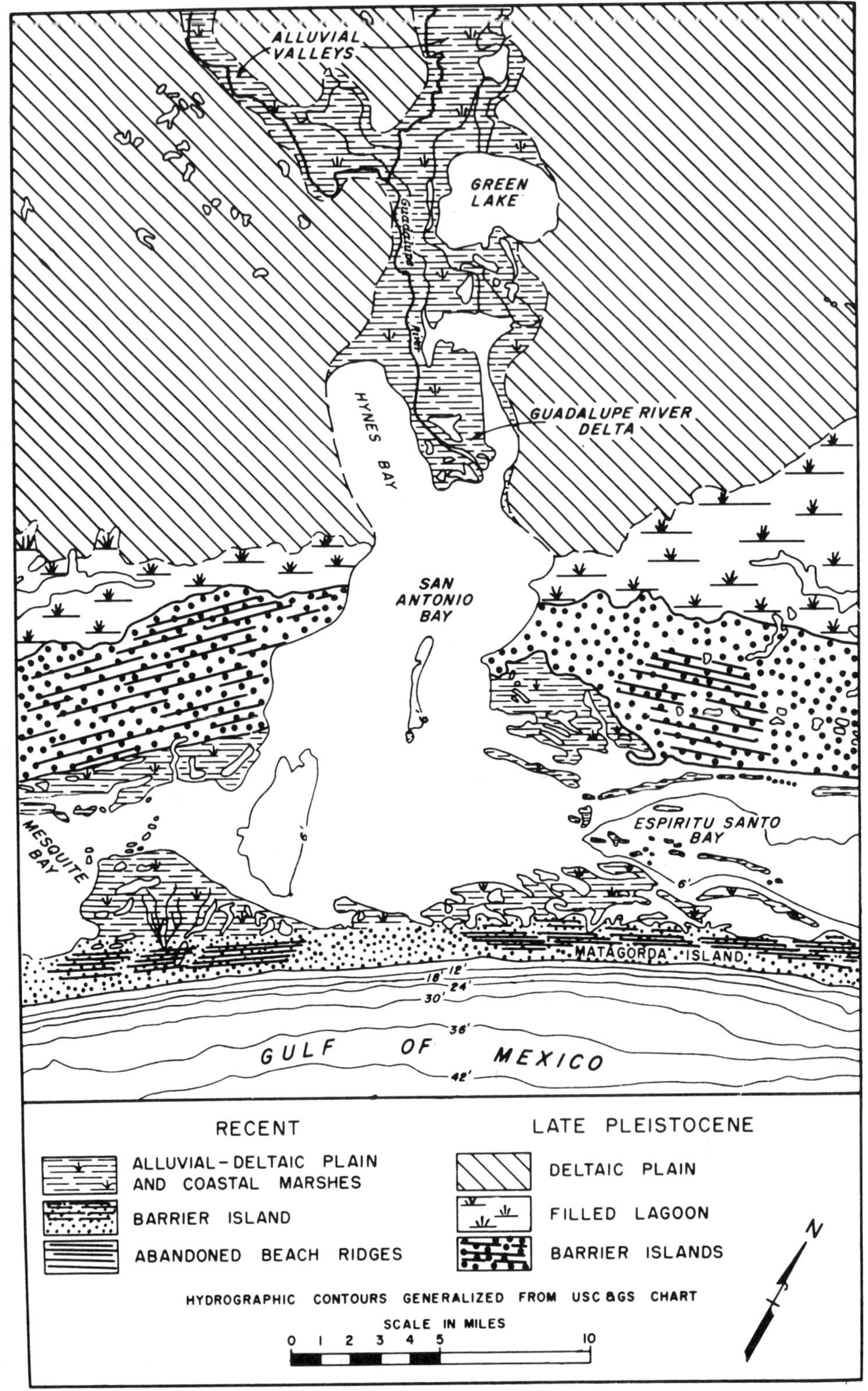

Figure 19. San Antonio Bay segment of the Texas shoreline.

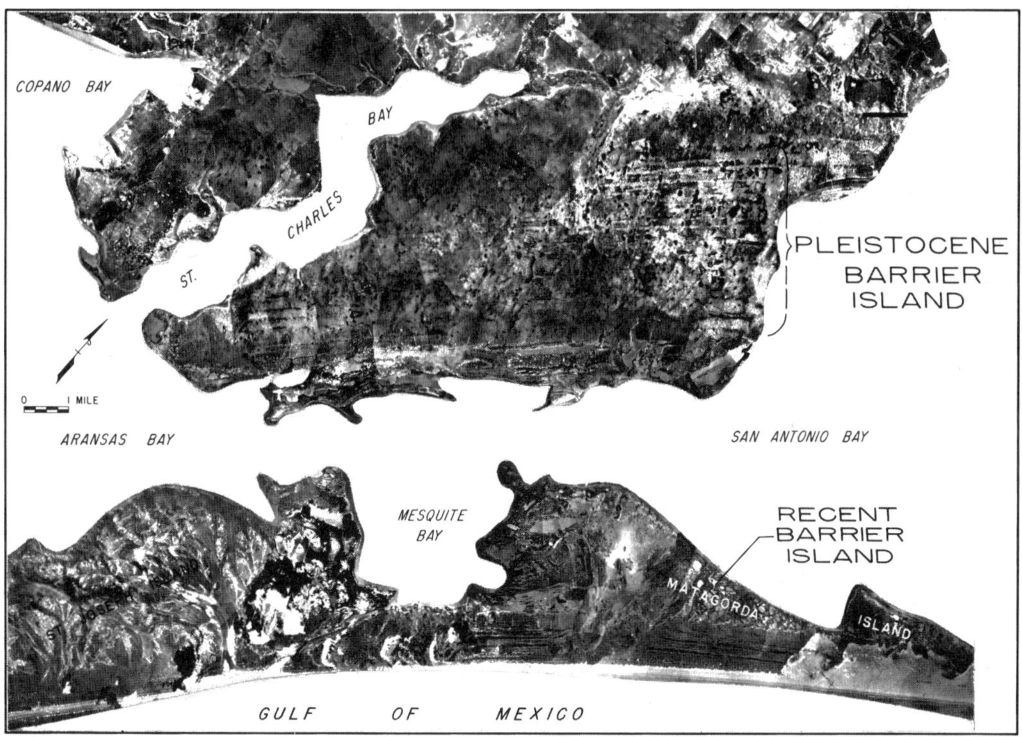

Figure 20. Aerial mosaic of a portion of Matagorda Island and the Pleistocene barrier island adjacent to San Antonio Bay (Courtesy, Tobin Aerial Surveys, Inc.).

The late Quaternary history of the San Antonio Bay region is very similar to that of the Galveston Bay region discussed above. The Guadalupe River entrenched its valley during the last Pleistocene glacial stage. The lower (seaward) part of this valley was drowned during the rising sea level stage. Since sea level reached its present stand further significant changes in the shoreline have occurred. These changes were related to three processes: (1) seaward growth of the Guadalupe delta in upper San Antonio Bay, (2) erosion of the San Antonio Bay shoreline by wave action, and (3) accretions to Matagorda Island by longshore currents and washover fans.

Remnants of a late Pleistocene barrier island which occur between the modern barrier island (Matagorda Island) and the late Pleistocene deltaic plain (figs. 19 and 20) extend along much of the Texas coast. The best development of this Pleistocene barrier island occurs in the region between the vicinity of Baffin Bay and the southwestern margin of the Colorado - Brazos deltaic plain. East of the Brazos deltaic plain remnants of this barrier island occur near Chocolate Bay and between Trinity Bay and East Bay (Smith Point), and also in the vicinity of Beaumont. Price (1947) recognized this feature several years ago and discussed its origin and its relationship to other physiographic features of the coastal region. In the San Antonio Bay region this barrier island is locally referred to as "Live Oak Barrier." It reaches elevations of about 30 feet above sea level and is flanked on the landward side by a filled lagoon, the surface of which has been termed the "Ingleside" by Price (1947).

Since its initial development about 5000 years ago, Matagorda Island has grown both gulfward and landward. The seaward growth of the island has been in the form of beach accretions. The sands and fine-grained sediments associated with the beach ridges and intervening low swales were probably derived from several sources, including the Brazos and Colorado deltas and erosion of the Gulf bottom. Growth on the back side of the island has been in the form of washover fans such as those shown on the aerial mosaic (fig. 20).

The processes of sedimentation which have prevailed in this region during the past few thousand years should continue relatively unchanged as long as sea level remains constant. Eventually San Antonio Bay should be completely filled by the Guadalupe delta. Mesquite and Espiritu Santo bays will eventually be filled, primarily with sediments washed over the barrier island by hurricane forces. During the nonhurricane seasons longshore currents will continue to transport and deposit sands along the seaward face of Matagorda Island, and the seaward prograding of the island should continue.

Figure 21. Aerial mosaic of the Sabine Pass region (mosaic by U. S. Department of Agriculture).

THE SABINE PASS SEGMENT OF THE SHORELINE

An interesting example of local shoreline migration associated with the mouth of a small river occurs in the Sabine Pass region of southeastern Texas (figs. 6 and 21). During the standing sea level stage the Gulf shoreline has regressed in a southeasterly direction for at least 12 miles as a result of beach and mud flat accretions. Longshore currents have played a major role in the development of the shoreline of this region.

LOCAL TRANSGRESSIVE SHORELINE

It has been demonstrated above that most of the Gulf shoreline of Texas has regressed seaward since sea level reached its present position about 5,000 years ago. There are a few notable local exceptions, however, where the Gulf shoreline has transgressed over the coastal plain during the past few hundred years. One of the best examples of this type of marine transgression lies along a 25-mile segment of the coast east of Bolivar Peninsula and west of Sabine Pass (figs. 6 and 22). A study of the shoreline in this area reveals that the modern transgressive beach deposits consist of only a thin veneer (2 or 3 feet) of sand overlying organic-rich clays. As a result of beach erosion, it has been necessary for the Texas Highway Department to relocate segments of Texas Highway 87 further inland a number of times during the past 20 years. It is quite probable that the shoreline of this segment has receded several thousand feet during the past few thousand years.

Another example of local marine transgression occurs along the western portion of the Colorado-Brazos deltaic plain, east of Matagorda Peninsula and west of Cedar

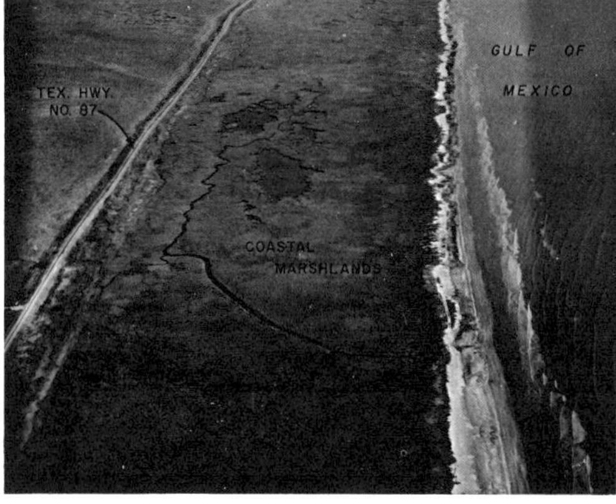

Figure 22. Oblique view of the transgressive shoreline about 11 miles west of the Sabine Pass along Texas Highway 87.

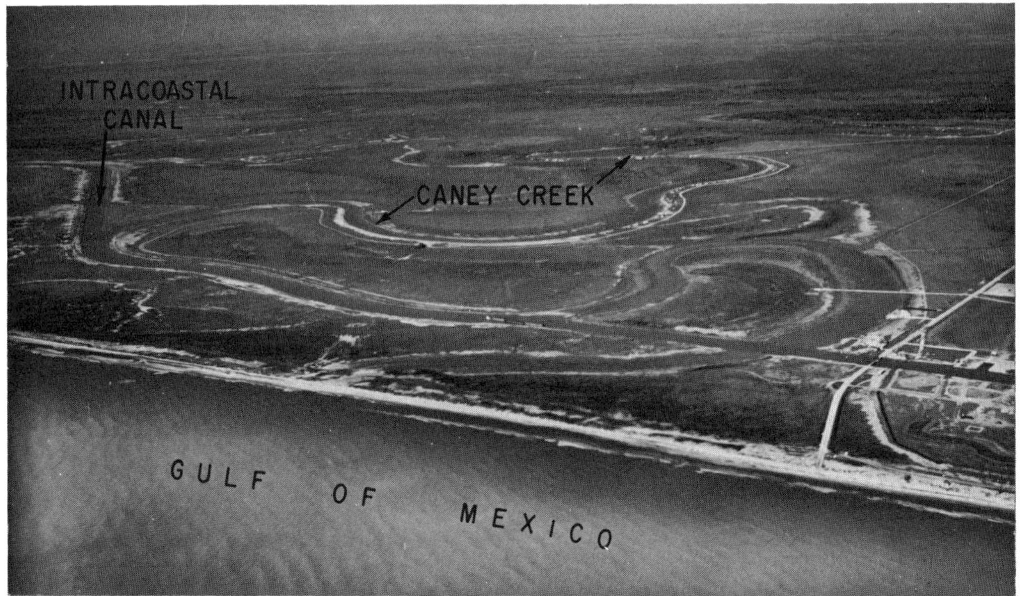

Figure 23. Oblique view of the transgressive shoreline in the vicinity of Caney Creek, approximately 23 miles east of Matagorda, Texas.

Lake (fig. 11). The Caney Creek meander belt of the Colorado River is believed to have extended gulfward from the modern shoreline for quite some distance while it was active. Since the abandonment of this meander belt the Gulf has been encroaching on this part of the coastal plain (fig. 23).

A similar situation exists east of Freeport where the Oyster Creek meander belt of the Brazos River is being truncated by a transgressive shoreline (fig. 11). Although marine transgression has occurred locally in the Colorado-Brazos deltaic plain region, the net result of deltaic sedimentation in this region has been a prograding of the shoreline (fig. 6).

ACKNOWLEDGMENTS

The writers are deeply indebted to Shell Development Company for permission to publish this paper and to E. H. Rainwater for critical review of the manuscript and numerous useful suggestions. Appreciation is also expressed to D. M. Graham, Bruce Gray, N. W. Kusakis, and H. V. Goehrs for assistance in preparing the illustrations, and to Miss N. R. Olson and Mrs. J. G. Breeding, who ably assisted in preparing and editing the manuscript, respectively.

The valuable assistance rendered by Mr. Albert B. Davis of the U. S. Engineers, Galveston, Texas, is gratefully acknowledged. He provided much useful information concerning the general geology of the Galveston Island area.

The senior author is particularly indebted to his colleague, H. A. Bernard. Many of the concepts discussed herein were formulated as a result of numerous field trips and discussions with him.

BIBLIOGRAPHY

Barden, W. J., et al., 1934, Beach erosion at Galveston, Texas: House Doc. 400, 73rd Congress, 2nd Session, 9 p.

Barton, D. C., 1930, Deltaic coastal plain of southeastern Texas: Geol. Soc. Amer. Bull., v. 41, no. 3, p. 359-382.

Bullard, F. M., 1942, Source of beach and river sands on Gulf Coast of Texas: Geol Soc. Amer. Bull., v. 53, no. 7, p. 1021-1043.

Carsey, J. B., 1950, Geology of Gulf coastal area and continental shelf: Amer. Assoc. Petroleum Geol. Bull., v. 34, no. 3, p. 361-385.

Collier, A. and Hedgpeth, J. W., 1950, An introduction to the hydrography of tidal waters of Texas: Pub. Inst. Marine Sci., Univ. of Texas, v. 1, no. 2, p. 121-194.

Deussen, A., 1924, Geology of the coastal plain of Texas west of Brazos River: U. S. Geol. Survey Prof. Paper 126, 139 p.

Doering, J. A., 1935, Post-Fleming surface formations of coastal Southeast Texas and South Louisiana: Amer. Assoc. Petroleum Geol. Bull., v. 19, no. 5, p. 651-688.

———, 1956, Review of Quaternary surface formations of Gulf coastal plain: Amer. Assoc. Petroleum Geol. Bull., v. 40, no. 8, p. 1816-1862.

Fariş, O. A., 1933, The silt load of Texas streams: U. S. Dept. Agriculture Bull. 382.

Fisk, H. N., 1944, Geological investigation of the alluvial valley of the lower Mississippi River: Mississippi River Commission.

———, 1947, Fine-grained alluvial deposits and their effects on Mississippi River activity: Mississippi River Commission.

———, 1952, Geological investigation of the Atchafalaya Basin and the problem of Mississippi River diversion: Mississippi River Commission.

———, 1959, Padre Island and Laguna Madre mud flats, south coastal Texas: Second Coastal Geography Conference, proceedings, p. 103-151.

Fisk, H. N. and McFarlan, E., Jr., 1955, Late Quaternary deltaic deposits of the Mississippi River, *in* Crust of the earth: Geol. Soc. America Spec. Paper 62, 279 p.

Gealy, B. L., 1955, Topography of the continental slope in northwest Gulf of Mexico: Geol. Soc. Amer. Bull., v. 66, no. 2, p. 203-228.

Geyer, R. A., 1951, Bibliography of oceanography, marine biology, geology, geophysics, and meteorology of the Gulf of Mexico: Texas Jur. Sci., v. 2, no. 1, p. 44-92.

Goldstein, A., Jr., 1942, Sedimentary petrologic provinces of the northern Gulf of Mexico: Jour. Sed. Petrology, v. 12, no. 2, p. 77-84.

Greenman, N. N., and LeBlanc, R. J., 1956, Recent marine sediments and environments of northwest Gulf of Mexico: Amer. Assoc. Petroleum Geol. Bull., v. 40, no. 5, p. 813-847.

Gulf Coast Association of Geological Societies, 1958, Sedimentology of South Texas: Field Trip Guidebook, 114 p.

Hedgpeth, J. W., 1947, The Laguna Madre of Texas: Trans. 12th North American Wildlife Conf., p. 364-380.

———, 1953, An introduction to the zoogeography of the northwestern Gulf of Mexico with reference to the invertebrate fauna: Pub. Inst. of Marine Sci., Univ. of Texas, v. 3, no. 1, p. 111-224.

Henry, V. J., 1956, Investigation of shoreline-like features in the Galveston Bay region, Texas, *in* Oceanographic survey of Gulf of Mexico: Texas A&M Research Foundation, A&M Project 24, Ref. 56-12T, April 1956, 76 p.

Huffman, G. G. and Price, W. A., 1949, Clay dune formation near Corpus Christi, Texas: Jour. Sed. Petrology, v. 19, no. 3, p. 118-127.

Johnson, D. W., 1919, Shore processes and shoreline development: John Wiley and Sons, New York.

LeBlanc, R. J. and Bernard, H. A., 1954, Résumé of late Recent geologic history of the Gulf Coast: Geologie en Mijnbouw, n. 6, Nw. Serie, 16e Jaargang, p. 185-194.

Leipper, D. F., 1951, Nature of ocean currents in the Gulf of Mexico: Texas Jour. Sci., v. 3, no. 1, p. 41-44.

———, 1954, Marine meteorology of the Gulf of Mexico, A brief review, *in* Gulf of Mexico, its origin, waters, and marine life: U. S. Dept. Interior, Fishery Bull. 89 of Fish and Wildlife Service, v. 55, p. 89-98.

Lohse, E. A., 1955, Dynamic geology of the modern coastal region, northwest Gulf of Mexico, *in* Finding ancient shorelines: Soc. Econ. Paleon. and Mineral. Spec. Pub. no. 3, p. 99-103.

———, 1958, Mouth of Rio Grande, *in* Sedimentology of South Texas: Gulf Coast Association of Geological Societies, Field Trip Guidebook, p. 55-56.

Lowman, S. W., 1947, Project 5, Recent and near-Recent sediments of the northern Gulf of Mexico—Gulf Coast region, *in* Amer. Assoc. Petroleum Geol. Research Committee, 1946-1947, Repts. on Projects 1-12, p. 52-97.

———, 1949, Sedimentary facies in Gulf Coast: Amer. Assoc. Petroleum Geol. Bull., v. 33, no. 12, p. 1939-1997.

Lynch, S. A., 1954, Geology of the Gulf of Mexico, *in* Gulf of Mexico, Its origin, waters and marine life: U. S. Dept. Interior, Fishery Bull. 89 of Fish and Wildlife Service, v. 55, p. 67-86.

Marmer, H. A., 1954, Tides and sea level in the Gulf of Mexico, *in* Gulf of Mexico, its origin, waters, and marine life: U. S. Dept. Interior, Fishery Bull. 89 of Fish and Wildlife Service, v. 55, p. 101-118.

McFarlan, E., Jr., 1955, Radiocarbon dating of the Late Quaternary in southern Louisiana, [abs.]: Geol. Soc. Amer. Bull., v. 66, no. 12, p. 1594-1595.

Metcalf, R. J., 1940, Deposition of Lissie and Beaumont formations of Gulf Coast of Texas: Amer. Assoc. Petroleum Geol. Bull., v. 24, no. 4, p. 693-700.

Nanz, R. H., 1954, Genesis of Oligocene sandstone reservoir, Seeligson Field, Jim Wells and Kleberg counties, Texas: Amer. Assoc. Petroleum Geol. Bull., v. 38, no. 1, p. 96-117.

Norris, R. M., 1953, Buried oyster reefs in some Texas bays: Jour. Paleontology, v. 27, no. 4, p. 569-576.

Penrose, R. A .F., Jr., 1890, A preliminary report of the geology of the Gulf Tertiary from the Red River to the Rio Grande, Texas: U. S. Geol. Survey, 1st. Ann. Rept., p. 3-101.

Price, W. A., 1933, Role of diastrophism in topography of Corpus Christi area, South Texas: Amer. Assoc. Petroleum Geol. Bull., v. 17, no. 8, p. 907-962.

———, 1947, Equilibrium of form and forces in tidal basins of coast of Texas and Louisiana: Amer. Assoc. Petroleum Geol. Bull., v. 31, no. 9, p. 1619-1663.

———, 1951a, Barrier island, not "offshore bar": Science, v. 113, no. 2939, p. 487-488.

———, 1951b, Building of Gulf of Mexico: secondary events in a regionally concordant basin: Gulf Coast Assoc. Geol. Societies, Trans., v. 1, p. 7-39.

———, 1954a, Dynamic environments: reconnaissance mapping, geologic and geomorphic, of continental shelf of Gulf of Mexico: Gulf Coast Assoc. Geol. Societies, Trans., v. 4, p. 75-107.

———, 1954b, Shorelines and coasts of the Gulf of Mexico, *in* Gulf of Mexico, its origin, waters and marine life: U. S. Dept. Interior, Fishery Bull. 89 of Fish and Wildlife Service, v. 55, p. 39-86.

———, 1958, Sedimentology and Quaternary geomorphology of South Texas: Gulf Coast Assoc. Geol. Societies Trans., v. 8, p. 41.

Russell, R. J., 1936, Physiography of the lower Mississippi River delta: Louisiana Dept. Cons. Geol. Bull. 8.

———, 1940, Quaternary history of Louisiana: Geol. Soc. Amer. Bull., v. 51, no. 8, p. 1199-1233.

———, 1945, Climates of Texas: Annals of the Assoc. of Amer. Geographers, v. 35, no. 2, p. 37-52.

Shepard, F. P., 1937, Revised classification of marine shorelines: Jour. Geol., v. 45, no. 6, p 602-624.

———, 1952, Revised nomenclature for depositional coastal features: Amer. Assoc. Petroleum Geol. Bull., v. 36, no. 10, p. 1902-1912.

———, 1953a, Sedimentation rates in Texas estuaries and lagoons: Amer. Assoc. Petroleum Geol. Bull., v. 37, no. 8, p. 1919-1934.

———, 1953b, Sediment zones bordering the barrier islands of central Texas coast, *in* Finding ancient shorelines: Soc. Econ. Paleon. Miner. Spec. Publ. no. 3, p. 78-96.

———, 1956, Late Pleistocene and Recent history of the central Texas coast: Jour. Geol., v. 64, p. 56-69.

———, and Moore, D. G., 1955, Central Texas coast sedimentation: characteristics of sedimentary environment, Recent history and diagenesis: Amer. Assoc. Petroleum Geol. Bull., v. 39, no. 8, p. 1463-1593.

Stetson, H. C., 1953, The sediments of the western Gulf of Mexico, Part I—the continental terrace of the western Gulf of Mexico: its surface sediments, origin, and development: Papers in Phys. Oceanog. and Meteorol., M.I.T. and Woods Hole Oceanog. Inst., v. 12, no. 4, 45 p.

Stevens, C. S., 1951, The silt load of Texas streams: Texas Jour. Sci., v. 3, no. 2, p. 162-173.

Storm, L. W., 1945, Résumé of facts and opinions on sedimentation in Gulf Coast region of Texas and Louisiana: Amer. Assoc. Petroleum Geol. Bull., v. 29, no. 9, p. 1304-1335.

Texas Board of Water Engineers and U. S. Dept. of Agriculture, Soil Conservation Service, Div. of Irrigation and Water Conservation, 1955, Progress Report No. 16 of the silt load of Texas streams (1953-1954), 54 p.

Thornthwaite, C. W., 1948, An approach toward a rational classification of climate: Geogr. Rev., v. 38, no. 1, p. 55-94.

Trowbridge, A. C., 1923, A geological reconnaissance of the Gulf Coastal Plain of Texas near the Rio Grande: U. S. Geol. Surv. Prof. Paper 131, p. 85-107.

———, 1932, Tertiary and Quaternary geology of the lower Rio Grande region, Texas: U. S. Geol. Survey Bull. 837, 260 p.

U. S. Department of Commerce, 1949, United States Coastal Pilot, Gulf Coast, Key West to Rio Grande, Washington, D. C.

U. S. Congress, 1953, Gulf shore of Galveston Island, Texas: Beach Erosion Control Study, House Doc. 218, 83rd Congr., 1st Session.

Weaver, P., 1950, Variations in history of continental shelves: Amer. Assoc. Petroleum Geol. Bull., v. 34, no. 3, p. 351-360.

Weeks, A. W., 1933, Lissie, Reynosa, and upland terrace deposits of coastal Texas between Brazos River and Rio Grande: Amer. Assoc. Petroleum Geol. Bull., v. 17, no. 5, p. 453-487.

———, 1945, Quaternary deposits of Texas Coastal Plain between Brazos River and Rio Grande: Amer. Assoc. Petroleum Geol., v. 29, no. 12, p. 1693-1720.

Williams, H. F., 1951, The Gulf of Mexico adjacent to Texas: Texas Jour. Sci., v. 3, no. 2, p. 237-250.

Editor's Comments on Paper 6

6 **Zeigler:** *Origin of the Sea Islands of Southeastern United States*

At the same time that Le Blanc and Hodgson were publishing their report on the Texas shoreline, John M. Zeigler, then a research associate in marine geology at the Woods Hole Oceanographic Institution, reported on the sea islands off the coasts of South Carolina and Georgia. Zeigler identified three types of islands in the region: erosion remnant islands, marsh islands, and beach-ridge islands. According to him, the erosion remnant islands originated by the drowning of cuestas; the marsh islands grew up through, but not much above, sea level; and the beach-ridge islands developed by littoral drift accompanying the destruction of headlands, during both a fall and a rise in sea level. These separate origins are somewhat similar to those advocated by Hoyt, Otvos, and Fisher, respectively, in the debate included later in this volume. Zeigler's view, however, appears to be an antecedent (unknown at the time of writing) to the editor's own paper on the multiple causality of barrier islands.

Zeigler is presently Assistant Director and Head of the Division of Physical, Chemical, and Geological Oceanography at the Virginia Institute of Marine Science at Gloucester Point, Virginia. His research interests include causes of erosion, tropical continental shelf studies, and nearshore circulation in bays and near beaches; and he has published very extensively over the last fifteen years. Zeigler was Associate Scientist at Woods Hole Oceanographic Institution from 1953 to 1967, and Professor of Marine Geology at the University of Puerto Rico from 1967 to 1971. At times during these years he was a lecturer at the University of Chicago; Director of Marine Laboratory, Barranguilla, Colombia; and President of Scientific Marine, Inc., Woods Hole, Massachusetts. Prior to that he had been a field geologist in Wyoming and Afghanistan. Zeigler received his B.A. at the University of Colorado in 1947, and his Ph.D. at Harvard University in 1954. He was born in St. Augustine, Florida, in 1922.

Reprinted from the *Geograph. Rev.*, Vol. 49, 222–237, 1959, copyrighted by the American Geographical Society of New York

6 ORIGIN OF THE SEA ISLANDS OF THE SOUTHEASTERN UNITED STATES*

JOHN M. ZEIGLER

THE coast of South Carolina and Georgia is characterized by a chain of islands parallel to the mainland and separated from it by salt marshes; passes or sounds between the islands lie approximately opposite the mouths of coastal rivers. This sea-island coast was of special interest to the participants of photographic flights carried out in 1953, 1956, 1957, and 1958 as part of coastal geographic studies conducted by the Woods Hole Oceanographic Institution and sponsored by the Office of Naval Research, Geography Branch.[1] On these flights motion pictures were taken of the east coast of the United States. Many of the islands were examined from the air much more closely than the shore line itself, and the writer's curiosity led him later to make two trips by land into the sea-island area to examine features that had been seen from the air. Field work was not detailed, but coverage extended from Winyah Bay, S. C., to the Florida boundary. The films, of course, were available for restudy of the features seen in the field.

Nature of the Sea-Island Coast

It is evident that two generations of rivers drain the coastal plain and form the sounds that separate the islands (Fig. 1): (1) pre-Wisconsin rivers, which cross the coastal plain and form part of the drainage of the Piedmont (for example, the Santee, the Edisto, the Savannah, the Ogeechee, the Altamaha, and the Satilla); and (2) post-Wisconsin rivers (such as the Medway, the Newport, and the Turtle), which terminate at or near the foot of the Wisconsin-age Pamlico scarp.[2]

**Contribution No. 988 from the Woods Hole Oceanographic Institution.* The writer is grateful to the Geography Branch of the Office of Naval Research for supporting coastal photographic flights that made it possible to see the entire sea-island coast several times. The ideas set forth in this paper more or less grew out of observations made during those flights. Dr. Robert A. Ragotzkie, director of the University of Georgia Marine Institute at Sapelo Island, encouraged the writer to examine his earlier ideas more closely and report them to the Marsh Conference held at Sapelo Island, March, 1958, and sponsored jointly by the National Science Foundation and the Marine Institute. The writer also wishes to thank Professors Donald C. Scott and Wilbur Duncan of the University of Georgia for many helpful suggestions. Encouragement from, and discussions with, Professor J. A. Steers of the University of Cambridge are gratefully acknowledged, as is criticism of the manuscript by Professors Alfred C. Redfield, Sherwood D. Tuttle, and Wilbur Duncan.

[1] See J. M. Zeigler and F. C. Ronne: Time-Lapse Photography—An Aid to Studies of the Shoreline, *Research Reviews,* Office of Naval Research, Washington, D. C., April, 1957, pp. 1–6.

[2] These relationships can be seen on Plate 19 of F. S. MacNeil: Pleistocene Shore Lines in Florida and Georgia, *U. S. Geol. Survey Professional Paper 221-F,* 1950.

➤ Dr. Zeigler is a research associate in marine geology at the Woods Hole Oceanographic Institution.

FIG. 1—Sea-island coast of Georgia and South Carolina, showing marsh types and locations. Ruled pattern indicates marsh areas localized by a zone of less resistant sediments. Dotted pattern shows location of marshes behind constructional features—beach ridges and beach ridge islands.

The sea-island coast is not uniform throughout its length either in appearance or in origin. South of the Edisto River in South Carolina it is composed of sea islands separated by sounds. North of the Edisto it is characterized by vast marshes, narrow beach-ridge islands, and fewer rivers. The location and pattern of the marshes also change from one end of the sea-island coast to the other. A well-defined band of marshes 2½ to 3 miles wide, developed in a long, narrow valley, extends continuously for 100 miles between Cumberland Island on the south and Skidaway Island on the north. The origin of the valley is the main concern of this paper.

Northward from Wassaw Island, Ga., the valley changes gradually into a zone of marshes and scattered, ragged-edge islands, both large and small (Fig. 1). This zone of marshes and islands terminates at Edisto Island; however, several large rivers that flow parallel to the coast line extend the marshes along a series of waterways to a point just north of Charleston. Above Charleston the coast is the marsh and beach-ridge island complex mentioned above.

Fenneman identifies a part of the Coastal Plain province of the southeastern United States as the "Sea Island section" and describes it as "young to mature terraced coastal plain with submerged border."[3] Following Veatch,[4] he suggests[5] that the islands of the coast of Georgia and South Carolina are parts of the Pamlico terrace,[6] "isolated by a slight submergence." The passageways between the islands were cut by streams during post-Pamlico emergence and were enlarged by tidal scour after postglacial submergence.

Flint[7] observed that many of the streams on the coastal plain flow parallel to the modern coast line in areas where the elevation is not more than 100 feet. He felt that the drainage pattern of the eastern part of the coastal plain was determined in part by the irregularities of an emerging sea floor. Before emergence numerous bars had been built on the sea floor, and these later controlled the drainage on the emerged coastal plain.

---

[3] N. M. Fenneman: [Map of] Physical Divisions of the United States, 1:7,000,000 (U. S. Geological Survey, 1946; reprinted 1949), reference in legend.

[4] Otto Veatch and L. W. Stephenson: Preliminary Report on the Geology of the Coastal Plain of Georgia, *Georgia Geol. Survey Bull. No. 26,* 1911, pp. 37–38.

[5] N. M. Fenneman: Physiography of Eastern United States (New York and London, 1938), p. 44.

[6] According to Cooke (C. W. Cooke: Correlation of Coastal Terraces, *Journ. of Geol.*, Vol. 38, 1930, pp. 577–589; reference on p. 578), the Pamlico terrace was named for North Carolina by Stephenson (L. W. Stephenson: The Quaternary Formations, *in* The Coastal Plain of North Carolina [by W. B. Clark and others], *North Carolina Geol. and Econ. Survey,* Vol. 3, 1912, pp. 266–290; reference on p. 286).

[7] R. F. Flint: Pleistocene Features of the Atlantic Coastal Plain, *Amer. Journ. of Sci.*, Vol. 238, 1940, pp. 757–787; reference on pp. 772–773.

The "bedrock" of the coastal area of Georgia and South Carolina is usually considered to be Quaternary sediments (Pleistocene and Recent). These deposits, for the most part relatively thin, also blanket the older formations.[8] Cooke[9] named these blanketing sediments Pamlico, as making up the terrace of the same name. The Pamlico formation as defined is a thin body of marine silt, clay, and sand containing a marine fauna that lived when temperatures were slightly higher than those of today. It was presumably deposited during the Sangamon interglacial. Its surface is a well-preserved sea-floor plain that rises gently landward from the present shore line to an elevation of about 25 feet, where it ends at the Suffolk strand line.[10] Everywhere within the area under discussion the Pleistocene sediments are less than 150 feet thick and dip seaward at less than 100 feet per mile.[11] However, stratigraphic succession is extremely difficult to determine because outcrops are infrequent and inaccessible.

Davis[12] long ago developed a genetic terminology for the erosional development of a coastal plain. Such a region is ideally of low elevation and is composed of sedimentary strata that have a gentle seaward dip. During the development of a drainage system, subsequent tributaries to the master consequent streams may erode headward along the outcrops of less resistant beds. This action produces a rectangular drainage pattern with the subsequent streams flowing in the strike valleys. The interfluves produced between these subsequent streams are elongate ridges called "cuestas." The southern part of the Atlantic Coastal Plain is underlain along the shore by relatively unconsolidated marine strata of Pleistocene age that dip seaward at less than 1°.[13]

Erosion Remnant Islands

There seem to be three different types of islands along the Georgia-South Carolina coast: erosion remnant islands, marsh islands, and beach-ridge

[8] A 2001-foot well at Charleston penetrated 75 feet of Pleistocene sediments described as dark-gray, finely arenaceous and micaceous clay, very coarse quartz sand, and shells and shell fragments. Below the Pleistocene were about two feet of Pliocene greenish phosphatic pebbles, and below that to 220 feet the Cooper marl of the uppermost Eocene or the Oligocene, composed of sandy and argillaceous limestone and nodules of phosphatic sandstone (L. W. Stephenson: A Deep Well at Charleston, South Carolina, *U. S. Geol. Survey Professional Paper 90-H*, 1914).

[9] *Op. cit.* [see footnote 6 above].

[10] R. F. Flint: Glacial and Pleistocene Geology (New York and London, 1957), pp. 266 and 363.

[11] H. G. Richards: Subsurface Stratigraphy of Atlantic Coastal Plain between New Jersey and Georgia, *Bull. Amer. Assn. of Petroleum Geologists*, Vol. 29, 1945, pp. 885-955; reference on p. 945.

[12] W. M. Davis: The Drainage of Cuestas, *Proc. Geologists' Assn.*, Vol. 16, 1899, pp. 75-93. See also O. D. von Engeln: Geomorphology (New York, 1942), pp. 106-132.

[13] Richards, *op. cit.* [see footnote 11 above], pp. 940 and 945.

FIG. 2—The north end of Wassaw Island, a typical beach-ridge island, looking east.

FIG. 3—View south over Sapelo Island—an erosion remnant island.

FIG. 4—Typical marsh islands, lying behind Sapelo Island.

islands. In general, the erosion remnant islands are sand-covered and irregular in outline (Fig. 3), the marsh islands are lower and are formed of detritus intermixed with vegetation (Fig. 4), and the beach-ridge islands have a characteristic topography of subparallel and arcuate ridges (Fig. 2). If the three types are present in the same area, they always occur in the same relative position to one another.

Erosion remnant islands are found from Charleston south to Florida (Figs. 5–7). The northern group, including Edisto, Port Royal, and Wadmalaw Islands, have conspicuously reticulate outlines and remind one of scattered pieces of a jigsaw puzzle. The southern group—for example, Cumberland and Sapelo Islands—are simpler in outline. The characteristic irregularity of the outline is apparently caused by the variable resistance of the sediments to erosion by streams. Erosion of the edges is primarily by marsh streams except along the sides facing the open coast. The pattern of the present topography is not uniform, and the resulting remnants have ragged edges. It is the writer's contention that the long, narrow, marsh-filled valley between the Florida border and Skidaway Island, described above, was eroded in a less resistant member of the coastal-plain sediments, which changes character northward, becoming not greatly different from the surrounding sediments, and finally disappears (Fig. 1). When it does, the zone of marshes and remnant islands also disappears. This explanation supports the thesis that many of the sea islands originated by the drowning of cuestas. Other criteria supporting erosional origin are types of sediment, types of bedding, and topographic form. If a fragment of the mainland was isolated by erosion, one would expect the island to have a similar lithology. Moreover, one would expect to be able to see the difference between typical coastal-plain sediments and the sediments making up beach ridges or barrier islands. In this part of the coastal plain the sediments are mostly sands and clayey or limy sands with scattered pockets of clay.[14] The beach ridges on this part of the Atlantic seaboard are made up of well-sorted fine sands.

The writer examined outcrops on some of the large islands—Sullivans, James, Johns, Wadmalaw, St. Helena, Hilton Head, Port Royal, Ladies, Sapelo, and Anastasia (south of the sea-island area in Florida)—and also visited scores of smaller islands; the results of his investigations are significant. Clay deposits were found on several of the remnant islands, and evidence for clay was found on others. There is a reddish clay deposit on Sapelo north of Marsh

[14] L. W. Stephenson: Major Marine Transgressions and Regressions and Structural Features of the Gulf Coastal Plain, *Amer. Journ. of Sci.*, Ser. 5, Vol. 16, 1928, pp. 281–298; Veatch and Stephenson, *loc. cit.* [see footnote 4 above]; and Richards, *op. cit.* [see footnote 11 above], pp. 919 and 922.

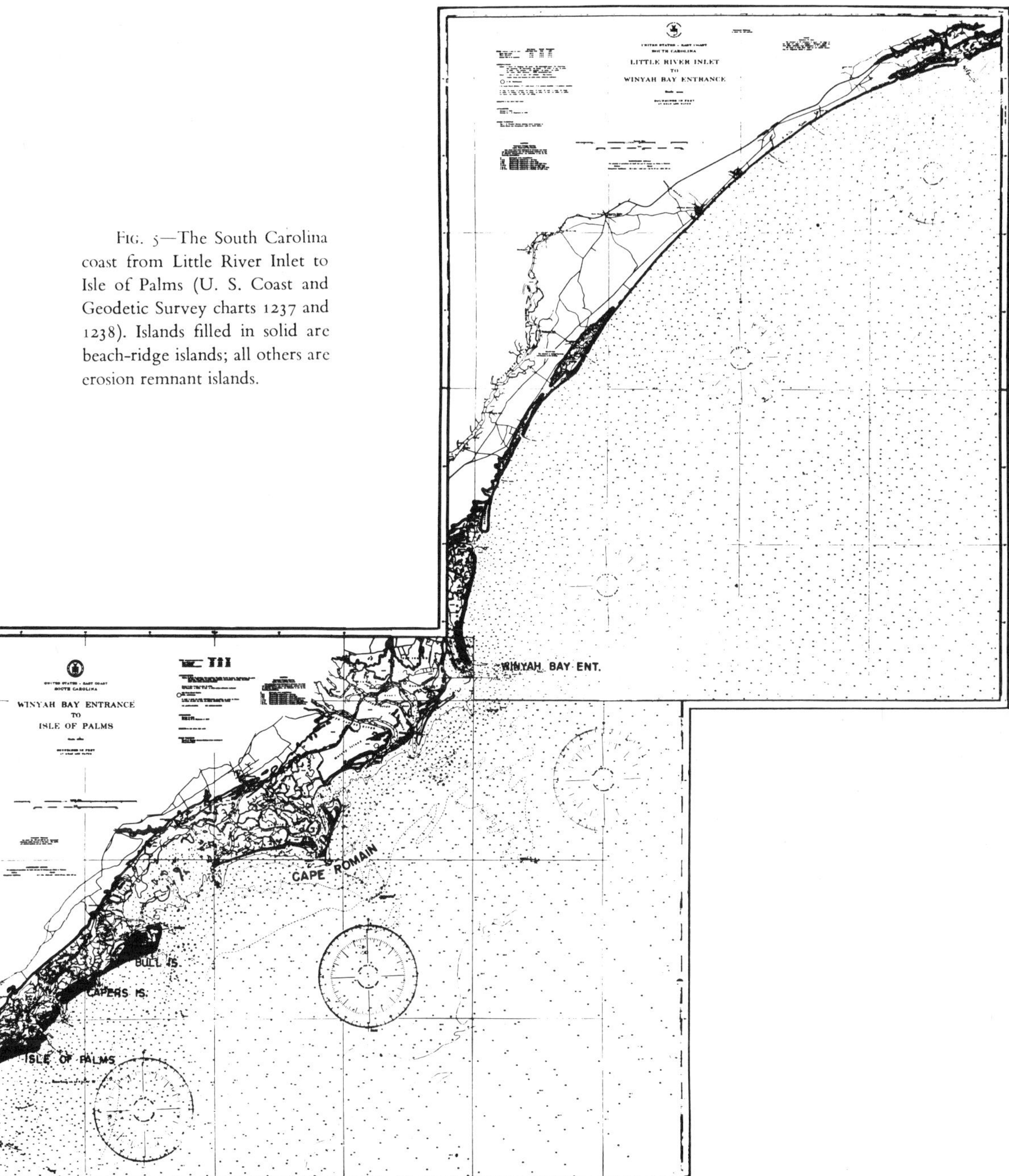

FIG. 5—The South Carolina coast from Little River Inlet to Isle of Palms (U. S. Coast and Geodetic Survey charts 1237 and 1238). Islands filled in solid are beach-ridge islands; all others are erosion remnant islands.

FIG. 6—The South Carolina coast from Isle of Palms to the Savannah River (U. S. Coast and Geodetic Survey charts 1239 and 1240). Islands filled in solid are beach-ridge islands; all others are erosion remnant islands.

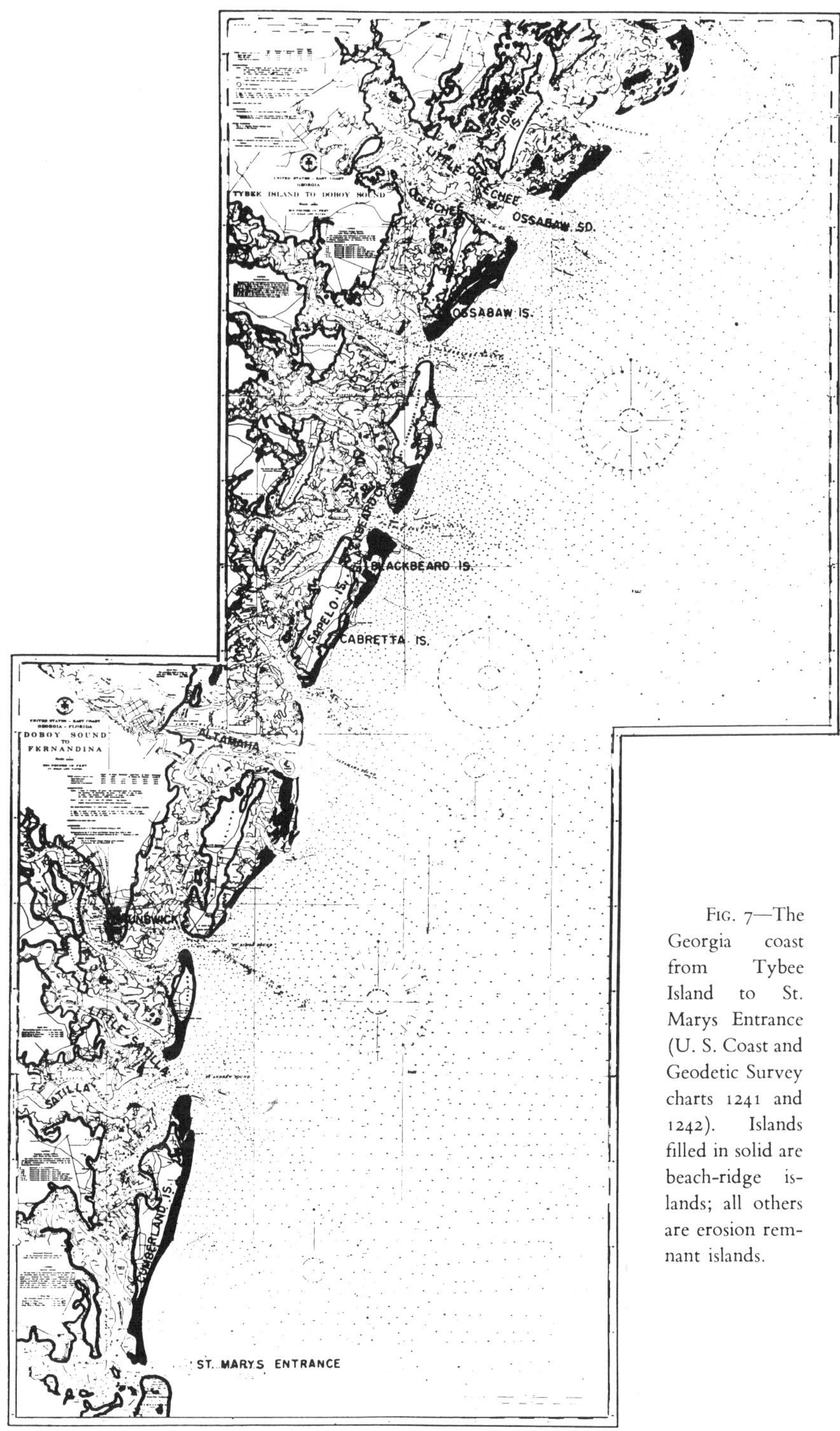

FIG. 7—The Georgia coast from Tybee Island to St. Marys Entrance (U. S. Coast and Geodetic Survey charts 1241 and 1242). Islands filled in solid are beach-ridge islands; all others are erosion remnant islands.

Landing. Extensive clay deposits, interbedded with fine tan-colored sand, crop out at the Bears Bluff Laboratory on Wadmalaw Island, and bricks are made today from a large clay deposit on Johns Island. No clay was found on Port Royal Island, but a brickyard creek was noted there. None of the deposits were exposed enough to indicate the total extent or the origin of the clay. The presence of clay, which would be unlikely on sand bars or barrier islands, indicates that the larger islands are indeed similar lithologically to the mainland.

A comparison shows differences in bedding between dunes, beach-ridge areas, and erosional remnants of coastal-plain formations. An excellent area for comparison is Raccoon Bluffs on Sapelo Island. Here Blackbeard Creek has cut into evenly bedded, fossiliferous sands, but on the other side of the creek one can see beautiful cross-bedded sands on Blackbeard Island, a beach-ridge island. Uniform bedding is also visible at Bears Bluff on Wadmalaw Island, an erosion remnant island. It is difficult to find bedding of any kind in many exposures which are sand from top to bottom, but at no place did the writer find cross-bedding on either large islands or small except those which were topographically obviously beach ridges. The presence of regular, even bedding is taken to mean that the cores of many of the sea islands are erosion remnants and not constructional marine features.

The different appearance of the erosion remnant islands from beach-ridge islands as seen from the air first attracted the writer's attention. The topography of the erosion remnant islands in no way resembles beach-ridge or barrier-beach topography. Even though the surface cover of the islands is sand, ancient dune topography is absent. The surficial appearance is like that of the mainland, even to the presence of a few small Carolina bays on Sapelo Island. Of course, there is no reason why the erosion remnant islands should not have superimposed marine features dating from the time when sea level fell from a level above them. Pleistocene bars and beaches have been recognized on much of the coastal plain of Georgia and South Carolina.[15]

Suggestions of old bars were seen from the air on Johns Island and Wadmalaw Island and questionably on Hilton Head Island, but these were not examined on the ground. It is certain that they are not beach ridges, nor have they trapped flooded areas behind them; they are still at much higher elevations than the bluffs bordering the marsh on Wadmalaw and Johns Islands. Linear features could, of course, aid the drainage to establish a trellis pattern. If this is the case in Georgia, the linear features themselves were also cut away when the long valley was formed.

[15] Flint, Pleistocene Features of the Atlantic Coastal Plain [see footnote 7 above]; and MacNeil, *op. cit.* [see footnote 2 above].

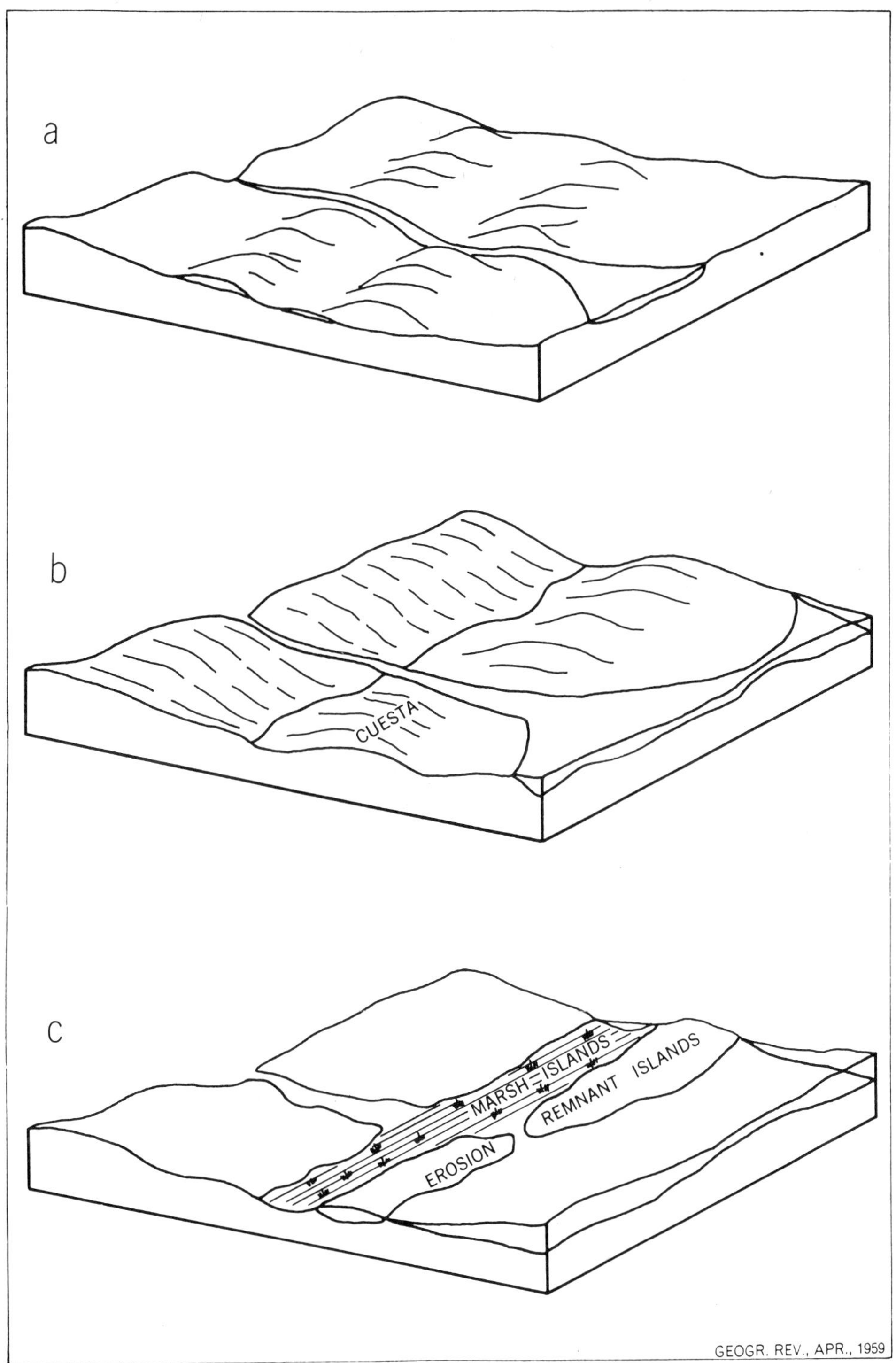

FIG. 8—Physiographic history of marsh islands and erosion remnant islands: (a) consequent stream cutting across the coastal plain at time of low sea level; (b) development of subsequent tributaries producing strike valleys and cuestas; (c) formation of erosion remnant islands by drowning; the marsh islands are now located in the position of the subsequent streams, while the cuesta tops remain as erosion remnant islands.

The sea covered coastal Georgia and South Carolina during the last interglacial period (Sangamon). As the Wisconsin glaciers advanced, sea level dropped and various marine features were left behind on the land (spits, sand bars, wave-cut scarps, and the like). Marine features were probably present all the way across the now submerged continental shelf to the Wisconsin shore line, 150 feet below the present sea level.[16] At the time of lowest sea level the present-day erosion remnant islands were part of the divides between the watersheds of rivers that drained the emerged portion of the coastal plain (Fig. 8a). Part of the developing drainage system followed the edge of a stratum of less resistant sediments roughly parallel to the present-day coast, and a valley that extended from Cumberland Island on the south to Skidaway Island on the north was cut into this less resistant stratum, and it truncated the main divides (Fig. 8b). As the subsequent valleys became flooded by the rising sea, sediments accumulated and marshes could and did develop on them between the mainland and the sea islands (Fig. 8c). The less resistant stratum is presumed to change its lithologic character northward until it gradually becomes no different from the surrounding sediments, its differential resistance to erosion decreases and becomes less uniform, and the resulting valley becomes less regular and finally is not found north of Edisto Island.

Marsh Islands

Marsh islands consist of various plants, mostly *Spartina* sp., growing on clay, silt, and peat. The general argument is that the marsh vegetation begins to grow on sand or mud flats exposed at low tide. Silt and clay are trapped by the stems and roots of the plants, and the level of the marsh is raised. Furthermore, the level will continue to rise as sea level rises. The rate of growth of the marsh vegetation does not exceed the rate at which sea level rises, because *Spartina* can flourish only in environments that are flooded by sea water from time to time; therefore the *Spartina* marshes are not built much above the average sea level. This summary of the formation of marsh islands is highly simplified; many complicating factors must be considered if one works with a specific marsh island, but in general the Carolina-Georgia marshes evolved in this manner.

The marsh islands form where marsh vegetation can survive, as in a bay or on a protected mud flat. Great marshes are forming in some of the drowned river valleys—for example, the Ogeechee—of this coastal area. One does not

[16] Flint, Glacial and Pleistocene Geology [see footnote 10 above].

expect vigorous marsh growth on the beaches exposed to the open sea on the Georgia–South Carolina coast.

Beach-Ridge Islands

Beach-ridge islands are unmistakable from the air (Fig. 2). They all show clear-cut lineation of ridges and swales, often accentuated by different vegetation on the ridges from that in the swales. They are formed entirely of well-sorted sand, predominantly quartz. The islands may be single long, thin strips or groups of parallel strips protruding through the marsh, or they may be sizable islands of ridges and swales facing the open sea, in which case marsh has grown up on the protected side.

The beach ridges of this part of the coast have several ages. Some are forming at present, such as those on the south end of Cabretta Island or on the south end of St. Catherines Island (Fig. 7), two older beach-ridge islands. Some formed during Wisconsin time as sea level dropped, and, of course, many are post-Wisconsin, formed during the subsequent rise in sea level.

There is some disagreement concerning the origin of beach ridges. A summary of the leading theories was given by Johnson.[17] The present writer believes that beach ridges indicate abundance of sand for transport. They are constructional features and indicate a net gain in elevation to the beach or coast on which they are built. When they are built at the tip of a pre-existing spit or island, as is often the case, they provide a lee or quiet area behind them where marshes can develop. The great marshes between Charleston and Winyah Bay developed in the lee of beach-ridge islands.

Beach-ridge islands are found along the entire coastal area under study; they are particularly well developed seaward of St. Helena Island, S. C., which is an erosion remnant island. An area of marsh 3½ miles wide has developed between St. Helena Island and the beach-ridge islands of Hunting, Fripp, and Ray Point. Many island groups show this complex structure of beach-ridge islands facing the sea, marshes behind them, and, finally, closest to the mainland, an erosion remnant island. Sapelo Island is bordered on the east by a marsh that developed behind the beach-ridge islands of Blackbeard and Cabretta (Fig. 7). St. Simons, Ossabaw, Skidaway, Hilton Head, Johns, and James Islands also show this complexity. To what extent the older erosion remnant islands influenced beach-ridge development seaward of them is not known, but if these erosion remnant islands represent fragments of higher

[17] D. W. Johnson: Shore Processes and Shoreline Development (New York and London, 1919), pp. 297–299.

land, divides between rivers, then the divides themselves extended farther seaward and would have represented shoal areas in the sea. Submarine contours showing the seaward extent of divides are well illustrated off Winyah Bay, Cape Romain, and Bull Bay, northeast of Charleston (Fig. 5). The shoals would have been places where wave energy was focused by wave refraction and thus increased the tendency for erosion. Erosion means removal of sand, and unless the sea is able to carry away the eroded sediment offshore, it must be transported by littoral drift alongshore. Since erosion would have been most intense on the submarine extensions of the divides, the eroded sand would indeed represent an excess of sediment near the edge of the divides. Wave attack would have tended to focus on headlands at a time of falling sea level (Wisconsin) just as it does on headlands today, but destruction of the headlands would have been delayed because even as they were being destroyed the sea dropped, and emerging contours would tend to keep the shoal areas emerging. Consequently, the retreating sea would have been continually piling up excess sediment, most likely to the sides of the shoals or headlands.

With the rise of sea level after the Wisconsin, the beach-ridge areas were modified. We do not know how far seaward they may have extended. They have all been destroyed except those which now face the sea. In general, the central coasts of the islands erode, and the ends of the islands grow as new material from the center is added at the ends.

The Silver Bluff shore line, so called by Cooke,[18] roughly coincides with the western edge of a marsh-filled former valley. Its scarp stands about eight feet above sea level and is well developed on the mainland west of Ossabaw Island. It can be followed with little difficulty along most of the Georgia and South Carolina coast, even though it has been considerably dissected by river meanders in most places. MacNeil[19] thinks the scarp was cut on the landward side of a sound during the climatic optimum, 4000–6000 years ago, when sea level was presumably higher than it is now. It should be clear that the existence of a Silver Bluff shore line does not explain the formation of the valley which became flooded to form the sound in the first place.

Erosional Control

We seek, then, some control that will explain a long, narrow valley. If the control could also explain why the valley ends as it does, this would

[18] C. W. Cooke: Geology of Florida, *Florida Geol. Survey Geol. Bull. No. 29*, 1945, p. 248.

[19] *Op. cit.* [see footnote 2 above], pp. 100 and 104.

contribute additional strength to the argument. Of the choices available for this erosional control, the writer chooses stratigraphic control as the most likely. It is in no way unusual for coastal-plain formations to be lenslike in aspect, nor is it unexpected for coastal-plain sediments to change lithologic character. Detailed lithologic descriptions of this coastal area are scarce, partly because outcrops that expose more than 20 or 30 feet of surface deposits are rare and partly because much field work remains to be done.

Structural control would also explain a narrow valley, but the coastal plain is not in a geologic province where one would expect large-scale block faulting since the Wisconsin. Thus the process of elimination supports the theory indicated by the sedimentary data and physiographic observations that the long, narrow valley between Cumberland Island and Skidaway Island owes its origin to erosion of a less resistant sedimentary unit. For if the valley is not structural, it must have been formed either by erosion or by trapping behind constructional marine features such as beach-ridge islands or barrier islands. Although many of the marshes north of the valley originated by trapping behind beach-ridge islands, the valley behind erosion remnant islands could not have had such an origin.

Editor's Comments on Paper 7

Returning to a study of the Gulf Coast area, we present the following paper by Francis P. Shepard. Shepard agrees with Johnson that barrier islands may form on a neutral, or steady-state shore. However, he does not believe, as does Johnson, that barrier islands develop during emergence, but rather at times of slow submergence. His reasons for the latter are set within the framework of an asymptotic rise in sea level.

Shepard's view of barrier island origin includes two possibilities: de Beaumont's wave-building through sea-level origin (Shepard's discussion of the Mississippi islands bears some resemblance, in fact, to the work of Otvos, which follows several years later) and Gilbert's longshore drift origin. As Shepard puts it, ". . . there is little necessity for deciding between the two, because both are probably important." The editor, in his thinly veiled bias, tends to agree with Shepard's catholic viewpoint.

Francis P. Shepard was born in Brookline, Massachusetts, in 1897. He received the A.B. degree from Harvard in 1919 and the Ph.D. from Chicago University in 1922. He then taught geology at the University of Illinois for a number of years. First Research Associate, then Professor of Submarine Geology at the Scripps Institution of Oceanography, he became Professor Emeritus in 1967. Shepard has specialized in submarine geology, continental shelf and bay sedimentation, submarine canyons, and coastal changes. He is a Fellow or honorary member of many professional organizations, and was president, from 1958 to 1963, of the International Association of Sedimentology. His textbooks are well known and he has published many articles on coastal processes.

Reprinted from *Recent Sediments, Northwest Gulf of Mexico,* F. P. Shepard et al. eds., American Association of Petroleum Geologists, Tulsa, Oklahoma, 197–220, 368–381 (1960)

GULF COAST BARRIERS[1]

7

FRANCIS P. SHEPARD[2]
La Jolla, California

ABSTRACT

The barriers which skirt the greater part of the northern Gulf Coast constitute sand bodies with widths up to several miles and thicknesses of 20 to 60 feet. In most places the sand bodies are bordered on both sides by muddy sediments. The larger barriers have at least four facies—beaches, dune belts, barrier flats or marshes, and inlets. Each of these has sediment characteristics which are usually distinctive, one from the other. The beach sands can generally be separated from the dune sands by lower grain roundness, lower silt content, and more even stratification. The barrier flats and marshes have higher silt and clay content than the other environments and commonly contain calcareous aggregates. The inlets are intermediate in sand content and can be recognized by their mixture of bay and open-gulf organic remains. The barrier islands have formed either during or since the rise in sea level at the end of the last glacial epoch. Contrary to a long-held opinion, they are not related to coast lines of emergence but either to slow submergence or to a steady state. Accordingly, it is reasonable that the ancient barriers should have been preserved in the stratigraphic column of a subsiding area like the Gulf Coast.

The source of sand necessary to maintain the barriers is to a great extent the sand deposits of the continental shelf, supplemented by the sands of the few rivers which enter directly into the gulf rather than into bays. In places where the shelf sand has been covered by mud, the barriers have been known to be completely eroded. Some of the barriers are growing seaward and others have retreated, while still others have been quite stable in position during historical times.

Barriers are common along other coastal lowlands around the various continents, although they may develop to a moderate extent along mountainous coasts. They are particularly common along the flanks of the large deltas of the world, occurring especially at points where the delta deposition has temporarily ceased and where the area is subsiding and being reworked by the waves.

INTRODUCTION

In the past the sandy islands and spits which border many coastal lowlands have been referred to as "offshore bars" (Johnson, 1919).[3] The inappropriateness of this name has been discussed elsewhere (Price, 1951; Shepard, 1952). Briefly speaking, the connotation of "bar" is a shallow, submerged sand body; whereas, the features under discussion are land masses which may be as much as 20 miles across and 100 miles or more in length, and may have hills 100 feet or more in height. The names "barrier island," or "barrier spit" if the sand mass is connected at one end to the land, or "bay barrier" if connected at both ends, seem to be more appropriate. Collectively (following Gilbert, 1885) they may be referred to as "barriers," or if narrow as "barrier beaches." In many cases these features are alternately separated from and then connected to the mainland, depending on the recency of storms which breach the barriers.

The importance of barriers to sedimentation studies is twofold. In the first place they are so common along lowland coasts where active sedimentation is occurring today as to lead one to suspect that they also have been common in the past. Secondly, most barrier islands represent a sand facies lying between two mud facies so that, if buried, they would provide potential traps for the future concentration of petroleum, as emphasized by Bass (1934).

During the seven years in which members of American Petroleum Institute Project 51 were engaged in the study of sediments along the Gulf Coast a number of barrier islands were investigated and a series of shallow borings were made to check the foundations of the islands. Much of the information has been published (Shepard and Moore, 1955; and Shepard and Rusnak, 1957). The purpose of the present paper is to attempt to coordinate the information obtained in the Texas and Louisiana coastal studies with data from other barriers, particularly from those east of the Mississippi Delta. Extensive use of charts and air photographs has been made in this study. The latter include photographs obtained during a series of flights along the Gulf Coast in 1951 made through the courtesy of the Gulf Research and Development Company.

[1] Manuscript received, October 5, 1958.

[2] Scripps Institution of Oceanography, University of California.

[3] See pages 368–81 for list of references.

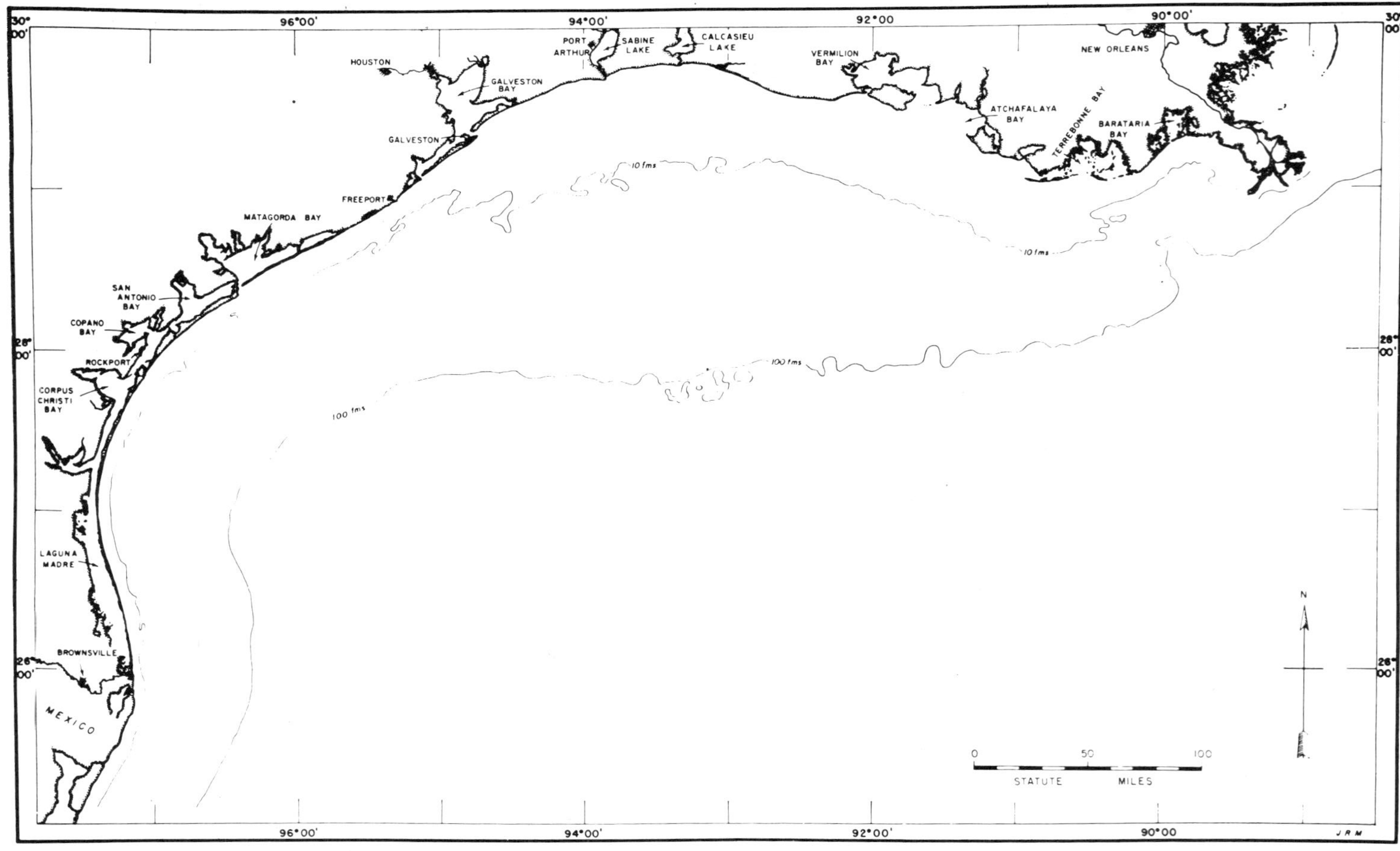

FIG. 1.—Showing the extent of the almost continuous barrier islands along the coast of Texas and the intermittent barriers along the Louisiana coast.

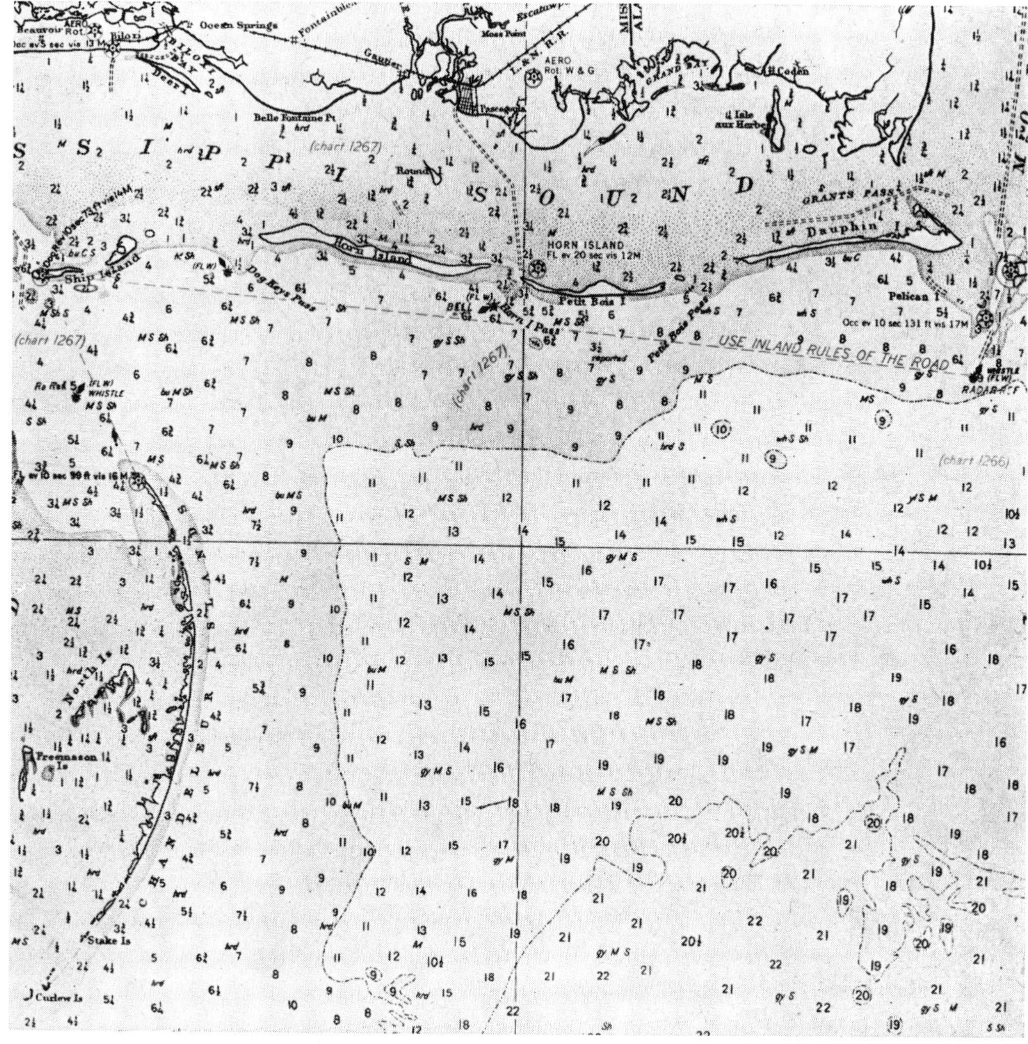

FIG. 2.—The discontinuous barriers bordering Mississippi Sound and the closely spaced Chandeleur Islands.

TYPES OF GULF COAST BARRIERS

Four types of barriers are described with gradations in between. All these are represented by islands and spits along the Gulf Coast. The following barrier types are of possible genetic significance.

1. Long straight or smoothly curving barriers, such as the almost continuous islands along the Texas coast (Fig. 1).
2. Segmented chains of islands with passes having width comparable to the length of the islands. For this type the Mississippi-Alabama offshore islands are representative (Fig. 2).
3. Cuspate islands or spits such as Cape San Blas in northwest Florida (Fig. 3).
4. Lobate or crescentic islands and spits (Fig. 4), such as the small islands along the southwest Florida coast.

The barriers also differ in respect to separation from the mainland. For one type, the lagoon inside some barriers has become largely filled, as in the central Laguna Madre where, except for the Intracoastal Waterway, the entire width of the former lagoon is represented by a sand or algal flat and in places is covered with migrating dunes. Elsewhere the barriers join the land at one or both ends but have an appreciable body of water inside. This is true of many of the barriers along the length of the Gulf Coast. The other type of

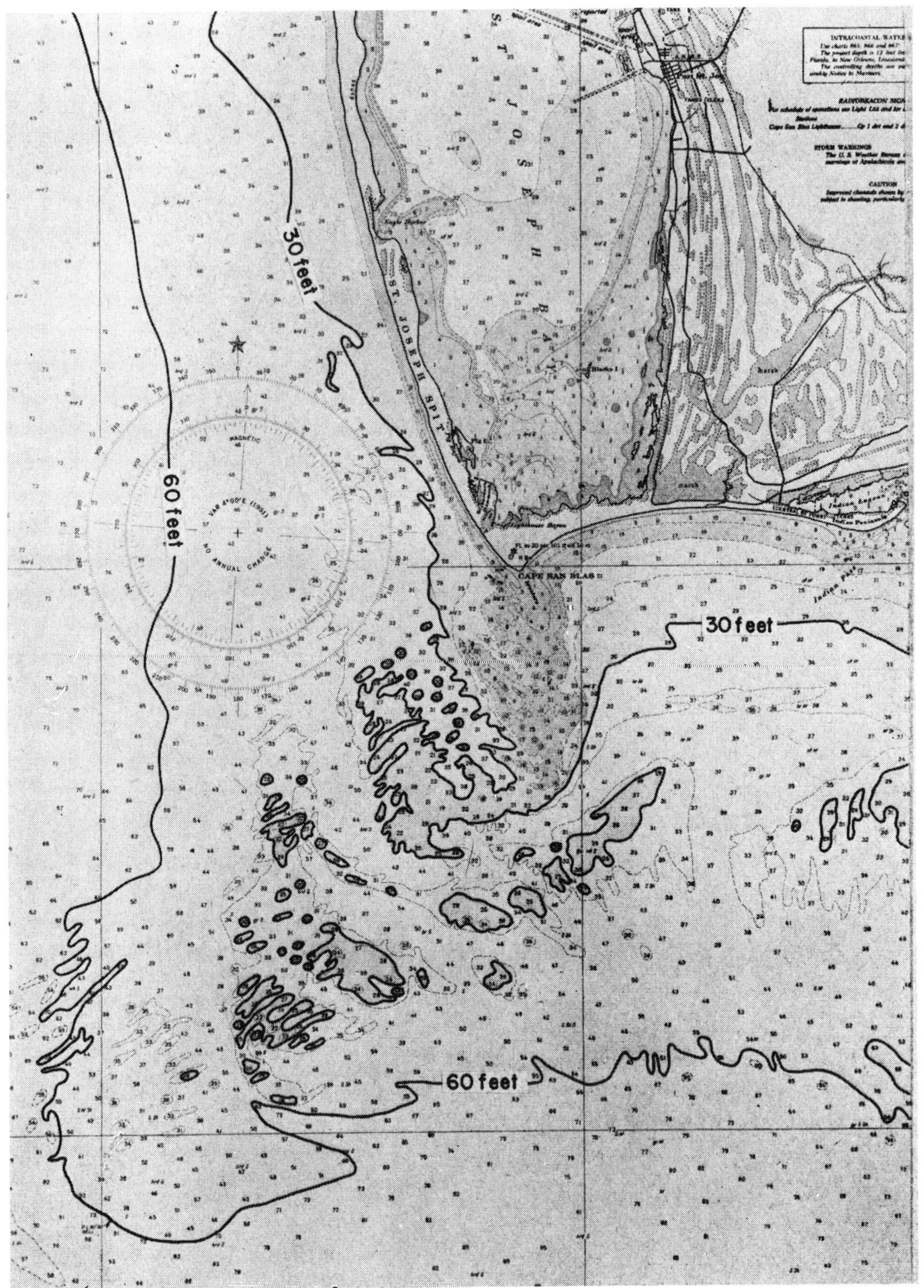

FIG. 3.—Showing the shoals which border the cuspate foreland at Cape San Blas, Florida.

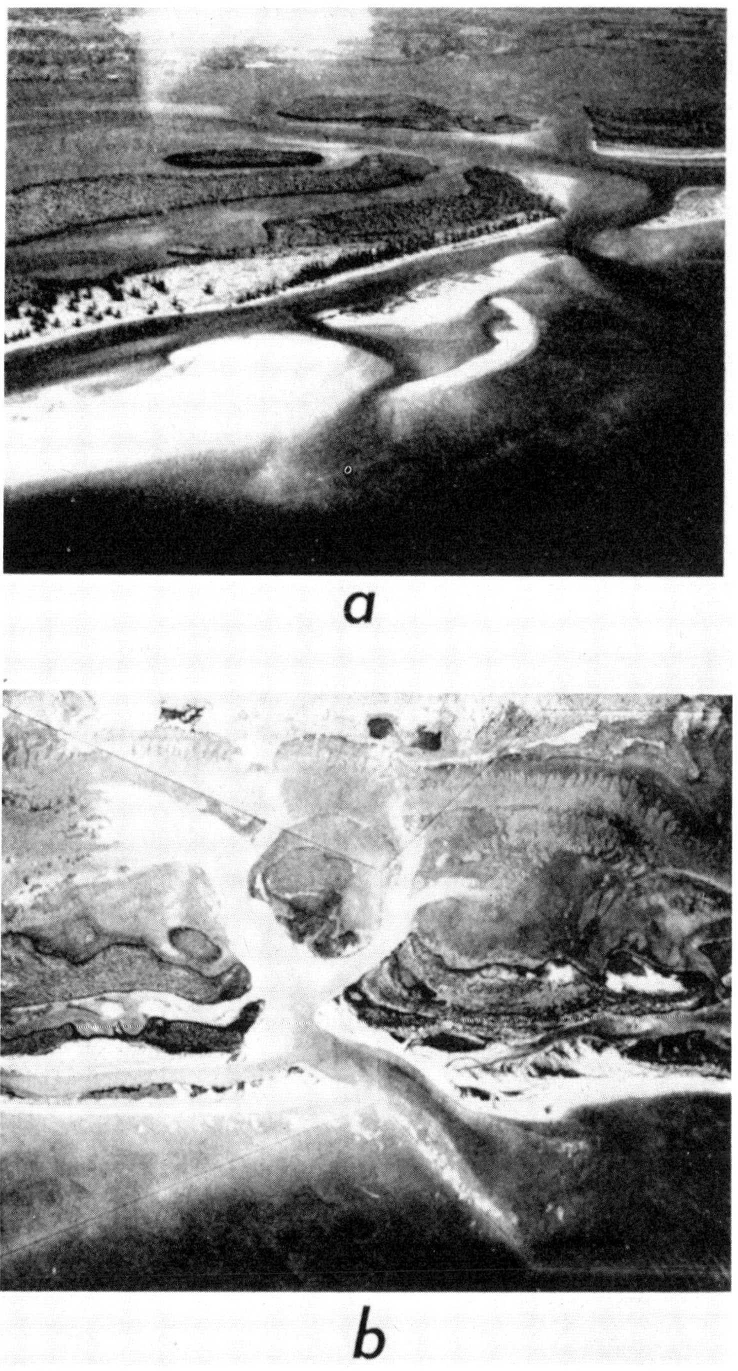

Fig. 4.—Showing the development of a lobate or crescentic island along the southwest Florida coast. Photograph *a*—by D. L. Inman, taken in 1951; photograph *b*—by U. S. Coast and Geodetic Survey, taken in 1939.

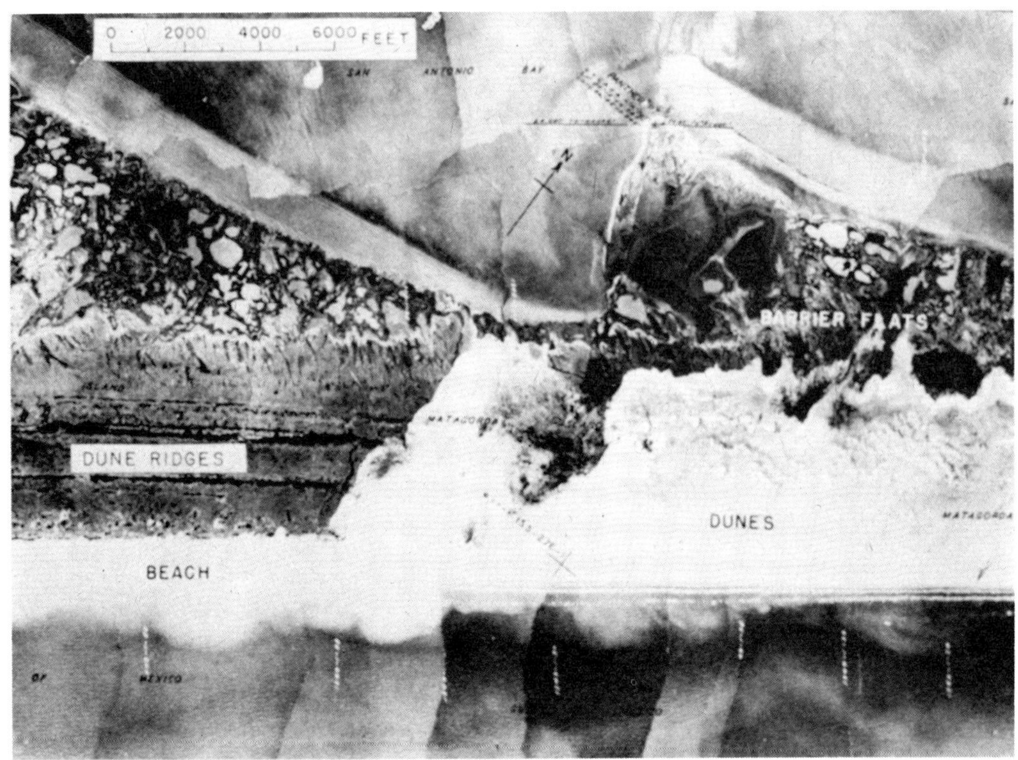

Fig. 5.—Showing the major physiographic divisions of a large barrier island. The foreground shows the Gulf of Mexico and the background a bay shore along the Texas coast. Two types of dunes are illustrated. Underwater bars are shown on the outside by the breaker lines, and on the inside, bars can be seen through the temporarily clear water. Photograph courtesy of Edgar Tobin Surveys.

barrier is a true island having no connection with the mainland.

CHARACTERISTICS OF BARRIERS

Barriers are characterized by (1) an outlying belt of sand separated from the mainland by a shallow body of water, (2) a much greater length than width, and (3) a straight seaward margin in contrast to a lobate, crenulate, or cuspate lagoonal shore line. The typical barrier has three major divisions (1) an outer beach with a broad berm, (2) a belt of dunes, and (3) an inner flat or marsh (Fig. 5). The dune belt may include a series of low ridges, commonly 5 to 20 feet high, which are not easily differentiated from beach ridges, or the dunes may consist of a broadly encroaching belt. Figure 5 illustrates both types. Some of the dune belts are cut by inlet channels.

The inner flats and marshes are commonly lobate or fan shaped and intersected with channels through which sea water flows during exceptionally high tides (Fig. 6). Where these channels have become blocked by high dunes, the flats are generally partly submerged and have numerous lakes (Fig. 7). These lakes are commonly oval or round, as are most lakes on coastal lowlands. Arcuate spits, which generally point away from the nearest inlet, fringe many of the barrier flats. In some places cuspate spits are found extending into the lagoon from the inside of the barrier island (Fig. 8). These spits generally have counterparts on the mainland side of the lagoon, and in some places the two form continuous barriers across the lagoon. Along the Gulf Coast these are best developed in northwest Florida.

A somewhat different type of barrier is found in the Chandeleur Islands. Here, according to Treadwell (1955), inside the beach (called the

FIG. 6.—Showing one side of a fan-like projection into Aransas Bay made by storm-tide overwashes crossing St. Joseph Island.

foreshore) there are dunes and beyond the dunes a slightly elevated shell-sand flat (the last two called the backshore) and beyond are mangrove swamps which extend to the Sound. The Mississippi Sound barrier islands consist mostly of beaches and dunes, with beaches developed on both sides and only narrow swamp and pond areas in the interior. To the east some of the barriers are all terrigenous sand, with either dunes inside the beach or shell-sand flats due to washovers.

SEAWARD SLOPE

On the exposed sides of the barriers beyond a narrow fringe of longshore bars and troughs, the sea floor descends at an angle which is considerably in excess of the general inclination of the

FIG. 7.—Showing the numerous lakes of the barrier flat on the inside of Matagorda Island and the arcuate spits pointing away from the nearest inlet (to the left).

FIG. 8.—A barrier island and lagoon along the southeast coast of Alabama showing the cuspate spits extending into the lagoon which have counterparts on the mainland side. Underwater shoals and bars are indicated by the shoaling in the lagoon and cuspate bars can be seen on the seaward side (in the foreground). Photography by U. S. Coast and Geodetic Survey.

continental shelf beyond (Fig. 9). An examination of soundings along the Gulf Coast shows that these slopes are mostly between 0.5 and 1.5 per cent. This compares with an average shelf slope of 0.07 per cent for the shelf in the Gulf Coast area, or about $\frac{1}{10}$th of the slope next to the barrier islands. The steeper base of the slope along the Gulf Coast occurs in most places at depths between 20 and 30 feet, although in some wave-protected areas the slope flattens at depths around 10 to 12 feet; notably on the west side, the lee, of the Mississippi Delta (profile *G*).

Examination of the outer slope of barriers in areas more exposed to waves than the Gulf of Mexico indicates that the slopes go to considerably greater depths. Where the large breakers come into the northern California and Washington coasts the slope continues to water depths of at least 50–60 feet and on the coast of the Carolinas, where waves have intermediate values, the slopes commonly flatten at about 45 feet. There are many exceptions to this relation of depth to wave size because of local factors such as the occurrence of rock bottom, but the general rule that depth is related to maximum breaker height appears to be a good one.

Slopes off cliffed shore lines where there are no barriers are commonly steep near shore also

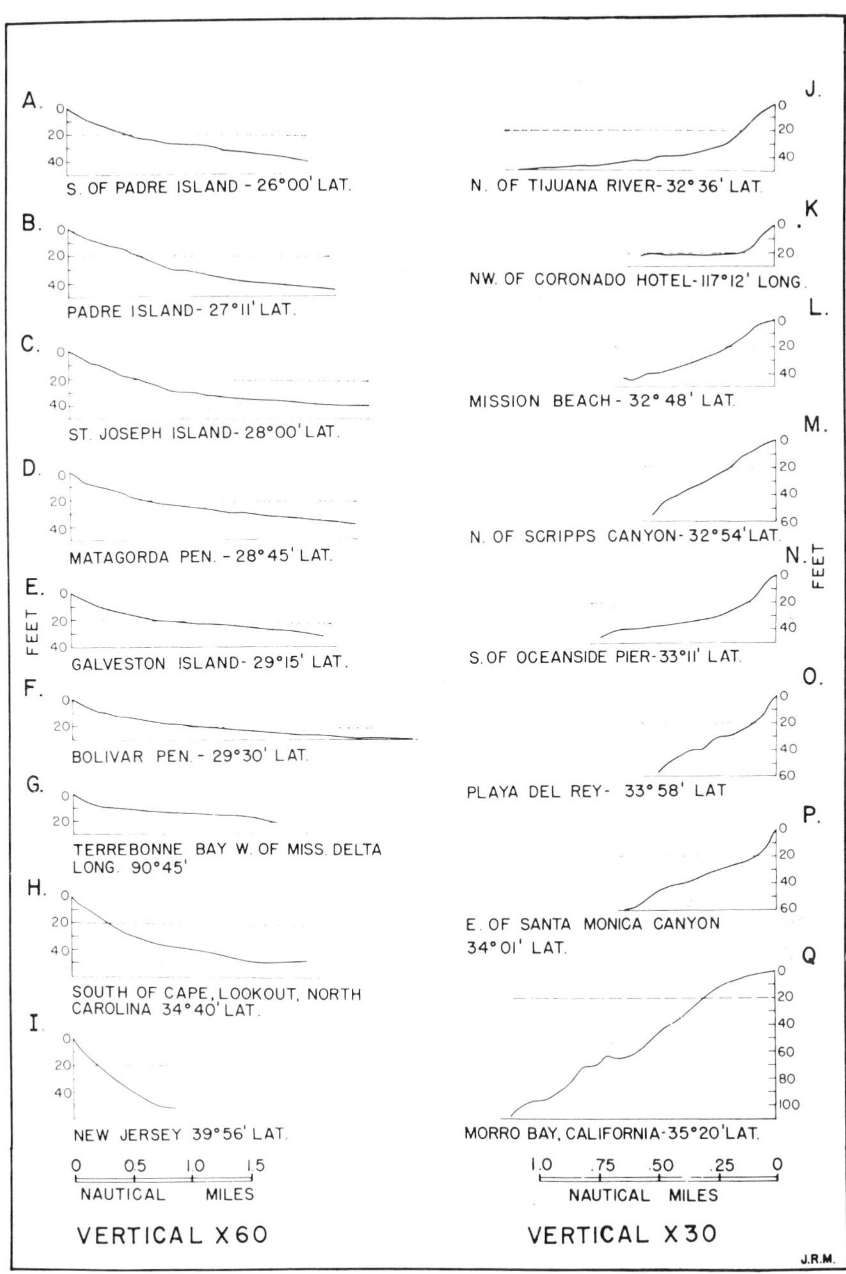

FIG. 9.—Showing the slopes off various barrier islands and a few other coasts of the United States. Illustrates the usual decrease in slope at a point which is related to the breaker heights along the coast. Profiles *M* to *P* are located off cliffed coasts. There appears to be no definite decrease of slope in profile *Q* where the coast has a barrier. Soundings taken from unpublished hydrographic sheets of U. S. Coast and Geodetic Survey. *A–F*, northwestern Gulf of Mexico; *J–Q*, southern California.

TABLE I. COMPARISON OF ROUNDNESS VALUES AND SILT CONTENT BETWEEN ADJACENT BEACH AND DUNE PAIRS

Location		*Roundness*[a]			*Silt Content (per cent)*		
		Beach	*Dune*	*Dif.*[b]	*Beach*	*Dune*	*Dif.*[b]
Texas (27 pairs)	(Ave.)	32.91	33.47	+ 0.56	.05	.06	+ .01
(Offshore wind during collection)							
Mustang Island (4)	(Ave.)	31.56	32.89	+ 1.33	.07	.11	+ .04
M-101		30.67	33.66	+ 2.99	.13	.18	+ .06
M-103		33.03	34.07	+ 1.04	.04	.15	+ .09
M-106		31.30	33.57	+ 2.27	.08	.03	− .05
M-107		31.25	30.26	− 0.99	.04	.08	+ .04
Padre Island (23)	(Ave.)	33.14	33.57	+ 0.43	.05	.06	+ .01
P- 8		33.72	34.62	+ 0.90	.08	.15	+ .07
P-10		30.84	32.30	+ 1.46	.02	.07	+ .05
P-13		32.03	33.65	+ 1.62	.12	.13	+ .01
P-14		34.22	34.61	+ 0.39	.07	.06	− .01
P-15		32.84	31.35	− 1.49	.09	.03	− .06
P-16		34.73	35.08	+ 0.35	.09	.06	− .03
P-18		30.31	34.20	+ 3.89	.03	.05	+ .02
P-20		32.25	33.70	+ 1.45	.04	.07	+ .03
P-22		29.76	32.82	+ 3.06	.04	.05	+ .01
P-25		33.25	32.39	− 0.86	.04	.07	+ .03
P-27		34.14	32.63	− 1.51	.03	.07	+ .04
P-30		32.54	34.50	+ 1.96	.01	.03	+ .02
P-32		33.57	35.23	+ 1.66	.02	.02	0
P-33		33.63	33.22	− 0.41	Tr.	Tr.	0
P-35		33.74	34.19	+ 0.45	.02	.06	+ .04
P-37		33.59	32.06	− 1.53	.02	.04	+ .02
P-39		32.90	32.33	− 0.57	Tr.	.05	+ .05
P-42		33.90	32.37	− 1.53	.02	.01	− .01
P-45		31.65	33.04	+ 1.39	.03	.03	0
P-48		33.65	35.86	+ 2.21	.02	.04	+ .02
P-50		35.17	32.32	− 2.85	.10	.07	− .03
P-53		33.89	33.82	− 0.07	.02	.03	+ .01
P-57		35.94	35.93	− 0.01	.15	.09	− .06
U. S. West Coast (11 pairs) (All onshore-wind areas)	(Ave.)	28.27	30.78	+ 2.51	.05	.13	+ .08
Pismo Beach, Calif.		29.92	31.73	+ 1.81	.20	.68	+ .48
Bodago, Calif.		29.73	30.17	+ 0.44	.04	.10	+ .06
Eureka, Calif.		29.94	31.20	+ 1.26	.01	.03	+ .02
Morro Bay, Calif.		31.06	34.23	+ 3.17	.02	.18	+ .16
Moss Landing, Calif.		28.64	28.53	− 0.12	.02	.01	− .01
Crescent City, Calif.		27.78	39.06	+11.28	.03	.01	− .02
Coos Bay, Oregon		26.49	26.75	+ 0.26	.02	.03	+ .01
Westport, Wash.		28.84	29.56	+ 0.72	.02	.03	+ .01
La Push, Wash.		25.82	29.15	+ 3.33	.02	.19	+ .17
Bay City, Wash.		25.47	28.15	+ 2.68	.14	.03	− .11
Copalis, Wash.		27.27	30.06	+ 2.79	.04	.16	+ .12
Baja California (7 pairs)	(Ave.)	25.91	27.43	+ 1.52	.11	.14	+ .03
Scammons		26.67	30.40	+ 3.73	.03	.05	+ .02
Las Cabras		24.12	23.00	− 1.12	.02	.04	+ .02
Magdalena Bay		34.36	29.87	− 4.49	.63	.44	− .19
Ensenada sta. 1		22.46	25.61	+ 3.15	.03	.11	+ .08
Ensenada sta. 2		23.41	26.76	+ 3.35	.02	.08	+ .06
La Mission		25.79	28.69	+ 2.90	.03	.12	+ .09
Rosarita		24.56	27.71	+ 3.15	.04	.11	+ .07

[a] Powers' scale × 100.
[b] + = Dune higher than beach.
− = Dune lower than beach.

TABLE I (*continued*)

Location		Roundness[a] Beach	Dune	Dif.[b]	Silt Content (per cent) Beach	Dune	Dif.[b]
Misc. Foreign (16 pairs)	(Ave.)	28.65	29.39	+ 0.74	.19	1.04	+ .85
Japan (6)	(Ave.)	26.37	27.93	+ 1.56	.06	.21	+ .15
Onuki #1		28.58	26.08	− 2.50	.01	.03	+ .02
Onuki #2		28.60	30.10	+ 1.50	.12	.60	+ .48
Sagami Bay		26.30	27.40	+ 1.10	.02	.23	+ .21
Eboshi Iwa		25.63	29.24	+ 3.61	.12	.24	+ .12
Oiso #1		23.95	27.03	+ 3.08	.05	.11	+ .06
Oiso #2		25.18	27.73	+ 2.55	.01	.02	+ .01
Dunes to west of beach in areas of prevailing westerly wind (3)	(Ave.)	34.59	31.98	− 2.61	.17	3.13	+2.96
Kiel, Germany		34.93	31.95	− 2.98	.02	2.16	+2.14
Dhahran, Saudi Arabia		39.16	36.89	− 2.27	.03	.14	+ .11
Lake Geneva		29.69	27.10	− 2.59	.45	7.08	+6.63
Miscellaneous (7)	(Ave.)	28.05	29.53	+ 1.48	.40	.73	+ .33
Puerto Rico #1		25.99	26.84	+ 0.85		Not measured	
Puerto Rico #2		27.05	30.15	+ 3.10		Not measured	
Puerto Rico #3		28.40	27.76	− 0.64		Not measured	
Karachi, Pakistan		19.54	21.47	+ 1.93	1.49	2.80	+1.31
Pyla, S.W. France		33.32	34.60	+ 1.28	Tr.	.02	+ .02
Skallingen, Denmark		29.81	32.43	+ 2.62	.05	.01	− .04
Fanø, Denmark		32.26	33.47	+ 1.21	.05	.09	+ .04
Total average (61 pairs)		30.15	31.22	+ 1.07	.09	.30	+ .21

(Fig. 9, profile *P*). It seems, therefore, as though this phenomenon may be an indication of a wave profile of equilibrium which exists along various types of coasts, although the steepening is particularly marked off barriers.

SEDIMENT CHARACTERISTICS OF BARRIER FACIES

Beaches.—Along the Gulf Coast to the west of the Mississippi Delta most of the beaches consist of clean, well-sorted sand[4] with a median diameter in the fine sand or very fine sand size, averaging about 0.12 mm. Locally shells are abundant, and along portions of the southern half of Padre Island they form the principal constituent of the berm. Elsewhere the beaches are so smooth that an ordinary car can be driven along the hard-packed sand. East of the Mississippi Delta the barrier beaches consist of coarser sand, commonly with medians around 0.3 mm.

The western Gulf beaches have only about one per cent silt. They consist predominantly of quartz sand, although a small percentage of feldspar is included. Terrigenous minerals make up about 95 percent of the whole, with shells approximately 2 per cent, and less than 1 per cent of Foraminifera, glauconite, and echinoids, respectively. Aside from unpublished oil company investigations, little study has been made of the eastern Gulf beaches, but the abundance of quartz is impressive, commonly more than 99 per cent (Martens, 1931, p. 82). The mineral content in the beaches around the Mississippi Delta is much more varied, quartz constituting only about 65 per cent. Beaches around southwest Florida are high in quartz, but the shell content is much greater than along the more northerly Gulf Coast.

Dunes.—The dune sands of the western Gulf barriers are quite similar to the beach sands. Their median diameters are virtually identical except where the beaches have abundant shells. The sorting, skewness, and kurtosis show no consistent differences, nor do the dune sands appear to be more frosted than the beach sands. Silt content, on the other hand, although small, is generally higher in the dunes (Table I). The constituents differ

[4] Along parts of the west Delta, mud shores are found locally.

slightly from those of the beach sands in having a higher content of plant material. Also the shells of all types are generally scarcer, except in blowouts where they have become locally concentrated.

Distinction between beach and dune sands.—The analyses of Allan Beal (Beal and Shepard, 1956) indicated that the dune sands have a greater degree of grain roundness than the beach sands on the Gulf Coast. Further studies (Shepard and Young, in press) have indicated that this finding applies in virtually all areas where there are onshore winds, but is not applicable to areas with offshore winds. Samples for these studies have been obtained from a large number of areas in different parts of the world (Table I).

The method used is somewhat of a modification of that of Powers (1953). Grains are judged on the basis of their angularity in three dimensions, and the highest grade (6) is given to those which would roll the best and the lowest (1) to those which could be rolled or dislodged with the greatest difficulty because of their platy surface and their projecting points which interlock with surrounding grains. Only quartz grains were used, mostly from the $\frac{1}{8}$–$\frac{1}{16}$ mm-size fraction. A total of 100 grains are classified using a Denominator Counter. All samples were run as unknowns to prevent bias and all samples were run by two investigators. Where rather different results were obtained by the two, the samples were rerun.

Table I lists the results from a comparison of paired sand samples from adjacent beach and dune locations for sand roundness and for silt content. Of 61 pairs compared for roundness, 42 pairs (69 per cent) showed the dune sand to be rounder; of 58 pairs examined for silt content, the dune sand contained more silt in 43 cases (74 per cent) (Table I).

The exceptions to the roundness rule occur largely in Padre Island, Texas, where the collections were made during a strong offshore wind. Other exceptions were found in areas which have either a predominant offshore wind or where the wind blows along the coast so that the sand in the dunes may not have been derived from the adjacent beaches.

A higher content of heavy minerals was found in the silt-size fraction of the dunes in about 80 per cent of the comparisons between beach and dune sands. On the other hand, shells were found to be more common in the beach sands although a considerable number of exceptions were found in the Padre Island area. Similarly mica is more common in beach sands than in adjacent dunes. Plant fibers are more abundant in dunes although the number of cases with plant fibers was not large enough for good statistical comparison. Comparing all of the criteria, we found in the areas where there are onshore winds, beach and dune sands could almost invariably be differentiated.

Structural contrasts, dunes and beaches.—Wherever it is possible to find bedding in the sands of a barrier, the structure is the best means of distinguishing dunes and beaches. Unfortunately, bedding was rarely seen in the cores from the many borings made in the barrier islands, although it might have become visible if the sands had been more consolidated. In the dunes typical aeolian cross-bedding is visible after a windstorm has etched it out, and the even foreshore laminations can be seen where a cut has been made into the beach. The scarcity of dark minerals in the Gulf beaches, however, prevents the usual dark layers of beaches from being found in the borings.

Barrier flats.—The sediments of the barrier flats and marshes are much more variable in composition than either the beach or dune sands. Where the flats are the result of recent hurricane washovers, the sand content of the sediments on the flats is almost as high as in the beaches. However, where the flats have not grown actively from the ocean, fine sediments have accumulated with the sand. In the portions which are partly submerged and have extensive lakes connected with channels to the lagoons (Fig. 7), very fine sediments have been introduced and sand may be less abundant than clay. The average analysis in the central Texas barrier flats gave 20 per cent silt and clay.

Because of the alternate wetting and drying of much of the flat surface, there has been extensive evaporation of the interstitial water and hence the production of calcareous aggregates and veinlets of calcareous material. Because vegetation partially covers the flats, plant fibers are more abundant than in the dunes.

Inlets.—Along most of the Gulf Coast the inlets occupy only a small fraction of the barrier length. However, some of these inlets change position frequently, and there are clear indications that as the islands have grown many of the inlets have been gradually eliminated by fill. As a result, a large portion of barrier island deposits, considered stratigraphically, may be included in the inlet category. Where inlets are being maintained at approximately their same depth by the tidal currents, their sediments may be only a surface lag concentrate and, hence, quite different from the deposits formed after the mouth of the inlet has been closed. On the floor of the open inlets, shells are abundant and consist of a combination of open gulf and bay species with the former somewhat more abundant. The sediments are largely sand, although muddy materials can be found in small pockets. Deposition in the blocked inlets results from occasional washovers from the sea, and as the result of slump from the dunes. In addition, slow circulation of water from the lagoon introduces fine sediment. Because of the latter, the sediments that blocked inlets are higher in mud content than those found on the floor of open inlets (Fig. 10–*A*). Inlets that are migrating slowly along the coast, as at Aransas Pass before it was controlled, may have quite thick sand and shell deposits formed on the depositing side (Fig. 10–*B*), as this deposition would occur while the inlet was still open and under the influence of tidal currents rather than after the inlet was blocked. Another variety of inlet deposit, described to me by M. A. Hanna, is formed where a hurricane tide has cut a deep hole in an inlet. This hole is rapidly filled with a very soft mud while sand is being carried a short distance into the marginal zone (Fig. 10–*C*). This will produce wedges of clayey sediments cutting across the longitudinal trend of the barrier.

Contrasts between barrier island and nearshore gulf deposits.—The sand content of samples taken within the 30-foot depth curve along the front of the barrier is rather comparable to typical barrier island deposits. In one respect, however, there is a decided contrast which is found all along the Texas coast and along at least a portion of the Louisiana coast. The barrier sands have, almost without exception, comparable quantities of the $\frac{1}{4}$ to $\frac{1}{8}$th and the $\frac{1}{8}$ to $\frac{1}{16}$th mm grade sizes, whereas the nearshore gulf sands consist predominantly of the $\frac{1}{8}$ to $\frac{1}{16}$th size. The best dividing line between the two is 70 per cent for the $\frac{1}{8}$ to $\frac{1}{16}$th mm size. The change between the two types takes place somewhere within the 12-foot depth line, inasmuch as the great majority of samples at that and greater depths have high percentages of the $\frac{1}{8}$ to $\frac{1}{16}$th mm sand. The difference is evidently a matter of sorting. The waves carry the coarser sand shoreward and carry the fine sand seaward. It is notable that reworking of Mis-

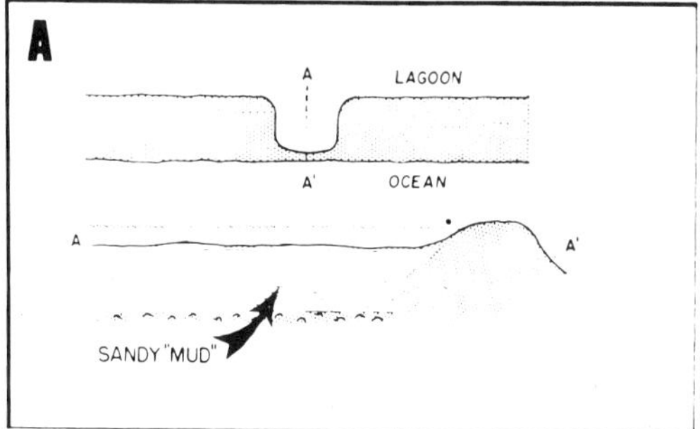

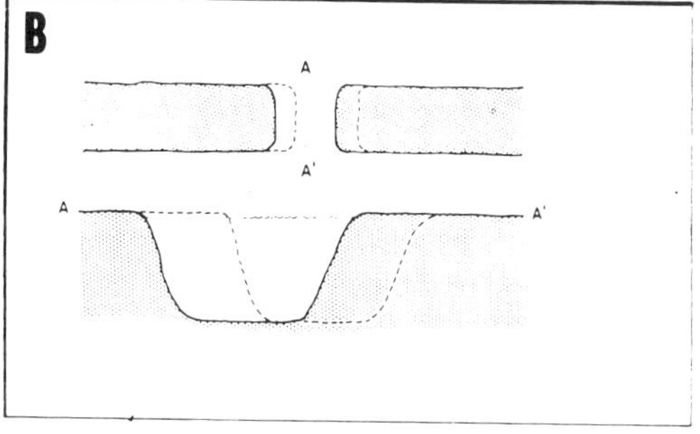

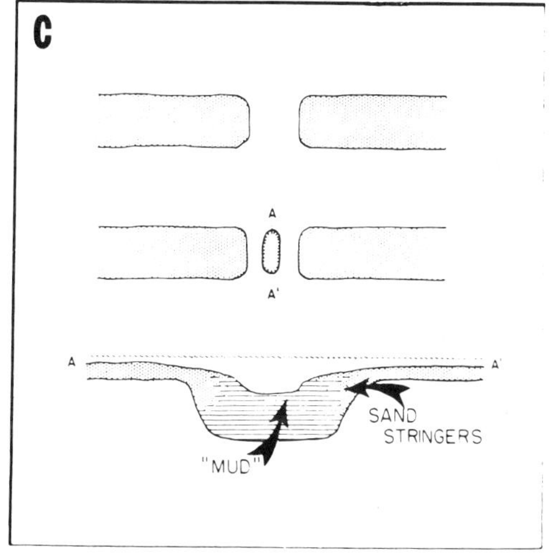

FIG. 10.—Illustrating various developments in the history of inlets. *A*—showing the blocking of an inlet on the ocean side and the development of relatively muddy sediment on the inside. *B*—illustrating the migration of an inlet along the coast under the influence of the longshore currents. *C*—the development of a deep hole in an inlet during a hurricane tide; later filled with a soft mud leaving a clayey sediment cutting across an otherwise sandy barrier island. Pointed out to the author by M. A. Hanna.

sissippi Delta deposits by the waves, resulting in the formation of barriers, produces the same change in the relative percentages of fine and very fine sands, so that the river deposits with their predominance of very fine sands are sorted to concentrate the fine sand on the barriers (Shepard, 1956a, p. 2562). Old drowned barriers out on the continental shelf off Texas similarly have abundant $1/4$ to $1/8$ mm sands in contrast to low percentages in typical shelf deposits (Curray, this volume, p. 235).

The nearshore gulf sands also differ from the beaches and dunes in having more Foraminifera and somewhat more glauconite. On the average, the shells are more abundant in the gulf sands, but locally there are exceptions where shells are concentrated on the beach or in blowouts among the dunes.

SHORE LINE CHANGES

The barriers are relatively stable in position along the western Gulf, as is shown by the charts of 1870 to 1880, which place the gulf shore line of the islands in approximately the same place as they are now found. A comparison made by D. L. Inman from Coast and Geodetic Survey original surveys showed that changes of the order of 100 to 500 feet in the shore line had occurred in about 60 years, with cut in some places and fill in others. Such changes result quite rapidly from hurricanes or other unusually large storms which generally produce recession, as determined in a recent Gulf Coast hurricane (Morgan, 1958) and in the East Coast hurricane of 1938 (Howard, 1939). The outward growth occurs during periods of small waves and at a much slower rate than the storm cut. Probably the changes between surveys of the western Gulf are largely the result of whether the survey was made after a period of cut or a period of fill.

In the 1869 survey of the Mississippi Delta a group of barrier islands, the Bird Islands, are shown where Baptiste Collette subdelta now exists. Muddy sediments from the advancing Baptiste Collette subdelta filled the strait inside Breton Island and covered the sand which had previously been the source for the islands. Subsequently, storm erosion gradually removed the islands and no sand was available for replacement (Shepard, this volume, p. 56), so that the islands have disappeared.

The Chandeleur Islands have changed in shape, particularly at the south end, and here the charts have shown a net retreat averaging 25 feet per year. This change in shape is probably authentic because the early charts have recognizable islands on the inside of the arc which can be assumed to have remained relatively constant in position. The distance to the outer barrier from these islands was used as an index of the change, rather than relying on the somewhat doubtful positions of the outer islands of the old survey.

According to old charts, the islands of Mississippi Sound have remained at approximately the same distance from the mainland during the past century (J. C. Ludwick, personal communication). The islands, however, have changed considerably in position along their length. All of them have been growing to the westward in the direction of the predominant current and are being eroded on the east side. The growth on the west, however, has exceeded the loss on the east, so that the islands are apparently increasing in size along with the gradual narrowing of the deep passes in between.

Along the northwesternmost Florida coast no chart evidence of change exists, although it can be seen that new barrier beaches have grown out beyond the old islands. To the east, near Panama City, definite evidence of change was found in the growth of a cuspate island. This island was photographed by the writer in 1951, and when seen again in 1953 it had become connected with the mainland on the southeast side. Farther south along the west coast of Florida several new islands could be recognized during the flights made in 1951 in a Gulf Research and Development Company plane (Fig. 11). One had formed just outside a break in a slightly older barrier island (Fig. 4). The barriers on the inside along this coast have become overgrown with mangroves, which are advancing into the lagoons and hence tying the islands to the mainland.

BARRIER ISLAND BORINGS

In the course of exploring the Gulf Coast sediments, we put down a total of 18 borings into the barrier islands to depths of from 23 to 60 feet.[5] In addition, samples from five borings in

[5] Eight of these obtained 3-inch cores with a drilling rig and the other ten obtained 1-inch cores with a Porter soil sampler (Shepard and Moore, 1955, Fig. 57).

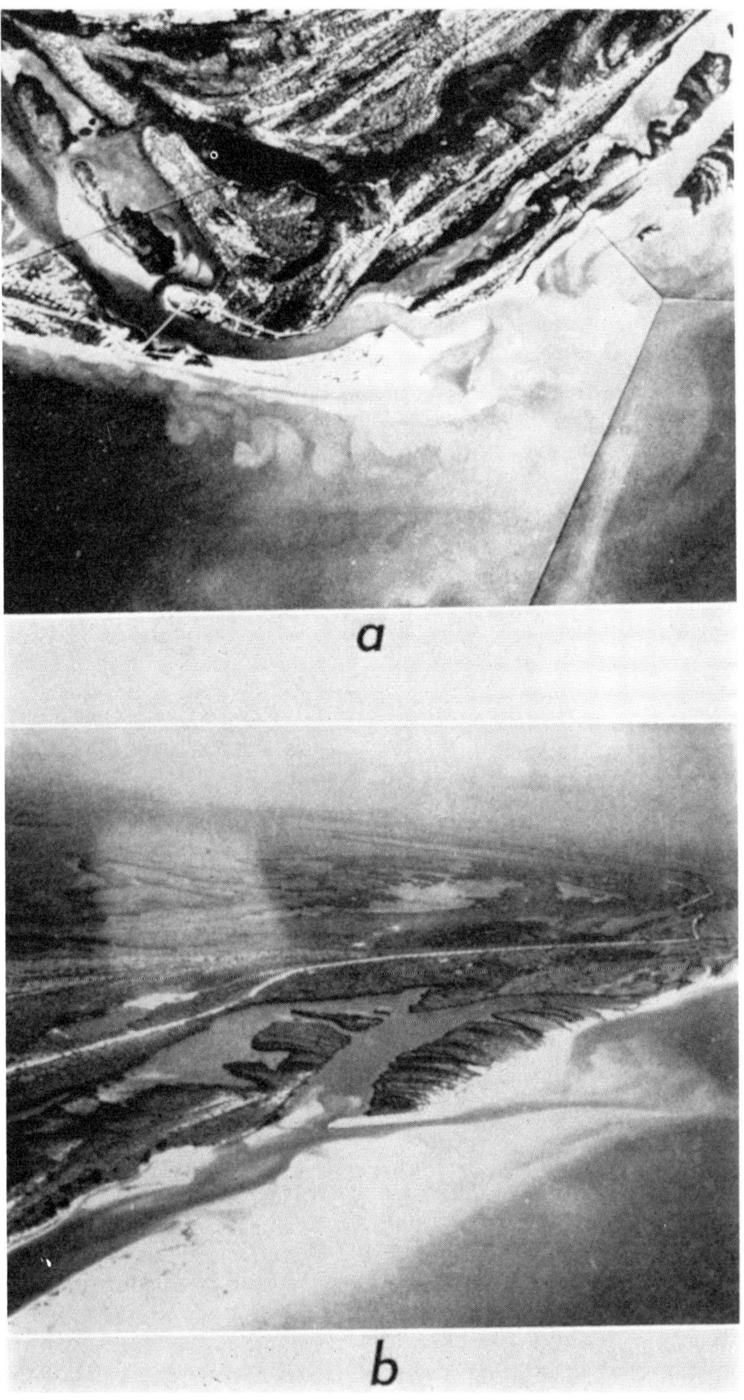

FIG. 11.—Showing the development of a new barrier island along the west coast of Florida between 1939 and 1951. Photograph *a*—by U. S. Coast and Geodetic Survey; photograph *b*—by D. L. Inman.

Galveston Island were supplied to us by Shell Development Company, and from two borings in Padre Island by Humble Oil Company. The driller's log of a large number of shot holes in St. Joseph Island was supplied to us by Perry Bass, of Richardson and Bass Oil Company. Using the grain-size, sorting, roundness and coarse-fraction criteria developed during the intensive study of sediment environments along this part of the Gulf Coast (Shepard and Moore, 1955), it has been possible to obtain considerable information on the origin and history of these barrier islands.

The most complete information comes from St. Joseph and Matagorda Islands near Rockport along the central Texas coast. These islands are separated by a narrow inlet which often becomes closed. Here five borings have been made in the beach, four inside the coastal dunes, two on a large overwash fan, and one on a curved spit on the lagoon side of the island. In addition, the generalized information from the shot holes drilled in St. Joseph Island gave some notion of the thickness of the sand formations.

All but one of the drillings appear to have reached the Pleistocene (Shepard and Moore, this volume, Figure 16, p. 137), which could generally be recognized by the appearance of an oxidized clay presumed to be the Pleistocene Beaumont formation. The depth to what was considered to be Pleistocene ranged from 25 to 60 feet, in general decreasing inland and to the northeast. The Holocene formations above the Pleistocene appear to be almost entirely barrier island in origin. They are primarily sands. The almost complete absence of continental-shelf sediments was judged largely on the basis of the relatively low percentages in the ⅛ to $1/16$th mm size; only two samples exceeded the 70 per cent dividing line, and even these had somewhat lower percentages than typical samples from the inner shelf. In addition, some of the samples at depths to about 45 feet in the holes have roundness values which are comparable to the present-day dunes. The boring at the tip of a large overwash fan on St. Joseph Island (XSJ 61), instead of encountering bay formations at a shallow depth, as was expected, encountered barrier island formations, including dunes, down to 30 feet where a greenish cast to the sand may indicate an older generation of sediment, exposed to weathering Similarly the boring on the spit called Mud Island (XMUD 11) showed barrier island (including dune) formations down to 25 feet, where the same greenish cast is found in the sediment. On the other hand, the borings on Matagorda Island along the shore of San Antonio Bay (XMT 103) extend down into bay deposits at a depth of about 30 feet.

A carbon-14 date was obtained from shells at 46 feet in a beach boring on St. Joseph Island (XSJ 56) showing an age of about 6,500 years (according to determinations made by both the U. S. Geological Survey and Magnolia Petroleum Laboratory). These shells were probably deposited in an inlet, because the formation is distinctly not open gulf and the shells include bay and gulf types.

Farther south, at the north end of Padre Island, a drilling showed continuous sand along its entire length of 23 feet. The analyzed samples show no sand with more than 54 per cent in the ⅛ to $1/16$th mm fraction, quite clearly excluding deposition in the nearshore gulf. At 19 feet, however, the greenish cast appears in the sand which may indicate an older generation. In southern Padre Island the Pleistocene Beaumont comes gradually nearer the surface, actually cropping out near the shore in one place (Rusnak, personal communication).

Humble Oil Company samples supplied by Dr. Harold N. Fisk from the middle section of Padre Island are not as clearly differentiated, as they represent an aggregate sample from 5- or 10-foot sections. The size analysis, however, indicates higher percentages in the ¼ to ⅛th than in the ⅛ to $1/16$ mm fraction. Also, the shells in the lower portion at 30–40 feet are indicative of bay conditions rather than open gulf. It is conceivable that the section passed through the top of the Pleistocene, although this was not believed to be the case by the Humble geologists.

In Matagorda Peninsula, north of the Rockport area and near the present mouth of the Colorado of Texas, a beach boring to 25 feet goes through what appears to be barrier island sands, judging from grain size, and reaches a bay deposit at the bottom of the hole.

Galveston Island has been drilled by the U. S. Corps of Engineers. Samples from several of these holes were sent to us by Shell Development Company of Houston. Analyses have shown a considerable difference at the two ends of the island, although the difference may be due to the samples coming from shallow depths, 0 to 18 feet in the southwest end of the island and 19 to 39 feet 8 miles from the northeast end of the island. The boring at the southwest end yielded samples with 80–90 per cent sand down to 12 feet but with the $\frac{1}{8}$ to $\frac{1}{16}$ mm fraction at less than 60 per cent, so presumably they had a barrier island origin. The two samples from 12 to 18 feet show a decrease in sand content downward, the lower one having only 35 per cent sand but with a fauna that is more related to bays than to the open gulf. Both these samples could have been deposited in an inlet.

The trans-island line of borings 8 miles from the northeast end of Galveston Island has samples which are much more likely to represent shelf deposits as they have $\frac{1}{8}$ to $\frac{1}{16}$th mm fractions above 85 per cent. In the first four samples, from 19 to 28 feet, the sand shows a progressive increase from 91 to 98 per cent. This suggests, although it does not prove, that during deposition the water was deepening or the distance from the island increasing, which may mean that the sea level was still rising from the glacial melting at the end of the Pleistocene. The glauconite and echinoids in these four samples are all relatively high, adding more probability to a shelf origin. The proximity to a barrier island is indicated, however, by the large sand content and the organisms, which show many forms suggestive of the nearness of an inlet through which bay forms, found sparingly, could come, to some extent. Below 28 feet, the deposits have a much lower sand content and are relatively high in clay. These lower deposits may represent an inlet in a barrier island. Below 36 feet along the shore, the material has been interpreted as Pleistocene by the engineers. It apparently represents a bay deposit.

Three borings with a Porter sampler were made on Breton Island, the island south of the Chandeleur group. This island differs from the Chandeleurs in that it seems not to have been moved toward the Mississippi Delta during the past 100 years. The chart of 1839 indicated an island with almost the same shape as the present one and having the same distance from the main channel of the Mississippi, which has also remained virtually constant. The pass through the island, however, has changed position, as shown by various charts. The borings encountered about 30 feet of sand before reaching clayey delta deposits. In one boring on the north side of Breton Island, a sand with a roundness comparable to that of the present island dunes and considerably greater than that of the beaches was found at a depth of 8 feet (Beal and Shepard, 1956, Fig. 7). At a depth of 20 feet in the same boring, the sand has 80 per cent in the $\frac{1}{8}$ to $\frac{1}{16}$ mm fraction, suggesting a shallow gulf origin. As this sand differs from Breton Sound deposits and from prodelta deposits because of the small percentage of silt and clay, its lack of plant fragments, and its relatively high content of glauconite, it can be presumed that it was deposited on the open gulf side of a barrier island.

The Corps of Engineers bored several holes to the north and east of Breton Island. These show a sandy zone going down to about 30 feet in the area near Breton Island, but to the north along the Chandeleurs the sand layer is thinner. Shallow borings by the staff of Louisiana State University (Treadwell, 1955, Fig. 7) show that the barrier sands overlie marsh or sound deposits at a depth of a few feet. The overlap of the barrier deposits onto sound deposits in the Chandeleurs and the absence of such overlap at Breton Island give further support to the different sequence of events in these islands, mentioned elsewhere.

The barrier islands which border Mississippi Sound have variable but large thicknesses of sand, according to well logs reported by the Mississippi Geological Survey (Brown, *et al.*, 1944). In one well on Ship Island, sand was logged to a depth of about 250 feet. Others show sand to about 50 feet with sand masses at still greater depth separated from each other by mud. The recent borings by Gulf Research and Development Company (Ludwick, personal communication) yielded thickness of not greater than 40 feet of barrier sand overlying a marine clay, which in turn overlies oxidized nonfossiliferous Pleistocene clay (Prairie or Beaumont). It is believed by Ludwick that this mass of sand is all quite recent and deposited after sea level reached its present height. The source of the sand is almost certainly from the east and south, judging from the minerals of Appalachian type.

ORIGIN AND HISTORY OF BARRIER ISLANDS

Longshore drift versus local derivation.—Considerable speculation as to the origin of coastal barriers preceded Douglas Johnson's classic work (1919), but this has been so well summarized by Johnson that most of it needs little review here.

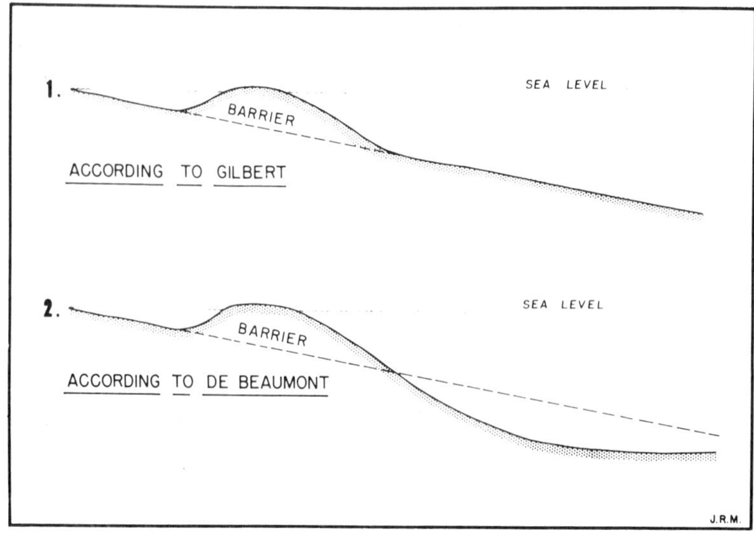

FIG. 12.—Illustrating the hypotheses of Gilbert and de Beaumont on the origin of barrier islands. Gilbert believed the barrier was simply added to the slope having been introduced by longshore currents, and de Beaumont believed the waves excavated the sea floor on the outside and built the barrier on the inside.

Johnson compared two theories of origin—one, attributed to Gilbert (1885), considered barriers as due to longshore drift forming a spit along the shore or across the mouth of a bay; the other attributed to de Beaumont (1845), considered the barriers as built up by waves which eroded the bottom in the breaker zone, and piled up their products as a ridge on the inside. Johnson tested these two concepts by drawing profiles across the islands and comparing them with a theoretical construction showing (1) the simple addition of material to an even slope, and (2) the pre-existing even slope cut by wave erosion outside, and covered by wave deposition inside (Fig. 12). His profiles (drawn by Miss B. M. Merrill) favored the second concept. However, Johnson did not realize that barrier islands are not simple modification of an even marine slope, but, as is now known from Texas and Louisiana coast borings, represent barrier deposits formed between bay sediment on one side and marine sediments on the other (Fig. 13). This would appear to cast much doubt on the evidence from the Johnson profiles.

The recent Gulf Coast studies have not solved the question of local *versus* longshore drift origin of the barriers, but there is little necessity for deciding between the two, because both are probably important. For some barriers there appears to be no good land source of sand. For example, the Mississippi islands have no nearby bluffs to supply the sediments, and the chief river source enters the head of the lengthy Mobile Bay. Transportation of sand across the long muddy stretches of the bay seems unlikely, although a little sand may move along the bay shore. Because of the small waves, this is a doubtful source of the large quantities required. The easterly winds and currents are carrying the sand from the present islands to the west. Therefore, unless there is an eastern source, the sand should give out on the eastern side, but the maps of the past 100 years do not show a net loss. In fact the islands are somewhat more continuous than is indicated on the old charts, which shows there must be an adequate source. The only sand available seems to be from the continental shelf. The presence of a mud zone outside the islands and a mud-filled trench along part of the island front (J. C. Ludwick, personal communication) apparently indicates that the shelf source is east of Mobile Bay where, according to the chart, sand appears to be continuous out from the shore over the shoal bottom. Thus the evidence from the Mississippi islands favors an offshore source, but also indicates the importance of longshore drift.

Emergence versus submergence.—Johnson also

considered alternative hypotheses that barriers formed either on shore lines of submergence or on shore lines of emergence. He concluded that emergence was the common type, but thought barriers might form along some neutral shore lines. He was convinced that barriers would not form along shore lines of submergence. His chief argument for barriers developing along shore lines of emergence was based on the assumption that the sea floor represents a very gently sloping plain. Emergence, bringing this plain into shallow water, would allow the waves to break well out from the shore and hence build a submerged bar and later a barrier island.

Johnson seems to have overlooked several points. One is that many flat plains exist on land, especially delta plains, that are even flatter than those of the continental shelf. The submergence of such plains would provide the gentle submarine slope which he considered necessary for the development of the barrier on the outside. Also, the emergence of numerous irregular shelves would leave conditions quite unfavorable for the supposed gently sloping bottom. Even more important, he failed to give much weight to the fact that the typical barrier islands are found along coasts with abundant estuaries. As already indicated, these estuaries along the Texas coast represent recent submergence (Shepard and Moore, 1955; Shepard; 1956a), and there is fairly substantial evidence that sea level has stopped rising, if at all, only in very recent times (Shepard and Suess, 1956). Furthermore, the barriers around deltas such as Breton Island and the Chandeleurs have clearly grown upward apace with the subsidence of an abandoned distributary (Russell, 1936). Finally, the succession in virtually all the borings described here seems to call for subsidence, or more likely rise in sea level, during the deposition of the barrier deposits. The 6,500-year age of shells in the St. Joseph Island boring adds further weight to the argument, as the sand directly above the shell bed appears to be beach sand. Thus, the evidence points overwhelmingly in favor of the formation of barriers, in the Gulf Coast area studied, to have been during, as well as directly following, inundation of the land.

A complication related to the growth of a barrier island during rising sea levels at the end of the last glacial stage comes from the source of sand. If sand comes from the shelf outside, the growth of the island upward should decrease the availability of the sand from the adjacent shelf, because it would be constantly getting into greater depth and presumably would become covered with muddy sediment, making it less available. The barrier might continue to grow, however, if sediments were transported toward it from either side by longshore drift or if the shelf currents prevented the sand from being covered. Also the barrier might migrate landward during the rise. Various ridges and older-generation sand masses are found

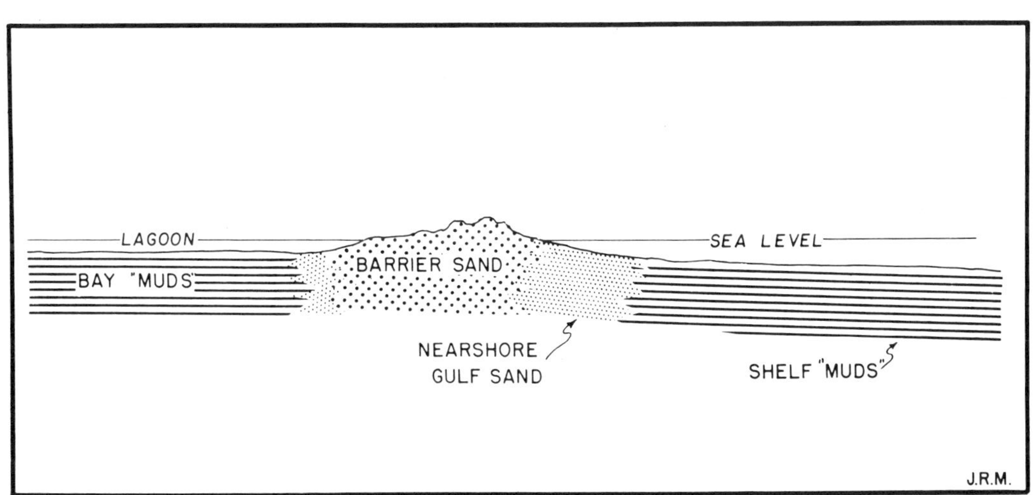

Fig. 13.—Showing the upgrowth of barrier sands between shelf muds on the seaward side and bay muds on the lagoon side.

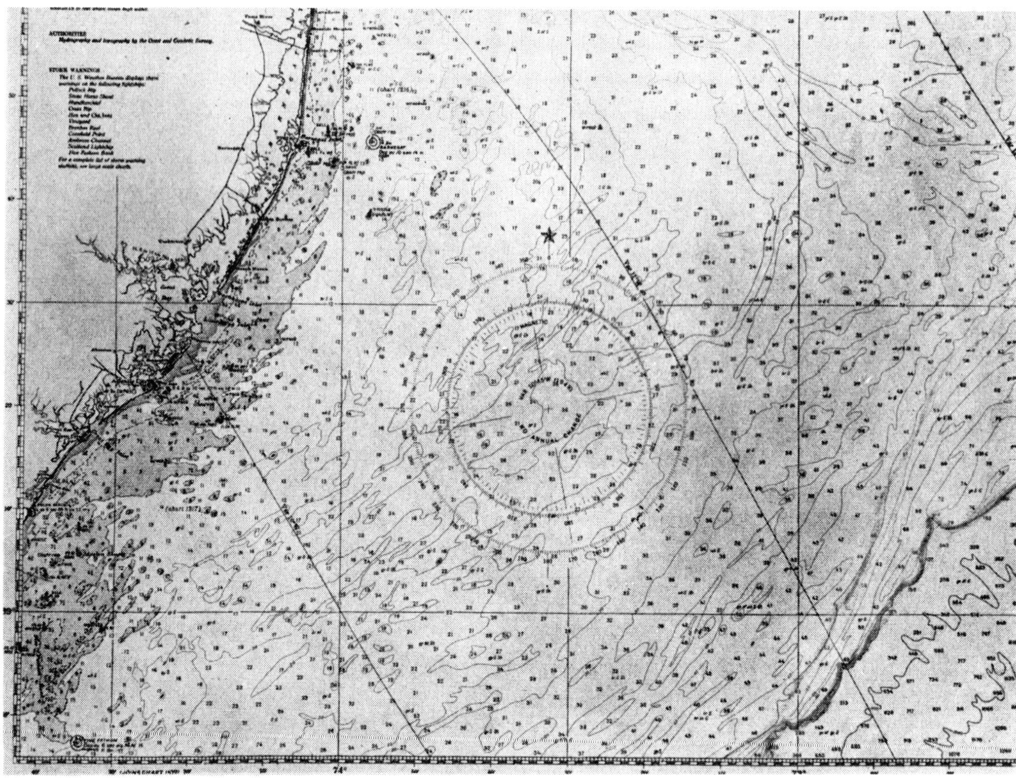

FIG. 14.—Showing the numerous drowned barriers on the continental shelf off the northern New Jersey coast. From U. S. Coast and Geodetic Survey Chart 1108.

out on the continental shelf (Curray, this volume, p. 256). Reworked faunas and carbon-14 dates show that these represent barriers formed during temporary slow downs in the rise in sea level. The 6,500-year date found in inlet deposits at 46 feet under St. Joseph Island beach suggests the barrier has moved slightly landward during the interval (an inlet must be inside the line of the beach).

In view of the definite evidence of sinking of the Chandeleur Islands in what has been dated as approximately the past 1,000 years (Treadwell, 1955, pp. 66-69), the reality of upgrowth of barrier islands despite sinking can scarcely be denied. Nor can one overlook the myriad of drowned barriers left behind on the East Coast continental shelf as the sea level rose when the glaciers melted (Fig. 14). In all probability these were not formed during a time of emergence.

Landward migration of barriers.—Johnson (following Davis, 1898) considered barriers as an early stage of a cycle, and in the later stages the islands are supposed to move up onto the land. We have seen that so far as historical changes are concerned, barriers are moving both landward and seaward. The Davis idea of a landward-migrating barrier could have been derived by a short-period series of observations of a longshore bar. These submerged bars, formed during winter storms, migrate landward during the small-wave periods of summer and are eventually added to the beach, where they are completely exposed at low tide (Shepard, 1950). However, these same small waves actually cause the barriers to grow seaward because the sand that is brought into the shore along with the longshore bar encroachment is added to the barrier. The storm waves, on the other hand, cut back the ocean side and, whenever they are accompanied by high sea level, develop washovers with the result that the barrier migrates landward. If there is insufficient sand supply on the outside, as may be true of the Chandeleurs, the islands cannot recover their former outer margin during the small wave periods following storms. Thus, the landward or seaward migration of a barrier depends on the sand supply and on wave conditions. A good balance between growth and retreat seems to exist along much of the Texas coast.

Factors controlling inlets.—Johnson (1919, pp. 367–374) gave considerable attention to the number and spacing of inlets through barriers, showing their apparent relation to the nearness to a headland source as well as to the tidal forces which keep inlets open. He assumed, quite reasonably, that the nearer the source the more any inlet, opened temporarily by waves, would tend to close because of being blocked by deposition from longshore drift. He also pointed to the greater likelihood of the lagoons filling where there were numerous inlets, because sediment is introduced and marsh grass can grow better under the condition of uniform salinity which would exist with closely spaced inlets. The case is somewhat oversimplified, however, because the filling of lagoons is also due, to a marked degree, to streams which are better able to deposit their load where inlets are missing. Also, sediments are introduced from the barrier by wind and washovers, neither of which are dependent on inlets.

It seems likely that Johnson's interpretation of inlets is more applicable to the East Coast, north of Florida, than to other areas such as the Gulf Coast where the situation is complicated because of more sediment-laden rivers entering the bays, smaller tides, and because of a considerable supply of sediment from the open gulf. The barriers of northwest Florida certainly are far too continuous to agree with Johnson's interpretation, as there is no headland or delta to supply the sand which, according to Johnson, should be necessary to keep the inlets closed. However, the small tide, combined with the small waves on the protected west side of the gulf, and the probable source of the sand from the shallow continental shelf outside may account for the continuity of these barriers.

ORIGIN OF CUSPATE BARRIER SHORE LINES

The barrier islands of the Texas coast south of Galveston have, for the most part, concave shore lines. To the east of Galveston the barriers show increasingly greater convexity, with the Chandeleur Islands as the first major convex arc. This convexity is in turn replaced by cuspate forelands farther east, such as Cape San Blas (Fig. 3) and Cape St. George. Along the Florida coast farther south there are several convexities, notably at the entrance to harbors. The best development of cuspate forelands is found on the southeast coast of the United States, with Cape Canaveral as the southernmost example, followed to the north by a series of capes terminating in Cape Hatteras. All these cuspate forelands are bordered by extensive shoals (Fig. 3), some of which can be traced entirely across the continental shelf and as much as 30 miles from the shore line. It seems possible that in these shoals we may have at least part of the explanation for the cuspate form. If the shoals existed as hills on the shelf prior to the cusp formation, the currents moving along the coast should set up gyrols on either side, and this in turn would favor deposition between the gyrols and hence the building out of the cuspate points. It is possible that the original existence of the shoals is related to coral growth on the shelf; at least, it is known that coral knolls exist on the shelf off Florida and some type of bioherm may extend as far north as Cape Hatteras, which is the point where the Gulf Stream leaves the coast and the shelf.

Cuspate spits found on the inside of the barrier islands (Figs. 8, 16) are not explicable on the same basis. Similar cusps in northeastern Nantucket Island were investigated by Johnson (1925, pp. 441–445). He explained them as a successive development of recurved hooked spits, which were formed as the barrier was extended to the southwest. The cuspate shape, Johnson believed, was formed later by minor tidal currents in the lagoon acting on the hooks. The lagoonal cusps were described by Fisher (1955) from St. Lawrence Island in Bering Sea, who explained them by lagoonal currents which gradually altered washover fans into a cusp shape. They are also found along portions of the coasts of Russia and Siberia; these have recently been explained by Zenkovitch (1959) as the result of longshore currents in the lagoons, which he thought were quite independent of washovers.

The cuspate spits along the Gulf Coast appear to have resulted from a combination of these several processes. They are clearly not the result of any strong tidal-current action, as tides are

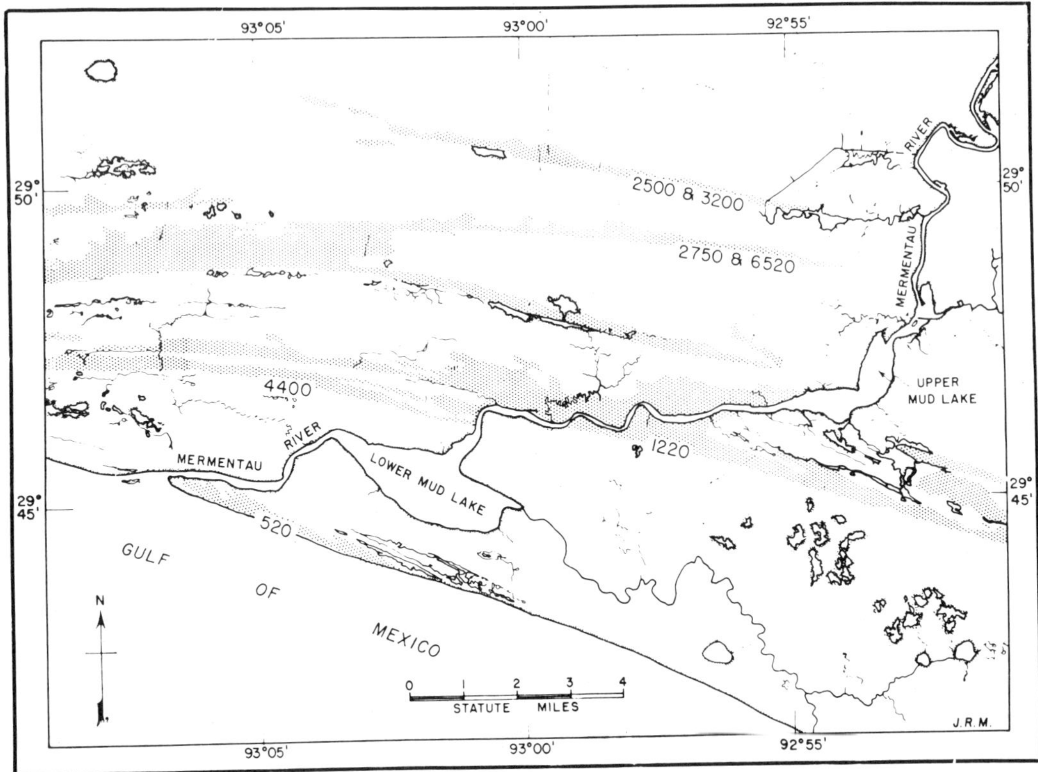

FIG. 15.—Showing the carbon-14 dates reported by Humble Oil Co. from shells taken from the cheniers (beach ridges) just to the west of the Mississippi Delta. The conflicts in ages may be due to the reworking of old shell deposits and introduction of shells during the rise in sea level. See Gould and McFarlan (1959).

very weak in this part of the Gulf of Mexico. Those shown in Figure 8 appear to be related to washovers because they can be traced across the entire island.

BARRIER RIDGES

Many barrier islands have a series of ridges extending parallel to the shore line (see Figs. 5, 8). These have generally been interpreted as beach ridges, built by a prograding shore line. Many authors have realized, however, that they are at least in part dune ridges. These also are formed near the shore. The significance of these ridges as regards sea level or land level changes depends to a considerable extent on whether the ridges are dunes or beaches. Storm beaches are thrown up for a number of feet above the high tide sea level, but all significant storm beaches, so far as the writer knows, are made of coarse debris, gravel, or shells. Sand beaches are not built to any great height above a plain because the sand is too easily transported inland across the plain by the waves. Accordingly, the storm beaches of sand are raised only some 2 or 3 feet above the tide level which produces them. Also, these beaches are very broad, quite in contrast to the steep-sided ridges of gravel and shell.

Ridges such as those from St. Joseph Island near Rockport, Texas (Fig. 5) are made of fine and very fine sand. They rise some 15 feet above surroundings, which would seem to rule them out as beach ridges formed above the present sea level. A study of sand roundness of samples from the ridges has shown that they are probably dunes (Beal and Shepard, 1956, Fig. 2). The height, therefore, does not tell us much about the sea level at which they were formed. The shell-sand ridges known as "cheniers" along the Louisiana coast, on the other hand, are much lower, rarely over 10 feet above mean sea level. Equally high ridges are found piled up by the waves during storms along the present coast (J. P. Morgan, personal communication). They are, therefore, indicative of formation at present sea level. The carbon-14 dating of shells in the ridges well inside the present shore line shows that some are as old as about 6,500 years. This has been used as an argument that sea level has been stationary during the past 5,000 years. Examination of the currently published data, however, (Fig. 15) shows that there is no progressive increase in age with distance from the shore. In fact, very striking age differences can be found in the same ridge. This rather suggests that in the formation of the ridges old shells have been reworked.[6] Possibly an age of about 2,500 years for the inner ridge may be indicated, as that is the youngest date found for this ridge. This may or may not mean a steady sea level during the past 2,500 years, as the area is located just west of the "hinge line" (Fisk and McFarlan, 1955), and therefore may have been slightly uplifted, keeping pace with a very slowly rising sea level.[7]

An alternative origin of barrier island ridges is illustrated in Figure 11. Here a longshore bar is being built up above sea level outside the beach. Another slightly submerged bar occurs farther out. This in turn may be brought up to sea level. A slight drop in sea level or a small uplift of these islands and bars would produce a series of more or less parallel ridges such as are found on the islands. To differentiate ridges formed in this way from dune ridges, a careful study of the sand is necessary, preferably including roundness determinations and comparison with the present beach and dune sands. Sands of ridges due to uplifted longshore bars would not ordinarily show dune roundness.

COMPARISON WITH OTHER BARRIERS

Barriers are by no means a monopoly of the Gulf Coast shore lines. They are found along the greater part of the East Coast of the United

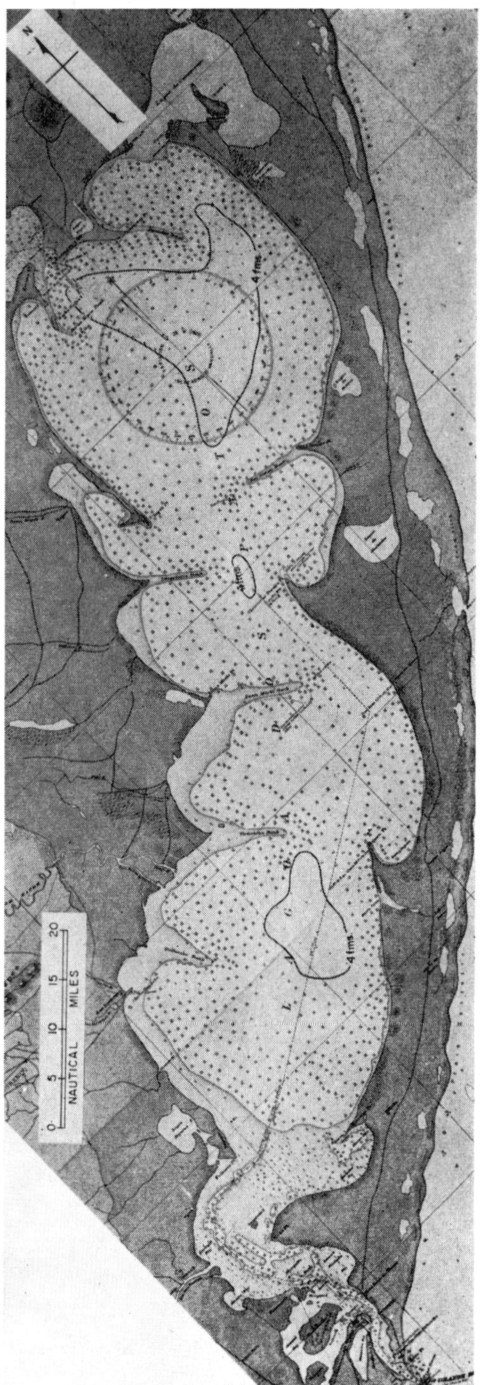

FIG. 16.—Showing the extremely wide barrier islands along the southeast coast of Brazil. The lagoonal depths are somewhat greater than along the Gulf Coast.

[6] According to R. H. Parker, examination of the cheniers has shown that the shells are very much reworked.

[7] More complete data have just been published (Gould and McFarlan, 1959) as this goes to press. Their information essentially confirms the data supplied here, although they consider that sea level has been constant for about 3,000 years.

States from Miami to easternmost Long Island. On the West Coast, barriers are quite scarce, but they are found locally, as for example at Morro Bay and Humboldt Bay in California and along the coast of southern Washington, notably at Willapa Bay and Grays Harbor. Europe has an extensive barrier development on the east side of the North Sea, including most of the Dutch Coast and adjacent Germany and Denmark. The south coast of France, west of the Rhone Delta, has a series of barriers with the long lagoons known as "Etangs." The Nile Delta is bordered by a row of barriers. Southeast Africa also has a series of these sand islands. The eastern Indian coast and east coast of Ceylon have many barriers along with areas where the lagoons inside former barriers have been filled. The features occur extensively along the coast of southeast Australia. The east coast of Brazil has what may be the widest barriers in the world, with a maximum width of 20 miles (Fig. 16).

In generalizing, it appears that barriers are found along practically all lowland coasts. They are common around deltas, particularly along the portions of the deltas which are not actively prograding. They are lacking, however, where the submerged deltaic plain is so gentle that the waves do not break but gradually lose their energy in approaching the shore, as for example, off the Orinoco (Tj. H. van Andel, personal quotation). It seems probable that wherever plains are slowly sinking, barriers will form at the margin and at times may grow upward, keeping pace with the sinking.

CONSOLIDATED BIBLIOGRAPHY

ABBOTT, R. T. (1954), *American Seashells*, D. Van Nostrand Co., Inc., New York.

ADAMS, R. M.; AND SORGNIT, F. F. (1951), "Comparison of Summer and Winter Temperatures, Gulf of Mexico," *Texas Agr. and Mech. Coll. Tech. Rept. 3*, Dept. of Oceanog., pp. 1–4.

ADKINS, W. S.; AND LOZO, F. E., JR. (1951), "Stratigraphy of the Woodbine and Eagle Ford, Waco Area, Texas," *Fondren Sci. Ser.*, No. 4, pp. 101–64.

AKERS, W. H. (1952), "General Ecology of the Foraminiferal Genus *Eponidella* With Description of a Recent Species," *Jour. Paleon.*, Vol. 26, No. 4, pp. 645–49.

——— (1954), "Ecologic Aspects and Stratigraphic Significance of the Foraminifer *Cyclammina cancellata* Brady," *ibid.*, Vol. 28, No. 1, pp. 132–52.

———, AND HOLCK, A. J. (1957), "Pleistocene Beds Near the Edge of the Continental Shelf, Southeastern Louisiana," *Bull. Geol. Soc. America*, Vol. 68, No. 8, pp. 983–91.

ALLEN, W. E.; STANLEY, L.; AND VINING, T. F. (1956), "Geophysical and Geological Investigation of Sea Mounds in the Gulf of Mexico" (Program Abst.), *Soc. Explor. Geophys.*, New Orleans Meeting, Oct., 1956.

ANDERSEN, H. V. (1951), "Two New Genera of Foraminifera from Recent Deposits in Louisiana," *Jour. Paleon.*, Vol. 25, No. 1, pp. 31–34.

——— (1953), "Two New Species of *Haplophragmoides* from the Louisiana Coast," *Contr. Cushman Found. Foram. Research*, Vol. 4, Pt. 1, pp. 21, 22.

ANDERSON, D. H.; AND ROBINSON, R. J. (1946), "Rapid Electrometric Determination of the Alkalinity of Sea Water, Using a Glass Electrode," *Indust. and Eng. Chem.* (Anal. Ed.), Vol. 18, pp. 767–73.

APPLIN, P. L. (1951), "Preliminary Report on Buried Pre-Mesozoic Rocks in Florida and Adjacent States, *U. S. Geol. Survey Circ. 91*.

——— (1952), "Sedimentary Volumes in Gulf Coastal Plain of United States and Mexico: Part I, Volume of Mesozoic Sediments in Florida and Georgia," *Bull. Geol. Soc. America*, Vol. 63, No. 12, pp. 1159–64.

———, AND APPLIN, E. R. (1944), "Regional Subsurface Stratigraphy and Structure of Florida and Southern Georgia," *Bull. Amer. Assoc. Petrol. Geol.*, Vol. 28, No. 12, pp. 1673–1753; *Dallas Digest* (Abst.) pp. 76–77; *Oil and Gas Journal*, Vol. 42, No. 46, pp. 92, 94.

———, AND ——— (1947), "Regional Subsurface Stratigraphy, Structure, and Correlation of Middle and Early Cretaceous Rocks in Alabama, Georgia, and North Florida," *U. S. Geol. Survey Prelim. Chart 26*, Oil and Gas Inves. Ser.

———, AND ——— (1953), "The Cored Section in George Basen's Fee Well 1, Stone County, Mississippi," *ibid.*, *Circ. 298*.

ARRHENIUS, G. O. (1952), "Sediment Cores from the East Pacific," *Repts. Swedish Deep-Sea Exped. 1947–48*, Vol. 5, Part 1, p. 85.

ATWATER, G. I. (1959), "Geology and Petroleum Development of the Continental Shelf of the Gulf of Mexico," *Proc. 5th World Petrol. Cong., Section I, Paper 21*.

———, AND FORMAN, M. J. (1959), "Nature of Growth of Southern Louisiana Salt Domes and its Effect on Petroleum Accumulation," *Bull. Amer. Assoc. Petrol. Geol.*, Vol. 43, No. 11, pp. 2592–2622.

AUSTIN, G. B. (1954), "On the Circulation and Tidal Flushing of Mobile Bay, Alabama, Part I," *Texas Agr. and Mech. Coll. Research Found. Proj. 24, Tech. Rept. 12*, pp. 1–28.

BAKKER, J. P. (1954), "Relative Sea-Level Changes in Northwest Friesland (Netherlands) since Prehistoric Times," *Geologie en Mijnbouw*, Nw. Ser. 16, pp. 232–46.

BALL, M. W. [EDITOR] (1951), *Possible Future Petroleum Provinces of North America*, Amer. Assoc. Petrol. Geol.

BANDY, O. L. (1954), "Distribution of Some Shallow-Water Foraminifera in the Gulf of Mexico," *U. S. Geol. Survey Prof. Paper 254-F*.

——— (1956), "Ecology of Foraminifera in Northeastern Gulf of Mexico," *ibid.*, *Prof. Paper 274-G*, pp. 179–204.

BARRELL, JOSEPH (1912), "Criteria for the Recognition of Ancient Delta Deposits," *Bull. Geol. Soc. America*, Vol. 23, No. 3, pp. 377–446.

BARTON, D. C. (1930), "Deltaic Coastal Plain of Southeastern Texas," *ibid.*, Vol. 41, No. 3, pp. 359–82.

BASS, N. W. (1934), "Origin of Bartlesville Shoestring Sands," *Bull. Amer. Assoc. Petrol. Geol.*, Vol. 18, No. 10, pp. 1313–45.

BATES, C. C. (1953), "Rational Theory of Delta Formation," *ibid.*, Vol. 37, No. 9, pp. 2119–62.

BEAL, M. A.; AND SHEPARD, F. P. (1956), "A Use of Roundness To Determine Depositional Environments," *Jour. Sed. Petrology*, Vol. 26, No. 1, pp. 49–60.

BEHRE, ELLINOR, H. (1950), "Annotated List of the Fauna of the Grand Isle Region, 1928–46," *Marine Lab., Louisiana State Univ.*, Baton Rouge, *Occasional Papers No. 6*.

BENNEMA, J. (1954), "Bodem-en Zeespiegelbewegingen in het Nederlands Kustegebied (Holocene Movements of Land and Sea Level in the Coastal Area of the Netherlands)," *Boor en Spade*, Vol. 7, pp. 1–96, Wageningen.

BERNARD, HUGH A. (1950), "Quaternary Geology of Southeast Texas," Unpub. Ph.D. dissertation, Louisiana State Univ., Baton Rouge.

BIEN, G. S.; CONTOIS, D. E.; AND THOMAS, W. H. (1958), "The Removal of Soluble Silica from Fresh Water Entering the Sea," *Geochim. et Cosmochim. Acta*, Vol. 14, pp. 35–54. Reprinted *in* Silica in Sediments, *Soc. Econ. Paleon. Min. Spec. Pub. No. 7*, pp. 20–35, 1959.

BLANKENSHIP, R. R. (1956), "Heavy Mineral Suites in Unconsolidated Paleocene and Younger Sands, Western Tennessee," *Jour. Sed. Petrology*, Vol. 26, No. 4, pp. 356–62.

BODENLOS, A. J.; AND BONET, FEDERICO (1956) "Geology of the Sierra Madre Oriental in the Tamazunchale Area, Mexico" (Abst.), *Resumenes de los Trabajos Presentados, XX Cong. Geol. Internac.*, pp. 25, 26.

BOKMAN, JOHN (1955), "Sandstone Classification: Relation to Composition and Texture," *Jour. Sed. Petrology*, Vol. 25, No. 3, pp. 201–6.

BOLLI, H. M.; AND SAUNDERS, J. B. (1954), "Discussion of some Thecamoebina Described Erroneously as Foraminifera," *Contr Cushman Found. Foram. Research*, Vol. 5, pp. 45–52.

BORNHAUSER, M. (1940), "Heavy Mineral Associations in Quaternary and Late Tertiary Sediments of the Gulf Coast of Louisiana and Texas," *Jour. Sed. Petrology*, Vol. 10, No. 3, pp. 125–35.

——— (1947), "Marine Sedimentary Cycles of Tertiary in Mississippi Embayment and Central Gulf Coast Area," *Bull. Amer. Assoc. of Petrol. Geol.*, Vol. 31, No. 4, pp. 698–712; also, *Oil and Gas Jour.*, Vol. 44, No. 48, pp. 88 (1946).

——— (1958), "Gulf Coast Tectonics," *Bull. Amer. Assoc. Petrol. Geol.*, Vol. 42, No. 2, pp. 339–70.

BRADSHAW, J. S. (1955), "Preliminary Laboratory Experiments on Ecology of Foraminiferal Populations," *Micropaleontology*, Vol. 1, pp. 31–58.

——— (1957), "Laboratory Studies on the Rate of Growth of the Foraminifer '*Streblus becarii* (Linné) var. *tepida* (Cushman)'," *Jour. Paleon.*, Vol. 31, No. 6, pp. 1138–47.

BRAUNSTEIN, J. (1950), "Subsurface Stratigraphy of the Upper Cretaceous in Mississippi," *Mississippi Geol. Soc. Guidebook, 8th Field Trip*, pp. 13–21.

——— (1957), "The Habitat of Oil in the Cretaceous of Mississippi and Alabama," in *Mesozoic-Paleozoic Producing Areas of Mississippi and Alabama*, Vol. 1, Mississippi Geol. Soc., pp. 1–11.

——— (1958), "Habitat of Oil in Eastern Gulf Coast," in *Habitat of Oil*, Amer. Assoc. Petrol. Geol., pp. 511–22.

BRETSCHNEIDER, CHARLES L. (1954), "Generation of Wind Waves Over a Shallow Bottom," *Beach Erosion Board Tech. Mem. 51*, p. 21.

———, AND GAUL, ROY D. (1956a), "Wave Statistics for the Gulf of Mexico off Brownsville Texas," *ibid.*, *Mem. 85*.

———, AND ——— (1956b) "Wave Statistics for the Gulf of Mexico off Caplen, Texas," *ibid.*, *Mem. 86*.

———, AND ——— (1956c), "Wave Statistics for the Gulf of Mexico off Burrwood, Louisiana," *ibid.*, *Mem. 87*.

BREUER, J. P. (1957), "An Ecological Survey of Baffin and Alazan Bays, Texas," *Pub. Inst. Marine Sci.*, Vol. 4, No. 2, pp. 134–55.

BROTHERHOOD, G. R.; AND GRIFFITHS, J. C. (1947), "Mathematical Derivation of the Unique Frequency Curve," *Jour. Sed. Petrology*, Vol. 17, No. 2, pp. 77–82.

BROWN, G. F., ET AL. (1944), "Geology and Ground-Water Resources of the Coastal Areas in Mississippi," *Mississippi Geol. Survey Bull. 60*.

BUCH, K. (1933), "On Boric Acid in the Sea and its Influence on the Carbonic Acid Equilibrium," *Jour. Cons. Internat. Explor. Mer.*, Vol. 8, pp. 309–25.

———, HARVEY, H. W.; WATTENBERG, H.; AND GRIPENBERG, S. (1932), "Über das Kohlensäuresystem in Meerwasser," *Rapp. et Process.-Verb. Cons. Perm. Internat. Explor. Mer.*, Vol. 79, pp. 1–70.

BUCHANAN, J. B. (1958), "The Bottom Fauna Communities Across the Continental Shelf of Accra, Ghana (Gold Coast)," *Proc. Zool. Soc. London*, Vol. 130, No. 1, pp. 1–56.

BULLARD, F. M. (1942), "Source of Beach and River Sands on Gulf Coast of Texas," *Bull. Geol. Soc. America*, Vol. 53, No. 7, pp. 1021–43.

BURST, J. F. (1958), "'Glauconite' Pellets: Their Mineral Nature and Applications to Stratigraphic Interpretations," *Bull. Amer. Assoc. Petrol. Geol.*, Vol. 42, No. 2, pp. 310–27.

CANADA, W. R. [CHAIRMAN] (1944–45), *Progress Report of the Geological Names and Correlations Committee*, South Louisiana Geol. Soc., Lake Charles.

CARROZI, A. (1957), "Contribution a l'Etude des Proprietes Geometriques de Oolithes, l'Example du Grand Lac Sale, Utah, U.S.A.," *Bulletin de l'Institut Natl. Genevois*, Vol. 58, pp. 4–52.

CARSEY, J. B. (1950), "Geology of Gulf Coastal Area and Continental Shelf," *Bull. Amer. Assoc. Petrol. Geol.*, Vol. 34, No. 3, pp. 361–85.

CARY, L. R. (1905), "A Contribution to the Fauna of the Coast of Louisiana," *Bull. Gulf Biol. Sta.*, No. 6, pp. 50–59.

———, AND SPAULDING, H. (1909), "Further Contributions to the Marine Fauna of the Louisiana Coast," *ibid.*, No. 12.

COE, W. R. (1957), "Fluctuations in Littoral Populations," *in* Treatise on Marine Ecology and Paleontology, *Geol. Soc. America Mem. 67*, Vol. 1, pp. 935–40.

COFFEY, G. N. (1909), "Clay Dunes," *Jour. Geol.*, Vol. 17, pp. 754–55.

COGEN, W. M. (1940), "Heavy-Mineral Zones of Louisiana and Texas Gulf Coast Sediments," *Bull. Amer. Assoc. Petrol. Geol.*, Vol. 24, No. 12, pp. 2069–2101.

COHEE, G. V., ET AL., *Tectonic Map of the United States*, Amer. Assoc. Petrol. Geol. and U. S. Geol Survey; second edition, in preparation.

COLLE, J., ET AL. (1952), "Sedimentary Volumes in Gulf Coastal Plain of United States and Mexico: Part IV, Volume of Mesozoic and Cenezoic Sediments in Western Gulf Coastal Plain of United States," *Bull. Geol. Soc. America*, Vol. 63, No. 12, pp. 1193–99.

COLLIER, ALBERT W.; AND HEDGPETH, J. W. (1950), "An Introduction to the Hydrography of Tidal Waters of Texas," *Pub. Inst. Marine Sci.*, Vol. 1, No. 2, pp. 120–94.

CURTIS, DORIS M. (1960), "Relation of Environmental Energy Levels and Ostracod Biofacies in East Mississippi Delta Area," *Bull. Amer. Assoc. Petrol. Geol.*, Vol. 44, No. 4, pp. 471–94.

CUSHMAN, J. A. (1918–31), "The Foraminifera of the Atlantic Ocean," *U. S. Natl. Mus. Bull. 104*, Pts. 1–8.

——— (1927), "Shallow-Water Foraminifera of the Tortugas Region," *Carnegie Inst. Washington Pub. 311*.

———, AND HENBEST, L. G. (1940), "Foraminifera," Pt. 2, *of* Geology and Biology of North Atlantic Deep-Sea Cores Between Newfoundland and Ireland, *U. S. Geol. Survey Prof. Paper 196-A*, pp. 35–50.

DALL, W. H. (1886), "Brachiopoda and Pelecypoda," Part I *of* Report on the Molluska, *Harvard Coll. Mus. Comp. Zool. Bull.*, Vol. 12, No. 6, pp. 171–318.

——— (1889), "Gastropoda and Scaphopoda," Part II *of* Report on the Molluska, *ibid.*, Vol. 18, pp. 1–492.

——— (1903), "A Preliminary Catalogue of the Shell-Bearing Mollusks and Brachiopods of the South-eastern Coast of the United States," *U. S. Natl Mus. Bull. 37*, 2d Ed., pp. 1–232.

DALLAS GEOLOGICAL SOCIETY AND DALLAS GEOPHYSICAL SOCIETY (1957), *The Geology and Geophysics of Cooke and Grayson Counties, Texas*.

DALY, R. A. (1920), "A Recent World-Wide Sinking of Ocean Level," *Geol. Mag.*, Vol. 57, pp. 246–61.

DAPPLES, E. C. (1942), "The Effect of Macro-Organisms Upon Near-Shore Marine Sediments," *Jour. Sed. Petrology,* Vol. 12, No. 3, pp. 118–26.

———; KRUMBEIN, W. C.; AND SLOSS, L. L. (1953), "Petrographic and Lithologic Attributes of Sandstones," *Jour. Geol.,* Vol. 61, No. 4, pp. 291–317.

DARNELL, R. M. (1958), "Food Habits of Fishes and Larger Invertebrates of Lake Ponchartrain, Louisiana, an Estuarine Community," *Pub. Inst. Marine Sci.,* Vol. 5, pp. 353–416.

DAVIS, W. M. (1898), *Physical Geography,* Boston.

DAY, P. R. (1950), "Physical Basis of Particle Size Analysis by the Hydrometer Method," *Soil Science,* Vol. 70, No. 5, pp. 363–74.

DE BEAUMONT, ELIE (1845), *Leçons de Geologie Pratique,* Paris, pp. 223–52.

DEGENS, E. T.; WILLIAMS, E. G.; AND KEITH, M. L. (1957), "Environmental Studies of Carboniferous Sediments. Part I: Geochemical Criteria for Differentiating Marine from Fresh-Water Shales," *Bull. Amer. Assoc. Petrol. Geol.,* Vol. 41, No. 11, pp. 2427–55.

DEUSSEN, ALEXANDER (1924), "Geology of the Coastal Plain of Texas West of the Brazos River," *U. S. Geol. Survey Prof. Paper 126.*

DE VRIES, H. L.; AND BARENDSEN, G. W. (1954), "Measurements of Age by the Carbon-14 Technique," *Nature,* Vol. 174, pp. 1138–41.

DIENERT, F.; AND WANDENBULCKE, F. (1923), "Sur le Dosage de la Silice Dans les Eaux," *Compt. Rend. Acad. Sci. Paris,* pp. 1478–80.

DIETZ, ROBERT S. (1954), "Marine Geology of Northwestern Pacific: Description of Japanese Bathymetric Chart 6901," *Bull. Geol. Soc. America,* Vol. 65, No. 12, pp. 1199–1224.

———, AND MENARD, HENRY W. (1951), "Origin of Abrupt Change in Slope at Continental Shelf Margin," *Bull. Amer. Assoc. Petrol. Geol.,* Vol. 35, No. 9, pp. 1994–2016.

DOERING, JOHN A. (1935), "Post-Fleming Surface Formations of Coastal Southeast Texas and South Louisiana," *Bull. Amer. Assoc. Petrol. Geol.,* Vol. 19, No. 5, pp. 651–88.

——— (1956), "Review of Quaternary Surface Formations of Gulf Coast Region," *ibid.,* Vol. 40, No. 8, pp. 1816–22.

——— (1958), "Citronelle Age Problem," *ibid.,* Vol. 42, No. 4, pp. 764–86.

DOHM, C. F. (1936), "Petrography of Two Mississippi River Subdeltas," *Louisiana Dept. Conserv., Geol. Bull. 8,* pp. 339–96.

DOTY, M.; AND OGURI, M. (1957), "Evidence for a Photosynthetic Daily Periodicity," *Limnol. and Oceanog.,* Vol. 2, pp. 37–40.

DROOGER, C. M.; AND KAASSCHIETER, J. H. P. (1958), "Foraminifera of the Orinoco-Trinidad-Paria Shelf," *in* Reports of the Orinoco Shelf Exped., Vol. 4, *Kon. Nederl. Acad. Wetencsch,* Part 22, pp. 1–108.

DUBAR, JULES R. (1958), "Stratigraphy and Paleontology of the Late Neogene Strata of the Caloosahatchee River Area of Southern Florida," *Florida Geol. Survey Bull. 40,* pp. 1–267.

DURHAM, C. O., JR. (1957), "The Austin Group in Central Texas," Unpub. Ph.D. dissertation, Columbia Univ., New York.

———, AND PEEPLES, E. M., III (1956), "Pleistocene Fault Zone in Southeastern Louisiana (Abst.)," *Trans. Gulf Coast Assoc. Geol. Soc.,* Vol. 6, p. 65.

EARDLEY, A. J. (1938), "Sediments of Great Salt Lake, Utah," *Bull. Amer. Assoc. Petrol. Geol.,* Vol. 22, No. 10, pp. 1359–87.

EMERY, K. O. (1938), "Rapid Method of Mechanical Analysis of Sands," *Jour. Sed. Petrology,* Vol. 8, No. 3, pp. 105–11.

———, AND RITTENBERG, S. C. (1952), "Early Diagenesis of California Basin Sediments in Relation to Origin of Oil," *Bull. Amer. Assoc. Petrol. Geol.,* Vol. 36, No. 5, pp. 735–806.

———, AND STEVENSON, R. E. (1957), "Estuaries and Lagoons," *in* Treatise on Marine Ecology and Paleoecology, *Geol. Soc. Amer. Mem. 67,* Vol. 1, pp. 673–749.

———; TRACEY, J. I.; AND LADD, H. S. (1954), "Geology of Bikini and Nearby Atolls," *U. S. Geol. Survey Prof. Paper 260-A.*

EMILIANI, CESARE (1954), "Depth Habitats of Some Species of Pelagic Foraminifera as Indicated by Oxygen Isotope Ratios," *Amer. Jour. Sci.,* Vol. 252, No. 3, pp. 149–58.

——— (1955), "Pleistocene Temperatures," *Jour. Geol.,* Vol. 63, No. 6, pp. 538–78.

——— (1958), "Paleotemperature Analysis of Core 280 and Pleistocene Correlations," *ibid.,* Vol. 66, No. 3, pp. 264–75.

ERBEN, H. K. (1956a), "New Biostratigraphic Correlations in the Jurassic of Eastern and South-Central Mexico" (Abst.), *Resumenes de los Trabajos Presentados XX Cong. Geol. Internac.,* pp. 26, 27.

——— (1956b), "Paleogeographic Reconstructions for the Lower and Middle Jurassic and for the Callovian of Mexico" (Abst.), *ibid.,* p. 27.

ERICSON, DAVID B.; BROECKER, WALLACE S.; KULP, J. L.; AND WOLLIN, G. (1956), "Late-Pleistocene Climates and Deep-Sea Sediments," *Science,* Vol. 124, No. 3218, pp. 385–89.

———, AND WOLLIN, GOESTA (1955), "Lithological Descriptions and Micropaleontological Analyses of Arctic Cores," Lamont Geol. Observ. (Columbia Univ.), New York. Mimeo. rept.

———, AND ——— (1956), "Correlation of Six Cores from the Equatorial Atlantic and the Carribbean," *Deep-Sea Research,* Vol. 3, No. 2, pp. 104–25.

EWING, GIFFORD C.; AND PHLEGER, FRED B (1958), "Tidal Influence on the Spacing of Inlets into Coastal Lagoons" (Abst.), *Trans. Amer. Geophys. Union,* Vol. 39, No. 3, p. 540.

EWING, MAURICE; AND DONN, WILLIAM L. (1956), "A Theory of Ice Ages, Part I," *Science,* Vol. 123, pp. 1061–66.

———, AND ——— (1958), "A Theory of Ice Ages, Part II," *ibid.,* Vol. 127, pp. 1159–62.

———; ERICSON, DAVID B., AND HEEZEN, BRUCE (1958), "Sediments and Topography of the Gulf of Mexico," in *Habitat of Oil,* Amer. Assoc. Petrol. Geol., pp. 995–1053.

FAGER, E. W. (1957), "Determination and Analysis of Recurrent Groups," *Ecology,* Vol. 38, pp. 586–95.

FAGG, DAVID B. (1957), "The Recent Marine Sediments and Pleistocene Surface of Matagorda Bay, Texas," *Trans. Gulf Coast Assoc. Geol. Soc.,* Vol. 7, pp. 119–33; also, *Texas Agr. and Mech. Coll. Research Found. Proj. 24, Ref. 57–17T,* pp. 119–33.

FAIRBRIDGE, R. W. (1958), "Dating the Latest Movements of the Quaternary Sea Level," *Trans. N. Y. Acad. Sci.,* Ser. 2, Vol. 20, No. 6, pp. 471–82.

FISHER, R. L. (1955), "Cuspate Spits of St. Lawrence Island, Alaska," *Jour. Geol.,* Vol. 63, No. 2, pp. 133–42.

FISK, H. N. (1938), "Geology of Grant and LaSalle Parishes," *Louisiana Dept. Conserv., Geol. Bull. 10.*

——— (1939), "Depositional Terrace Slopes in Louisiana," *Jour. Geomorphology,* Vol. 2, No. 3, pp. 181–99.

——— (1940), "Geology of Avoyelles and Rapides Parishes, Louisiana," *Louisiana Dept. Conserv., Geol. Bull. 18,* pp. 3–240.

——— (1944), "Geological Investigation of the Alluvial Valley of the Lower Mississippi River," *Mississippi River Comm.,* Vicksburg; also (Abst.), *Tulsa Geol. Soc. Digest,* Vol. 15, pp. 50–55 (1947).

——— (1947), "Fine-Grained Alluvial Deposits and Their Effect on Mississippi River Activity," *Mississippi River Comm.,* Vicksburg.

——— (1948), "Geological Investigation of the Lower Mermentau River Basin and Adjacent Areas in Coastal Louisiana," *U. S. Army, Corps of Engineers, Def. Project Rept.,* Appendix II.

——— (1952), *Geological Investigation of the Atchafalaya Basin and the Problem of Mississippi River Diversion,* U. S. Corps of Engineers, Waterways Expt. Sta., Vicksburg, pp. 1–145.

——— (1955), "Sand Facies of Recent Mississippi Delta Deposits," *Proc. 4th World Petrol. Cong.,* Sec. 1-C, pp. 377–98.

——— (1956), "Nearsurface Sediments of the Continental Shelf off Louisiana," *Proc. 8th Texas Conf. Soil Mech. and Found. Eng.,* Austin, Sept. 14–15.

——— (1959), "Padre Island and the Laguna Madre Flats, Coastal South Texas," *2d Coastal Geog. Conf., Coastal Studies Inst., Louisiana State Univ.,* Baton Rouge, pp. 103–52.

———, AND MCCLELLAND, BRAMLETTE (1959), "Geology of Continental Shelf off Louisiana: Its Influence on Offshore Foundation Design," *Bull. Geol. Soc. America,* Vol. 70, No. 10, pp. 1369–94.

———, AND MCFARLAN, E., JR. (1955), "Late Quaternary Deltaic Deposits of the Mississippi River," *in* Crust of the Earth, *Geol. Soc. America Spec. Paper 62,* pp. 279–302.

———; MCFARLAN, E., JR.; KOLB, C. R.; AND WILBERT, L. J., JR. (1954), "Sedimentary Framework of the Modern Mississippi Delta," *Jour. Sed. Petrology,* Vol. 24, No. 2, pp. 76–99.

FLAWN, PETER T. (1954), "Texas Basement Rocks: A Progress Report," *Bull. Amer. Assoc. Petrol. Geol.,* Vol. 38, No. 5, pp. 900–12.

——— (1956), "Basement Rocks of Texas and Southeast New Mexico," *Univ. Texas Bull. 5605.*

———, AND DÍAZ, TEODORO (1959), "Problems of Paleozoic Tectonics in North-Central and Northeastern Mexico," *Bull. Amer. Assoc. Petrol. Geol.,* Vol. 43, No. 1, pp. 224–30.

FLEMING, R. H. (1938), "Tides and Tidal Currents in the Gulf of Panama," *Jour. Marine Research,* Vol. 1, pp. 192–206.

———, AND REVELLE, ROGER (1939), "Physical Processes in the Ocean," *Recent Marine Sediments,* Amer. Assoc. Petrol. Geol. pp. 48–141; reprinted by Soc. Econ. Paleon. Min. as *Spec. Pub. No. 4* (1955).

FLINT, J. M. (1899), "Recent Foraminifera," *U. S. Natl. Mus. Ann. Rept. 1897,* pp. 249–349.

FLINT, R. F. (1957), *Glacial and Pleistocene Geology,* John Wiley & Sons, Inc., New York.

FOLK, ROBERT L. (1951), "Stages of Textural Maturity in Sedimentary Rocks," *Jour. Sed. Petrology,* Vol. 21, No. 3, pp. 127–30.

——— (1954), "The Distinction Between Grain Size and Mineral Composition in Sedimentary-Rock Nomenclature," *Jour. Geol.,* Vol. 62, No. 4, pp. 344–59.

——— (1956), "The Role of Texture and Composition in Sandstone Classification," *Jour. Sed. Petrology,* Vol. 26, No. 2, pp. 166–71.

FORGOTSON, J. M., JR. (1954), "Regional Stratigraphic Analysis of Cotton Vâlley Group of Upper Gulf Coastal Plain," *Bull. Amer. Assoc. Petrol. Geol.,* Vol. 38, No. 12, pp. 2476–99; also, *Trans. Gulf Coast Assoc. Geol. Soc.,* Vol. 4, pp. 143–54.

——— (1956), "A Correlation and Regional Stratigraphic Analysis of the Formations of the Trinity Group and the Genesis and Petrography of the Ferry Lake Anhydrite" (Abst.), *Trans. Gulf Coast Assoc. Geol. Soc.,* Vol. 6, p. 91.

——— (1957), "Stratigraphy of Comanchean Cretaceous Trinity Group," *Bull. Amer. Assoc. Petrol. Geol.,* Vol. 41, No. 10, pp. 2328–63.

FORMAN, M. J.; AND SCHLANGER, S. O. (1957), "Tertiary Reef and Associated Limestone Facies from Louisiana and Guam," *Jour. Geol.,* Vol. 65, No. 6, pp. 611–27.

FÜCHTBAUER, H. (1954), "Transport und Sedimentation der Westlichen Alpenvorlands-Molasse," *Heidelberger Beitr, Miner. u. Petrog.,* Vol. 4, Nos. 1–2, pp. 26–53.

FUGLISTER, F. C. (1947), "Average Monthly Sea Surface Temperatures of the Western North Atlantic Ocean," *Papers in Phys. Oceanog. and Meteorol.,* Mass. Inst. Tech. & Woods Hole Oceanog. Inst., Vol. 10, No. 2, pp. 1–25.

——— (1954), "Average Temperature and Salinity at a Depth of 200 Meters in the North Atlantic," *Tellus,* Vol. 6, pp. 46–58.

GALSTOFF, P. S. (1931), "Surveys of Oyster Bottoms in Texas," *U. S. Dept. Interior, Fish and Wildlife Serv., Fishery Inves. Rept.,* Vol. 6, pp. 1–30.

———, ET AL. (1954), "The Gulf of Mexico, Its Origin, Waters, and Marine Life," *ibid., Fishery Bull. 89,* Vol. 55.

GARDNER, JULIA (1926–47), "The Molluscan Fauna of the Alum Bluff Group of Florida," *U. S. Geol. Survey Prof. Paper 142,* pp. 1–656.

GILBERT, G. K. (1885), "The Topographic Features of Lake Shores," *ibid., 5th Ann. Rept.,* p. 87.

GINSBURG, R. N. (1953), "Beachrock in South Florida," *Jour. Sed. Petrology,* Vol. 23, No. 2, pp. 85–92.

GODWIN, H.; SUGGATE, R. P.; AND WILLIS, E. H. (1958), "Radiocarbon Dating of the Eustatic Rise in Ocean-Level," *Nature,* Vol. 181, pp. 1518–19.

GOLDBERG, E. D.; WALKER, T. J.; AND WHISENAND, A. (1951), "Phosphate Utilization by Diatoms," *Biol. Bull.,* Vol. 101, pp. 274–84.

GOLDSTEIN, AUGUST, JR. (1942), "Sedimentary Petrologic Provinces of the Northern Gulf of Mexico," *Jour. Sed. Petrology,* Vol. 12, No. 2, pp. 77–84.

———, AND FLAWN, P. T. (1958), "Oil and Gas Possibilities of Ouachita Structural Belt in Texas and Oklahoma," *Bull. Amer. Assoc. Petrol. Geol.,* Vol. 42, No. 3, pp. 876–81.

———, AND RENO, D. H. (1952), "Petrography and Metamorphism of Sediments of Ouachita Facies," *ibid.,* Vol. 36, No. 12, pp. 2275–90.

GOULD, H. R.; AND MCFARLAN, E., JR. (1959), "Geologic History of the Chenier Plain, Southwestern Louisiana," *Trans. Gulf Coast Assoc. Geol. Soc.,* Vol 9, pp. 261–70.

———, AND STEWART, R. H. (1955), "Continental

Terrace Sediments in the Northeastern Gulf of Mexico," *in* Finding Ancient Shorelines, *Soc. Econ. Paleon. Min. Spec. Pub. No. 3,* pp. 2–20 [Jan. 1956].

GREENMAN, NORMAN N.; AND LEBLANC, RUFUS J. (1956), "Recent Marine Sediments and Environments of Northwest Gulf of Mexico," *Bull. Amer. Assoc. Petrol. Geol.*, Vol. 40, No. 5, pp. 813–47.

GRIGG, R. P. (1956), "Key to the Nodosaria Embayment of South Louisiana," *Trans. Gulf Coast Assoc. Geol. Soc.*, Vol. 6, pp. 55–62.

GRIM, RALPH E. (1956a), "Clay-Mineral Studies," *A.P.I. Proj. 51, Rept. 19,* IMR Ref. 56–1, pp. 16–19.

——— (1956b), "Clay-Mineral Investigations," *ibid., Rept. 20,* IMR Ref. 56–3, pp. 23–27

——— (1958), "Concept of Diagenesis in Argillaceous Sediments," *Bull. Amer. Assoc. Petrol. Geol.*, Vol. 42, No. 2, pp. 246–53.

———, AND JOHNS, W. D. (1955), "Clay-Mineral Investigation of Sediments in the Northern Gulf of Mexico," *Natl. Research Council, Proc. 2d Ann. Clay Minerals Conf.*, pp. 81–103; also, *A.P.I. Proj. 51, Rept. 16,* IMR Ref. 55–4, pp. 23–27.

GULF COAST ASSOCIATION OF GEOLOGICAL SOCIETIES (1951–55), *Transactions,* Vols. 1–5

GUNTER, GORDON (1942), "Seasonal Condition in Texas Oysters," *Proc. and Trans. Texas Acad Sci.*, Vol. 25, pp. 89–93.

——— (1945), "Studies of Marine Fishes of Texas," *Pub. Inst. Marine Sci.*, Vol. 1, pp. 1–190.

——— (1950), "Seasonal Population Changes and Distributions as Related to Salinity of Certain Invertebrates of the Texas Coast, Including Commercial Shrimp," *ibid.*, Vol. 1, No. 2, pp. 7–51.

———, AND HILDEBRAND, H. H. (1951), "Destruction of Fishes and Other Organisms on the South Texas Coast by the Cold Wave of January 28-February 3, 1951," *Ecology,* Vol. 32, No. 4, pp. 731–36.

HALBOUTY, M. T.; AND HARDIN, GEO. C., JR. (1956), "Genesis of Salt Domes of Gulf Coastal Plain," *Bull. Amer. Assoc. Petrol. Geol.,* Vol. 40, No. 4, pp. 737–46.

HANNA, M. A. (1934), "Geology of the Gulf Coast Salt Domes," in *Problems of Petroleum Geology,* Amer. Assoc. Petrol. Geol., pp. 629–78.

HÄNTZSCHEL, W. (1939), "Tidal Flat Deposits (Wattenschlick)," in *Recent Marine Sediments,* Amer. Assoc. Petrol. Geol., pp. 195–206; reprinted by Soc. Econ. Paleon. Min. as *Spec. Pub. No. 4* (1955).

HARRY, H. W. (1942), "List of Mollusca of Grand Isle, La., Recorded From the Louisiana State Univ. Marine Laboratory, 1929–41," *Occasional Papers No. 1, Marine Lab., Louisiana State Univ.*, Baton Rouge, pp. 1–12.

HARVEY, H. W. (1955), *The Chemistry and Fertility of Sea Waters,* Cambridge Univ. Press.

HAZZARD, R. T.; BLANPIED, B. W.; AND SPOONER, W. C. (1947), "Notes on the Stratigraphy of the Formations Which Underlie the Smackover Limestone in South Arkansas, Northeast Texas, and North Louisiana," *Shreveport Geol. Soc., 1945 Ref. Rept.*, Vol. 2, pp. 483–503.

HEDGPETH, J. W. (1947), "The Laguna Madre of Texas," *Trans. 12th North Amer. Wildlife Conf.*, pp. 364–80.

——— (1953), "An Introduction to the Zoogeography of the Northwestern Gulf of Mexico with Reference to the Invertebrate Fauna," *Pub. Inst. Marine Sci.*, Vol. 3, No. 1, pp. 111–224.

——— (1957), "Estuaries and Lagoons, II. Biological Aspects," *in* Treatise on Marine Ecology and Paleoecology, *Geol. Soc. America Mem. 67,* Vol. 1, pp. 693–729.

HEEZEN, B. C. (1957), "1908 Messina Earthquake, Tsunami, and Turbidity Current" (Abst.), *Bull. Geol. Soc. America,* Vol. 68, No. 12, p. 1743.

HENRY, VERNON J. (1956), "Investigation of the Shoreline-Like Features in the Galveston Bay Region, Texas," Chap. IV, *Tectonic History of the Galveston Bay Region,* Texas Agr. and Mech. Coll. Research Found. Proj. 24 Tech. Rept., Ref. 56–12T.

HERALD, F. A. [EDITOR] (1951), Occurrence of Oil and Gas in Northeast Texas, *Univ. Texas Bur. Econ. Geol. Pub. 5116.*

HILDEBRAND, H. H. (1954), "A Study of the Fauna of the Brown Shrimp (*Penaeus aztectus* Ives) Grounds in the Western Gulf of Mexico," *Pub. Inst. Marine Sci.*, Vol. 3, No. 2, pp. 233–366.

——— (1958), "Estudios Biologicos Preliminares Sobre la Laguna Madre de Tamaulipas," *Ciencia* (Mex.), Vol. 17, Nos. 7–9, pp. 151–73.

HILL, R. T. (1937), "Paluxy Sands, with Further Notes on the Comanche Series" (Abst.), *Proc. Geol. Soc. America,* 1936, pp. 79–80.

HOLLAND, W. C.; HOUGH, L. W.; AND MURRAY, G. E. (1952), "Geology of Beauregard and Allen Parishes," *Louisiana Dept. Conserv. Geol. Bull. 27.*

HOLLE, C. G. (1952), "Sedimentation at the Mouth of the Mississippi River," *Proc. 2d Conf. on Coastal Eng., Council on Wave Research,* Univ. Calif., Berkeley, pp. 111–29.

HOLMES, H. W. (1955), *The Chemistry and Fertility of Sea Waters,* Cambridge Univ. Press.

HOLMES, R. W. (1958a), "Surface Chlorophyll A, Surface Primary Production, and Zooplankton Volumes in the Eastern Pacific Ocean," *Rapp. et Proces-Verb. Cons. Perm. Internat. Explor. Mer.*, Vol. 144, pp. 109–16.

——— [EDITOR] (1958b), "Physical, Chemical, and Biological Oceanographic Observations Obtained on Expeditions *Scope* in the Eastern Tropical Pacific, November-December, 1956," *U. S. Dept. Interior, Fish and Wildlife Serv., Spec. Sci. Rept. Fish. No. 279.*

———, AND REID, F., "The Preparation of Pelagic Marine Nannoplankton for Microscopic Examination and Enumeration with the Molecular Filter," Unpub. ms.

———; SCHAEFER, M. B.; AND SHIMADA, B. M. (1957), "Primary Production, Chlorophyll, and Zooplankton Volumes in the Tropical Eastern Pacific Ocean," *Bull. Interamer. Trop. Tuna Comm.*, Vol. 2, pp. 129–69.

HONEA, J. W. (1956), "Sam Fordyce-Vanderbilt Fault System of Southwest Texas," *Trans. Gulf Coast Assoc. Geol. Soc.*, Vol. 6, p. 51.

HOPKINS, D. M. (1959), "Cenozoic History of the Bering Land Bridge," *Science,* Vol. 129, No. 3362, pp. 1519–28.

HORBERG, C. L. (1955), "Radiocarbon Dates and Pleistocene Chronological Problems in the Mississippi Valley Region," *Jour. Geol.*, Vol. 63, No. 3, pp. 278–85.

HORRER, PAUL L. (1951), "Oceanographic Analysis of Marine Pipeline Problems (Atchafalaya Bay, La., and Adjacent Continental Shelf Area), Sec. III, Waves," *Texas Agr. and Mech. Coll. Research Found. Proj. 25, Final Report.*

HOUGH, L. W. (1937), "Petrographic Comparison of Mississippi River and Tributary Bed Material Sands," Unpub. M.A. thesis, Louisiana State Univ., Baton Rouge.

HOUSTON GEOLOGICAL SOCIETY (1953), *AAPG-SEPM-SEG Guidebook: Field Trip Routes, Oil Fields, Geology,* Joint Ann. Mtg., Houston, 1953.

——— (1954a), "Cross Sections, Upper Gulf of Texas, Four Strike and Two Dip Sections."

——— (1954b), "Stratigraphy of the Upper Gulf Coast of Texas and Strike and Dip Cross Sections, Upper Gulf Coast of Texas," Study Group Report, 1953-54.

——— (1954c), "Study Group Cross Section and Stratigraphic Review Project on Texas Upper Gulf Coast."

——— (1958), "Middle Tertiary (Lower Oligocene to Middle Eocene) of Brazos River Valley," *Guidebook of Field Trip, Dec. 6, 1958.*

——— (1959), "The Frio Formation of the Upper Gulf Coast of Texas."

HOWARD, A. D. (1939), "Hurricane Modification of the Offshore Bar of Long Island, New York," *Geog. Rev.,* Vol. 29, No. 3, pp. 400–15.

HOWE, H. V. (1933), "Review of Tertiary Stratigraphy of Louisiana," *Bull. Amer. Assoc. Petrol. Geol.,* Vol. 17, No. 6, pp. 613–55; reprinted in *Gulf Coast Oil Fields,* Amer. Assoc. Petrol. Geol., pp. 383–424 (1936).

——— (1936), "Louisiana Petroleum Stratigraphy," *Amer. Petrol. Inst., Div. of Prod., Paper 901-12B;* also, *Oil and Gas Journal,* Vol. 34, No. 48, pp. 98–111, 124–28; *Louisiana Dept. Conserv. Geol. Bull. 27,* pp. 1–46; Amer. Petrol. Inst., *Drilling and Production Practice,* pp. 405–19 (1937).

HUFFMAN, G. G.; AND PRICE, W. A. (1949), "Clay Dune Formation near Corpus Christi, Texas," *Jour. Sed. Petrology,* Vol. 19, No. 3, pp. 118–27.

HUMPHREY, W. E. (1956), "Tectonic Framework of Northeastern Mexico," *Trans. Gulf Coast Assoc. Geol. Soc.,* Vol. 6, p. 25.

HUTCHINS, L. W. (1947), "The Bases for Temperature Zonation in Geographical Distribution," *Ecol. Mon.* Vol. 17, pp. 324–35.

ILLING, L. V. (1954), "Bahaman Calcareous Sands," *Bull. Amer. Assoc. Petrol. Geol.,* Vol. 38, No. 1, pp. 1–95.

IMLAY, R. W. (1940), "Lower Cretaceous and Jurassic Formations of Southern Arkansas and their Oil and Gas Possibilities," *Arkansas Geol. Survey, Info. Circ. 12.*

——— (1943), "Jurassic Formations of Gulf Region," *Bull. Amer. Assoc. Petrol. Geol.,* Vol. 27, No. 11, pp. 1407–1533.

——— (1945), "Subsurface Lower Cretaceous Formations of South Texas," *ibid.,* Vol. 29, No. 10, pp. 1416–69.

———, ET AL. (1948), "Stratigraphic Relations of Certain Jurassic Formations in Eastern Mexico," *ibid.,* Vol. 32, No. 9, pp. 1750–61.

INMAN, D. L. (1949), "Sorting of Sediments in the Light of Fluid Mechanics," *Jour. Sed. Petrology,* Vol. 19, No. 2, pp. 51–70.

——— (1952), "Measures for Describing the Size Distribution of Sediments," *ibid.,* Vol. 22, No 3, pp. 125–45.

———, AND CHAMBERLAIN, T. K. (1955), "Particle-Size Distribution in Nearshore Sediments," *in* Finding Ancient Shorelines, *Soc. Econ. Paleon. Min. Spec. Pub. No. 3,* pp. 78–96 [Jan. 1956].

———, AND NASU, N. (1956), "Orbital Velocity Associated with Wave Action near the Breaker Zone," *Beach Erosion Board Tech. Mem. 79.*

JOHNS, W. D.; AND GRIM, R. E. (1958), "Clay Mineral Composition of Recent Sediments from the Mississippi River Delta," *Jour. Sed. Petrology,* Vol. 28, No. 2, pp. 186–99.

JOHNSON, C. W. (1934), "List of the Marine Mollusca of the Atlantic Coast from Labrador to Texas," *Proc. Boston Soc. Nat. Hist.,* Vol. 40, pp. 1–204.

JOHNSON, D. W. (1919), *Shore Processes and Shore Line Development,* John Wiley & Sons, Inc., New York.

——— (1925), *The New England-Acadian Shore Line,* John Wiley & Sons, Inc., New York.

JOHNSTON, W. A. (1921), "Sedimentation of the Fraser River Delta," *Geol. Survey Canada Mem. 125.*

——— (1922), "The Character of the Stratification of the Sediments in the Recent Delta of Frazer River, British Columbia, Canada," *Jour. Geol.,* Vol. 30, No. 2, pp. 115–29.

JONES, G. E.; THOMAS, W. H.; AND HAXO, F. T. (1958), "Preliminary Studies of Bacterial Growth in Relation to Dark and Light Fixation of $C^{14}O_2$ During Productivity Determinations," in *U. S. Dept. Interior, Fish and Wildlife Spec. Sci. Rept. Fisheries No. 279,* pp. 79–86.

JONES, N. S. (1950), "Marine Bottom Communities," *Biol. Rev.,* Vol. 25, pp. 283–313.

JORDAN, G. F.; AND STEWART, H. B., JR. (1959), "Continental Slope off Southwest Florida," *Bull. Amer. Assoc. Petrol. Geol.,* Vol. 42, No. 4, pp. 974–91.

JORDAN, LOUISE (1952), "Preliminary Notes on the Mesozoic Rocks of Florida," *Guidebook, 44th Ann. Mtg. Assoc. Amer. State Geol.,* Florida Geol. Survey, pp. 39–45.

KELLER, W. D.; WESTCOTT, J. F.; AND BLEDSOE, A. O. (1954), "The Origin of Missouri Fireclays," *in* Clays and Clay Minerals; Proc. 2d Natl. Conf. on Clays and Clay Minerals, *Natl. Research Council Pub. 327,* pp. 7–46.

KELLOG, J. L. (1905), "Notes on Marine Food Molluska of Louisiana," *Bull. Gulf Biol. Sta. 3,* pp. 1–43.

KELLOUGH, G. R. (1956), "Distribution of Foraminifera Around a Submerged Hill in the Gulf of Mexico," *Trans. Gulf Coast Assoc. Geol. Soc.,* Vol. 6, pp. 205–16.

KETCHUM, B. H. (1939), "The Absorption of Phosphate and Nitrate by Illuminated Cultures of *Nitzchia closterium,*" *Amer. Jour. Bot.,* Vol. 26, pp. 309–407.

KIDWELL, A. L. (1951), "Mesozoic Igneous Activity in the Northern Gulf Coastal Plain," *Guidebook, Gulf Coast Assoc. Geol. Soc., 1st Ann. Mtg.,* pp. 182–99.

———, AND HUNT, J. A. (1958), "Migration of Oil in Recent Sediments of Pedernales, Venezuela," in *Habitat of Oil,* Amer. Assoc. Petrol. Geol., pp. 790–817.

KIMBALL, H. H. (1928), "Amount of Solar Radiation That Reaches the Surface of the Earth on the Land and on the Sea, and Methods By Which It Is Measured," *Mon. Weath. Rev., Washington,* Vol. 56, pp. 393–98.

KOLB, C. R. (1956), "Review of Petrographic Studies of Bed Material, Mississippi River, Its Tributaries, and Offshore Areas of Deposition," *U. S. Corps of Engineers, Waterways Expt. Sta. Tech. Rept. 3-4336,* Vicksburg.

———, AND VAN LOPIK, J. R. (1958), "Geology of the Mississippi River Deltaic Plain, Southeastern Louisiana," *ibid., Tech. Repts. 3-483* and *3-484,* 2 vols.

KOLDEWIJN, B. W. (1958), "Sediments of the Paria-Trinidad Shelf," *Reports of the Orinoco Shelf Expedition,* Vol. 3, Mouton & Co., The Hague, pp.1-109.

KORNFELD, M. M. (1931), "Recent Littoral Foraminifera from Texas and Louisiana," *Stanford Univ., Contr. Dept. Geol.,* Vol. 1, No. 3, pp. 77-101.

KRUIT, CORNELIUS (1955), *Sediments of the Rhone Delta: I, Grain Size and Microfauna,* Rijksuniv. te Groningen. Pub. by Mouton & Co., The Hague. Also, in *Kon. Nederlands Geol. Mijnbouwk. Gen. Verhand.,* Vol. 15, No. 3, pp. 357-499.

KRUMBEIN, W. C. (1934), "Size Frequency Distributions of Sediments," *Jour. Sed. Petrology,* Vol. 4, No. 2, pp. 65-77.

——— (1939), "Tidal Lagoon Sediments on the Mississippi Delta," in *Recent Marine Sediments,* Amer. Assoc. Petrol. Geol., pp. 178-94; reprinted by Soc. Econ. Paleon. Min. as *Spec. Pub. No. 4* (1955).

———, AND ABERDEEN, ESTHER (1937), "The Sediments of Barataria Bay," *Jour. Sed. Petrology,* Vol. 7, No. 1, pp. 3-17.

———, AND GARRELS, R. M. (1952), "Origin and Classification of Chemical Sediments in Terms of pH and Oxidation-Reduction Potentials," *Jour. Geol.,* Vol. 60, No. 1, pp. 1-33.

———, AND PETTIJOHN, F. J. (1938), *Manual of Sedimentary Petrography,* D. Appleton-Century Co., Inc., N. Y., London.

KRYNINE, P. D. (1942), "Differential Sedimentation and its Products During One Complete Geosynclinal Cycle," *Proc. 1st Pan Amer. Congr. Mining Eng. & Geol.,* Vol. 2, Pt. 1, pp. 537-61.

——— (1943), "Diastrophism and the Evolution of Sedimentary Rocks," *Pennsylvania State Min. Ind. Exper. Sta. Tech. Paper 84-A.*

——— (1948), "The Megascopic Study and Field Classification of Sedimentary Rocks," *Jour. Geol.,* Vol. 56, No. 2, pp. 130-65.

LADD, HARRY S. (1951), "Brackish-Water and Marine Assemblages of the Texas Coast, with Special Reference to Mollusks," *Pub. Inst. Marine Sci.,* Vol. 2, No. 1, pp. 125-63.

———; HEDGPETH, J. W.; AND POST, RITA (1957), "Environments and Facies of Existing Bays on the Central Texas Coast," *in* Treatise on Marine Ecology and Paleontology, *Geol. Soc. America Mem. 67,* Vol. 2, pp. 599-640.

LAFAYETTE GEOLOGICAL SOCIETY (1954), "Idealized Isometric Block Diagram of Southern Louisiana, Showing the Stratigraphy of the Miocene (post-*Margulina ascensionensis*)."

LAHEE, F. H. (1929), "Oil and Gas Fields of the Mexia and Tehuacana Fault Zones, Texas," in *Structure of Typical American Oil Fields,* Vol. 1, Amer. Assoc. Petrol. Geol., pp. 304-88.

LAMB, H. (1945), *Hydrodynamics,* Dover Pubs., New York.

LANKFORD, R. R. (1959), "Distribution and Ecology of Foraminifera from East Mississippi Delta Margin," *Bull. Amer. Assoc. Petrol. Geol.,* Vol. 43, No. 9. pp. 2068-99.

———, AND CURRAY, J. R. (1957), "Mid-Tertiary Rock Outcrop on Continental Shelf, Northwest Gulf of Mexico," *ibid.,* Vol. 41, No. 9, pp. 2113-17.

———, AND SHEPARD, F. P. (1960), "Facies Interpretation in Mississippi Delta Borings," *Jour. Geol.,* Vol. 68, No. 4, pp. 408-26.

LEBLANC, R. J.; AND BERNARD, H. A. (1954), "Résumé of Late Recent Geological History of the Gulf Coast," *Geologie en Mijnbouw,* Vol. 16e, No. 6, Nw. Ser., pp. 185-94.

———, AND HODGSON, W. D. (1959), "Origin and Development of the Texas Shoreline," *2d Coastal Geog. Conf., Coastal Studies Inst., Louisiana State Univ.,* Baton Rouge, pp. 57-101.

LEHMAN, E. P. (1957), "Statistical Study of Texas Gulf Coast Recent Foraminiferal Studies," *Micropaleontology,* Vol. 3, pp. 325-56.

LEIPPER, D. F. (1954), "Physical Oceanography of the Gulf of Mexico," *in* Gulf of Mexico, Its Origin, Waters, and Marine Life, *U.S. Dept. Interior, Fish and Wildlife Serv. Bull. 89,* pp. 119-37.

LIEBMAN, E. (1940), "River Discharges and Their Effect on the Cycles and Productivity of the Sea," *Proc. 6th Pac. Sci. Cong.,* Vol. 3, pp. 517-23.

LOEBLICH, A. R., JR.; AND TAPPAN, H. (1957), "Correlation of the Gulf and Atlantic Coastal Plain Paleocene and Lower Eocene Formations by Means of Planktonic Foraminifera," *Jour. Paleon.,* Vol. 31, No. 6, pp. 1109-37.

LOHSE, E. ALAN (1952), "Shallow-Marine Sediments of the Rio Grande Delta," Unpub. Ph.D. dissertation, Univ. of Texas, Austin.

——— (1955), "Dynamic Geology of the Modern Coastal Region, Northwest Gulf of Mexico," *in* Finding Ancient Shorelines, *Soc. Econ. Paleon. Min. Spec. Pub. No. 3,* pp. 99-104 [Jan. 1956].

——— (1958), "Geochronology of Mud Flats Through Varve Analysis," *in* Sedimentology of South Texas, *Corpus Christi Geol. Soc. Guidebook,* May, 1958, p. 47.

LOWMAN, S. W. (1949), "Sedimentary Facies in Gulf Coast," *Bull. Amer. Assoc. Petrol. Geol.,* Vol. 33, No. 12, pp. 1939-97.

——— (1951), "The Relationship of the Biotic and the Lithic Facies in Recent Gulf Coast Sedimentation," *Jour. Sed. Petrology,* Vol. 21, No. 4, pp. 233-37.

LOZO, F. E., JR. (1949), "Stratigraphic Relations of Fredericksburg Limestones, North Central Texas," *Shreveport Geol. Soc. Guidebook, 17th Ann. Field Trip,* pp. 85-91.

——— [EDITOR] (1951), "The Woodbine and Adjacent Strata of the Waco Area of Central Texas," East Texas Geol. Soc. Symposium, *Fondren Sci. Ser., No. 4.*

———, AND STRICKLIN, F. L., JR. (1956), "Stratigraphic Notes on the Outcrop Basal Cretaceous, Central Texas," *Trans. Gulf Coast Assoc. Geol. Soc.,* Vol. 6, pp. 67-78.

LUCAS, GABRIEL (1955), "Oolithes Marines Actualles et Calcaires Oolithiques Recents sur la Rivage Africain de la Mediterranée Orientale (Egypte et Sud Tunisien)," *Bull. Sta. Oceanog. de Salammbô* (Tunisie), *No. 52.*

LUDWICK, J. C. (1950), "Deep Water Sand Layers off San Diego," Unpub. Ph.D. dissertation, Univ. Calif., Los Angeles.

———, and Walton, W. R. (1957), "Shelf-Edge, Calcareous Prominences in Northeastern Gulf of Mexico," *Bull. Amer. Assoc. Petrol. Geol.,* Vol. 41, No. 9, pp. 2054–2101.

McClure, C. D.; Nelson, H. F.; and Huckabay, W. B. (1958), "Marine Sonoprobe System, New Tool for Geologic Mapping," *ibid.,* Vol. 42, No. 4, pp. 701–16.

McFarlan, E., Jr.; and Thomson, M. R. (1957), "Subsurface Quaternary Stratigraphy in Coastal Louisiana and Adjacent Continental Shelf" (Program Abst.), *Amer. Assoc. Petrol. Geol.* 42nd Ann. Mtg., St. Louis, pp. 28–29.

McIntire, W. G. (1954), "Correlation of Prehistoric Settlements and Delta Development," *Louisiana State Univ.,* Baton Rouge, *Tech. Rept. No. 5.*

McLean, C. M. (1957), "Miocene Geology of Southeastern Louisiana," *Trans. Gulf Coast Assoc. Geol. Soc.,* Vol. 7, pp. 241–45.

Marland, F. C. (1958), "An Ecological Study of the Benthic Macro-Fauna of Matagorda Bay, Texas," Unpub. M.S. thesis, Texas Agr. and Mech Coll., College Station, Dept. of Oceanog., pp. 1–75.

Marmer, H. A. (1954), "Tides and Sea Level in the Gulf of Mexico," *in* Gulf of Mexico, Its Origin, Waters, and Marine Life, *U. S. Dept. Interior, Fish and Wildlife Serv. Bull. 89,* pp. 101–18.

Martens, J. H. C. (1931), "Beaches of Florida," *Florida Geol. Survey 21st–22d Ann. Repts.,* pp. 67–119.

Mason, M. A. (1952), "Pertinent Factors in the Protection of the Gulf Coast," *Proc. 2d Conf. Coastal Eng.,* Houston, 1951, pp. 217–25.

Masson, P. H. (1955), "An Occurrence of Gypsum in Southwest Texas," *Jour. Sed. Petrology,* Vol. 25, No. 1, pp. 72–77.

Mathis, R. W. (1944), "Heavy Minerals of Colorado River Terraces of Texas," *ibid.,* Vol. 14, No. 2, pp. 86–93.

Mattison, G. C. (1948), "Bottom Configuration in the Gulf of Mexico," *Jour. Coast and Geod. Survey,* Vol. 1, pp. 78–82.

Mellen, F. F. (1940), "Geology [Part I] of Yazoo County Mineral Resources," *Mississippi Geol. Survey Bull. 39,* pp. 9–72.

——— (1941), "Warren County Mineral Resources," *ibid., Bull. 43,* pp. 9–88.

Menzel, R. W. (1956), "Annotated Check-List of the Marine Fauna and Flora of the St. George Sound-Apalachee Bay Region, Florida Gulf Coast," *Florida State Univ. Contrib. 61,* Oceanog. Inst., Tallahassee. Mimeo. rept., pp. 1–78.

Milne, I. H.; and Earley, J. W. (1958), "Effect of Source and Environment on Clay Minerals," *Bull. Amer. Assoc. Petrol. Geol.,* Vol. 42, No. 2, pp. 328–38.

Mitchell, J. D. (1894), "List of Texas Mollusca," *Times Steam Print,* Victoria, Texas, Vol. 22.

Mixon, R. B.; Murray, G. E.; and Diaz, T. (1959), "Age and Correlation of Huizachal Group (Mesozoic), State of Tamaulipas, Mexico," *Bull. Amer. Assoc. Petrol. Geol.,* Vol. 43, No. 4, pp. 757–71.

Monroe, W. H. (1955), "Reverse Faulting in the Coastal Plain of Alabama" (Abst.), *Bull. Geol. Soc. America,* Vol. 66, No. 12, Pt. 2, p. 1598.

Moody, C. L. (1949), "Mesozoic Igneous Rocks of Northern Gulf Coastal Plain," *Bull. Amer. Assoc. Petrol. Geol.,* Vol. 33, No. 8, pp. 1410–28.

Moore, David G. (1955), "Rate of Deposition Shown by Relative Abundance of Foraminifera," *ibid.,* Vol. 39, No. 8, pp. 1594–1600.

———, and Scruton, P. C. (1957), "Minor Internal Structures of Some Recent Unconsolidated Sediments," *ibid.,* Vol. 41, No. 12, pp. 2723–51.

Moore, Derek (1959), "Role of Deltas in the Formation of Some British Lower Carboniferous Cyclothems," *Jour. Geol.,* Vol. 67, No. 5, pp. 522–39.

Moore, H. F. (1907), "Survey of Oyster Bottoms in Matagorda Bay, Texas," *U. S. Bur. Fisheries Doc. No. 610, Rept. U. S. Fish Comm. and Spec. Papers, 1905,* pp. 1–86.

———, and Danglade, E. (1915), "Condition and Extent of the Natural Oyster Beds and Barren Bottoms of Lavaca Bay, Texas," *ibid., Doc. 809,* Appendix II, *Rept., 1914,* pp. 1–45.

Moore, R. C. (1949), "Meaning of Facies," *in* Sedimentary Facies in Geologic History," *Geol. Soc. America Mem. 39,* pp. 2–34.

Moore, W. E. (1957), "Ecology of Foraminifera in Northern Florida Keys," *Bull. Amer. Assoc. Petrol. Geol.,* Vol. 41, No. 4, pp. 727–41.

Morgan, H. J., Jr. (1952), "Paleozoic Beds South and East of Ouachita Folded Belt," *ibid.,* Vol. 36, No. 12, pp. 2266–74.

Morgan, J. P.; Nichols, L. G.; and Wright, M. (1958), "Morphological Effects of Hurricane Audrey on the Louisiana Coast," *Louisiana State Univ.,* Baton Rouge, *Coastal Studies Inst. Tech. Rept. 10.*

———; Van Lopik, J. R.; and Nichols, L. G. (1953), "Occurrence and Development of Mudflats Along the Western Louisiana Coast," *ibid., Tech. Rept. 2.*

Muehlberger, W. R. (1959), "Internal Structure of the Grand Saline Salt Dome, Van Zandt County, Texas," *Univ. Texas Bur. Econ. Geol. Rept. Inves. No. 38.*

Mullin, J. B.; and Riley, J. P. (1955), "The Spectrophotometric Determination of Nitrate in Natural Waters, with Particular Reference to Sea Water," *Anal. Chem. Acta.,* Vol. 12, pp. 464–80.

Murray, Grover E. (1952a), "Vicksburg Stage and Mosley Hill Formation," *Bull. Amer. Assoc. Petrol. Geol.,* Vol. 36, No. 4, pp. 700–07.

——— (1952b), "Sedimentary Volumes in Gulf Central Plain of United States and Mexico: Foreword and Summary," *Bull. Geol. Soc. America,* Vol. 63, No. 12, p. 1157.

——— (1952c), *Idem,* "Part III, Volume of Mesozoic and Cenozoic Sediments in Central Gulf Coastal Plain," *ibid.,* pp. 1177–92.

——— (1955), "Midway Stage, Sabine Stage, and Wilcox Group," *Bull. Amer. Assoc. Petrol. Geol.,* Vol. 39, No. 5, pp. 671–96.

——— (1956), "Relationships of Paleozoic Structure to Large Anomalies of Coastal Element of Eastern North America," *Trans. Gulf Coast Assoc. Geol. Soc.,* Vol. 6, pp. 13–24; also, *Resumenes de los Trabajos Presentados, XX Cong. Geol. Internac.,* p. 289.

——— (1957), "Geologic Occurrence of Hydrocarbons in Gulf Coastal Province of the United States," *Trans. Gulf. Coast Assoc. Geol. Soc.,* Vol. 7, pp. 253–300.

———, et al. (1952), "Sedimentary Volumes in Gulf Coastal Plain of United States and Mexico," *Bull. Geol. Soc. America,* Vol. 63, No. 12, pp. 1157–1228.

———, and Thomas, E. P. (1945), "Midway-Wilcox Surface Stratigraphy of Sabine Uplift, Louisiana and

Texas," *Bull. Amer. Assoc. Petrol. Geol.,* Vol. 29, No. 1, pp. 45–70.

———, AND WILBERT, L. J., JR. (1950), "Jacksonian Stage," *ibid.,* Vol. 34, No. 10, pp. 1990–97.

NANZ, ROBERT H., JR. (1954), "Genesis of Oligocene Sandstone Reservoir, Seeligson Field, Jim Wells, and Kleberg Counties, Texas," *ibid.,* Vol. 38, No. 1, pp. 96–117.

NATIONAL RESEARCH COUNCIL (1948), *Rock Color Chart;* reprinted by Geol. Soc. America (1951).

NEEDHAM, C. E. (1934), "The Petrology of the Tombigbee Sands of Eastern Mississippi," *Jour. Sed. Petrology,* Vol. 4, No. 2, pp. 55–59.

NETTLETON, L. L. (1952), "Sedimentary Volumes in Gulf Coastal Plain of the United States and Mexico: Part VI, Geophysical Aspects," *Bull. Geol. Soc. America,* Vol. 63, No. 12, pp. 1221–28.

——— (1957), "Gravity Survey Over a Gulf Coast Continental Shelf Mound," *Geophysics,* Vol. 22, pp. 630–42.

NEUMANN, ANDREW C. (1958), "The Configuration and Sediments of Stetson Bank, Northwestern Gulf of Mexico," *Texas Agr. and Mech. Coll. Research Found. Proj. 24,* IMR Ref. 58–FT.

NEW ORLEANS GEOLOGICAL SOCIETY (1954), "Cross Sections Through South Louisiana."

NIENABER, J. H. (1958), "Shallow Marine Sediments Offshore from the Brazos River, Texas," Unpub. Ph.D. dissertation, Univ. of Texas, Austin.

NORRIS, R. M. (1953), "Buried Oyster Reefs in Some Texas Bays," *Jour. Paleon.,* Vol. 27, No. 4, pp. 569–76.

NOTA, D. J. G. (1958), "Sediments of the Western Guiana Shelf," *in* Reports of the Orinoco Shelf Expedition, Vol. 2, *Utrecht. Meded. Landbouw. te Wageningen,* Vol. 58, No. 2, pp. 1–98.

NUNNALLY, J. D.; AND FOWLER, H. F. (1954), "Lower Cretaceous Stratigraphy of Mississippi," *Mississippi Geol. Survey Bull. 79.*

O'NEIL, T. (1949), "The Muskrat in Louisiana Coastal Marshes," *Louisiana Dept. of Wildlife and Fisheries,* New Orleans.

ORR, W. L.; EMERY, K. O.; AND GRADY, J. R. (1958), "Preservation of Chlorophyll Derivatives in Sediments off Southern California," *Bull. Amer. Assoc. Petrol. Geol.,* Vol. 42, No. 5, pp. 925–62.

PARKER, F. L. (1954), "Distribution of the Foraminifera in the Northeastern Gulf of Mexico," *Harvard Coll. Mus. Comp. Zool. Bull.,* Vol. 111, No. 10, pp. 453–588.

———; PHLEGER, F. B; AND PEIRSON, J. F. (1953), "Ecology of Foraminifera from San Antonio Bay and Environs, Southwest Texas," *Cushman Found. Foram. Research Spec. Pub. 2.*

PARKER, R. H. (1955), "Changes in the Invertebrate Fauna, Apparently Attributable to Salinity Changes, in the Bays of Central Texas," *Jour. Paleon.,* Vol. 29, No. 2, pp. 193–211.

——— (1956), "Macro-Invertebrate Assemblages as Indicators of Sedimentary Environments in East Mississippi Delta Region," *Bull. Amer. Assoc. Petrol. Geol.,* Vol. 40, No. 2, pp. 295–376.

——— (1959a), "Macro-Invertebrate Assemblages of Central Texas Coastal Bays and Laguna Madre," *ibid.,* Vol. 43, No. 9, pp. 2100–66.

——— (1959b), "Marine-Invertebrate Assemblages and their Relation to Nearshore Sedimentary Environments," Preprint vol., *Internat. Oceanog. Cong.,* New York, pp. 648–49.

———, AND CURRAY, JOSEPH R. (1956), "Fauna and Bathymetry of Banks on Continental Shelf, Northwest Gulf of Mexico," *Bull. Amer. Assoc. Petrol. Geol.,* Vol. 40, No. 10, pp. 2428–39.

PASSEGA, R. (1954), "Turbidity Currents and Petroleum Exploration," *ibid.,* Vol. 38, No. 9, pp. 1871–87.

PATRICK, W. W. (1953), "Salt Dome Statistics," in *AAPG-SEPM-SEG Guidebook, Joint Ann. Mtg., Houston,* 1953, pp. 13–20.

PEPPER, J. R.; DE WITT, WALLACE, JR.; AND DEMAREST, D. F. (1954), "Geology of the Bedford Shale and Berea Sandstone in the Appalachian Basin," *U. S. Geol. Survey Prof. Paper 259.*

PETERSEN, C. G. J. (1913), "Valuation of the Sea. II. The Animal Communities of the Sea Bottom and their Importance for Marine Zoogeography," *Repts. Danish Biol. Sta.,* Vol. 21, pp. 3–44.

——— (1915), "On the Animal Communities of the Sea Bottom in the Skagerrak, the Christianiafjord and Danish Waters," *ibid.,* Vol. 23, pp. 3–28.

PETTIJOHN, F. J. (1954), "Classification of Sandstones," *Jour. Geol.,* Vol. 62, No. 4, pp. 360–65.

——— (1957), *Sedimentary Rocks,* Harper & Bros., New York.

PHLEGER, FRED B (1951a), "Ecology of Foraminifera, Northwest Gulf of Mexico. Part I: Foraminifera Distribution," *Geol. Soc. America Mem. 46,* pp. 1–88.

——— (1951b), "Displaced Foraminifera Faunas," *in* Turbidity Currents and the Transportation of Coarse Sediments to Deep Water, *Soc. Econ. Paleon. Min. Spec. Pub. No. 2,* pp. 66–75.

——— (1954a), "Foraminifera and Deep-Sea Research," *Deep-Sea Research,* Vol. 2, pp. 1–23.

——— (1954b), "Ecology of Foraminifera and Associated Micro-Organisms from Mississippi Sound and Environs," *Bull. Amer. Assoc. Petrol. Geol.,* Vol. 38, No. 4, pp. 584–647.

——— (1955a), "Ecology of Foraminifera in Southeastern Mississippi Delta Area," *ibid.,* Vol. 39, No. 5, pp. 712–52.

——— (1955b), "Foraminiferal Faunas in Cores Offshore from the Mississippi Delta," *Deep-Sea Research, Papers on Marine Biol. and Oceanog.,* Supp. to Vol. 3, pp. 45–57.

——— (1956), "Significance of Living Foraminiferal Populations along the Central Texas Coast," *Contr. Cushman Found. Foram. Research,* Vol. 7, Pt. 4, pp. 106–51.

——— (1960) "Foraminiferal Populations in Laguna Madre, Texas," *Sci. Repts. Tohoku Univ.* 2d Ser. (Geology), *Spec. Vol. No. 4* (Hanzawa Memorial Volume), pp. 83–91.

———, AND LANKFORD, R. R. (1957), "Seasonal Occurrences of Living Benthonic Foraminifera in Some Texas Bays," *Contr. Cushman Found. Foram. Research,* Vol. 8, Pt. 3, pp. 93–105.

———, AND PARKER, F. L. (1951), "Ecology of Foraminifera, Northwest Gulf of Mexico. Part II: Foraminifera Species," *Geol. Soc. America Mem. 46,* pp. 1–64.

———; PARKER, F. L.; AND PEIRSON, J. F. (1953), "North Atlantic Core Foraminifera," *Repts. Swedish Deep-Sea Exped.,* Vol. 7, No. 1, pp. 1–122.

PINSAK, A. P. (1958), "A Regional, Chemical, and Mineralogical Study of Surficial Sediments in the Gulf of Mexico," Unpub. Ph.D. dissertation, Univ. of Indiana, Bloomington.

POOLE, D. M. (1958), "Heavy Mineral Variation in San Antonio and Mesquite Bays of the Central

Texas Coast," *Jour. Sed. Petrology,* Vol. 28, No. 1, pp. 65–74.

———; Butcher, W. S.; and Fisher, R. L. (1951), "The Use and Accuracy of the Emery Settling Tube for Sand Analysis," *Beach Erosion Board Tech. Mem. 23.*

Powers, M. C. (1953), "A New Roundness Scale for Sedimentary Particles," *Jour. Sed. Petrology,* Vol. 23, No. 2, pp. 117–19.

Price, W. A. (1933), "Role of Diastrophism in Topography of Corpus Christi Area, South Texas," *Bull. Amer. Assoc. Petrol. Geol.,* Vol. 17, No. 8, pp. 907–62.

——— (1934), "Lissie Formation and Beaumont Clay in South Texas," *ibid.,* Vol. 18, No. 7, pp. 948–59.

——— (1947), "Equilibrium of Form and Forces in Tidal Basins of Coast of Texas and Louisiana," *ibid.,* Vol. 31, No. 9, pp. 1619–63.

——— (1951), "Barrier Island, not 'Offshore Bar'," *Science,* Vol. 113, No. 2939, pp. 487–88.

——— (1952), "Reduction of Maintenance by Proper Orientation of Ship Channels Through Tidal Inlets," *Proc. 2d Conf. Coastal Eng.,* pp. 243–55.

——— (1954a), "Dynamic Environments: Reconnaissance Mapping, Geologic and Geomorphic, of Continental Shelf of Gulf of Mexico," *Trans. Gulf Coast Assoc. Geol. Soc.,* Vol. 4, pp. 75–107.

——— (1945b), "Environment and Formation of the Chenier Plain," *Texas Agr. and Mech. Coll. Research Found. Proj. 63,* Ref. 54–64T, p. 15.

——— (1954c), "Shorelines and Coasts of the Gulf of Mexico," *in* Gulf of Mexico, Its Origin, Waters, and Marine Life, *U. S. Dept. Interior, Fish and Wildlife Serv. Bull. 89,* pp. 39–62.

——— (1956), "Hurricanes Affecting the Coast of Texas from Galveston to Rio Grande," *Beach Erosion Board Tech. Mem. 78.*

——— (1958), "Sedimentology and Quaternary Geomorphology of South Texas," *Trans. Gulf Coast Assoc. Geol. Soc.,* Vol. 8, pp. 41–75.

———, and Gunter, G. (1942), "Certain Recent Geological and Biological Changes in South Texas with Consideration of Probable Causes," *Texas Acad. Sci. Proc.,* Vol. 26, pp. 138–56.

Pritchard, D. W. (1952), "Estuarine Hydrography," *Advances in Geophysics,* Vol. 1, pp. 243–80.

Puffer, E. L.; and Emerson, W. K. (1953), "The Molluscan Community of the Oyster-Reef Biotope on the Central Texas Coast," *Jour. Paleon.,* Vol. 27, No. 4, pp. 536–54.

Pulley, T. E. (1953), "A Zoogeographic Study Based on the Bivalves of the Gulf of Mexico," Unpub. Ph.D. dissertation, Harvard Univ., Cambridge, pp. 1–215.

Puri, H. S.; and Vernon, R. O. (1959), "Summary of the Geology of Florida and a Guidebook to the Classic Exposures," *Florida Geol. Survey Spec. Pub. 5.*

Razavet, C. D. (1956), "Contribution a l'Etude Geologique et Sedimentologique du Delta du Rhone," *Soc. Geol. France Mem. 76.*

Reedy, M. F., Jr. (1949), "Stratigraphy of Frio Formation, Orange and Jefferson Counties, Texas," *Bull. Amer. Assoc. Petrol. Geol.,* Vol. 33, No. 11, pp. 1830–58; also [Abstract], *ibid.,* No. 1, pp. 108–109.

Reid, G. K., Jr. (1955), "A Summer Study of the Biology and Ecology of East Bay, Texas, Part I," *Texas Jour. Sci.,* Vol. 7, No. 3, pp. 316–43.

Revelle, R.; and Fairbridge, R. W. (1957), "Carbonates and Carbon Dioxide," *in* Treatise on Marine Ecology and Paleoecology, *Geol. Soc. America Mem. 67,* Vol. 1, pp. 239–95.

Richards, H. G. (1939), "Marine Pleistocene of Texas," *Bull. Geol. Soc. America,* pp. 1885–98.

——— (1954), "Mollusks from the Mississippi Delta," *Notulae Naturae,* Acad. Sci. Philadelphia, No. 263.

Riley, G. A. (1937), "The Significance of the Mississippi River for Biological Conditions in the Northern Gulf of Mexico," *Jour. Marine Research,* Vol. 1, pp. 60–74.

———; Stommel, H.; and Bumpus, D. F. (1949), "Quantitative Ecology of the Plankton of the Western North Atlantic," *Bull. Bingham Oceanog. Colln.,* Vol. 12, pp. 1–169.

Riley, J. P. (1953), "The Spectrophotometric Determination of Ammonia in Natural Waters with Particular Reference to Sea Water," *Anal. Chem. Acta,* Vol. 9, pp. 575–89.

Rittenhouse, G. (1944), "Sources of Modern Sands in the Middle Rio Grande Valley, New Mexico," *Jour. Geol.,* Vol. 52, No. 3, pp. 145–83.

Ritter, H. P. (1896), "Report of a Reconnaissance of the Oyster Beds of Mobile Bay and Mississippi Sound, Alabama," *Bull. U. S. Fishery Comm. for 1895,* Vol. 15, No. 6, pp. 325–39.

Rodhe, W.; Vollenweider, R. A.; and Nauwerck, A. (1958), "The Primary Production and Standing Crop of Phytoplankton," in *Perspectives in Marine Biology,* A. Buzzatti-Traverso, Ed., Univ. of Calif. Press, pp. 299–322.

Rogers, J. J. W.; and Dawson, R. E., Jr. (1958), "Size Distribution of Zircon and Tourmaline Grains in Some Samples of the Lissie Formation," *Jour. Sed. Petrology,* Vol. 28, No. 3, pp. 361–65.

———, and Powell, W. F. (1958), "Size Distribution of Zircon Grains in Some Samples of Lower Beaumont Clay," *Jour. Sed Petrology,* Vol. 28, No. 1, pp. 36–39.

Rukhin, L. B. (1958), *Grundzüge der Lithologie,* Akademie Verlag, Berlin.

Rusnak, Gene A. (1957), "A Fabric and Petrologic Study of the Pleasantview Sandstone," *Jour. Sed. Petrology,* Vol. 27, No. 1, pp. 41–55.

Russell, R. D. (1937), "Mineral Composition of Mississippi River Sands," *Bull. Geol. Soc. America,* Vol. 48, No. 9, pp. 1307–48.

Russell, R. J. (1936), "Physiography of the Lower Mississippi River Delta," *in* Reports on the Geology of Plaquemines and St. Bernard Parishes, *Louisiana Dept. Conserv. Geol. Bull. 8,* pp. 3–199.

——— (1942), "Geomorphology of the Rhone Delta," *Assoc. Amer. Geog. Ann.,* Vol. 32, No. 1, pp. 149–254.

———, and Russell, R. D. (1939), "Mississippi River Delta Sedimentation," in *Recent Marine Sediments,* Amer. Assoc. Petrol. Geol., pp. 153–77; reprinted by Soc. Econ. Paleon. Min as *Spec. Pub. No. 4* (1955).

Rutsch, R. F. (1957), "Mollusk Report," Unpub. ms., Gulf Research and Devel. Corp., Houston.

Ryther, J. H. (1956a), "Photosynthesis in the Ocean as a Function of Light Intensity," *Limnol. and Oceanog.,* Vol. 1, pp. 61–70.

——— (1956b), "The Measurement of Primary Production," *ibid.,* Vol. 1, pp. 72–84.

Saunders, H. L. (1956), "Oceanography of Long

Island Sound, 1952–54. Part X: The Biology of Marine Bottom Communities," *Bull. Bingham Oceanog. Colln.*, Vol. 15, pp. 345–414.

SAWTELLE, GEORGE (1936), "Salt Dome Statistics," in *Gulf Coast Oil Fields,* Amer. Assoc. Petrol. Geol., pp. 109–18.

SCHOTT, W. (1935), "Die Foraminiferan in den Aquatorial Teil des Atlantischen Ozeans," *Deutsche Atlantische Exped.*, Vol. 11, No. 6, pp. 411–616.

SCRUTON, P. C. (1952), "Preliminary Report to the Project Committee," *A.P.I. Proj. 51, Rept. 7,* IMR Ref. 53–1.

——— (1955). "Sediments of the Eastern Mississippi Delta," *in* Finding Ancient Shorelines, *Soc. Econ. Paleon. Min. Spec. Pub. No. 3,* pp. 21–50 [Jan. 1956].

——— (1956), "Oceanography of Mississippi Delta Sedimentary Environments," *Bull. Amer. Assoc. Petrol. Geol.,* Vol. 40, No. 12, pp. 2864–2952.

——— (1957), "Oceanography of Mississippi Delta Sedimentary Environments: A Correction," *ibid.*, Vol. 41, No. 3, p. 566.

———, AND MOORE, D. G. (1953), "Distribution of Surface Turbidity off Mississippi Delta," *ibid.*, Vol. 37, No. 5, pp. 1067–74.

SELLARDS, E. H. (1940), "Pleistocene Artifacts and Associated Fossils from Bee County, Texas," *Bull. Geol. Soc. America,* Vol. 51, No. 11, pp. 1627–57.

———; ADKINS, W. S.; AND PLUMMER, F. B. (1933), "Stratigraphy," Vol. 1 *of* The Geology of Texas, *Univ. Texas Bull. 3232.*

———; BAKER, C. L.; ET AL. (1934), "Structural and Economic Geology," Vol. II, *ibid., Bull. 3401.*

———, AND HENDRICKS, LEO (1946), *Structural Map of Texas,* 3d Ed., Univ. Texas Bur. Econ. Geol.

SHAW, E. W. (1913), "The Mudlumps at the Mouths of the Mississippi," *U. S. Geol. Survey Prof. Paper 85-B,* pp. 11–27.

SHENTON, E. H. (1957), "A Study of the Foraminifera and Sediments of Matagorda Bay, Texas," *Trans. Gulf Coast Assoc. Geol. Soc.*, Vol. 7, pp. 135–50.

SHEPARD, FRANCIS P. (1937), "'Salt' Domes Related to Mississippi Submarine Trough," *Bull. Geol. Soc. America,* Vol. 48, No. 9, pp. 1349–62.

——— (1948), *Submarine Geology,* Harper & Bros. Pub., New York.

——— (1950), "Longshore-Bars and Longshore-Troughs," *Beach Erosion Board Tech. Mem. 15.*

——— (1952), "Revised Nomenclature for Depositional Coastal Features," *Bull. Amer. Assoc. Petrol. Geol.*, Vol. 36, No. 10, pp. 1902–12.

——— (1953), "Sedimentation Rates in Texas Estuaries and Lagoons," *ibid.*, Vol. 37, No. 8, pp. 1919–34.

——— (1954), "Nomenclature Based on Sand-Silt-Clay Ratios," *Jour. Sed. Petrology,* Vol. 24, No. 3, pp. 151–58.

——— (1955), "Delta-Front Valleys Bordering the Mississippi Distributaries," *Bull. Geol. Soc. America,* Vol. 66, No. 12, pp. 1489–98.

——— (1956a), "Marginal Sediments of Mississippi Delta," *Bull. Amer. Assoc. Petrol. Geol.,* Vol. 40, No. 11, pp. 2537–2623.

——— (1956b), "Late Pleistocene and Recent History of the Central Texas Coast," *Jour. Geol.,* Vol. 64, No. 11, pp. 56–69.

——— (1957), "Postglacial Sea Level," *A.P.I. Proj. 51, Rept. 25,* IMR Ref. 57–3, p. 5.

——— (1958), "Sedimentation of the Northwestern Gulf of Mexico," *Geol. Rundschau,* Vol. 47, No. 1, pp. 150–67.

———; EMERY, K. O.; AND GOULD, H. R. (1949), "Distribution of Sediments on East Asiatic Continental Shelf," *Allan Hancock Found. Pub., Occasional Paper No. 9,* pp. 1–64.

———, AND LANKFORD, R. R. (1959), "Sedimentary Facies from Shallow Borings in Lower Mississippi Delta," *Bull. Amer. Assoc. Petrol. Geol.,* Vol. 43, No. 9, pp. 2051–67.

———, AND MOORE, D. G. (1954), "Sedimentary Environments Differentiated by Coarse-Fraction Studies," *ibid.*, Vol. 38, No. 8, pp. 1792–1802.

———, AND ——— (1955), "Central Texas Coast Sedimentation: Characteristics of Sedimentary Environment, Recent History, and Diagenesis," *ibid.*, Vol. 39, No. 8, pp. 1463–1593.

———, AND RUSNAK, G. A. (1957), "Texas Bay Sediments," *Texas Inst. Marine Sci.,* Vol. 4, No. 2, pp. 5–13.

———, AND SUESS, H. E. (1956), "Rate of Postglacial Rise of Sea Level," *Science,* Vol. 123, No. 3207, pp. 1082–83.

———, AND WRATH, W. F. (1937), "Marine Sediments Around Catalina Island," *Jour. Sed. Petrology,* Vol. 7, No. 2, pp. 41–50.

———, AND YOUNG, R., "Distinguishing Between Beach and Dune Sands," (in press, *Jour. Sed. Petrology*).

SHREVEPORT GEOLOGICAL SOCIETY (1946–47; 1951; 1953), "Reference Reports on Certain Oil and Gas Fields of North Louisiana, South Arkansas, Mississippi, and Alabama," Vol. I (1946); Vol. II (1947); Vol. III, Pt. 1 (1951); Vol. III, Pt. 2 (1953).

SHUKRI, N. M. (1950), "The Mineralogy of Some Nile Sediments," *Quart. Jour. Geol. Soc. London,* Vol. 105, Pt. 4, pp. 511–34.

SIBUL, O. (1955), "Laboratory Study of Wind Tides in Shallow Water," *Beach Erosion Board Tech. Mem. 61,* pp. 1–50.

SIDWELL, R. G. (1941), "Sediments of Pecos River, New Mexico," *Jour. Sed. Petrology,* Vol. 11, No. 2, pp. 80–84.

——— (1947), "Trinity Sediments of North and Central Texas," *ibid.*, Vol. 17, No. 2, pp. 68–72.

———, AND COLE, C. A. (1937), "Sedimentation of Colorado River in Runnels and Coleman Counties, Texas," *ibid.*, Vol. 7, No. 3, pp. 104–07.

SIEGEL, S. (1956), *Nonparametric Statistics for the Behavioral Sciences,* McGraw-Hill Book Co., New York.

SIMMONS, E. G. (1957), "An Ecological Survey of the Upper Laguna Madre of Texas," *Pub. Inst. Marine Sci.*, Vol. 4, No. 2, pp. 156–200.

———, AND THOMAS, W. H., "Phytoplankton of the Mississippi Delta," Unpub. ms.

SMITH, F. G.; MEDINA, A. F.; AND ABELLA, A. F. (1951), "Distribution of Vertical Water Movement Calculated From Surface Drift Vectors," *Bull. Marine Sci. Gulf and Caribbean,* Vol. 1, pp. 187–95.

SMITH, P. V., JR. (1954), "Studies on Origin of Petroleum: Occurrence of Hydrocarbons in Recent Sediments," *Bull. Amer. Assoc. Petrol. Geol.*, Vol. 38, No. 3, pp. 377–404.

SOUTHEASTERN GEOLOGICAL SOCIETY (1949), "Mesozoic ·Cross Section, Southeast Alabama, South Georgia, and Florida," 4 charts.

SOUTH TEXAS GEOLOGICAL SOCIETY (1941), "Tertiary Cross Section, South Texas (Gillespie County to Aransas County), No. 17, Ordovician, Pennsylvanian, Cretaceous, Tertiary."

——— (1949), "Mesozoic Cross Section, South Texas (Robertson County to Maverick County), No. 20a, Jurassic, Cretaceous."

——— (1951a), "Nueces River Cross Section, South Texas (Maverick County to Kleberg County), No. 22, Cretaceous, Tertiary."

——— (1951b), "Downdip Mesozoic Rocks of South Texas," *Bull. Amer. Assoc. Petrol. Geol.*, Vol. 35, No. 2, pp. 357–60.

——— (1952), "Guadalupe River Cross Section, South Texas (Guadalupe County to Calhoun County), No. 24, Cretaceous, Tertiary."

SPOONER, W. C. (1935), "Oil and Gas Geology of the Gulf Costal Plain in Arkansas," *Arkansas Geol. Survey Bull. 2*, pp. 1–474.

SPRINGER, S., AND BULLIS, H. R. (1952), "Exploratory Shrimp Fishing in the Gulf of Mexico, 1950–51," *U. S. Dept. Interior, Fish and Wildlife Serv., Fishery Leaflet 406*, pp. 1–34.

———, AND ——— (1954), "Exploratory Shrimp Fishing in the Gulf of Mexico, Summary Report for 1952–54," *Commercial Fishery Review*, Vol. 16, No. 10, pp. 1–16.

———, AND ——— (1956), "Collections for the *Oregon* in the Gulf of Mexico," *U. S. Dept. Interior, Fish and Wildlife Serv. Spec. Sci. Rept. Fisheries, No. 196*, pp. 1–134.

STEEMAN-NIELSEN, E. (1952), "The Use of Radioactive Carbon (C^{14}) for Measuring Organic Production in the Sea," *Jour. Cons. Internat. Explor. Mer.*, Vol. 18, pp. 117–40.

——— (1954), "On Organic Production in the Oceans," *ibid.*, Vol. 19, pp. 309–28.

STEINBERG, D.; AND UDENFRIEND, S. (1957), "The Measurement of Radioisotopes," in *Methods of Enzymology*, Colowick, S. P., and Kaplan, N. O., Eds., Academic Press, New York, Vol. 4, pp. 425–72.

STENZEL, H. B. (1952a), "Boundary Problems," *Mississippi Geol. Soc. Guidebook, 9th Ann. Field Trip*, pp. 11–31.

——— (1952b), "Notes on Surface Correlation Chart," *ibid.*, pp. 32–33.

——— (1952c), "Transgression of the Jackson Group," *ibid.*, pp. 36–41.

STEPHENSON, L. W. (1928), "Major Marine Transgressions and Regressions, and Structural Features of Gulf Coastal Plain," *Amer. Jour. Sci.*, Ser. 5, Vol. 16, pp. 281–98.

STETSON, HENRY C. (1953), "The Sediments of the Western Gulf of Mexico: Part I—the Continental Terrace of the Western Gulf of Mexico; its Surface Sediments, Origin, and Development," *Papers in Phys. Oceanog. and Meteorol.*, Mass. Inst. Tech. & Woods Hole Oceanog. Inst., Vol. 12, No. 4, pp. 3–45.

STEVENSON, R. E.; AND EMERY, K. O. (1958), "Marshlands at Newport Bay, California," *Allan Hancock Found. Pub. 20*, Univ. Southern Calif. Press.

STEWART, H. B., JR. (1958), "Sedimentary Reflections of Depositional Environment in San Miguel Lagoon, Baja California, Mexico," *Bull. Amer. Assoc. Petrol. Geol.*, Vol. 42, No. 11, pp. 2567–2618.

STORM, L. W. (1945), "Résumé of Facts and Opinions on Sedimentation in Gulf Coast Region of Texas and Louisiana," *ibid.*, Vol. 29, No. 9, pp. 1304–35.

SUN, MING-SHAN (1954), "Heavy Minerals of the Jacksonian Sediments of Mississippi and Adjacent Areas," *Jour. Sed. Petrology*, Vol. 24, No. 3, pp. 200–06.

SUNDBORG, ÅKE (1956), "The River Klarälven—A Study of Fluvial Processes," *Geografiska Ann.*, Stockholm, Vol. 38, Nos. 2–3, pp. 127–316.

SUTTKUS, R. D.; DARNELL, R. M.; AND DARNELL, J. H. (1953–54), "Biological Study of Lake Ponchartrain," Progress Rept. to Louisiana Wildlife and Fish. Comm., Nos. 1–8, Tulane Univ., New Orleans, Zool. Dept. Mimeo. rept.

SVERDRUP, H. U.; JOHNSON, M. W.; AND FLEMING, R. H. (1942), *The Oceans: Their Physics, Chemistry, and General Biology*, Prentice-Hall, Inc., New York.

SWAIN, F. M., JR. (1944), "Stratigraphy of Cotton Valley Beds of Northern Gulf Coastal Plain," *Bull. Amer. Assoc. Petrol. Geol.*, Vol. 28, No. 5, pp. 577–614.

——— (1949), "Upper Jurassic of Northeastern Texas," *ibid.*, Vol. 33, No. 7, pp. 1206–50.

——— (1955), "Ostracoda of San Antonio Bay, Texas," *Jour. Paleon.*, Vol. 29, No. 4, pp. 561–646.

SWEITZER, N. B., JR. (1898), "Origin of the Gulf Stream and Circulation of Waters in the Gulf of Mexico, with Special Reference to the Effect on Jetty Construction," *Trans. Amer. Soc. Civil Eng.*, Vol. 40, pp. 86–98.

SYKES, G. (1937), "The Colorado Delta," *Carnegie Inst. Washington Pub. 460*.

TALLMAN, S. L. (1949), "Sandstone Types: Their Abundance and Cementing Agents," *Jour. Geol.*, Vol. 57, No. 6, pp. 582–91.

TERZAGHI, KARL (1956), "Varieties of Submarine Slope Failures," *Proc. 8th Texas Conf. on Soil Mech. and Found. Eng.*

TEXAS AGRICULTURAL AND MECHANICAL COLLEGE RESEARCH FOUNDATION (1956), "Hydrographic Studies in the Regions of Port Isabel and Port O'Connor, Texas," Proj. 140, Ref. 56–22F.

TEXAS BOARD OF WATER ENGINEERS AND U. S. SOIL CONSERVATION SERVICE (1952), "Silt Load of Texas Streams (1950–51)," *Progress Rept. 13, Div. of Irrig. and Water Conserv.*

THOMAS, W. A. (1953) [CHAIRMAN], "Geology of the Gulf Coast," Guidebook, Joint Ann. Mtg., AAPG-SEPM-SEG, Houston, 1953; summarized in *Oil and Gas Jour.*, Vol. 52, No. 7 (June 22), 1953, pp. 249–51; 274.

THOMAS, W. H. (1958), "Comment," in *Perspectives in Marine Biology*, Buzzati-Traverso, A., Ed., Univ. of Calif. Press, p. 322.

THOMPSON, W. C. (1955), "Sandless Coastal Terrain of the Atchafalaya Bay Area, Louisiana," *in* Finding Ancient Shorelines, *Soc. Econ. Paleon. Min. Spec. Pub. No. 3*, pp. 52–77 [Jan. 1956].

THORNTHWAITE, C. W. (1948), "An Approach Toward a Rational Classification of Climate," *Geog. Rev.*, Vol. 38, No. 1, pp. 55–94.

THORSON, G. (1957), "Bottom Communities (Sublittoral or Shallow Shelf)," *in* Treatise on Marine Ecology and Paleontology, *Geol. Soc. America Mem. 67*, Vol. 1, pp. 461–534.

TOULMIN, L. D. (1952), "Sedimentary Volumes in Gulf Coastal Plain of the United States and Mexico: Part II, Volume of Cenezoic Sediments in Florida and Georgia," *ibid.*, *Bull.*, Vol. 63, No. 12, pp. 1165–76.

——— (1955), "Cenezoic Geology of Southeastern Alabama, Florida, and Georgia," *Bull. Amer. Assoc. Petrol. Geol.*, Vol. 39, No. 2, pp. 207–35.

TRASK, P. D. (1932), *Origin and Environment of Source Sediments of Petroleum*, Gulf Pub. Co., Houston.

TREADWELL, R. C. (1955), "Sedimentology and Ecology of Southeast Coastal Louisiana," *Louisiana State Univ.*, Baton Rouge, *Tech. Rept. 6*, Proj. No. N7, ONR, pp. 1–78.

TROWBRIDGE, A. C. (1923), "A Geological Reconnaissance in the Gulf Coastal Plain of Texas Near the Rio Grande," *U. S. Geol. Survey Prof. Paper 131*, pp. 85–107.

——— (1930), "Building of Mississippi Delta," *Bull. Amer. Assoc. Petrol. Geol.*, Vol. 14, No. 7, pp. 867–901.

——— (1932), "Tertiary and Quaternary Geology of the Lower Rio Grande Region, Texas," *U. S. Geol. Survey Bull. 837*.

——— (1954), "Mississippi River and Gulf Coast Terraces and Sediments as Related to Pleistocene History—A Problem," *Bull. Geol. Soc. America*, Vol. 62, No. 8, pp. 793–812.

U. S. ARMY, CORPS OF ENGINEERS, MISSISSPPI RIVER COMMISSION (1958), "Geology of the Mississippi River Deltaic Plain, Southeastern Louisiana," *Tech. Rept. 3-483*, Vicksburg.

U. S. BEACH EROSION BOARD (1954), "Shore Protection, Planning, and Design," *U. S. Corps of Engineers Tech. Rept. 4*.

U. S. COAST AND GEODETIC SURVEY (1949), *United States Coast Pilot Gulf Coast, Key West to Rio Grande*, 3d Ed., U. S. Dept. of Commerce, Ser. No. 725.

U. S. DEPARTMENT OF AGRICULTURE (1938), "Atlas of Climatic Charts of the Oceans," Washington.

VAN ANDEL, TJ. H. (1950), "Provenance, Transport, and Deposition of Rhine Sediments," Ph.D. dissertation, Wageningen, Utrecht, Meded. Landbouwh. te Netherlands.

——— (1955), "Sediments of the Rhone Delta; II, Sources and Deposition of Heavy Minerals," *Geol. Mijnbouwk. Gen. Neder. Verh.*, Vol. 15, Pt. 3, pp. 516–56.

——— (1958), "Origin and Classification of Cretaceous, Paleocene, and Eocene Sandstones of Western Venezuela," *Bull. Amer. Assoc. Petrol. Geol.*, Vol. 42, No. 4, pp. 734–63.

———, AND POOLE, D. H. (1960), "Sources of Holocene Sediments in the Northern Gulf of Mexico," *Jour. Sed. Petrology*, Vol. 30, No. 1, pp. 91–122.

———, POSTMA, H., ET AL. (1954), "Recent Sediments of the Gulf of Paria," *in* Reports of the Orinoco Shelf Expedition, Vol. I, *Kon. Nederl. Akad. Wetensch. Verh.*, Vol. 20, No. 5.

VAN DORN, W. G. (1956), "Large-Volume Water Samplers," *Trans. Amer. Geophys. Union*, Vol. 37, pp. 682–84.

VAN LOPIK, J. R. (1955), "Recent Geology and Geomorphic History of Central Coastal Louisiana," *Louisiana State Univ.*, Baton Rouge, *Tech. Rept. 7*, pp. 1–88.

VANN, J. H. (1959), "The Geomorphology of the Guayana Coast," *Proc. 2d Coastal Geog. Conf.*, Louisiana State Univ., Baton Rouge, pp. 153–88.

VAN STRAATEN, L. M. J. U. (1954), "Radiocarbon Dating and Changes of Sea Level at Velzen (Netherlands)," *Geologie en Mijnbouw*, Nw. Ser., Vol. 16, pp. 247–53.

——— (1959a), "Littoral and Submarine Morphology of the Rhone Delta," *Proc. 2d Coastal Geog. Conf.* Louisiana State Univ., Baton Rouge, pp. 233–64.

——— (1959b), "Minor Structures of Some Recent Littoral and Neritic Sediments," *Geologie en Mijnbouw*, No. 21e, Nw. Ser., pp. 197-216.

VERNON, R. O. (1952), "The Cenozoic Rocks of the Northern Peninsula and the Panhandle of Florida," *Assoc. Amer. State Geol. Guidebook 44th Ann. Mtg.*, Florida Geol. Survey, pp. 46–61.

WALLACE, W. E., JR. (1957), "Fault Map of South Louisiana," *Trans. Gulf Coast Assoc. Geol. Soc.*, Vol. 7, p. 240.

WALTON, W. R. (1952), "Techniques for Recognition of Living Foraminifera," *Contr. Cushman Found. Foram. Research*, Vol. 3, pp. 56–60.

WARREN, A. D. (1956), "Ecology of Foraminifera of the Buras-Scofield Bayou Region, Southeast Louisiana," *Trans. Gulf Coast Assoc. Geol. Soc.*, Vol. 6, pp. 131–52.

——— (1957), "Foraminifera of the Buras-Scofield Bayou Region, Southeast Louisiana," *Contr. Cushman Found. Foram. Research*, Vol. 8, pp. 29–40.

WATERS, J. A.; MCFARLAND, P. W.; AND LEA, J. W. (1955), "Geologic Framework of Gulf Coastal Plain of Texas," *Bull. Amer. Assoc. Petrol. Geol.*, Vol. 39, No. 9, pp. 1821–50.

WEAVER, C. E. (1958), "Geologic Interpretation of Argillaceous Sediments; Part 1, Origin and Significance of Clay Minerals in Sedimentary Rocks," *ibid.*, Vol. 42, No. 2, pp. 254–71.

WEAVER, PAUL (1950), "Variations in History of Continental Shelves," *ibid.*, Vol. 34, No. 3, pp. 351–60; [Discussion], *ibid.*, No. 7, pp. 1589–92.

——— (1951), "Continental Shelf of Gulf of Mexico," *ibid.*, Vol. 35, No. 2, pp. 393–98.

——— (1955), "Gulf of Mexico," *in* Crust of the Earth, *Geol. Soc. America Spec. Paper 62*, pp. 269–78.

WEEKS, A. W. (1945a), "Balcones, Luling, and Mexia Fault Zones in Texas," *Bull. Amer. Assoc. Petrol. Geol.*, Vol. 29, No. 12, pp. 1733 37.

——— (1945b), "Quaternary Deposits of Texas Coastal Plain Between Brazos River and Rio Grande," *ibid.*, Vol. 29, No. 12, pp. 1693–1720.

WELDER, F. A. (1959), "Processes of Deltaic Sedimentation in the Lower Mississippi River," *Coastal Studies Inst., Louisiana State Univ.*, Baton Rouge, *Tech. Rept. 12*.

WENTWORTH, C. K. (1922), "A Scale of Grade and Class Terms for Clastic Sediments," *Jour Geol.*, Vol. 30, No. 5, pp. 377–92.

WHITE, W. A. (1950), "Blue Ridge Front—A Fault Scarp," *Bull. Geol. Soc. America*, Vol. 61, No. 12, pp. 1309–46.

WHITEHOUSE, U. G. (1953), "Chemistry of Sedimentation," *Amer. Petr. Inst. Proj. 51, Rept. 9*, Univ. of Calif., Scripps Inst. Oceanog., IMR ref. 53–43, pp. 31–38.

WHITTEN, H. L.; ROSENE, H. F.; AND HEDGPETH, J. W. (1950), "The Invertebrate Fauna of Texas Coast Jetties: A Preliminary Survey," *Pub. Inst. Marine Sci.*, Vol. 1, No. 2, pp. 53–87.

WILLIAMSON, J. D. M. (1959), "Gulf Coast Cenozoic History," *Trans. Gulf Coast Assoc. Geol. Soc.*, Vol. 9, pp. 15–29.

WILSON, BASIL W. (1957), "Hurricane Wave Statistics for the Gulf of Mexico," *Beach Erosion Board Tech. Mem. 98*, U. S. Corps of Engineers.

WOODS, R. D. (1956), "The Northern Structural Rim of the Gulf Basin," *Trans. Gulf Coast Assoc. Geol. Soc.*, Vol. 6, pp. 3–12.

WOOLLARD, GEORGE P. (1956), "Pre-Cretaceous (Basement) Map of Atlantic Coastal Region," Unpub.

WOOSTER, W. S.; AND RAKESTRAW, N. W. (1951), "The Estimation of Dissolved Phosphate in Sea Water," *Jour. Marine Research*, Vol. 10, pp. 91–100.

ZENKOVITCH, V. P. (1959), "On the Genesis of Cuspate Spits Along Lagoon Shores," *Jour. Geol.*, Vol. 67, pp. 269–77.

ZOBELL, C. E. (1946), "Studies on Redox Potential of Marine Sediments," *Bull. Amer. Assoc. Petrol. Geol.* Vol. 30, No. 4, pp. 477–513.

ZONNEVELD, J. I. S. (1953), "Waarnemingen Langs de Kust van Suriname," *Tijdschrift Kon. Neder. Aardrij. Genootschap*, Vol. 71, pp. 18–31.

Editor's Comments on Paper 8

8 **Zenkovitch:** *Some New Exploration Results About Sand Shores Development During the Sea Transgression*

In this, the first of two papers by V. P. Zenkovich included in this volume, we have an investigation of barrier formation along the shores of the Black Sea. Basic to the study are Zenkovich's observations on the transport of sediment normal to the shore, somewhat along the lines described by R. L. Miller and J. M. Zeigler (*Journal of Geology*, **66**, 417–441, 1958). Transversal debris drifting, as Zenkovich calls it, moves coarse sediment shoreward on the nearshore bottom.

The development of barriers and lagoons require two basic conditions: a gentle bottom slope of 0.001 to 0.005, and an abundant sediment supply. Transversal debris drifting under these conditions generates shore ridges in shallow water or wave-built terraces at the edge of the land. With the partial submergence of these features accompanying a transgression of the sea, lagoons are formed in the low area behind the barriers. As would be expected, lagoons developed in this manner behind wave-built terraces would exhibit lagoonal muds over a terrestrial surface. Zenkovich's group, utilizing a vibro-piston corer, found stratigraphic evidence for all the conditions outlined above. In a later paper by Zenkovich, we shall find further details of his continuing investigation into the origin of barrier islands.

Vsevolod Pavlovich Zenkovich, who has authored several books and more than 200 papers on coastal processes, is presently Head of the Seashore Laboratory at the Institute of Geography. He was born in Moscow in 1910 and studied at the Geological Institute there. In 1943 Zenkovich received a doctorate, and he was named Professor in 1947. In 1944 he was elected a member of the Coastal Geomorphology Commission of the International Geographical Union. From 1944 to 1970 Zenkovich was Head of the Coastal Dynamics Laboratory of the Institute of Oceanology, Academy of Science, USSR which, in 1971, was transferred to the Institute of Geography in Moscow.

Zenkovich has investigated the shores of all the seas surrounding his native country, as well as the coasts of Poland, China, Vietnam, Cuba, Mexico, Egypt, and Yugoslavia. He is a member of the Oceanographic Commission, Academy of Science, USSR, and is on the editorial board of *Oceanology* and *Geomorphology* magazines.

Reprinted from *De Ingenieur*, **9**, 113–121 (1962)

De Ingenieur

Bouw- en Waterbouwkunde 9

8

Some new exploration results about sand shores development during the sea transgression [1])

627.52

by prof. dr. V. P. Zenkovitch. Institute of Oceanology of the Academy of Sciences of the U.S.S.R.

Summary: By means of thorough analysis of the findings in the near-shore zones of the Black Sea author gives a general picture of the geological history of different types of morphological features.

After a survey of the practical importance of similar researches and a brief description of the dynamical processes in the near shore zone author treats the various observed deposits in considerable detail and draws conclusions of general validity.

Introduction

It is well known, that the overwhelming majority of useful minerals of sedimentary origin were deposited in the nearshore zone of seas, that is in the very upper part of a shelf and in the lagoons. Coals and combustible shales, manganese, aluminium and iron ores, different salts, phosphorites, siliceous rocks and purer types of lime-stores – all were formed there. The sediments in which petroleum could be formed, spread wider, but its productive accumulations are concentrated in sand bodies, the majority of which have the nearshore genesis too.

Therefore it is quite natural that there arouse among geologists the high interest to study the structure and composition of recent nearshore sediments. Unfortunately, as there are some technical difficulties in geological researches of the near-shore sea zone till now, nearly all these questions were studied by means of the fossil materials. Only during last years some regularities of forming of the recent near-shore strata became clear.

In some parts of the U.S.S.R. sea shores it was quite possible to state the fact that during the sea transgression over the low-lying plains consisting of loose sediments the submarine strata of rather compound structure are formed. For the understanding of the structure it was essential to get an idea about the process of debris-shifting in two opposite directions in the normal to the shore, the 'transversal drift'. The study of the forms of the submerging shores was also useful for this purpose.

[1]) Laatste van de drie voordrachten door prof. Zenkovitch gehouden voor de Afdeling voor Bouw- en Waterbouwkunde van het K.I.v.I. en het Kon. Geol. Mijnbouwkundig Genootschap op 20 en 22 juni 1961, resp. te Scheveningen en te Utrecht. Zie *De Ingenieur* 1961 No. 23, blz. A 341 en 1962 No. 13, blz. B 81 en No. 15, blz. B 95.

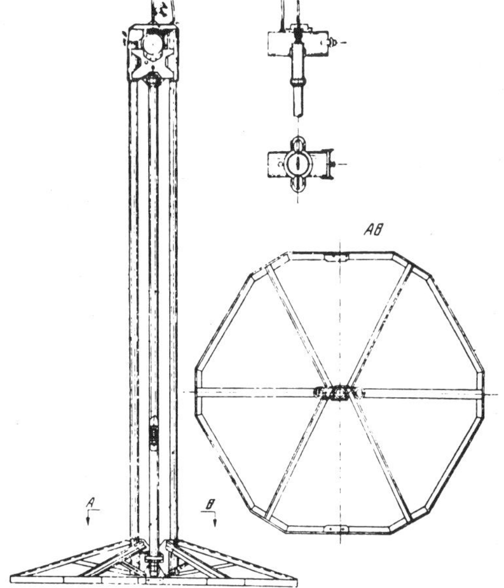

Fig. 1. The construction scheme of the vibro-piston core sampler.
The section AB is shown.

Fig. 2. The vibro-piston core sampler is being drawn from the ship.

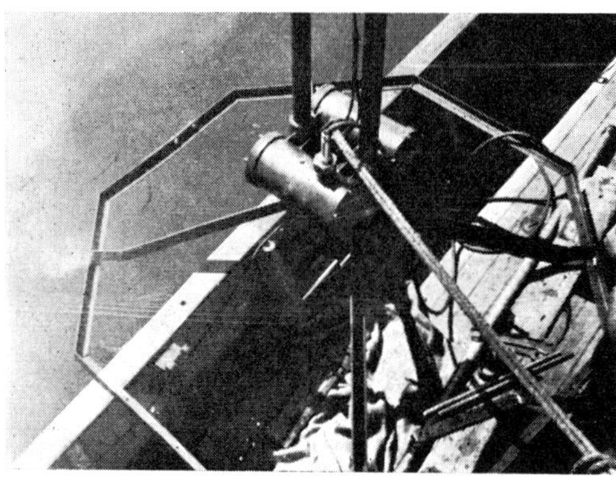

Fig. 3. The basement frame of the sampler with its vibrators lowered down.
The travelling tube is laid out on the deck.

The majority of sea shores of the globe have been submerged during the late-glacial transgression and the level of the ocean became relatively stable 5000 years ago. We knew about the structure of the peripheral part of this transgressive strata only after a few borings made on the land and even to a smaller degree on the sea aquatorium. Bottom borings are usually carried out in those parts where it is important for practical purposes e.g. in connection with the searches of oil deposits or in the course of seaport prospections. As to the structure of transgressive strata, such material is only incidental at hand. Therefore we attach great importance to special sea researches carried out by means of vibro-piston core sampler, when explorers could have chosen the most typical shore sections for their work.

The vibro-piston core sampler

The set-up of a vibro-piston corer is rather simple (fig. 1). Its mechanical basement consists of a frame which looks like a metal polygon 3 m diameter, over which two 'directing' tubes 5 m long rise vertically. On the sliding muffs between them a travelling tube is mounted in which core enteres. On the top of the tube the vibrator is fixed. It is supplied with electricity through a cable from a ship. At the lower opening of a travelling tube there is a piston connected to the top of the 'directing' tubes by the wire rope. This keeps its fixed position relatively to the sea bottom surface. The lower opening of the travelling tube is equipped with a valve which is automatically sealed when the instrument's lifting begins. This instrument can be used from a little ship at a depth till 100 m, as well as at land surface using a lorry for this aim (fig. 2, 3 and 4). In very shallow aquatorium near the shore, work can be carried out from a raft.

After the instrument sunk to the bottom, the electricity is switched on and it gives the vibrator very frequent oscillations. So the travelling tube is plunging into the bottom at full length under its own weight. Usually the penetration into sediment lasts not more than one minute. Thanks to the piston, the tube is filled with core at a little vacuum which secures the receiving of long monolith without any changes of its structure. The instrument can be used successfully even on such coarse sediments as shingle and shell beds.

Fig. 4. The use of the vibro-piston core sampler on the surface of the land.

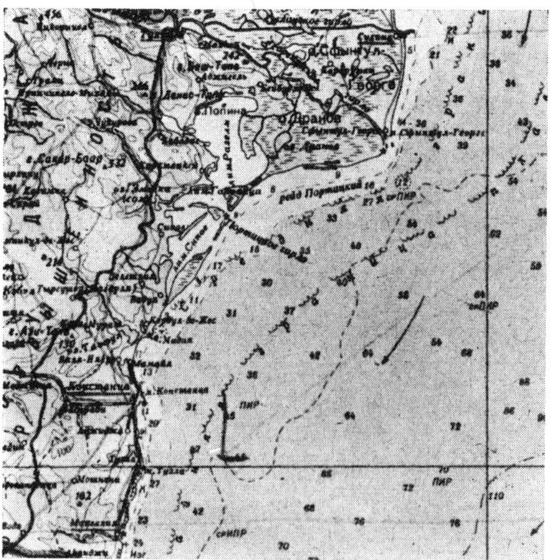

Fig. 5. The North-West part of the Black Sea.

Processes of the near-shore zone

Before dealing with the material received, it is necessary to dwell on some typical processes of the near-shore zone development with a slightly sloping bottom ($< 0{,}01$) because all the sand shores in particular belong to them. In Russian they are called 'otmely' shores. The debris shifting up and down this slope is of great importance in these processes. Even one of the first explorers E. De Beaumont enunciated the hypothesis that on the flat bottom waves cast shoreward the alluvia occurring on it and build up, above the water, the stretched strips of debris called offshore bars or barrier islands. The lagoons appear behind these barriers.

Although this hypothesis is accepted by the majority of the scientists, up till now no reliable quantative data were obtained concerning this process and there was no certainty that it really plays a significant role in the dynamics of the shore. As this question is very important we tried to study it in different ways.

In the previous article some results have been given about the researches of the influence of the waves asymmetries in the near-shore zone and their influence on the bottom changes. Conclusions from this research are based on natural observations made by means of horizontal wave-pressure recorders. The analysis of records gave the opportunity to calculate and summarize impulses of the transversal debris drifting grain sizes for a certain point, or for a certain part of profile, where several instruments were set simultaneously.

As a result, we have got a series of typical curves. Near the shores with a slightly sloping submarine part, for example the sand shores, such impulses for coarse particles are almost always positive, that is they are directed towards the shore. As to fine sand particles, the corresponding impulse can have different directions at a given place depending on the phase of hydrodynamic regime.

Then the comparison of these curves with a real sand drifting was made. For this purpose the artificial mixtures of sand were used, which have been dyed by luminiscent substances. Compounds of one and the same colour (yellow, red, blue) were dumped on different bottom points of the same profile and then the direction and velocity of their displacement were traced. Simultaneous records of near-bottom wave pressures were made on the basis of which the corresponding impulses were calculated. The comparison of the data received have convinced us that the transversal debris-drifting takes place in accordance with the character of wave movements and accompanying piling up or outflow of near-bottom waters.

The velocities measured of the fine sand transversal drifting have the order up to 25 cm/min (in some cases up to 50 cm/min). For the coarse sand grains these figures are much lower. Therefore, during some fluctuations of the hydrodynamical regime impulses for the fine sand drift may change their direction, but the coarse particles keep their movement shore-wards stable. So the process

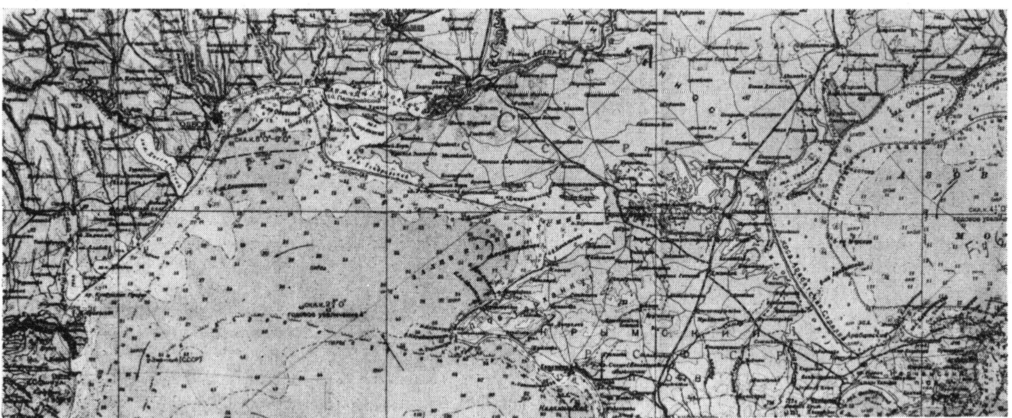

Fig. 6. The Azov Sea, Kertch Strait and Anapa region.

Fig. 7. The North-East shore of the Black Sea. 5-7 from the U.S.S.R. Marine Sea Atlas.

of transversal drifting itself was traced, although the data received do not allow yet to state the quantative dependencies.

Observations in nature

We succeeded in studying the other aspects of the process with the regional material analysis. The first question was how powerful this process was and whether the large shore structures could be formed as a result of it (fig. 5, 6 and 7).

On the Black Sea, the Asov Sea and the Caspian Sea we know vast areas in which there is no sand supply from the land and the shore cliffs consist of clay and loesses only. In such places all the shore-debris are composed of shell fragments, real thalassogenetic material. It is casted to the shore and accumulated in such large quantities, that it forms rather long (tens of km) and voluminous embankments of different types.

The Arabat barrier-iceland is the largest of them (120 km long, 200 m wide and up to 6 m thick) containing more than 1 cubic km of debris. At present this embankment continues to be active and its formation is connected with the present sea level.

Biologists discovered that the average annual shell productivity (Cardium edule) is about 400 tons per one square kilometer in many near-shore regions of the Asov Sea. If we consider: 1) that the weight of a shell is 40 percents of living mollusk weight; 2) that in the shell-debris the porosity is 30 procent; 3) that the shell is cast to the shore not more than from 5 meters depth and correspondingly from the bottom belt only of five km wide, we have the result that during one year the shore ridge of 0.6 square meter section could be formed. Such calculations have been corroborated by the real data of some shell spits growing, for instance that of Atchuev.

It may be proved in some cases that the bottom material has a terrestrial origin as well. This can be made clear if the heavy minerals of the sand bar essentially differ from that of the continental ones at the region considered, and if one looks at the primary sources of its genesis at the bottom. As an example, I may point out that the large underwater bar in the north-west part of the Black Sea is composed of the sand which includes some types of hornblendes characteristic to the Danube alluvion. The details will be reported later.

The second problem, which we tried to solve is to define the depth from which rather coarse material could be cast to the shore. In the region between the Tarchankut peninsula and the Evpatorian bay we succeeded to answer the question from what dephts the shells and their fragments can be thrown out at present time. The beaches of the region to a large extent consist of shells. The near-shore sand deposits are descending to 2-3 depth only. Lower, to a depth of 20 m, there is a bench of tertiary limestone, and below the depth mentioned muddy deposits begin. On the bench surface the poor faunistic complex of molluscs is developed (Mytilus, Cerithium etc.). The richer complex exists on the muddy bottom (Pecten, Mactra, Tapes etc.). It was found out that the beach sediments include the forms not only of rock and sand biocoenosises but in large amount a muddy one as well. It means that the 20 m depth is not a lower limit for casting up of the material even in a comparatively small basin as this part of the Black Sea is where the hight of waves rarely surpasses 3 m and the length of 80 m. The belt of the bottom which is crossed by shells is about 7 km wide.

Developments of barriers and lagoons

The process of the barrier development is going on under certain favourable conditions only. The most significant of them is a bottom slope being within the limits of 0.001-0.005. The reason of this are some regularities of wave energy expenditure when passing over the shoal zones. The specific energy of the waves remains stable

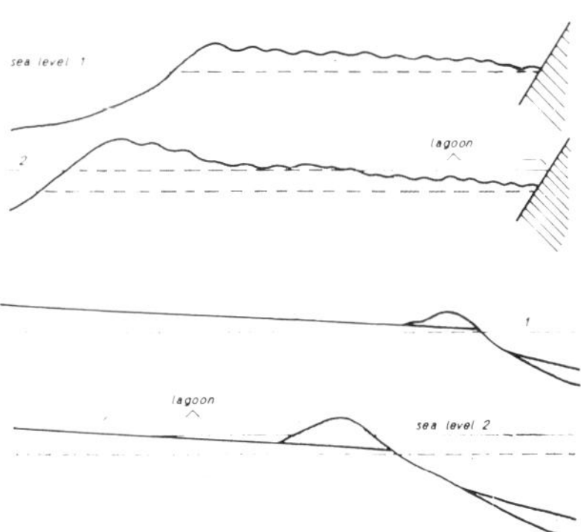

Fig. 8. Two particular cases of the offshore bars formation. Upper-as a result of the wave-built terrace sinking. Below-during the submersion of an alluvial plain.

nearly over the whole length of its run if the bottom slope is about 0.01. If the waves cause the shoreward drift of debris from a relatively large depths, this would go on up to the beach in the same quantities. With a larger bottom slope this process even intensifies as the dephts decrease. But if the bottom slope is less than 0.005, there exists a sufficient loss of waves energy while their going over the shoal. So, some part of the loose materials is inevitably settling on the bottom, thus giving rise to an underwater bar and stimulating it to grow up. With bottom slopes less than 0.001, the energy exhausts so slowly that waves are not able to move the sand shorewards at all.

The second condition is the presence of a thick layer of loose sediments. If the debris are not in sufficient quantities and they cover the bench surface with a comparatively thin layer, the whole material is thrown right to the shore. So instead of a barrier island in such a case the above-water wave-built terrace is formed, its surface being crossed with a large series of shore ridges. The terrace of a given structure is also formed in those cases when the loose bottom has a slope about 0.005-0.01. This pecularity is also connected with wave deformation processes which I have told about in some more details in my previous article.

Then, it is known that many off-shore bars can move themselves landwards by all their bodies and they can crawl over the lagoon deposits (muds, peat) accumulated behind them. Such formations were repeatedly described in the literature and we know a lot of such examples on the U.S.S.R. shores too. However, more detailed investigations proved that barrier island's shifting chiefly occurs in the regions of the recent submergence. If the sea level is stable, the bars having gone up to a shoal zone, usually enlarge in width and come to the state of equilibrium. Their position becomes fixed.

The lagoons are widely spread along the submerging boarders of lower plains. However, many of them have entirely different genesis from that was supposed according to the theory of E. De Beaumont supported by Johnson.

If the front edge of the low plain is composed of loose sediments, it often undergoes the washing away, notwithstanding some other dynamical processes of the second order. The fine material is carried away to a deep sea, but coarse components are accumulated in form of a large shore ridge. As further submergence and washing over of the new and new sediments is going on, such a ridge is shifted landwards. However as its volume increases the process becomes slower. Meanwhile the prolonging submergence may become the cause of the fact that the surface of the plain will get lower than the sea level and will form a lagoon in the place which have never been a part of the sea aquatorium (fig. 8). There occurs the common processes of the lagoon mud accumulation of the terrestrial run-off origin or the material being carried in from the sea with the tides.

The borings on some lagoon bottoms of the Black Sea and Tchukot Sea showed that there is no recent marine sediments at all under the cover of lagoonal muds. Immediately beneath them there are diluvial accumulations (the first case) or gravels of piedmont terrace, consolidated with permafrost.

The lagoons may arise as a result of the wave-built

terrace sinking too. This type differs in some relations from one described above. If the shore prograding is in progress during the submergence, the terrace surface acquires a peculiar profile declined landwards. During further submergence the back part of such a terrace becomes lower than the sea level and also turns into a lagoon. Such forms were observed on the shores of the Camchatka peninsula.

In the first case described the analogues of barrier islands are formed as well although they consist not of material thrown up from the sea bottom but of that having its origin from the submerging land. These formations being deliberately of terrestrial origin, as it was proved, can be concealed during further transgressive run and be buried beneath the bottom strata between purly sea sediments. I shall give you an example of these below.

The last feature of lagoon shores which is worth to mention is following. According to the classical scheme of barrier islands formation the predecessors of bars were to be some kind of voluminous submarine ridges or banks. Only after a certain period of their growing up and shifting towards the shore, the top of such bank rises over the sea level and turns into an off-shore bar. However, such formations have never been described in literature and on this account there were the deductive suppositions only.

The suggested considerations and examples proved that we are to expect sufficiently compound and various complex of sediments in the transgressive series. So, beginning our works at the Black Sea with a vibro-piston corer, we have chosen the regions with different sedimentational environments and recent vertical movements. In particular the regions studied were characterized by a very poor run off from the land and by a slow sedimentation process correspondingly. The subwater banks of the western part of the sea marked on the bathimetric maps were of special interest to us. Supposedly they were to be considered as the formations resulted from the transversal drift of debris according to the scheme mentioned before.

The sediments investigation in the region chosen gave very interesting results. Subsequently we proceeded with our studying on other seas (the Japanese Sea, the Baltic Sea etc.) and displayed a lot of common traits in transgressive strata composition. In all the cases studied the relics of ancient shore structures were revealed on the open sea bottom. Some of them were manifested in the relief as subaqual banks, but sometimes they were absolutely leveled under the cover of later muddy sediments. Between them there can be met the lagoonal muds or the open sea muddy sediments. Thus, on the whole, the transgressive series obtain the stepped structure.

The brief sketch of the geological history of the Black Sea for the last milleniums should precede the description of this material. The analysis of the sea history was given even between 1920 and 1940 by Arkchangelsky and Strakchov as a result of studying the open sea cores covering almost 4 m.

In the so-called New-Euxinus period corresponding to the last glaciation the sea level was about 40 m lower than the present one. The sea turned then into brackish water lake with rather specific fauna. When the late glacial transgression took place and when the ocean level rose over the Bosphor threshold the water became more salty and the impoverished Mediterranean fauna penetrated the basin.

Then came the ancient Black Sea period which was replaced by the recent one about 5000 years ago. From that time one cannot trace any significant changes in the water salinity or the molluscs population, although the sea level apparently continued to rise slowly. Today in the regions of tectonic depressions the relative shore submergence occurs with the velocity from 30 cm (Odessa) to 50 cm (Kolkhida Lowland) in a century. In the stable and rising tectonic regions level changes are not marked/ Anapa/during the last decades.

Cores of the north-west shallow part of the sea revealed that under the fine bed of recent shell and sand sediments there are fluviatil and moor formations. Therefore, on this part of the sea existed the old alluvial plain created by the Dneper, the Buge and the Dnester rivers discharge, connected in the South with that of the Danube river.

Sedimentological features around the Black Sea

The analysis of near-shore transgressive strata will follow the series sections described in a number of E. N. Nevessky's articles. The sections were taken in the open part of the sea and in two large bays in the North-west and in the Anapa region as well. All of them have much in common.

The larger form of the submarine relief is a great Odessa bank (fig. 9). This submarine ridge is about 60 km long, and has about 15 m relative excess above the bottom hollow, lying behind it shorewards. The bank profile resembles that of a dune, having a steep back edge. Its crest lies in the depth of 5-7 m. A very similar form but smaller in dimensions has one of the submarine ridges of the Karkinitis gulf, which we shall name the Northern one (see A_a, fig. 9).

In both cases the crest of the bank consists of the pure shell valves, which form a steep inner scarp of the bank as well. This layer is more than 5 m thick. The front bank declivity having the slope about 0.001 is covered with the sandy-shell material which is buried under the muddy sediments as the depth increases. The lithological studying of the surface layer showed that the selection of the material takes place here at present. In its upward drifting the relatively coarse material (shells) overruns the finer one (sand) and having reached the crest it falls down the bank's back side. Cores at the foot of the declined shell layer have passed through its train and cut into the mud.

Studying the large bank material showed that under the recent sand-shell layer the sands are deposited, which do not include any fossil molluscs and are characterized by rather high sorting. The lens of fluvial alluvia were found among them, and down the slope these strata contact with the typical fluviatil material. All the above mentioned facts permits us to consider the main body of the bank to be an ancient shore dune or a large off-shore bar, the material of which has undergone an eolian reworking. As the sea level rose the whole form turns into submerged one. The transgressive sea did not destroy essentially the constitution of this relic shore form. Nowa-

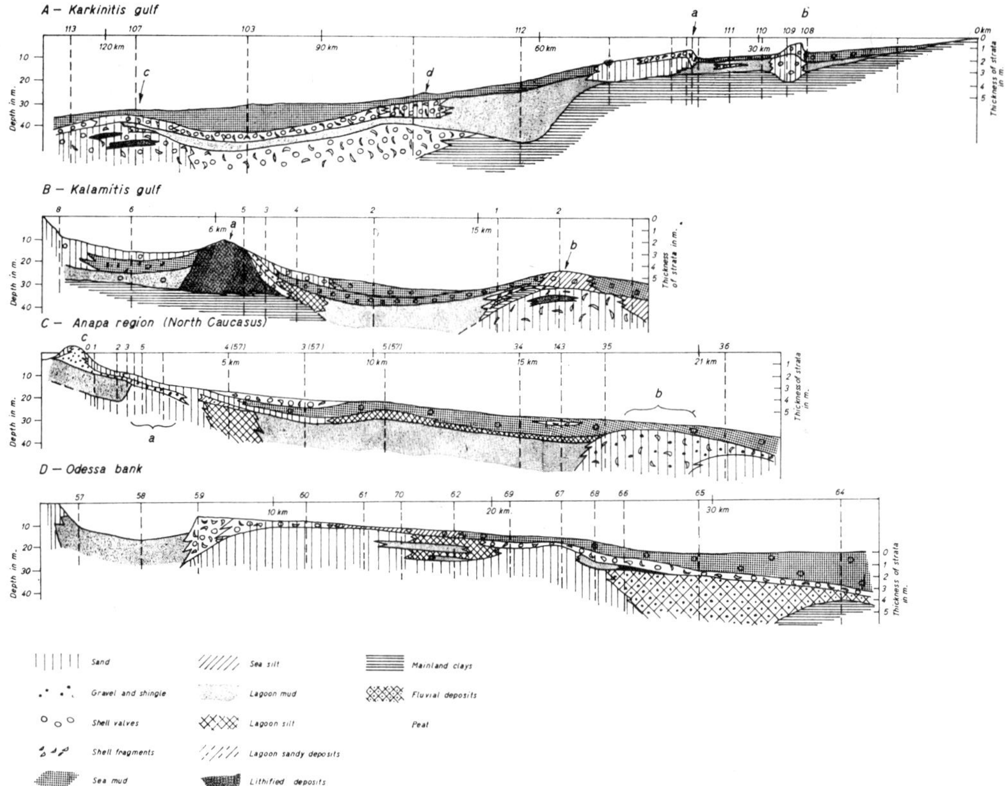

Fig. 9. The sections of the near-shore Black Sea sediments (after E. Nevessky).

days the bank is becoming deformed and undoubtedly continues its shifting towards the North.

The other bank mentioned (A_a in the north gulf) has apparently subaqueous origin on the whole. Everywhere its strata contain shell valves well preserved. The inner bank of the northern (A_b) bay consists mostly of sand. Because of the specific hydrological conditions at the head of the gulf (oxygen deficiency in winter times) the fauna productivity is much smaller, than in the open sea. On this bank a series of borings were made. They passed through the sand body of the bank, the underlaying mud bed and penetrated into mainland clay subtratum. Probably it is also of the bottom origin, now being quickly shifted towards the East.

On the whole in the northern gulf four banklike relics are displayed. The outer two banks are wholly buried under the most recent muds and the bottom surface is leveled. They both also consist of sand with some shells. The absolute depth of the outer banks top is of 35 (A_c) and 23 m (A_d). The faunistic analyses showed that first of them was already formed during the New-Euxinus and the second one at the Ancient Black Sea period. Their origin as two off-shore bars is proved by the fact that fine muds of lagoonal habit, almost free from fossils, are deposited between and beneath them.

In the southern (Kalamitis) gulf there are three shore-barriers (see fig. 9, B). The first of them is a wide recent bar partly, connected with the main shore. Along the axis of two large bays the coarse bar sediments thickness exceeds 20 m, and beneath them the lagoonal muds are buried. The sea part of the section is carried out opposite the watershed between the bays, and the thickness of lagoonal muds is far less here.

In the sea are two relic forms which are none-active at present. The shell material of the bank nearest to the shore (B_a) is lithified now and exposed over the bottom as a nacked rocky ridge. The second barrier of a loose materials is disposed on the depth about 25 m (B_b) and covered with a muddy layer.

The shell bank is deliberately of a subaqual formation, that is the predecessor of an offshore bar which have not completed its development. Both the recent bar and the bank under large depths have quite another origin. In the structure of the first the gravel and shingle layers are of great significance besides sea shells. The coarse terrigenous material is drifting along the shore from the South, where the cliffs are abraded intensively.

The similar material was displayed in cores of the deep-water bank (B_b). Hence we can conclude that it had its feeding from the South as well and was a supraqual beach or spit which separated the southern bay being considerably larger than nowadays.

The section made opposite the accumulative shore of Anapa (fig. 9, C) has approximately the same constitution. Here one recent shore barrier and two ancient ones were met. Between them the lagoonal muds with a distinct annual microlamination were discovered. However the underwater barriers are situated here on considerably smaller depths than those of North-West region. It should be kept in mind that in the Anapa region the shore submergence probably had a smaller amplitude, as it was not accompanied by the tectonic sinking of the land.

For comparing the sections of the Anapa region with those of the north-western part of the sea, the detailed studying of molluscs fauna was carried out (L. Nevesskaya). She succeeded in subdividing the total complex into four separate horizons and comparing them with each other for these remote districts.

On the ground of this analysis, it was established for example, that the first underwater barrier of the Anapa section lying at the depth of about 10 m (C_a) is of the same age as the third one (A_d) in the northern gulf. The latter is 23 m deep. Both of them were formed near ancient shores about 5000 years ago. The difference between their recent depth about 13 m is connected with the tectonic submergence of the northwestern region. The deep-water barrier of the southern gulf corresponds to this period too.

The most ancient formations are the deep-water barrier of the Anapa section (C_b) and the remotest one in the northern gulf (A_e). Their formation took place 6 till 7.5 thousand years ago in the end of New-Euxinus period. The difference between their absolute position is less than 10 m. Probably the Anapa barrier was an underwater one, but in the northern gulf a barrier was originated as a shore form. The second (beginning from the shore) barriers of the southern (B_a) and northern (A_a) gulfs are of 2-3 thousand years old.

An estimation of absolute age of nearshore relics was approximately made on the ground of calculation of annual microlayers in lagoonal muds and related to their thickness.

The whole complex of the data received led to the conclusion that the transgression run at both regions was nearly the same, but the sea level rose with different relative velocity due to heterogeneity of the tectonic structure. Because of that the rythmic constitution of these sedimentary formations was not caused by some local factors; it reflects the general run of the transgression during which periods of decelerations and accelerations were marked.

While analysing the strata of the structures described sometimes it is very difficult to determine sources of the origin of the coarse material composing the relics of shore forms. The sections given before were made in the parts where the alongshore debris drifting was of secondary significance and promoted in forming of two barriers in the southern bay only. The material of most forms was completely of bottom origin. It arose as a result of rewashing of ancient sediments of the alluvial plain, partly of mainland clay rocks on the bottom and to a considerable extent it represents the remains of recent fauna. Thus, in this case we could see the importance of the transversal drifting process in its pure aspect.

However, following along sea shores in the common case, we see the alternation of different types. There are some parts of substantial material accumulation beside those of abraded or washed away loose shores, as well as the parts of the alluvium supply from the land concentrated in several points. So the alongshore debris drifting brings some significant complications in the constitution of transgressive strata. They are especially appearing on the embayed shores.

Having a desire to study the origin of the near-shore sand sediments, we try to compare the sand mass accumulated in the form of terraces of barrier islands with the possible reserves which represent the sea bottom within the limits influenced by waves at present level. Taking the minimum surface slope of the submerging land and exploiting it towards the sea we can calculate the general sand quantity having been cast to the land. In most cases, even when there is no long-shore drift, the quantity of the deposits near the shore is many times greater than could be transported from the bottom under the permanent sea level.

In searching additional sources of debris supply on submerging shores we should draw our attention to deeper parts of the see which are covered now with the mud. For example, in the real case of the Great bank we display the strata of ancient alluvial sediments, containing sandy material on a large space of the shelf with the depths not more than 50 m. The comparison of heavy minerals complex of the bank sands with the river alluvium of this region demonstrates an interesting fact. Along with minerals typical for the alluvium of the norther rivers (Garnet, Zircon etc.) there were found a great quantity of amphiboles inherent to the Danube alluvium. Hence, a conclusion can be made that even the Danube alluvium was partly washed away by the sea and this sand passed northwards the distance more than 150 km during the transgression. Thus for the bank forming there was an enormous reserve of the sand.

In other less favourable conditions, for instance when the sea bottom slopes are relatively high, even taking 100 m as a possible transgressive hight, we cannot suppose the presence of such a reserve on the bottom. In such a case we should suggest either long-shore-drift from adjacent parts of the shore or repeated sandy supply from the land with high river floods in the course of level oscillations. This material should be distributed then over a large distance along the shores with a debris-stream or as a result of opposed migrations.

Just the material collected on the Black Sea has been adduced because it was here, that a series of detailed sections in the parts of different shores structure were carried out. But simular materials were received on other seas, such as the Azov Sea, the Baltic Sea (the southern shores within Poland), the Japanese and the Yellow Sea as well. Everywhere we discovered on the bottom the ancient alluvial, moor or lake sediments, the rock benches drowned and as to the Baltic Sea the terrestrial morainic sediments also were found. Everywhere the relic shore formations as sandy bars were revealed, sometimes attended by lagoonal sediments. The available data allow to draw some general conclusions.

1. The sections made on different seas of the Atlantic and Pacific basins give once again a corroboration of the universal 'eustatic late-glacial transgression and emphasize its strong influence on the near-shore zone structure of recent seas.
2. If the shallow zone is wide, and the supply of the coarse materials from the land is rather slow, we can reveal by means of vibropiston corer the very complicated constitution of the transgressive strata including numerous ancient shore-formations.
3. The reality of the barrier islands formation mechanism was proved by the works accomplished. These forms arise as a result of transversal shorewards debris drifting from the sea bottom.
4. Notwithstanding this process, the occurrence of barrier islands of terrestrial material is proved. Their formation takes place if the sea invades the plaine land in which constitution there abounds the sandy or more coarse components. In this case lagoons were never a part of the sea aquatorium in the previous period.
5. In spite of the views stated before, the debris barriers shifting landwards with a permanent level has its limit. It is most pronounced during a longer period of the land submerging.
6. However, if the sea level rapidly rises, the sea can overflow the shore barriers formed before. In the shallow water conditions the influence of rather weak waves cannot destroy them and further they are conserved under the muddy sediments cover.
7. As a more interesting phenomenon it is considered the rhythmic structure of transgressive strata in which the shore barriers (subaqual) banks or offshore bars relics are interspaced horizontally by the muddy sediments (partly of lagoonal origin). Undoubtedly there ought to be a certain optimal velocity of the sea level rising, the most favourable to the formation of subaqual banks or offshore bars supplied with the bottom material. Therefore this rhythm may be most easily to explain as irregular transgressive run being increased or decreased under the influence of climate changes. However, it is quite probable that the barriers can be rhythmically formed during the same velocity of the level rising. The underwater banks shifting landwards can be slowed down as a result of their volume increasing. We hope that this question will be successfully solved in course of further investigations.
8. Finally we saw that in some cases, when bottom sections of the tectonic heterogeneous regions are compared with each other, it turns out to corroborate the existence of differential vertical movements of the land and to prove the considerable duration of this process.

At the end I should like to emphasize a great interest of the near-shore sea zone and its sediments, which up to the present time are not thoroughly studied. Oceanographical ships are not able to enter the shoals, and it is only during last years that the land-geologists and geographers began to penetrate beyond the limits of a shoreline. It is here that a lot of significant discoveries can be made. In particular, the important results from many points of view can be obtained by means of an instrument like a vibro-piston core sampler. Scientists of the Polish People's Republic, Chinese People's Republic and the German Democratic Republic were acquainted with this instrument. In the first two countries with the help of our specialists the investigations of the described character are already being carried out. I hope that here in Holland too such works can provoke an interest of geologists and give many valuable data.

Selected bibliography

[1] Arhangelsky, A. D., Strakhov, N. M. Geological Structure and History of the Black Sea. Ac. Sci., U.S.S.R. Press, 1938 (English Summary).

[2] Beaumont, E. de. Leçons de géologie pratique. Paris, 1845.

[3] Johnson, D. W. Shore processes and Shoreline development. New York, 1919.

[4] Kudinov, E. I. Vibropiston Core-Sampler. Trans. Inst. of Oceanol. Ac. Sci., U.S.S.R., v. 25. 1957 (in Russian).

[5] Nevesskaya, L. A. On the Change of the Black Sea Bivalve Mollusc Complex in the Latequarternary Period. Proc. (Doklady) Ac. Sci., U.S.S.R., v. 121, N 1, 1958 (in Russian).

[6] Nevesskaya, L. A. Mollusc Complex of the Upperquarternary Littoral Deposits in Anapa Region of the Black Sea. Trans. Oceanogr. Com., Ac. Sci., U.S.S.R., v. 4, 1959 (in Russian).

[7] Nevessky, E. N. Etude des Sediments Marins Littoraux a l'aide du Tube a Piston Vibreur. Bull. d'Inform. du Comité Central d'oceanographie et d'etude des Cotes, X, 6, juin 1958.

[8] Nevessky, E. N. On the Rhythmicity of Sea Transgressions. Journal 'Oceanology', v. 1, N 1, 1961 (in Russian).

[9] Zenkovitch, V. P. One Peculiar Mode of the Lagoon Formation. Proc. (Doklady) Ac. Sci., U.S.S.R., v. 75, N 4, 1950 (in Russian).

[10] Zenkovitch, V. P. On the Lagoonal Sea-Shores and offshore Bars origin. Trans. Inst. of Oceanol. Ac. Sci, U.S.S.R., v. 21, 1957 (in Russian).

[11] Zenkovitch, V. P. The Black and the Azov Sea Shores. Moscow Geograph. St. Publish. House, 1958 (in Russian).

[12] Zenkovitch, V. P. The Morphology and Dynamics of the Soviet Black Sea Shores. Publ. Ac. Sci., Moscow v. 1, 1958; v. II, 1960 (in Russian).

Editor's Comments on Paper 9

Oleg Konstantinovich Leontiev is Professor and Doctor of Geography, and Chief of the Department of Geomorphology, Faculty of Geography, Moscow State University, USSR. He specializes in coastal geomorphology and marine geology and has published more than 150 scientific works concerning these fields. He is presently the Scientific Manager of the Sea Geomorphology Laboratory. Leontiev has participated in numerous expeditions to the seas of the Far East and Southeast Asia and the Caspian and Adriatic seas, as well as oceanographic investigations in the Indian and Pacific oceans. His present research activities are concerned with coral reef geomorphology, geomorphic subdivisions of the oceans, and the coastal zone and floor relief of the Caspian Sea.

Leontiev is concerned in this paper with the world-wide erosion of barrier islands now occurring, a phenomenon caused, he maintains, by present conditions, which differ from those of the period of the islands' origin. In Leontiev's view, underwater bars form during a transgression of the sea and then become barrier islands following a regression. Such events, within the framework of an oscillating, or fluctuating (Fairbridge), sea-level curve, have occurred during the last 6,000 years. The recent rise in sea level, therefore, provides unfavorable conditions leading to the present erosion of barrier islands.

The next paper, and a later one, will further serve to elaborate on Leontiev's investigations of barrier islands.

Reprinted from *Coastal Research Notes*, No. 12, 5–7 (1965)

9

On the Cause of the Present-Day Erosion of Barrier Bars

O. K. LEONTYEV

Barrier bars are the biggest and most widespread accumulation forms on coasts. A great part of the Atlantic coast, as well as the coast of the Gulf of Mexico in the United States, are fringed with barrier bars. In the USSR the largest bars occur along the western coast of Kamchatka, on the Sakhalin coasts, and on the shores of the Black, Caspian, and Karskoye seas. The summary length of barrier bars is about 10 percent of the world's ocean shoreline (Vinogradov, 1963).

Barrier bars usually separate lagoons from the sea, from which, not infrequently, salt and some other commercial minerals are obtained; and in cases when lagoons are desalted, they serve as basins for cultivation of valuable kinds of fish. As major accumulation forms, barrier bars are thick deposits of building material (sand, gravel). Sea health resorts are located on many barrier bar beaches, and often large cities spring up on bars (e.g., Atlantic City and Ocean City on the Atlantic coast; Galveston on the Texas coast). Hence these accumulation forms are of great importance in the economic activity of man.

However, there are many facts indicating that the barrier bar shoreline is unstable. On all the U.S. coasts rimmed with barrier bars that are mentioned, coastal erosion surveys have led to a whole complex of expensive coast protection measures. Intensive barrier bar erosion is observed on the western coast of Kamchatka (Vladimirov, 1958) and on many other barrier bars.

The global occurrence of the processes of barrier bar erosion seems to be related to some general causes operating in quite different regions of the world's ocean coast. In our opinion the main cause is the fact that barrier bars are relict forms that originated under hydrodynamic conditions differing from those of the present day. This, of course, does not exclude the significance of purely local phenomena which can cause bar erosion in certain areas. Such occurrences are widely known, and there is no need to speak about them in this brief communication.

The formation of island and barrier bars was preceded by the formation of underwater bars (Leontyev, 1961). As was shown by V. P. Zenkovich, O. K. Leontyev, and E. N. Nevesski (1960), underwater coastal bars were formed in the course of the postglacial transgression of the world's ocean. The latter, as is known, was caused by huge masses of water, which earlier had been preserved in the form of ice sheets, returning to the ocean. The coastal continental plains, which were often formed of unconsolidated alluvial, fluvioglacial, or proluvial deposits, were submerged as a result of the transgression. The surfaces of those plains were of insignificant slope: about 0.001–0.003, i.e., far less than the gradients of the profile of dynamic equilibrium, adapted to the hydrodynamic conditions of sea swell. Following reconstruction of the initial profile in accordance with the undulation conditions, the masses of loose material began to move. The dislocation of great masses of deposits toward the shoreline caused underwater bars to be formed on the upper part of the underwater coastal slope.

As was pointed out in the book cited (Zenkovich, Leontyev, and Nevesski, 1960) and in the monographs by Leontyev (1961) and by Zenkovich (1962), the further development of underwater bars led to the formation of barrier islands, and, then, of barrier bars. However, as early as 1951 it was established by E. N. Egorov that certain forms relative to bars, namely, offshore bars, did not emerge above the surface when the sea level was constant, but remained underwater forms. In 1963–1964, at the Experimental Geomorphology Laboratory, Moscow University, tests were carried out on modeling the processes of bar formation (Leontyev and Nikiforov, 1965). The tests resulted in the establishment of the fact that an underwater bar can become a subsurface form only if, in the course of its formation, the sea level recedes and is lower than the bar crest, if only for a short period of time. With the sea level constant, or even more so with the level rising, the bar remains under water, and the depth above its crest remains not lower than h (the wave depth).

The cause of the phenomenon mentioned can, to our mind, be explained as follows. When, as a result of the upward growth of the bar, the depth above its crest reaches the value of h, the complete breaking of the wave will take place over the bar crest, and the maximum swash rates (with which the evacuation of the material takes place) will be observed right on the bar crest. The washout of the material from the crest down the shoreward slope of the bar is a factor that prevents the bar from growing farther upward. It is only when sea level recedes and the shoreline is lower than the crest that the crest can become a zone of accumulation and will grow in height as a result of drift deposition at the upper borderline of the lapping. If the level rises again to the previous mark, then the bar crest may find itself even higher than this mark, and so the bar will remain as a subsurface accumulation form. With the further slow rise of sea level, the seaward slope of the bar will be eroded, since the depths in front of it will be greater, and a new reconstruction of the profile will start in conformity with the new depths.

It is just such a succession of phenomena that was observed in the course of the test on modeling of the process of bar formation. It follows from the above that barrier bars, so abundant on the world's ocean coasts, became superaqueous forms as a result of the sea-level recession that took place not long ago. Most probably, it took place during Postatlantic time (i.e., immediately after the Holocene climatic optimum). In the history of the Holocene, this recession of sea level has left its trace in the form of the Flandrean Terrace.

The numerous definitions of the absolute ages of the deposits that constitute the barrier bars (Fisk, 1959; Bird, 1963; Shofield, 1962; Fedorov, 1963; and others) show that around 6000 B.P. the world's ocean level reached a mark 3–5 m higher than the present one, followed by recession of the level by about the same value. The rise of sea level that started again about 4000–4500 B.P. created conditions favorable for the subsequent slow erosion of barrier bars that is continuing.

Thus barrier bars are relict formations and subject to reconstruction and erosion as a result of the continuing development of the postglacial transgression. This conclusion is an additional confirmation of the fact that the course of the postglacial transgression, as was noted by Fairbridge (1961), Shnitnikov (1957), and others, is complicated by short-period recessions and subsequent rises of the sea level, reflecting cyclic changes of climate.

References

Bird, E. C. F., 1963. The physiography of the Gippsland lakes of Australia. *Z. Geomorphol.*, **7**, No. 3.

Fairbridge, R. W., 1961. La base eustatique de la geomorphologie. *Ann. Geogr.*, **70**, No. 381.

Fedorov, P. V., 1963. Stratigraphy of the Quaternary deposits in the Crimean-Caucasian coast and some aspects of the geological history of the Black Sea. *Proc. Geol. Inst., Acad. Sci., USSR*, **88**.

Fisk, H., 1959. Padre Island and the lagoon Madre flats, coastal south Texas. *2nd Coast. Geog. Conf.*

Leontyev, O. K., 1961. *The fundamentals of geomorphology of coasts.* Moscow Univ. Publ. House, Moscow.

Leontyev, O. K., and Nikiforov, L. G., 1965. On the causes of the global occurrence of barrier bars. *Oceanology*, No. 4.

Shnitnikov, A. V., 1957. The changeability of the general humidity of the continents in the Northern Hemisphere. *Proc. Geog. Soc., USSR*, **16**.

Shofield, I. C., 1962. Postglacial sea levels. *Nature*, **195**, No. 4847.

Vinogradov, O. N., 1963. Representation of the features of morphology, dynamics and origin of coasts on general geographic maps. *Acad. Sci., USSR.*

Vladimirov, A. T., 1958. An approach to the morphology and dynamics of the western Kamchatka coast. *Proc. Acad. Sci., USSR, Ser. Geog.*, No. 2.

Zenkovich, V. P., 1962. *The fundamentals of the theory of the development of coasts.* "Nauka" Publ. House, Moscow.

Zenkovich, V. P., Leontyev, O. K., and Nevesski, E. N., 1960. The influence of the Eustatic Postglacial Transgression on the development of the coastal zone in the USSR. Coll. *Soviet Geologists' Papers at the XXIst Intern. Geol. Cong.*, Acad. Sci., USSR.

Editor's Comments on Paper 10

An English translation of this article from *Oceanology* is reproduced here for the perusal of the reader. In it Leontiev and Nikiforov continue the theme of the two previous papers. Submarine bars are formed when (1) there is a gently sloping nearshore bottom, above the profile of equilibrium; (2) abundant sediment with which to build the bar is available; (3) transversal debris-drift transports the sediment shoreward; and (4) the sea level is stable or rising. The bar evolves into a barrier island when a recession in sea level follows these events. The present stage is one of rising sea level, as stated by Leontiev in his first paper, causing extensive barrier island erosion. The authors cite evidence from around the world, and along the Gulf of Mexico in particular, illustrating just such a sequence.

Reprinted from *Oceanology*, **5**, 61–67 (1965)

REASONS FOR THE WORLD-WIDE OCCURRENCE OF BARRIER BEACHES

10

O.K. Leont'yev and L.G. Nikiforov

All major summaries on coastal geomorphology [8, 10, 12, 19, 21, 42] note the extensive distribution of coasts sheltered by barriers. The coasts of this type in the USSR include the greater part of the Eastern Caspian, a considerable part of the shores of the Black Sea, the western shores of Kamchatka and the eastern shores of Sakhalin, and large areas of the coasts of Soviet northern waters: in all 11% of the Soviet coastline. Similar coasts in Western Europe include the Netherlands seaboard, the greater part of the southern coast of France, parts of the British Isles, the upper end of the Adriatic, andthe marshes in France; a total, according to Vinogradov [4], of 8.4% of the overall length of coast in Western Europe. In Asia outside of the Soviet Union there are the coasts of the Indian subcontinent and Ceylon, and parts of the coasts of Vietnam, China, Japan and Indonesia (4% of the total length of coastline); in North America there is the greater part of the Atlantic seaboard and the shores of the Gulf of Mexico, and some parts of the low Alaskan coast; in Africa there is almost the entire northern seaboard of the Gulf of Guinea and a number of other stretches (15.6% of the coastline of the African continent); in Australia almost the entire coast of the Great Australian Bight. Coasts fronted by barrier beaches account for 9% of the total length of coastline of the seas and oceans.

It may therefore be said that barrier beaches are a feature of world-wide importance, and that they are for some reasons typical of the recent geologic epoch.

Present views on barrier beaches are examined in a number of special studies, mainly in relation to the origin of lagoon coasts, with which they are genetically linked. Detailed bibliographies are to be found in Zenkovich [12] and Leont'yev [21], and in some articles [11, 15, 16, 20, 25, 43, 52].

Zenkovich [11] gives the following definition of a barrier beach: "... long narrow strips of detritus raised above sea level and extending at some distance from the original land parallel to the general trend of the coast." A barrier is usually backed by a lagoon. The material of which barrier beaches are formed reaches the coast from the sea floor. Unlike spits and other forms fed by long-shore drifts, barrier beaches are produced by movement of material normal to the coast, and derive their material from the sea floor around them. Although the long-shore movement of material may play a part in the dynamics of a barrier, it is not a factor determining its formation.

The stages in formation of a barrier have been considered by Zenkovich [11, 12] and Leont'yev [20, 21]. The three stages distinguished are submarine bar, barrier island, and barrier beach proper.

The main conditions for the development of barrier beaches are the presence of considerable reserves of unconsolidated material on the submarine beach slope, a gentle initial bottom gradient ensuring that the position of the initial submarine slope profile is "above" the profile of equilibrium for the given hydrodynamic conditions, and some bend in the submarine slope profile. It would appear that since barriers are produced by movement of material normal to the coast, that the prevailing direction of wave approach should also of necessity be at right angles to the initial shorelines, but this condition is in fact not essential, since if the initial gradient is gentler than the gradients of the profile of equilibrium, material will move shoreward from the sea floor whatever the angle of wave approach.

Unconsolidated material on the sea floor may originate in various ways. In some instances barriers are formed of a mixture of chemogenic (oolitic) material and material of organic origin (shells), as on the eastern shores of the Caspian, and on the coasts of Florida and the Bahamas; in other instances the material may be mainly shells. For example, the Arabat spit, which is a typical barrier beach separating the Sivash Lagoon from the Sea of Azov, is almost entirely composed of shells. In most instance, however, the material is terrigenous. The question of how large reserves of terrigenous material come to be present on the submarine slope in front of a developing barrier is intimately bound up with the concept of a gentle initial slope as one of the most important conditions of barrier formation.

The following explanation is given: as a result of liquidation of the continental ice sheet in

North America and Eurasia at the end of the Upper Quaternary a eustatic transgression developed. According to Fairbridge [37] sea level was approximately 100 m below the contemporary level at the end of the Wisconsin (approximately 16,000 years ago), which is in good agreement with the views of other authors who have studied this problem [5, 40, 53], but in the post Wisconsin (11,000 years ago) it had already reached levels of 30 m. In the ensuing 5,000 years sea level almost reached the contemporary level and has not altered since by more than ± 3 m.

In the course of this transgression the edges of the continents, which in most instances consisted of alluvial, proluvial or fluvioglacial accumulation plains, became submerged. Vast masses of unconsolidated material were therefore present on the sea floor, and these began to be moved shoreward by wave action as the transgression developed [13]. The slope profile had begun to adapt itself to the new dynamic conditions, which could be achieved only by "incising" a new profile (satisfying the new dynamic conditions of equilibrium) in the initial surface and by carrying the greater part of the erosion products shorewards, in accordance with the pattern of development of the submarine slope profile on a detrital slope [10, 12, 18].

There are a number of known facts to indicate that barrier beaches are in very many instances constructed from material of subaerial origin. The relevant information is given, in particular, by Zenkovich [11]. In referring to the Meyechken shingle barrier and other barriers on the Chukchi coast Zenkovich points out that they derive their material from sheets of rock fragments lying on a foothill plain, which is now submerged as a result of the postglacial transgression and has been converted into the submarine beach slope. Zenkovich notes that this fact is "very interesting, since no description of shingle barriers is to be found anywhere in the world literature. All references are to sand barriers" [11 (p. 12)].

In this latter context we would note that one of the authors of this paper was recently involved in the investigation of a barrier consisting largely of shingle on the mainland coast of Sakhalin Bay, where there is a shingle barrier approximately 40 km long between Capes Litke and Perovskiy supplied by a submerged Pre-Holocene shingle terrace.

It is noteworthy that, according to the extensive material cited in the works of Zenkovich and Leont'yev mentioned above, barriers form on a sandy bottom at gradients of approximately 0.005-0.003. Such gradients are common on subaerial accumulation plains.

In his 1946 monograph Zenkovich related the formation of barrier beaches directly with that of submarine bars, treating the barrier as the final stage in the development of a submarine bar. It has, however, been demonstrated in later works [6, 7] that the connection is no stronger than that both forms owe their origin to development of a profile of equilibrium of the submarine slope. Submarine bars and barriers are phenomena that are too different in scale for conversion from one to the other, although there are common features to their formation. The submarine forms that precede barrier beaches are evidently larger features from the very beginning of their evolution. The accumulation of material on the bottom that gives rise to the submarine feature preceding a barrier beach forms initially at considerably greater depth and is possibly associated with transition from an already assimilated part of the bottom profile to a part nearer the shore, which has not yet been reconstructed, but which still retains the initial gradient.

Attempts were made to prove this hypothesis at the experimental geomorphological laboratory of Moscow State University in connection with solution of the problem of the effect of the wave factor on the slope of a tectonic uplift forming in shallow water.

The experiment yielded very interesting information and, in particular, provided confirmation of the hypothesis that some of the barrier islands of the North Caspian owe their origin to tectonic factors [20, 25]. In the course of the experiment it was established that the top of the barrier never breaks the surface when sea level is constant or rising, but that it remains a submarine form. Ye. N. Yegorov had previously arrived at the same conclusion; many years of repeated surveys and observations of the dynamics of submarine bars convinced him that they could never become above-water forms at constant sea level.

It is apposite to recall that, when considering the conditions in which an avant-delta is converted into delta land, Samoylov [26] emphasized that a fall of level was an essential condition: only in this case could the various submarine accumulations become above-water forms of the delta. This condition is, however, always satisfied in the formation of a delta, since flooding of the river alternates with low water.

Yegorov noted that a coastal ridge, usually formed while one type of sea was operative, was an exception to his conclusion. This is a local relief form differing considerably in size and formation conditions from other submarine bars and developing close to the shoreline. When the sea calms such a ridge may be uncovered. Most known cases of description of a bar breaking the surface apparently relate to this type of ridge. Even here, however, a drop of level is involved: the ridge invariably breaks the surface towards the end of the stage during which the sea abates, and the piling up of water to the coast, which is a feature of all wave disturbances, is replaced by a decline of level.

Johnson [40] treated barriers as a sign of coastal elevation, .e., he assumed that the formation of a barrier as an above-water relief form was related to a relative fall of level.

It was, however, established in subsequent and more detailed investigations of the distribution of coasts fronted by barriers, and of the nature of vertical movements on these coasts (cf. e.g., [1, 14]) that, with few exceptions, coasts fronted by barriers exhibit all the features of recent submergence. The facts do not, therefore, support the view that barriers are a sign of coastal elevation.

A contradictory pattern develops: on the one hand, a submarine bar is incapable of becoming an above-water feature even at constant level (and all the more so under conditions of submergence); on the other hand, coasts fronted by a barrier show all the signs of recent submergence. It is generally held that this latter circumstance is related to the postglacial transgression of the seas and oceans. It has been established by processing extensive measurement records that sea level has risen by 1 mm annually on average in the last seventy years [37, 53].

In the authors' opinion it was possible for the tops of submarine bars formed under the conditions of this transgression to emerge only because the postglacial transgression was discontinuous and the rise in sea level alternated with brief falls, as a result of which the bars became subaerial features. The universal occurrence of barrier beaches indicates that at least one of these falls in level was general and sufficiently prolonged for the barrier beaches emerging at the surface to consolidate as above-water forms sufficiently to be able to withstand the renewed transgression for some time.

Before considering the reality of this event in postglacial history, however, we must first consider the following two questions: (1) why should the top of a submarine accumulation be unable to break the surface when sea level is constant or rising, (2) is it possible for a lagoon coast, with which the formation of barrier beaches is usually connected, to form without formation of a bar?

The first question may be answered by an analysis of wave destruction over a submarine bar in the course of formation. As the bar grows a situation may be reached in which depth above the top is similar to wave height. In this case partial destruction of the wave over the bar gives way to total destruction and the formation of swash. The effective swash zone will be limited to a narrow strip of the top of the submarine feature, beyond which depth again increases towards the shore. As a result the submarine ridge lies entirely in the zone of maximum swash velocities, i. e., entirely within that part of the swash zone which by virtue of the maximum velocities can only be eroded. A zone of accumulation, corresponding to the zone on the beach where the velocities of the swash decrease, is absent, and its place is taken by the beach slope of the submarine bar on to which material is carried from the crest by the swash. Material is carried from the crest on to this slope both under the effect of the vertical component of the swash directed downwards, and under the effect of gravity when the swash is greatly weakened as a result of the increase of depth. When sea level is constant (and especially when it is rising) the crest of a submarine bar is therefore unable to approach the surface, since the least depth to which it can rise in its growth is equal to wave height.

The answer to the second question is to be found in the writings of Zenkovich [11, 12] and Kaplin [16]. In the course of transgression the sea submerges that part of the coastal plain which lies immediately behind the coastal accumulation terrace. The surface of this terrace may be considerably raised (several meters) not only owing to beach accumulation, but also because of the formation of aeolian accumulations at the upper limit of the beach. The area submerged behind the terrace becomes a lagoon, and the high unsubmerged part of the terrace consisting of storm berms and dunes remains as a narrow strip of land that is morphologically similar to a barrier beach.

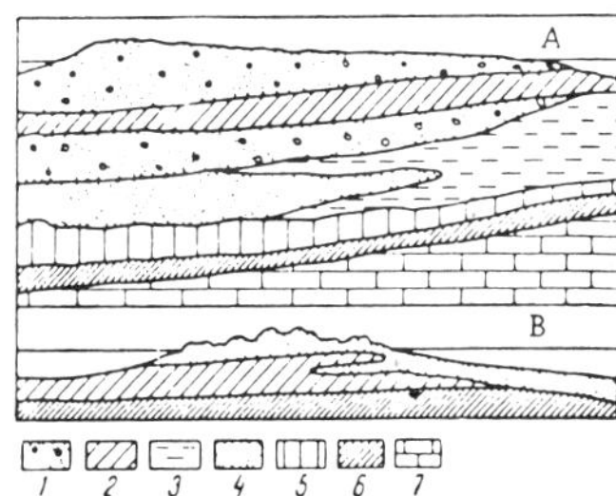

Fig. 1. A) through the Moynak barrier (according to A.I. Dzens-Litovskiy); B) section through the Alibey barrier near the Dniester liman (according to V. P. Zenkovich):

1) sand and shingle, 2) gray lagoon mud, 3) reddish-brown mud, 4) sand, 5) brown mud, 6) bedded clay and loesses, 7) limestones.

The case here described is quite possible and is frequently encountered on sea coasts. The lagoon deposits should clearly be bedded directly on the subaerial sediments of the plain. This is revealed in, for example, the geologic profile of the barrier beaches of the Dniester liman and the Moynak lagoon on the Black Sea coast (Fig. 1): here the deposits of the barrier are bedded on lagoon sediments, and the latter occur on subaerial deposits or on bedrock. The deposits of the open bay are absent from the profile. Consequently, a lagoon (or liman) formed in this instance in a place where prior to its formation there had been no sea. The sea water advanced over the coastal terrace and the lagoon formed only after the barrier had formed and not at the same time.

Nevertheless, not all lagoons are formed in this way; in fact this is not the most typical instance. Thus, if we analyze the geologic profile of one of the world's major barriers — the barrier of the Gulf of Mexico, which has been studied most thoroughly by Fisk [38] and by Le Blanc and Hodgson [44], we see the lagoon deposits are here underlain by the deposits of the open bay, and that a former submarine bar dating back to the first stages of the transgression, when sea level was 8-10 m below the present level, is clearly outlined in the lower part of the profile. This bar did not, however, break the surface at the time of its formation. Radiocarbon dating gave an age of 5,550-5,680 years for the bay sediments replacing the deposits of this barrier [38] (Fig. 2).

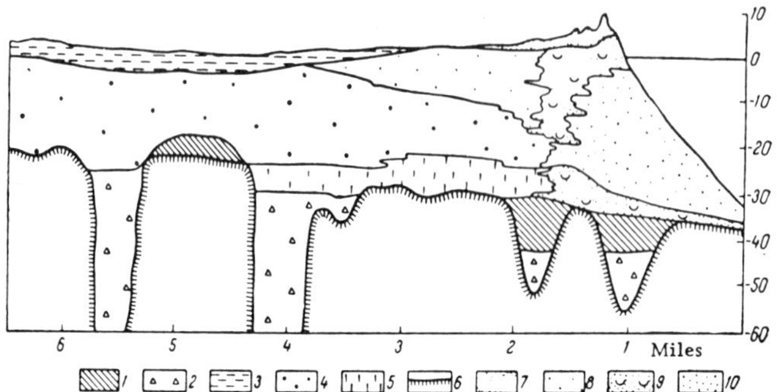

Fig. 2. Section through the barrier of the Gulf of Mexico (according to Fisk):

1) complex of aeolian plain deposits, 2) estuary deposits, 3) closed lagoon deposits, 4) open lagoon deposits, 5) open bay or gulf deposits, 6) saline zone, 7) dunes, 8) aeolian sands and proluvial train, 9) beach deposits, 10) bottom sand deposits.

It is evident from the profiles given by Fisk that the growth of this accumulation feature was interrupted approximately 5,000 years ago, apparently as a result of a rapid rise of level, but was subsequently renewed. Open lagoon sediments were deposited between the barrier and the coast (i.e., clearly the deposits of a lagoon not yet cut off by an above-water barrier). These deposits are pure, unsilted gray sands with a high content of sea shells. Approximately 4,000 years ago these sediments were replaced by closed lagoon deposits consisting of inter-layers of rust-red sands and gray silts containing many stunted mollusks. This was apparently the time of universal drying out of the Texas barrier in the Gulf of Mexico. It follows from what has been said that this drying out could only have taken place as a result of a fall of sea level. It would appear from the structure of the geologic profile that level fell by 3 m relative to the present level, which is in good agreement with the deductions of Fairbridge (see above), and gives a mean rate of 0.7 mm a year for the subsequent rise of level, i.e., a lesser rate than that observed in the last seventy years (1 mm annually).

In recent years many authors have described a marine terrace clearly of postglacial age 5-6 m above present level in the most varied regions of coast throughout the seas and oceans. Many authors compare it in time of formation with the Flandrian Holocene terrace of Western Europe or the Nicean terrace of the Mediterranean. This information is here presented in tabular form.

This table could be longer, but there is no need to extend it. It shows quite clearly that the Flandrian transgression (i.e., the stage of Holocene history when sea level was several meters above the present level) did in fact take place, and that the attempts of various authors to establish its absolute age yield quite comparable results. Most authors put this transgression at about the time of the climatic postglacial optimum and recognize its eustatic nature. It is evident from the information given that the transgression gave way approximately 4,000-4,500 years ago to a fall of sea level, as a result of which the tops of barrier beaches that had previously been submarine forms emerged as land.

Many barrier beaches have been directly dated, and it has been shown that their formation was related to the fall of sea level in post-Flandrian times. We have already referred to the

Data on distribution and height of postglacial (Flandrian) marine terrace

Coastal region	Author and year	Height of terrace, m	Dating
Netherlands, Friesian Islands	de Johng, 1960	2-3	Sub-Atlantic times
Scotland	Donner, 1963	7, 6	Postglacial
Nova Scotia, Acadia	Harrison and Lyon, 1963	3	4,500 years
Crete	Boekschoten, 1963	7	Flandrian transgression
New Zealand	Shofield, 1962	2, 1	3,900 B.C.
India	Chatterjee, 1963	5	Flandrian transgression
Gibraltar	Giermann, 1962	3	Postglacial
Gulf of St. Lawrence	Dionne, 1963	6	Postglacial
Western Australia	Committee for the Investigation of Changes of the Coastline, 1962	3	Flandrian transgression
Solway Bay (Irish Sea)	Marshall, 1962	6-7	Postglacial
Gippsland barriers (Bass Strait)	Bird, 1963	3-4	Before formation of the barriers, 6,000 years ago level was 3 m above the present level
Kent, England	Oldfield, 1960	5-8	5,200-3,000 years B.C
Region of the Moroccan coast	Markov, 1961	2-5	Flandrian transgression
La Rochelle	Verzhe, 1960	8	5,500 years
Eastern Australia	Ruszchzynska, 1961	2	5,000 years
Ghana	Ruszchzynska, 1961	3-6	5,000 years
Chile, between Valparaiso and Los Vilas	Paskoss, 1963	5-7	Holocene
Soviet Maritime Territory	Solov'yev, 1959	4	Holocene — transgression 3,000-4,000 years ago
Tanabe, Japan	Mii Hideo, 1962	6	Transgressive Holocene, retreat
McMurdo Sound, Antarctia	Speden, 1960	10	Maximum stage of last transgression 6,000 years ago
Northwest of the Okhotsk area	Zabelin, 1951	7-10	Postglacial
Matanzas area, Cuba	Duclos, 1963	4-9	Flandrian transgression
North Pacific seaboard	Chemekov, 1961	3	North Pacific terrace
Black Sea	Fedorov and Geptner, 1959		
	Fedorov, 1963	3-4	Flandrian

Note: One of the authors has also observed this terrace at a height of 7-10 m on the mainland coast of Sakhalin Bay, and the other author at a height of 5-8 m on the Dalmatian coast in Yugoslavia.

absolute age of the Texas barrier in the Gulf of Mexico. Bird [31] also gives information on the age of a coastal barrier. The Ghana Flandrian terrace [49] is also morphologically a barrier beach.

It is a characteristic feature of most barrier beaches that they are subject to erosion. Clear examples of this are to be found in the works of Vladimirov on the barriers of Western Kamchatka and the eastern coast of Sakhalin [2, 3]. Zenkovich [12], who describes many examples of the encroachment of barriers on to lagoon deposits, states that this process takes place because storm waves spill over the crest of the barrier and carry material from the crest to the rear, i.e., that encroachment results from erosion of the frontal surface of the barrier. This also confirms the suggestion made above that barrier beaches are relics, that they developed as above-water forms as a result of a fall of level after the Flandrian transgression, and now that sea level is once again rising they are unstable forms of coastal relief that do not correspond to the new hydrodynamic conditions of the coastal zone. It is noteworthy that even in the Caspian, where development of barriers was until recently favored by the fall of sea level, erosion of barrier beaches has commenced since 1956, when the level was more or less stabilized, especially in the eastern half of the sea, where the conditions of supply are not conducive to stability when sea level is stable.

REFERENCES

1. Budanov, V.I., A.S Ionin, P.A. Kaplin, and V.S. Medvedev. Recent vertical movements of the shores of Soviet seas. In: The 21st International Geological Congress. USSR Acad. Sci. Press, 1960.
2. Vladimirov, A.T. Morphology and dynamics of the shores of Western Kamchatka. Izv. Akad. nauk SSSR, ser. geogr., No. 2, 1958.
3. Vladimirov, A.T. Morphology and evolution of the lagoon coast of Sakhalin. Tr. In-ta okeanol. Akad. nauk SSSR, 48, 1961.
4. Vinogradov, O.N. Depiction of features of the morphology, dynamics and origin of coasts on general geographic maps on scales of 1:25,000, 1:50,000 and 1:100,000. Author's summary of thesis, Moscow, 1963.
5. Voronov, P.S. Procedure for paleogeographic and Cretaceous geographic reconstruction of the morphometry of the continents and ice sheets. Izv. Vses. geogr. ob-va, 96, No. 5, 1964.
6. Yegorov, Ye.N Observations of the dynamics of submarine sand bars. Tr. In-ta okeanol. Akad. nauk SSSR, 6. 1951.
7. Yegorov, Ye.N. Some dynamic features of a shallow coast of accumulation. Tr. In-ta geogr. Akad. nauk SSSR No. 68, 1956.
8. Guilcher, A. Essai sur la zonation et la distribution des formes littorales de dissolution du calcaire. Ann. de Géog. 62, 1953.
9. Zabelin, A.V. The recent uplift of the Northwestern shore of the Sea of Okhotsk. Priroda, No. 8, 1951.
10. Zenkovich, V.P Dinamika i morfologiya morskikh beregov. Ch. I. Volnovyye protsessy. (Dynamics and morphology of sea coasts, part 1. Wave processes.) Morskoy transport, 1946.
11. Zenkovich, V.P. The origin of barrier beaches and lagoon coasts. Tr. In-ta okeanol. Akad. nauk SSSR, 21, 1957.
12. Zenkovich, V.P. Osnovy ucheniya morskikh beregov. (Theoretical principles of coastal development.) Moscow, USSR Acad. Sci. Press, 1962.
13. Zenkovich, V.P., O.K. Leont'yev, and Ye.N. Nevesskiy. The effect of the eustatic post-Glacial transgression on development of the coastal zone in Soviet seas. In: 21st International Geological Congress. USSR Acad. Sci. Press 1960.
14. Ionin, A.S., P.A. Kaplin, and V S. Medvedev. Some results of regional research on the coasts of Soviet seas. Tr. In-ta okeanol. Akad. nauk SSSR, 48, 1961.
15. Kaplin, P A. Some features of the lagoons of the northeastern coast of the USSR. Tr. Okeanogr. komiss. Akad. nauk SSSR, 2 1957.
16. Kaplin, P A. Some features of lagoon formation. Okeanologiya, 4, No. 2, 1964.
17. King, C.A.M. Beaches and coasts. London, 1959. (Russian edition, For. Lit. Press, Moscow, 1963.)
18. Leont'yev, O.K. Modification of the profile of a coast of accumulation when sea level falls. Dokl. Akad. nauk SSSR, 66, No. 3, 1949.
19. Leont'yev, O.K. Geomorfologiya morskikh beregov i dna. (Geomorphology of sea coasts and the sea floor.) Moscow Univ. Press, 1955.
20. Leont'yev, O.K. Lagoon types and lagoon formation on recent coasts. In: 21st International Geological Congress, USSR Acad. Sci. Press, 1960.
21. Leont'yev, O.K. Osnovy geomorfologii morskikh beregov. (Principles of coastal geomorphology.) Moscow Univ. Press, 1961.
22. Markov, K.K. Paleogeographic problems of Morocco in human times. Byul. Komis. po izuch. chetvertichn. perioda Akad. nauk SSSR, No. 26, 1961.
23. Nikiforov, L.G. Notes on the formation of barrier beaches. Conference of the Caspian Commission, Abstracts of Proceedings, Baku, 1963.
24. Nikiforov, L.G The conditions under which barrier beaches form (taking the Island of Ogurchinskiy as an example.) Okeanologiya, 4, No. 4, 1964.
25. Nikiforov, L.G. and O.K. Leont'yev. Experimental studies of the interaction of endogenous and marine exogenous factors taking as an example the modelling of the Ukhta Platform uplift. Priroda, No. 10, 1965.

26. Samoylov, A. V. Ust'ya rek. (River mouths.) Gidrometeoizdat, 1952.
27. Solov'yev, A. V. The Holocene ingression in the southern Maritime Territory. Mat. Vses. n.-i. geol. in-ta, No. 2, 1959.
28. Fedorov, P. V. Stratigraphy of the Quaternary deposits of the Crimean and Caucasian coast and some aspects of the geologic history of the Black Sea. Tr. Geol. in-ta Akad. nauk SSSR No. 88, 1963.
29. Fedorov, P. V. Stratigraphy of the Quaternary deposits of the coastal zone in the northeastern Black Sea area. Tr. Geol. in-ta Akad. nauk SSSR No. 32, 1959.
30. Chemekov, Yu. F. Quaternary deposits and main stages in the development of the Soviet Far East. Mat. Vses. n.-i. geol. in-ta, No. 34, 1961.
31. Bird, E. C. F The physiography of the Gippsland lakes, Australia. Z. Geomorphol., 7, No. 3, 1963.
32. Boekschoten, G. J. Some geological observations on the coasts of Crete. Geol. en mijnbouw, 42, No. 8, 1963.
33. Chaterjee, S. P. Fluctuations of sea level around the coasts of India during the Quarternary period. Z. Geomorphol., suppl. No. 3, 1961.
34. Dionne, J. C Le problème de la terrasse et de la falaise. Mic. Mac. Rev. canad. geogr., 17, No. 1, 1963.
35. Donner, J J. The late- and postglacial raised beaches in Scotland. II. Suomalais tiedeakat. toimituks, ser. AII, No. 68, 1963.
36. Duclos, C. Etude géomorphologique de la région de Matanzas, Cuba. Arch. sci., 16, No. 2, 1963.
37. Fairbridge, R. W. La base eustatique de la géomorphologie. Ann. geogr., 70, No. 381, 1961.
38. Fisk, H. Padre island and the laguna Madre flats, coastal south Texas. Second Coast. Geogr. Conf. Louisiana State Univ., 1959.
39. Giermann, G. Meeresterrassen am Nordufer der Strasse von Gibraltar. Ber. Naturforsch. ges. Freiberg, 52, 1962.
40. Gutenberg, B. Changes in sea level, postglacial uplift and mobility of the earth interior. Bull. Geol. Soc. America, 52, No. 5, 1941.
41. Harrison, W. and C. J. Lyon. Sea level and crustal movements along the New England-Acadian shore, 4,500-3,000 B. C., J. Geol., 71, No. 1, 1963.
42. Johnson, D. W. Shore processes and shoreline development, New York, 1919.
43. de Johng, J. D. The morphological evolution of the Dutch coast. Geol. en mijnbouw, 39, No. 11, 1960.
44. Le Blanc, R. J. and W. D. Hodgson. Origin and development of the Texas shoreline. Second Coast. Geogr. Conf. Louisiana State Univ., 1959.
45. Marshall, J. R. The morphology of the Upper Solway salt marshes. Scott. Geogr. Mag., 78, No. 2, 1962.
46. Mii Hideo. Coastal geology of Tanabe Bay. Sci. Repts Tohoku Univ. ser. 2, 34, No. 1, 1962.
47. Oldfield, F. Late Quaternary changes in climate, vegetation and sea level in lowland Lonsdale. Publ. Inst. Brit. Geographers, No. 28, 1960.
48. Paskoff, R. Indices morphologiques d'un stationnement de l'océan Pacifique à 5-7 m au-dessus de son niveau moyen actuel sur le littoral du Chili central. Compt. rend. Soc. géol. France, No. 6, 1963.
49. Ruszvzynska, A. Z. bodan nad czwartorzedowymi zmianami poziomu oceanow. Przegl. geol. 8, No. 2, 1961.
50. Shofield, J. C. Postglacial sea levels. Nature, 195, No. 4847, 1962.
51. Speden, I. G. Postglacial terraces near Cape Chocolate, McMurdo Sound, Antarctica. N. Z. J. Geol. and Geophys., 3, No. 2, 1960.
52. Sanders, J. E. Effect of sea level rise on established barriers. Geol. soc. America Spec. Paper, No. 73, 1963.
53. Valentin, H. Die Küsten der Erde. Ergänzungsheft zu Petermanns Geogr. Mitt. No. 246, 1952.

Lomonosov State University,
Moscow

Received January 21, 1965

Editor's Comments on Paper 11

In this very comprehensive report L. M. J. U. van Straaten develops a sequence of barrier genesis similar to that put forth in the previous three papers by Zenkovich, Leontiev, and Nikiforov. In common with these scientists, he agrees upon the essential prerequisites of a shallow low slope, transverse sediment drift, and a rising sea level. Unlike the others, however, he recognizes a substantial component of longshore drift.

Van Straaten's stratigraphic evidence is most interesting, as is the interrelationship between the submarine bar and primary beach ridge that he describes. The reader is advised to examine, in particular, the text on (van Straaten's) pages 72 and 73 together with the figure on page 71.

L. M. J. U. van Straaten was born in 1920 in Rotterdam, Netherlands. He attended Leiden University; received the bachelor's degree (candidaats examen) in 1940, then the master's degree (doctoraal examen) and Ph.D. in 1946. He became teacher of geology at the Groningen University in 1947 and Professor in 1962. In 1954–1955 van Straaten was Visiting Associate Professor of Geological Oceanography at Texas A & M College, College Station, Texas. He has studied the geology of Europe and the surrounding seas since 1941 and has published 56 papers on his investigations. In 1972 both the Van Waterschoot van der Gracht Medal of the Koninklÿk Nederlands Geologisch Mÿnbouwkundig Genootschap, and the F. P. Shepard Award of the Society of Economic Paleontologists and Mineralogists were bestowed upon him.

Reprinted from *Mededel. Geol. Sticht.*, Nieuwe Serie No. 17, 41–75 (1965)

11

Coastal barrier deposits in South- and North-Holland

in particular in the areas around Scheveningen and IJmuiden

L. M. J. U. VAN STRAATEN

Geologisch Instituut, Melkweg 1, Groningen, Netherlands

4 Tables; 26 figures in text; Plates 10—14.

SUMMARY

This paper deals with the general lithology, grain size distributions, sediment structures and mollusc remains of the Dutch coastal barrier formation between Scheveningen and Voorburg (borings), at IJmuiden (excavation) and on the forelying sea floor (grab samples).

Some of the main conclusions (partly tentative) following from these data are:

1) The barrier system has formed during the Subboreal, in the course of not more than 2000 years.
2) More than half of the sediments were supplied by longshore drift, the remaining part by wave action transverse to the coast.
3) Longitudinal sediment transport was probably of the order of 3.10^6 m^3 per year.
4) Transverse supply was intense, owing to the small depths of the sea floor in front of the coast, a consequence of the rapid rise of sea level during the preceding Atlantic stage and the slight slope of the original land surface.
5) The chief sources of the sediment supplied by longshore drift were a) the nearshore and coastal sand masses that originally existed further to the southwest and b) the mouths of the main branch of the Rhine and of the Old Rhine distributary.
6) Transverse supply was the main factor leading to the formation of large submarine bars, on the leeward side of which fine-grained, clayey sediments could be deposited in relatively small depths.
7) When the innermost submarine bar had reached its maximum height, it protected the coast from heavy waves. No high ridges of sand could be thrown up on the backshore at this stage, and the beach remained of small elevation. A relatively low area was formed, consisting of minor ridges and swales. Later on tidal channels developed in the swales, and after filling up of the channels the area became covered by peat swamps.
8) The coarse sand of the innermost submarine bar was covered by fine-grained and clayey deposits formed in the shelter of a second bar. The seaward flow of ground water in the barrier complex was concentrated below the clayey sediments. Where it emerged on the sea bottom the dissolved iron compounds were oxidized and precipitated in and on mollusc shells. A similar precipitation from emerging ground water probably accounts for much of the brown coatings on the sand grains south of Bergen.
9) The mollusc fauna living in front of the coast during the Subboreal was much richer in *Macoma balthica* than the present fauna in front of the North Sea coast, in connection with the higher clay content of the bottom sediments deposited in the lee of the bars. The difference in fauna is reflected in the composition of the brown shell material found on the present beach, which is derived by the waves from the submarine outcrops of the coarse sand of the former bars.
10) The conclusions lead to a reconstruction of the coastal evolution at Scheveningen as shown in fig. 26.

CONTENTS

INTRODUCTION

The sand ridges in the coastal area of the provinces of South- and North-Holland which run parallel with or at a slight angle to the present coast-line, have been recognized already long ago as beach formations, covered at most places by wind-blown sand. Their distribution and surface geology have been studied in great detail (see e.g. ref. 5, 11, 16, 18). Much less, however, is known about their deeper parts. It is true that a great many borings have been made which penetrate the whole complex of coastal deposits, but practically all of them were bailer borings. They did not provide a sufficiently accurate picture of the successive strata to allow conclusions on the origin of the coastal barrier system.

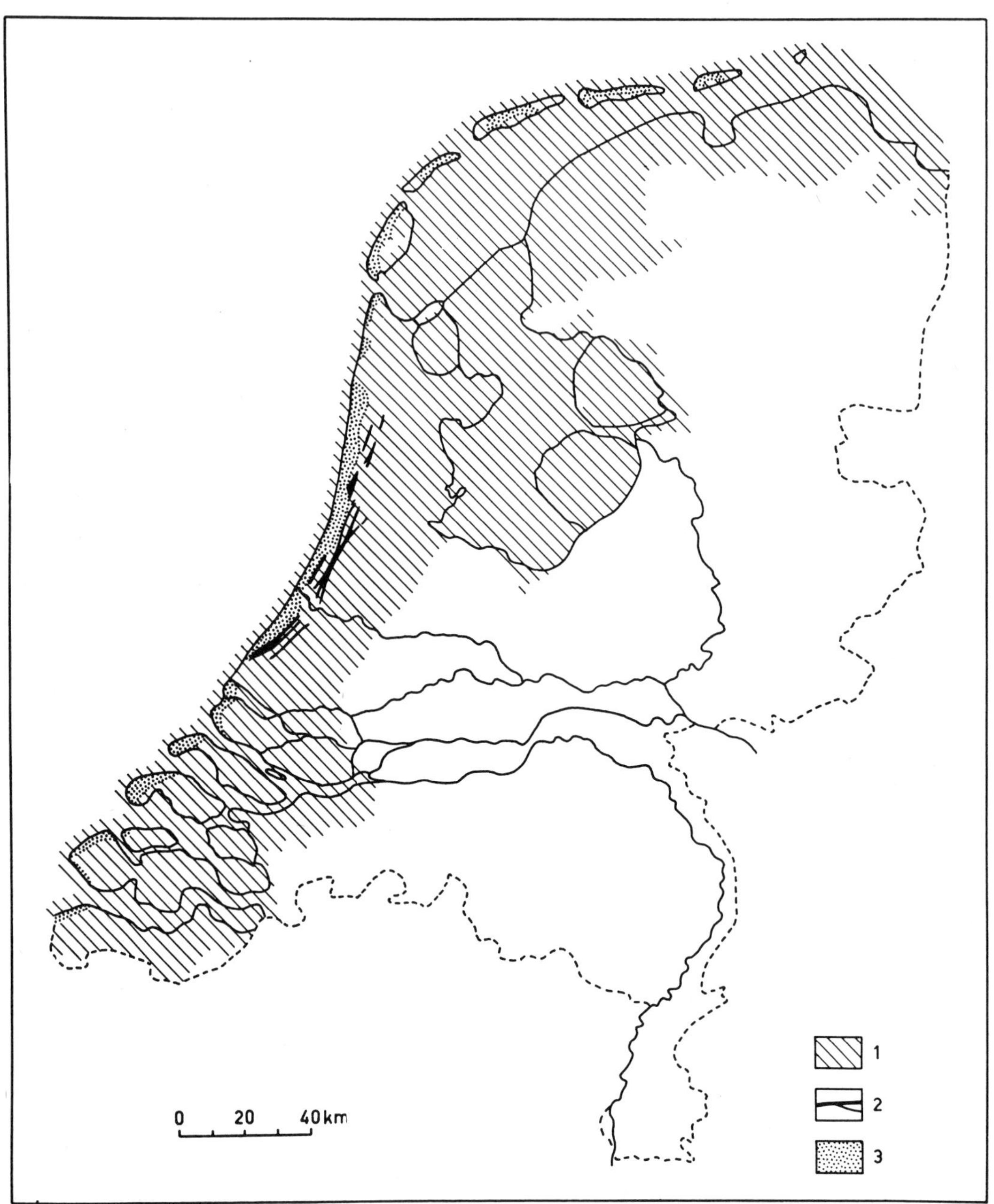

Fig. 1.

Distribution of Holocene marine (littoral s.l.) deposits in the Netherlands.

1. Sediments of beaches, inlets, estuaries, wadden areas, salt marshes and lagoons.
2. Subboreal barrier ridges.
3. Young dunes.

In the course of the last few years more precise data about these lower parts have become available. First, an excavation could be studied that had been opened for the enlargement of the fishing port of IJmuiden. Then, 10 completely cored borings were made in the coastal area near Den Haag, by the State Geological Survey. The sites of the borings were chosen by Dr. W. H. ZAGWIJN, paleobotanist at this institution, in conjunction with the present author. They lie along a line from Scheveningen to Voorburg. Additional data were collected by sampling the sea floor off Scheveningen and IJmuiden.

In the following pages the results of these investigations are given, as far as they are related to lithology, grain sizes, sediment structures and mollusc remains. The author wishes to acknowledge the great help he has received from the Geological Survey. This help did not only involve the above named boring activities, but also the making of numerous grain size analyses and the preparation of hundreds of lacquer peels from the cores.

RIDGES AND INLETS

The distribution of old barrier ridges in the coastal area between Hoek van Holland and Den Helder is shown on figs. 1 and 2. Along the coast itself, the topography of ridges and intervening swales is buried by high sand dunes of younger age. In the southwest the system of barrier ridges is partly cut off by the present coastline. At the other end the inner ridges terminate at and north of Alkmaar between intracoastal[1]) deposits. Here a large inlet has formerly existed. On the other side of this inlet the coast bent out far to the west, well beyond the present shore-line. Consequently, no old barrier ridges are found further northward, the deposits of tidal flats, salt marshes and peat swamps extending till below the dunes and the North Sea beach.

Cross sections through these two different parts of the coast are given in fig. 3. They show that at Scheveningen and IJmuiden the barrier system is well developed, with widths of 8 to 10 kilometres, and with thicknesses of up to 16 metres (below sea level). In the section at Callantsoog, on the other hand, the shore formations are limited to a thin wedge of beach sand and to young, mainly post-Roman dunes. Whereas the coast south of Alkmaar (-Petten) has advanced considerably, owing to the deposition of barrier sediments, the (net) result of coastal processes north of this place has been a recession of the shore-line.

The accretion of the coast south of Petten took place mainly in Subboreal times. During the preceding stages of the Holocene this part of the coast had been retreating, in consequence of the rapid postglacial rise of the sea. In that period it probably consisted of a narrow strip of beach sand, covered by low dunes. Thus the innermost barrier ridge does not represent the first stage of beach sand accumulation along the open sea coast. It only marks the landward limit of the transgressive open sea deposits of the earlier part of the Holocene.

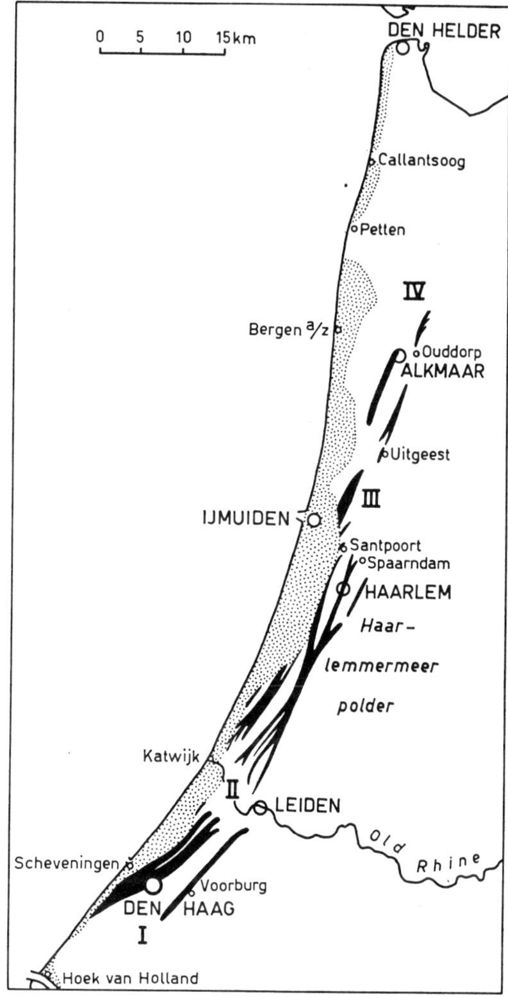

Fig. 2.
Distribution of Subboreal barrier ridges.
I—IV Major inlets or other interruptions.

Between Hoek van Holland and Alkmaar the barrier system was interrupted by three major inlets, which connected the areas behind the innermost ridges with the North Sea (fig. 2). Little is known about the

[1]) Two main categories of littoral sediments may be distinguished, viz. 1) those, formed along the open sea, and 2) those which are deposited on the landward side of coastal barriers. The latter sediments, as well as the environments in which they are formed (salt marshes, lagoons, estuaries, tidal flat areas etc.) are referred to in this paper by the term „intracoastal".

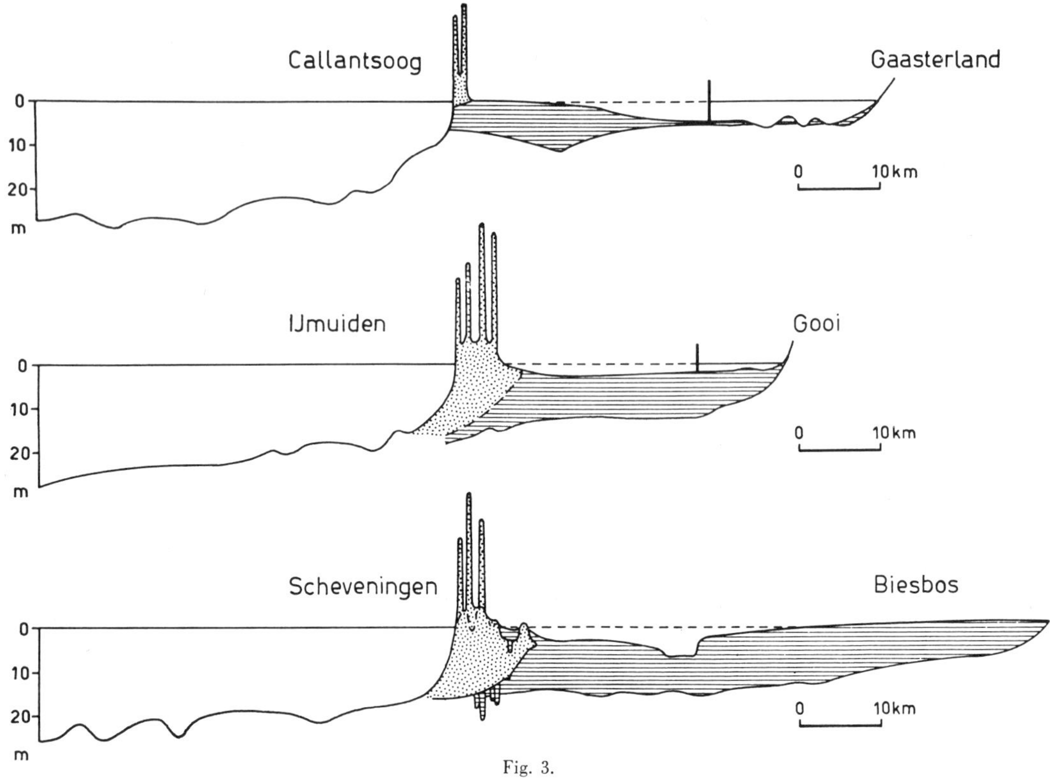

Fig. 3.
Cross sections transverse to coast.
Punctated: dunes, beach sands and submarine deposits of Subboreal and Subatlantic coastal barrier formations.
Horizontal striations: other Holocene deposits.

southernmost inlet, south of Den Haag. The area where it must have been situated has been covered completely by younger deposits (the „Westland-dek"). Its former existence is inferred from the distribution of sandy sediments of tidal flats and channels, exposed more to the east (ref. 17).

The next interruption was formed by the mouth of the Old Rhine distributary near Leiden. From the way the barrier ridges finger out towards this pass, notably on its northern side, and from the shifted position of the northern ridges relatively to the southern ones, it can be deduced that the opening was an active pass during the whole period of coastal barrier formation. The sense of shifting of the successive barrier ridges with respect to each other is the same as that shown by the present barrier islands in the northern part of the Netherlands. It reflects the direction of propagation of the tidal oscillations along our coast from the Belgian to the German side (VAN VEEN, 1936). In the course of its existence, the mouth of the Old Rhine pass has migrated towards the northeast over a distance of some $2\frac{1}{2}$ kilometres. A similar sense of migration is known from many inlets of the present Wadden Sea.

The third opening in the barrier system was located between Spaarndam (-Santpoort) and Uitgeest. Here, no such clear picture of the development of the consecutive ridges on both sides of the inlet is preserved. This is the result of later erosion by tidal channels, traversing the area when the coast had advanced more to the west. Possibly the mouth of this inlet has also migrated along the coast, from south to north.

Besides these major interruptions of the barrier system, minor ones are found in the innermost ridges, viz. south of the town centre of Haarlem, north of Uitgeest, and east of Alkmaar. The opening at Haarlem may be related to the inlet which served as the channel of supply of the intracoastal Hoofddorp deposits. This formation covers large surfaces of the Haarlemmermeerpolder (ref. 8). The opening at Alkmaar bears a similar relation to the Omval deposits, east and southeast of Ouddorp (ref. 16).

However, there is no exact correspondence between the distribution of these intracoastal formations and the openings in the barrier ridge, and the gap at Haarlem is far too small to account for the supply of the extensive Hoofddorp deposits. It can be deduced from the distribution of sandy ridges in this latter formation that the main supply came from a point that was situated a little to the north of the opening in the barrier ridge. The same applies to the inlet at Alkmaar. Sandy ridges of the Omval deposits are cut off by the barrier ridge of Ouddorp. Furthermore, if the Alkmaar inlet had been active during the formation of the inner parts of the barrier system, the ridges on both sides of it should show a slight shift in their relative position, owing to the above mentioned effect of longshore tide propagation. In reality the two lie exactly in line with each other.

It must therefore be concluded that the main activity of the inlets, through which the material of the Hoofddorp and Omval deposits was brought in, had terminated before the first advance of the coast in the beginning of the Subboreal. It is possible, though, that with the filling up of the inlets till about sea level, the beach ridges in this area still underwent periodical overflows during high floods, which prevented the formation of dunes. Then the interruptions in the outcrops of ridge sand would merely be the result of the absence of an eolian cover.

EVOLUTION OF INTRACOASTAL AREAS

Owing to the rapid rise of the sea during the earlier stages of the Holocene, the low areas on the landward side of the beach sand accumulations were covered with water. Much sediment was carried into these areas, from the open sea coast, by flood currents flowing in through the tidal inlets, and possibly also from the land, by rivers. Yet, the supply was insufficient to bring the bottom above the (rising) low tide level. Hence, most of the area had the character of a l a g o o n : the lagoonal environment in which the „Hydrobia-clay" (ref. 24) was deposited. On the inner side of the entrances to this lagoon inner tidal deltas must have developed. Their highest parts may have grown up above low water level, forming tidal flats. Along its eastern shores the lagoon was bordered by peat swamps. Such were the conditions during the early Atlantic stage, e.g. about 7000 years ago. The coast at that time was still lying to the west of its present position.

With continuing rise of the sea the whole sequence of environments, and of their deposits, migrated in landward direction. Thus the peat advanced over the land, the lagoon clays over the peat, the tidal delta deposits over the lagoon clays and the beach sands of the open sea coast over the tidal delta deposits. Meanwhile, the supply of sediment to the intracoastal areas gained on the rise of the water level, so that the tidal deltas grew out over the lagoon floor. The lagoonal parts became more and more reduced in size, and when the North Sea shore finally reached its easternmost position, about 5000 years ago, the beach and dune sands were bordered on their landward side by a wide t i d a l f l a t l a n d s c a p e.

This situation persisted during the first stages of accretion of the open sea coast, when the inner series of barrier ridges was formed. In these times the different sections of the barrier complex, separated from each other by inlets, were therefore normal barrier islands, like those along the present Wadden Sea.

When the barrier system was built out further seaward, the supply of sediment to the intracoastal areas gradually diminished again. Large surfaces of tidal flats, after having been silted up to marsh level, became covered with p e a t s w a m p s. From now on the (organic) upgrowth in most of the areas could keep ahead of the continuing (though slower) rise of the water level. The tidal flat environments became limited to a few separate estuaries. This situation lasted till the end of the period of coastal accretion.

The decrease of sediment supply through the inlets may have been largely caused by the rapid deposition of sand on the seaward side of the coast, which often must have led to a partial plugging of the inlets. At the same time, the reduced access of sea water to the intracoastal areas probably resulted in a freshening of the ground water. This of course, greatly improved the conditions for peat growth and the ensuing organic upgrowth of the surface may in its turn have diminished the possibilities of marine sediment supply.

It must be stressed that the above considerations apply especially to the areas lying directly behind the coastal barrier complexes between Den Haag and Alkmaar. More to the east, and in particular in the Zuiderzee region different circumstances prevailed. Owing to its greater distance from the coast this region never received enough sediment for the development of extensive tidal flats. Throughout the Atlantic and later Holocene times it remained an area of lagoonal, lacustrine and peat swamp environments.

THE EXCAVATION AT IJMUIDEN

The excavation at IJmuiden, mentioned in the introduction, was situated at a distance of about 1 kilometre east of the present North Sea beach, i.e. it exposed sediments of the outer parts of the coastal barrier complex. These sediments were formed along a normal open sea beach, well beyond the influence of tidal inlets. The nearest inlet to the north, at the time of their formation, was about 10 kilometres away, and the nearest one in the opposite direction about 30 kilometres (the mouth of the Old Rhine).

A photograph of an east-west section in the excavation is given on Pl. 10 A. It shows, among other things, that the beds have a slight inclination towards the west, in correspondence with the westward accretion of the coastal barrier system. The angle of inclination is on the average some $2\frac{1}{2}°$. Similar slopes occur on the present beach and the forelying parts of the sea floor.

In a representative section of the excavation the following succession was found (fig. 4):

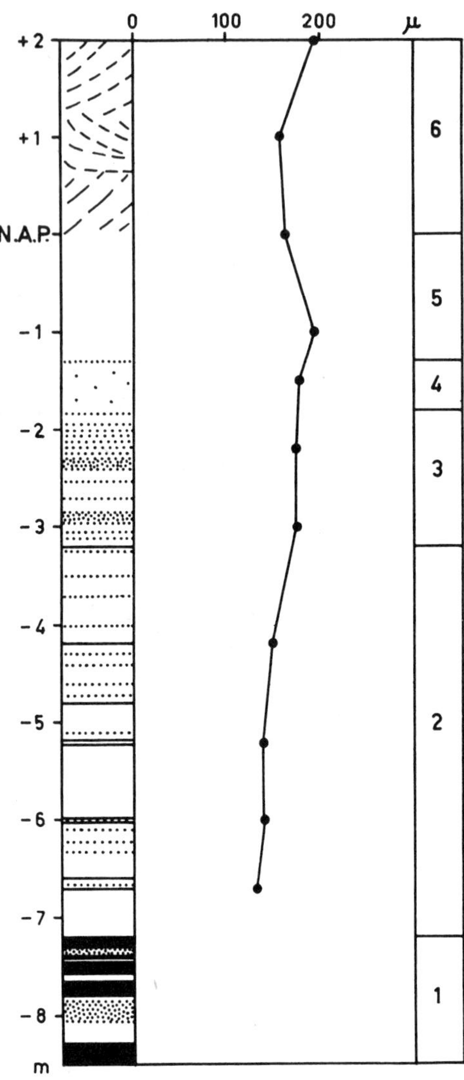

cross bedding

shell beds

clayey laminae

clay beds

median of grain size frequency distribution

Fig. 4.
Succession of strata in excavation at IJmuiden, with median diameters of sand.

6) 2.0—0.0 m above N.A.P. [2]: cross bedded dune sand;
5) 0.0—1.3 m below N.A.P. : sand with approximately horizontal laminations;
4) 1.3—1.8 m below N.A.P. : sand with shells, and with parallel and cross laminations;
3) 1.8—3.2 m below N.A.P. : sand with abundant shells, partly concentrated in separate beds;
2) 3.2—7.2 m below N.A.P. : sand with thin shell beds and with scarce, thin, clayey laminae;
1) 7.2—8.5 m below N.A.P. : alternations of clay and relatively fine-grained sand, with thick shell beds. This deposit was only partly exposed. The data given in fig. 4 are mainly baseu on a core taken from the floor of the excavation.

From 1 to 7 metres below N.A.P. the sand shows a more or less gradual downward decrease of the grain sizes. This must be due to the decreasing effect of the waves on the bottom with increase of water depth. The structures of series 5 are rather typical for beach environments. The scarcity or absence of shells in these sediments then probably indicates deposition on the backshore. The rich concentrations of shells between 2 and 3 metres depth most probably were formed on the lower foreshore and directly in front of the low water line. Taking together all evidence, one may conclude that the sediments corresponding to the original mean sea level lie somewhere between 1 and 3 metres below N.A.P., most probably around 2 metres below N.A.P.

The presence of a clayey formation below a depth of 7 metres in the area of IJmuiden was known already from boring data, and has been mentioned by BENNEMA and PONS (1957). These authors interpreted it as the fill of a former tidal inlet, mainly because they believed it had the outline of a funnel open to the west. However, according to S. JELGERSMA (personal communication), the clay is not at all limited to this funnel-shaped area. Moreover, neither the grain size distributions, nor the mollusc shell assemblages of the deposit in the excavation are in favour of this interpretation. The grain sizes are in general too small for deposition in an inlet, and the mollusc shells definitely point to formation in the open sea, washed shells of species typical for tidal flat areas or estuaries being almost completely absent. As a matter of fact, the presence of clayey sediments in barrier deposits at depths of 7 metres and more is a very normal phenomenon. It was also found in the borings at Scheveningen (fig. 6).

One of the most conspicuous features of the excavation at IJmuiden was the great abundance of ripple structures. The crests of the ripples have at most places an approximately north-south orientation, i.e. parallel to the beach. Evidently they were formed under influence of waves approaching from the open sea. One

[2] All depths and altitudes in this paper refer to N.A.P., the Dutch Ordnance Datum. Along the coast between Hoek van Holland and Den Helder this datum level lies from 3 to 8 cm above mean sea level.

may wonder why typical current ripples were not more commonly observed. At the present day the strength of the tides along the Dutch coast is considerable, and it can hardly be supposed that during the formation of the coastal barrier deposits the tides were much weaker. Since the tidal currents flow parallel to the shore (except around the time of slack water, when their velocities are small), one might expect the crests of the current ripples to be more or less perpendicular to those of the wave ripples.

However, it is known that where waves and a (unidirectional) current simultaneously influence the bottom, and the waves are propagated at a large angle to the current, the ripple marks which are formed have their crests not at right angles, but parallel to the current direction. These ripples have been described as „longitudinal wave-current-ripples" (VAN STRAATEN, 1951, 1953). While their orientation thus is determined by the current, the distance between their crests depends on the dimensions of the waves and the depth of the water. Probably the ripples exposed in the excavation are partly of this combination type. In deeper water, where the influence of the waves on the bottom is smaller, normal transverse current ripples can be expected. They may have been formed in the clayey series nr. 1, but their presence could not be demonstrated because of insufficient exposure of these deposits.

An east-west section through wave (or wave-current-) ripples in series 2, at a depth of about 5 metres, is shown in Pl. 10 B. The ripples were apparently produced during heavy storms, when the water was very rich in suspended mud. When the wave motion had subsided again, part of this mud was deposited as a cover on the rippled sand surface.

While in the lower parts of this series distinct ripple marks are limited to separate levels, they become more and more abundant in the higher parts of the succession, see e.g. Pl. 11. At some places almost the whole sediment is rippled. The majority of the ripples are of asymmetrical development, both as regards profile and internal structures. Their steeper side, and the inclination of their laminations are directed towards the east, i.e. up the general slope of the bedding planes. A similar asymmetry of wave ripples is commonly observed on recent beaches. It is due to the difference in strength between the forward and the backward movements of the water particles in the waves. In normal circumstances the forward motion (up the

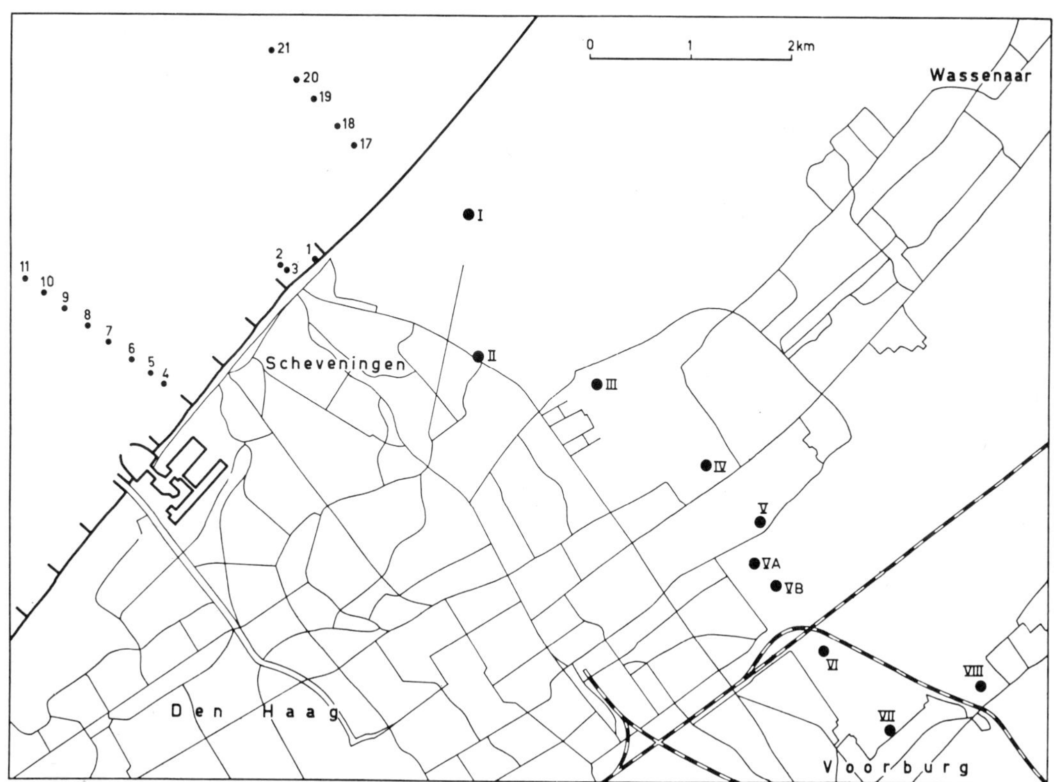

Fig. 5.
Locations of borings between Scheveningen and Voorburg (I—VIII).

beach) is the stronger of the two. It causes a slow upslope migration of the ripples, and hence a (residual) transport of sand in this direction.

THE BORINGS BETWEEN SCHEVENINGEN AND VOORBURG

The locations of the borings between Scheveningen and Voorburg are given in fig. 5. Combination of the data following from these borings (supplemented with data from bailer borings) results in the profile of fig. 6. The profile shows that the coastal barrier formation has more or less the shape of a wedge, pointing in landward direction. The base of the formation rises from about 16 metres below N.A.P. on the seaward side to less than 6 metres along its landward edge.

Most of the wedge rests on older Holocene intracoastal deposits of peat swamps, lakes, lagoons and tidal flat areas, except its outer portion, which is underlain by Pleistocene sand. Both the intracoastal Holocene and the Pleistocene sediments have been incised by fairly deep tidal channels.

The coastal barrier formation consists of open sea deposits (including beach sands), covered by dune sand. Accumulation of the dune sand occurred during the whole period of coastal accretion, but its intensity varied greatly. At many places the eolian sand transport was interrupted for long intervals. There soils developed, which in the depressions between successive barrier ridges became soon covered by peat layers (cf. JELGERSMA, ref. 10). The largest body of peat was formed in the wide depression between the innermost barrier ridge (boring VIII) and the ridge of Huis ten Bosch (boring V).

Before peat growth started in this area, it had passed through a stage of tidal flat conditions. One of the channels in this tidal flat landscape had a depth of 2 to 4 metres (boring VI). Probably the depth of the channels increased towards the southwest, where the area stood in connection with the inlet south of Den Haag (I in fig. 2). More to the northeast the depression must have been connected with the pass of the Old Rhine. The tidal divide presumably was located southeast of Wassenaar.

The strongest accumulation of dune sands took place in the Middle Ages. Thick masses of wind-blown sand then covered the whole seaward part of the barrier system, burying not only the older dune topography, but also most of the peat layers that had developed in the depressions. In that period the coast was retreating again, and it is very likely that this retreat was partly caused by the removal of sand from the beach and the forelying sea floor to these dune areas.

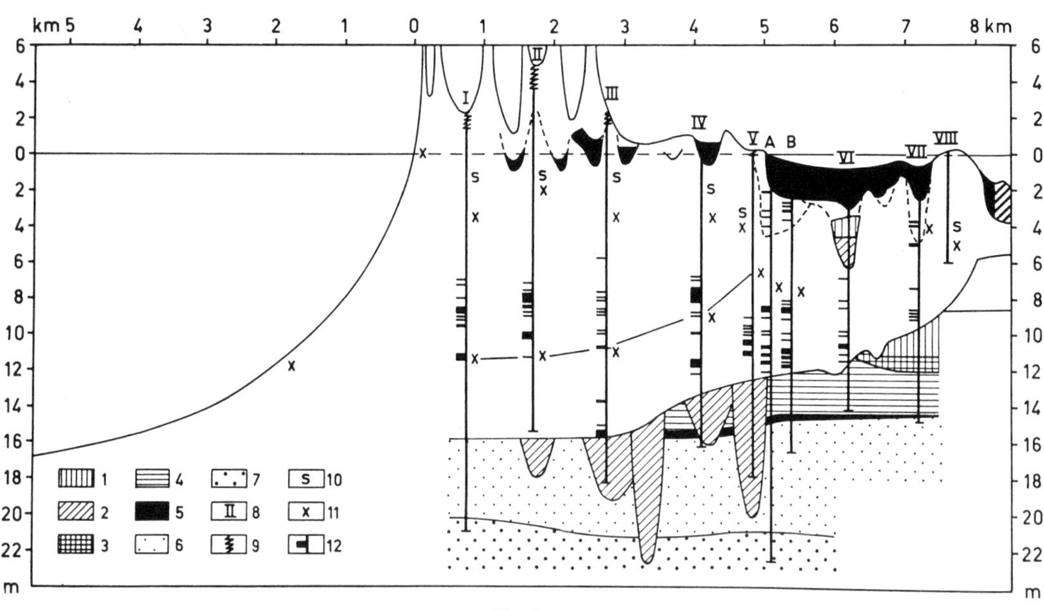

Fig. 6.

General profile of coastal area between Scheveningen and Voorburg.

1. Deposits of tidal flats.
2. Sediments of major tidal channels.
3. Lagoonal clay.
4. Lacustrine clays.
5. Peat.
6. Weichselian (last glacial stage).
7. Eemian (last interglacial stage).
8. Number of boring.
9. Structures disturbed by human activities.
10. Highest shell layers.
11. Maximum of median diameters of sand.
12. Clay layers (simplified).

MOLLUSC SHELL ASSEMBLAGES OF COASTAL BARRIER DEPOSITS

Among the organic remains enclosed in the barrier deposits along the Dutch coast, the bulk is formed of mollusc shells. These also constitute the most important criterion for distinguishing the barrier sediments from intracoastal deposits. Figures 7 and 8 give the percentage composition of 169 samples of mollusc shell material from marine coastal and intracoastal sediments in the Netherlands. The percentages are based on counts of specimens larger than 2.5 mm. Shell fragments larger than 2.5 mm were included in the counting, and their frequencies expressed as broken numbers, on the base of estimates of their relative size (compared to complete specimens). The various species have been grouped together as follows:

Group I (stippled in figs. 7 and 8): *Spisula subtruncata* (DA COSTA), *Spisula solida* L., and possibly other *Spisula* species. *Spisula subtruncata* is by far the most abundant mollusc of the near-shore association in this part of the North Sea. The other *Spisula* species, though much less common is (are) combined with *Spisula subtruncata* because their fragments can rarely be distinguished from each other.

Group II (white in figs. 7 and 8): *Polinices catena* (DA COSTA), *Polinices poliana* (DELLE CHIAJE), *Arca lactea* L., *Mysella bidentata* (MONTAGU), *Venus gallina* L., *Mactra corallina* (L.), *Lutraria lutraria* (L.), *Donax vittatus* (DA COSTA), *Abra alba* (W. WOOD), *Angulus fabula* (GMELIN), *Angulus tenuis* (DA COSTA), *Phaxas pellucidus* (PENNANT), *Ensis ensis* (L.), *Hiatella arctica* (L.), *Thracia papyracea* (POLI) and other species that are more or less characteristic for the North Sea associations along the coast of South- and North-Holland. Some of them also live (locally) in the Wadden Sea, e.g. *Mysella bidentata* and *Angulus tenuis*, but in much smaller numbers.

Group III (horizontal striations): *Macoma balthica* (L.) and *Barnea candida* (L.). These species live both in the coastal zone of the North Sea and in the estuaries and other tidal flat environments of the Netherlands. *Macoma balthica* is one of the main constituents of the mollusc fauna. *Barnea candida*, on the other hand, is a very minor component. It is a borer, and requires a firm bottom of peat or compact clay. It is therefore limited to places where old sediments are laid bare by erosion.

Group IV (oblique striations): *Mytilus edulis* L. This mollusc was originally mainly restricted to tidal flat areas, estuaries and lagoons. At the present day it also lives, in large quantities, along the North Sea shore, where it inhabits artificial structures such as wrecks, harbour moles and groynes.

Group V (cross striations): *Cardium edule* L. This is the dominant species of intracoastal marine environments in the Netherlands. Yet it also lives, in small numbers, along the North Sea beaches, at depths of a few metres (see e.g. VAN STRAATEN, 1961).

Group VI (vertical striations): *Scrobicularia plana* (DA COSTA), *Mya arenaria* L. and other species that are chiefly restricted [3]) to estuaries, tidal flat areas and lagoons.

The first two columns in fig. 7 give the group percentages in a number of grab samples from the North Sea bottom, near Scheveningen and near IJmuiden [4]). The third column shows the composition of mollusc shell assemblages sampled in the excavation at IJmuiden. The fourth, fifth and sixth columns refer to the sediments of large tidal channels exposed in an earlier excavation, at Velsen (see ref. 24). The channel deposits in this excavation were underlain, at depths of 14 to 15 metres below N.A.P., by old lagoonal sediments: the „Hydrobia"-clay (see above). The percentage distributions in columns 1-6 are plotted against depth. The seventh column, finally, gives examples of assemblages from tidal flat areas of the present day, and from the barrier islands along the Wadden Sea. The localities where these latter samples were collected are listed in table I [5]).

TABLE I

Mixed assemblages

A. North Sea beach, De Slufter, Texel.
B. North Sea beach, Nes, Ameland.
C. North Sea beach, Nes, Ameland.
D. Beach of Borndiep inlet, southwest of Hollum, Ameland.
E. Bottom of Vlie inlet, between Vlieland and Terschelling.
F. Bottom of Zoutkamperlaag inlet, southwest of Schiermonnikoog.
G. Bottom of Blauwe Slenk channel, south of Griend.
H. Tidal flat near Griend.
I. Beach, Griend.

Typical assemblages of tidal flat areas, Wadden Sea

J. Tidal flat, Noordpolderzijl.
K. Tidal flat, Lauwerszee.
L. Tidal flat, Noordpolderzijl.
M. Tidal flat, south of Schiermonnikoog.
N. Tidal flat, south of Schiermonnikoog.
O. Tidal flat, Lauwerszee.
P. Tidal flat, Lauwerszee.
Q. Tidal flat, south of Schiermonnikoog.
R. Tidal flat, south of Nes, Ameland.
S. Gully in tidal flat, south of Nes, Ameland.
T. Tidal flat, Oostmahorn.
U. Tidal flat, Lauwerszee.
V. Channel floor, Omdraai.
W. Gully in tidal flat, Eendracht, Texel.
X. Tidal flat, Lauwerszee.
Y. Tidal flat, Eendracht, Texel.

[3]) After a storm in early December 1964 the author found large bivalves of *Mya arenaria* with the animals still in them, washed ashore on the North Sea beach at Bakkum.

[4]) For the IJmuiden area see also ref. 26.

[5]) Most of these analyses have been published before, in another form (VAN STRAATEN, 1956).

Typical assemblages of tidal flat areas, Zeeland

I. Tidal flat near Woensdrecht.
II. Tidal flat near Bergen op Zoom.
III. Tidal flat near Ierseke.
IV. Tidal flat near Ierseke.
V. Tidal flat near Wilhelminadorp.

It is seen in fig. 7 that the quantitative composition of the assemblages from the recent sea floor off Scheveningen and IJmuiden and from the barrier deposits at IJmuiden (columns 1-3) is totally different from that of the material found in the recent tidal flat areas (column 7, middle and lower parts). The first named assemblages are characterized by high percentages of groups I and II, the latter by high percentages of groups V and VI.

Assemblages with a more intermediate composition are found in and along the inlets to the Wadden Sea, on the North Sea beaches of the barrier islands between these inlets (column 7, A-F), and in the old channel deposits at Velsen (columns 4-6). The intermediate character of the samples from Velsen is chiefly due to reworking of *Spisula* shells and other North Sea material from the coastal barrier deposits in which the channels were incised. A smaller quantity of these foreign admixtures may have been brought down from the open sea coast of that time, by flood currents flowing inward through the channels.

While at Velsen the North Sea shells are present as foreign elements in intracoastal assemblages, the opposite situation is found on the beaches of the barrier islands along the Wadden Sea. There appreciable quantities of *Cardium* shells are found in North Sea assemblages. At least a part of these shells comes from the Wadden Sea. They have been carried by ebb currents to the inlets, whence they were further displaced by longshore-drifting. Another part, however, must be of local provenance, i.e. it must have come from the forelying sea floor. Though the numbers of *Cardium edule*, inhabiting the nearshore zone along the Dutch North Sea coast are on the average very small, it seems probable that along the barrier islands somewhat richer populations can develop, at least locally and temporarily. Besides these living *Cardium* populations, there is perhaps another source of *Cardium* shells on the North Sea bottom, viz. outcrops of older deposits, formed in tidal inlets or more inward parts of tidal flat environments. There are indications that at some places in front of the islands such outcrops are present, or have been present in the last few centuries (e.g. off Ameland, see ref. 23) as the result of the landward migration of the coast.

The composition of the assemblages in 7 of the 10 borings between Scheveningen and Voorburg is shown in fig. 8. The high percentages of groups I and II make it at once clear that the greater part of the sediments in these borings, as far as they contain mollusc shells, have been formed in open sea environment. This is further proved by the presence of numerous bivalves of *Spisula subtruncata* (marked with S in fig. 8), which could not have withstood a long transport from the open sea to other environments without becoming unhinged. Moreover, though the majority of the *Spisula* shells are small and delicate, the percentages of broken material are mostly very low.

Comparison of the mollusc shell assemblages of these former open sea deposits with those formed on the recent North Sea floor shows that there is one important difference between the two, viz. the percentages of *Macoma balthica*. In the borings these are on the average much higher than in the recent material. This greater abundance of *Macoma* during the formation of the barrier deposits may have been caused by greater mud contents of the bottom. It is known that this species preferably inhabits muddy sediments (ref. 4; see also the distribution of living *Macoma* specimens off IJmuiden, ref. 26).

In the borings IV, V and VII the open sea deposits are underlain by sediments of tidal flat environments. Their mollusc shell assemblages are characterized by the complete absence of North Sea elements (Groups I and II) and by the high percentages of Group VI. In boring VII these deposits contain a few bivalves of *Scrobicularia plana*, standing in the position of life.

Another intracoastal marine deposit is present in boring VI, above the open sea series. Its lower part was formed in a small tidal channel. The upper part is composed of tidal flat muds, formed above low tide level. Its intracoastal nature is proved by the presence of numerous bivalves of *Scrobicularia* in the living position. The *Spisula* admixtures are probably reworked from the barrier sediments, eroded by this channel.

The subject of the composition of mollusc shell assemblages will be resumed in the section on colours of sand and shells.

LITHOLOGY AND TEXTURE

In the section dealing with the excavation at IJmuiden it was shown that the grain sizes of the coastal barrier sand at that place decrease rather gradually downwards, from a depth of about 1 metre below N.A.P. to the base of the exposure. The decrease is accompanied by an increase of the number and the thicknesses of the intercalated clay layers.

The same situation is found on the recent sea floor off IJmuiden, see figs. 9 and 10. Between the shore and the 7 metres depth contour, the median diameter of the analyzed samples (of this series) drops from about 200—220 μ to 125 μ. However, beyond a depth of 13 metres the bottom is again composed of relatively coarse sand, with median diameters of 210 to 240 μ, and with little or no mud. Between 7 and 13 metres the sediments have a typical mixed character, the grain size distributions showing two maxima instead of one. While the median diameters of the two components in these mixed deposits are approximately constant, their relative quantities vary, the percentages of the fine grained material decreasing, and those of the coarse grained fraction increasing towards greater depths.

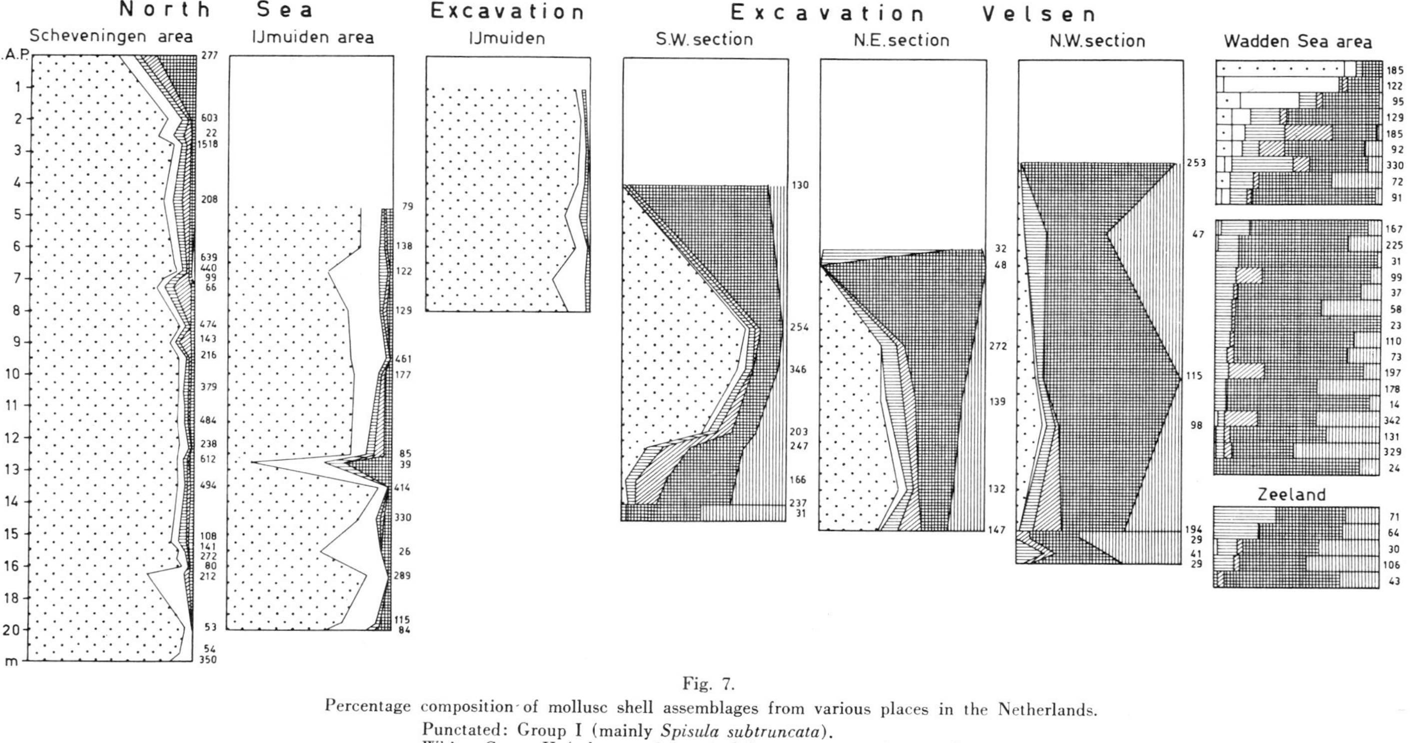

Fig. 7.

Percentage composition of mollusc shell assemblages from various places in the Netherlands.

Punctated: Group I (mainly *Spisula subtruncata*).
White: Group II (other species typical for open sea environment).
Horizontal striations: Group III (mainly *Macoma balthica*).
Oblique striations: Group IV (*Mytilus edulis*).
Cross striations: Group V (*Cardium edule*).
Vertical striations: Group VI (*Scrobicularia plana* and other species typical for intracoastal marine environments).
Numbers on right side of columns: total numbers of counted specimens.

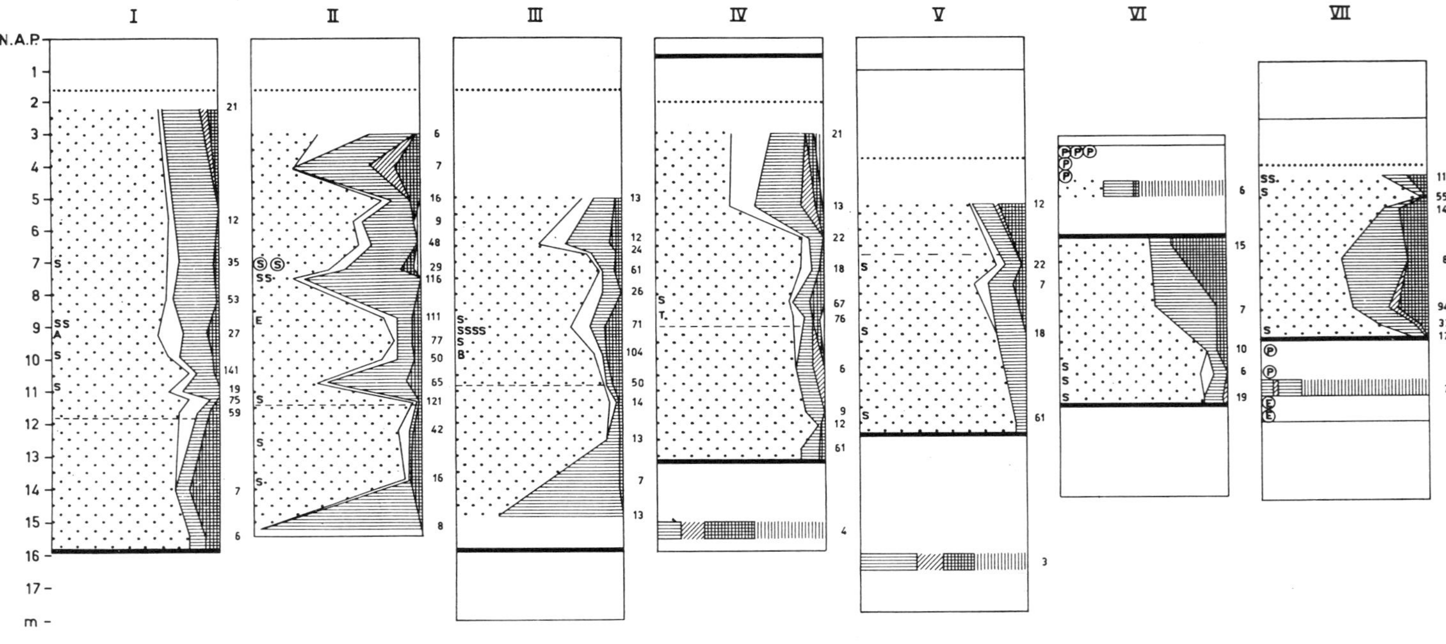

Fig. 8.

Percentage composition of mollusc shell assemblages in borings between Scheveningen and Voorburg.

Heavy black lines: Upper and lower boundaries of open sea-, beach- and dune deposits.
Broken lines: Top of lower coarse grained zone.
Dotted line: Highest shell layers.

S = Bivalves of *Spisula subtruncata*
A = Bivalves of *Abra alba*
E = Bivalves of *Cardium edule*
B = Bivalves of *Macoma balthica*
T = Bivalves of *Angulus tenuis*
P = Bivalves of *Scrobicularia plana*

Circles around S, P, and E: bivalves in the position of life.
For other explanation see fig. 7.

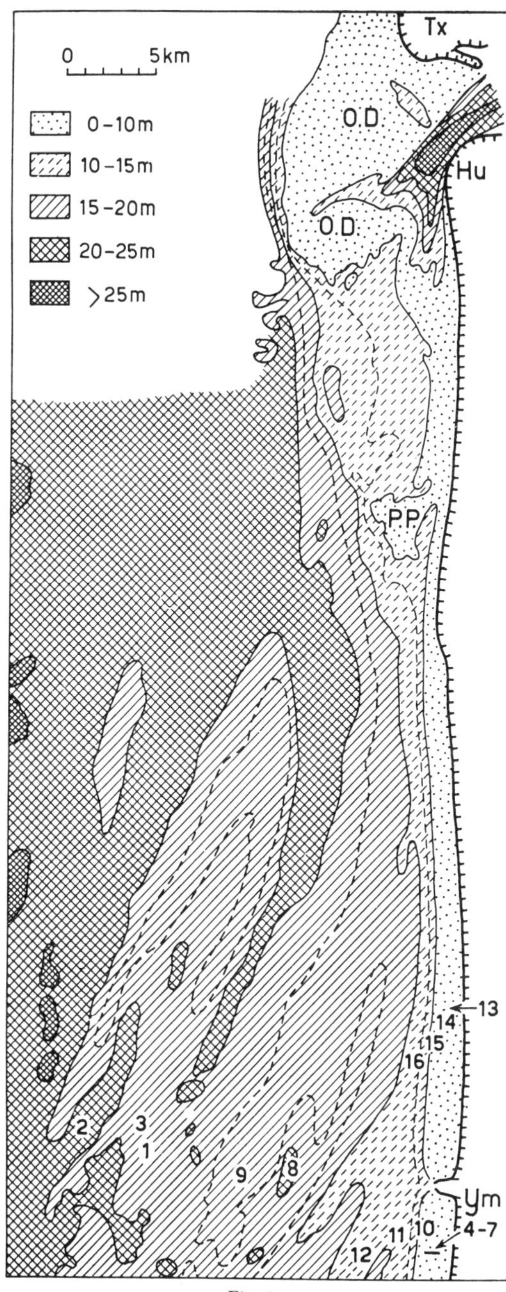

Fig. 9.
Bathymetry of sea floor off North-Holland between IJmuiden (IJm.) and Huisduinen-Den Helder (Hu). Depths in metres below N.A.P. After Chart of Rijkswaterstaat. 1—16 Samples (see figs. 7 and 10).

On the sea floor off Scheveningen (fig. 11) and in the borings near this place the grain size distributions show a similar zonation, see fig. 12. The absolute values of the median diameters are a little different, though, especially in the lower zone of relatively coarse sands. They are distinctly higher in the Scheveningen area than off IJmuiden. In the upper parts of this zone the medians reach values of up to 310 μ (in the borings) and 350 μ (on the sea floor near Scheveningen). Further downward they decrease again, rapidly in the borings, and less markedly on the sea floor. The finer sediments, underlying the lower coarse sands in the borings contain again intercalations of clay layers. They may be distinguished as a fourth zone.

Taken together the recent sea floor deposits and the older coastal barrier formation at Scheveningen thus consist of the following zones (table II):

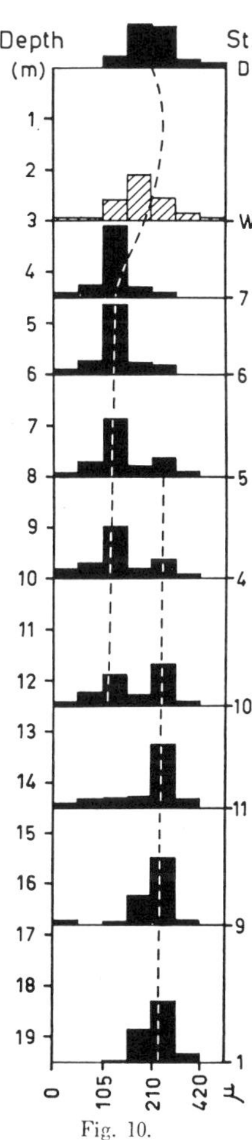

Fig. 10.
Grain size frequency histograms of samples from sea floor off IJmuiden (see fig. 9). Broken line connects median diameters. D = Sample DOEGLAS (ref. 6). W = Sample Boring I (for comparison). Depths in metres below N.A.P.

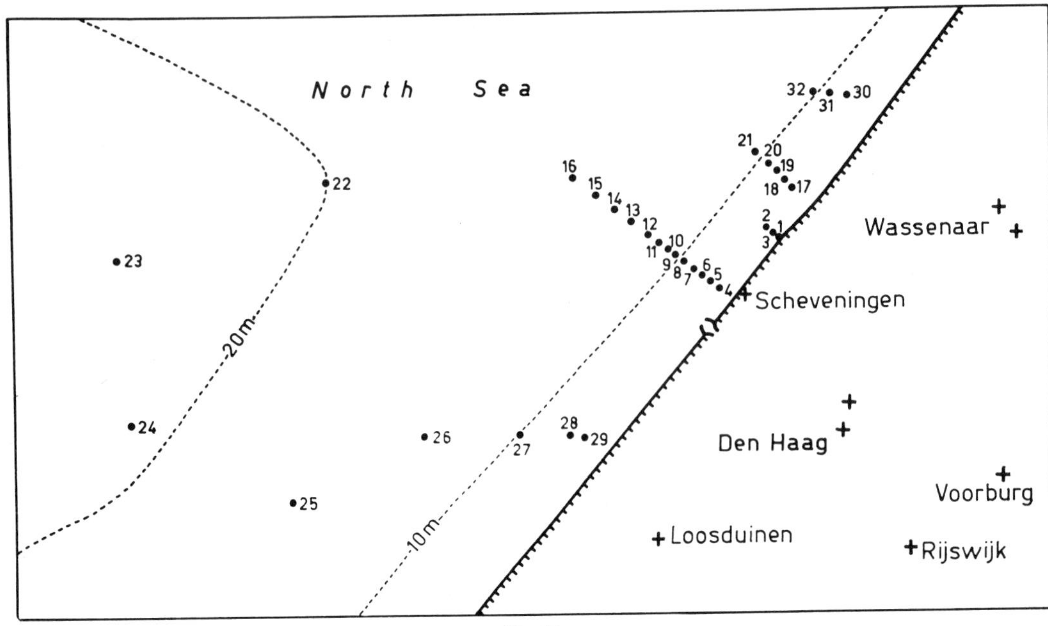

Fig. 11.
Approximate location of samples from sea floor off Scheveningen (see figs. 7, 12, 14 and 18).

TABLE II
Lithologic zones of sea floor and of coastal barrier formation at Scheveningen

I. Upper zone of relatively coarse grained sands, grading downward into:
II. Upper zone of fine sands and muds, which in its lower parts is mixed with material from:
III. Lower zone of relatively coarse grained sands, grading downward into:
IV. Lower zone of fine sands and muds (not clearly marked on the present sea bottom).

The variations in grain size distribution on the present sea floor, from the beach towards deeper water are mainly the result of the different effects of waves, tidal currents, supply of sedimentary material, and rate of deposition. The concentration of relatively coarse sand on the beach and in shallow water is primarily influenced by the waves. They keep the water in this environment almost continuously in strong motion, so that fine-grained particles are only rarely deposited. Most of the fine material is carried off along the coast. A part of it thereby accidentally comes in deeper water, further off-shore, where wave motion, because of its rapid decrease in intensity from the surface downward, no longer prevents its sedimentation. Another part may disappear through inlets into tidal flat areas (or lagoons) where it may be deposited more or less definitely.

Yet, the relative coarseness of the beach and „near-beach" sediments is not only due to the washing out of fine particles. The waves also actively concentrate the coarse elements in shoreward direction, in consequence of the asymmetry of their movements. It was mentioned already before that where waves pass from deeper into shallower water, the forward swing of the oscillating water particles is stronger than the backward motion. This does not necessarily lead to accumulation of water on the beach, since the strong forward movement is usually of shorter duration than the weaker seaward motion. However, in many cases these waves do cause a residual transport of bottom sediment, since the stronger forward motion may be of greater effect than the longer backward swing. This applies especially to coarse bottom material, that can just be taken along by the forward movement of the water, but not by the backward movement. Fine particles, on the other hand, may be carried back and forth over equal distances, or they may experience a slight residual seaward transport as the result of slow, wave-induced currents, superimposed on the oscillations of the water particles (undertow).

In deeper water, further from the beach (the upper parts of zone II of the above classification) the influence of waves on the bottom is mainly limited to occasional storms or periods of strong swell. Coarse sand is present only in subordinate amounts, since it is kept concentrated on and along the beach by the asymmetrical motion of the waves. Fine sand and mud, on the contrary, are supplied in considerable quantities. The mud is chiefly deposited at the turn of the tides, when the currents are weak. In many cases all mud that settles to the bottom at these moments of slack

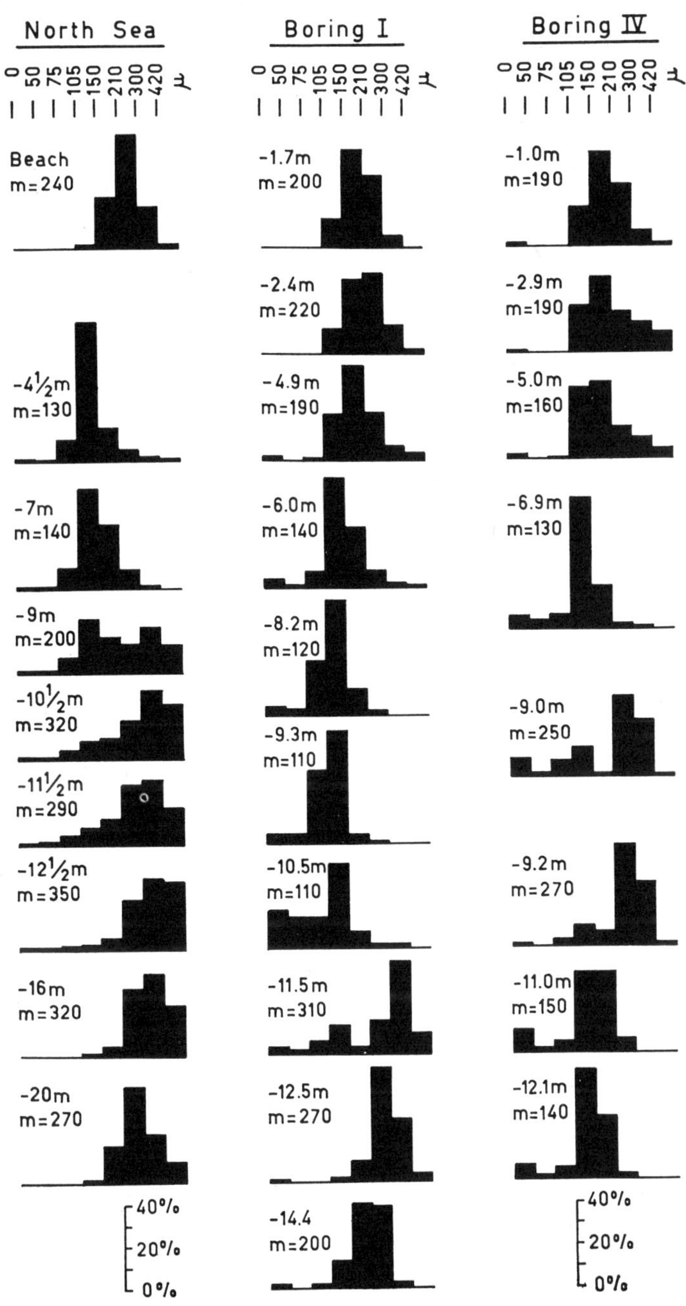

Fig. 12.

Grain size frequency histograms of samples from sea floor off Scheveningen (fig. 11) and from borings I and IV (fig. 5). Depths in metres below N.A.P.; m = median diameters.

water, is eroded again during the subsequent stages of higher current velocities. However, it also happens, from time to time, and from place to place, that these stronger currents bring on more sand than they carry off. Then part of the mud is preserved, because it is covered by sand before it can be eroded. In such conditions the mud layers are soon compacted a little, whereby their resistance to possible later erosion is strongly increased.

Thus the amounts of mud that are encorporated in the sediment depend to a large extent on the rates of sedimentation. Where these latter are high the resulting sediments may contain appreciable quantities of mud. On the contrary, where they are low practically all mud that is deposited during slack water is sooner or later removed during stages of increased current activity. The fact that the sediments of the fine grained zone are relatively rich in mud, notwithstanding the strength of the tidal currents, may therefore be ascribed to the high rates of deposition. This does not necessarily mean that the average, resulting upgrowth of the bottom has also been rapid. It only means that when sedimentation took place, it occurred swiftly. High mud contents may also be produced when the stages of deposition alternate with stages of erosion, now here, then there, so that the final result is a slow vertical accretion.

In the lower parts of this zone the fine sand and muds are mixed with sand of larger grain sizes. The appearance of coarse elements in these deposits has nothing to do with increased strength of the currents (or the waves). It is merely due to the vicinity of the coarse grained third zone of the succession, which lies directly below them, and which forms the sea floor further offshore. Part of the coarse sand admixtures may be of local provenance, having (originally) been reworked from the surface of the underlying formation before it was completely covered by fine grained material. Another part may come from further seaward, under the influence of large waves and tidal currents. This supply probably continues at the present day.

With increase of depth the relative amounts of fine sand and mud gradually diminish and in the following zone of coarse sand their percentages are generally quite small. To a certain extent this decrease may be caused by lessening of the supply. Most of the fine material that is transported in the sea, is originally supplied by rivers and by coastal erosion, i.e. it starts its marine peregrinations in the coastal zone. Since the tidal currents move very predominantly parallel to the shore, it is mainly spread out along the coast itself. Only relatively small quantities are dispersed further seaward, by „diffusion" and under the influence of secondary currents, such as undertow and rip currents.

However, it is not unlikely that this seaward dispersion, at least of part of the material, is counteracted by the asymmetrical character of the wave motion. Just as the waves in shallow water, owing to this asymmetry, tend to sort out the coarse sand and move it towards the beach, they may have a directive effect on the movements of fine sand on the lower parts of the subaqueous slope, so that this latter is prevented from spreading out further from the coast. Then only silt and clay floccules are supplied to the more distant parts of the sea floor. But in the absence of sufficient fine sand little of this mud can be enclosed in the sediments. While the tidal currents at these depths are strong enough to move large quantities of fine sand (if available) during a single stage of ebb or flood, they probably can transport in this time only minor amounts of the coarse sand which here forms the bottom. Hence most of the mud that is deposited at the turn of the tide beyond the outer limit of the fine sands, must disappear again in suspension during the following stages of the tide.

The above considerations apply especially to cases where the sea bottom slopes gradually down from the beach towards deeper water (fig. 13 A). Complications arise when in front of the coast a large subaqueous bar is present (fig. 13 B). Such a bar brings about a diminishing of the wave movements in the depression behind it. Large waves approaching the coast are caused to break over the bar. They lose much of their energy, and can have only little effect on the bottom of the deeper area behind it. Sometimes the formation of such a bar may lead to a considerable increase in strength of the tidal currents passing through the depression on its shoreward side. But if this does not happen, relatively fine grained sands and muds may be deposited in it.

Under certain conditions the bar may grow up to such a height that fine-grained, muddy sediments (zone II) are formed at much smaller depths than would have been possible without this sheltering effect. The outer, seaward slope of the bar, on the other hand, is exposed to the full strength of the waves, and the possibilities of mud deposition on that side are accordingly reduced. It the rates of deposition are not particularly high, or if the seaward slope is even slowly

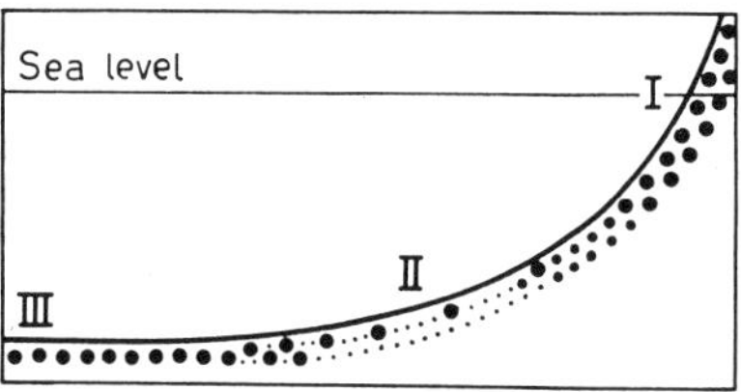

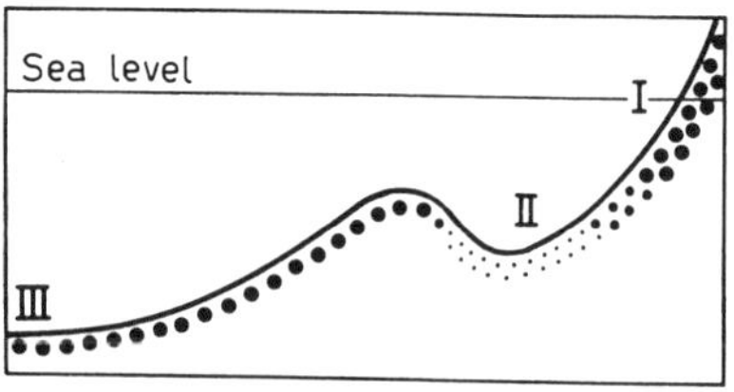

Fig. 13.
Disrtibution of coarse and fine sands
above: along coast without submarine bar, and
below: along coast with submarine bar.
I—III: Lithologic zones of table II.

eroded (e.g. when the bar migrates in landward direction), its surface will then be composed of relatively coarse sand, at depths at which elsewhere the second zone of the classification is found.

It is probable that the (older) coastal barrier deposits were formed under the complicating influence of such bars [6]). The origin of their lithological zonation will be dealt with in the section on correlation of strata.

In order to get a more continuous picture of the median diameter of the sand in the borings all lacquer peels and cores have been studied by means of a binocular microscope. The dominant grain sizes of the sand laminae were measured at intervals of a few centimetres. On account of these data estimates were made of the average median diameters for each successive 25 centimeters of core length. These averages were plotted in diagrams against depth, and connected by lines which were simplified a little by smoothing out minor irregularities.

The position of the lines was then checked by 27 sieve analyses. It was found that they showed a systematic error (to the coarser side), which was greater in coarse sand than in fine-grained material. The course of the median lines corrected for this error is given in fig. 14. Along the left side of the columns in this figure the average (estimated) clay contents of each successive 10 centimetres of core length is added, as far as there were no gaps between the cores.

It is seen in fig. 14 (cf. fig. 6) that the top of the lower zone of coarse sand rises in landward direction from boring I to boring V. From there it slopes down a little and disappears between borings V B and VI. These depth variations are more clearly illustrated by fig. 15, which gives the distribution of the median diameters in the whole profile, as deduced from the boring data. It follows from figures 14 and 15 that the fine grained sands and muds between the upper and the lower zones of comparatively coarse material are not limited to the present sea floor and borings I and II (fig. 12), but that they extend to borings IV and V. Thus the greater part of the beach and dune sands of the barrier formation are separated from the lower coarse sands by a continuous zone of fine grained sediments.

This situation originally led the author (ref. 26, 27) to conclude that the bulk of the beach and dune sand (and the fine grained sediments as well) had been supplied by longshore drift and not by transverse transport of sedimentary material from deeper water to the coast. Since then the author has come back from this opinion, though he still feels that longitudinal transport must have played a very important part in providing the sediment. The main argument that seemed to point to the dominating influence of longshore supply was based on the grain size distributions of the upper fine-grained zone and on the assumption that the bedding planes cut obliquely through this zone (fig. 16), in other words that is accretion has taken place both in vertical and in seaward direction. On account of what is known from other coasts this assumption appeared indeed very likely. But in that case it was impossible to explain the coarse sand, deposited in shallow water and in the beach and dune environments, as being derived from the lower zone of coarse sand on the deeper sea floor. For it should then have passed the areas where the fine-grained and muddy deposits were formed, and notable quantities of this coarse material should have become trapped between the finer sediments. In reality the percentages of coarse elements in this fine-grained zone are mostly extremely small, the fraction $> 210\ \mu$ amounting in many cases to less than 1%.

The argument is invalid, however, if the fine grained zone has been mainly formed by vertical accretion, and the layers have accordingly a more horizontal position. Then a much greater portion of the sand in the upper coarse-grained zone may have come from deeper water, viz. by transport over the top of the fine-grained formation, instead of by passing „through" it. As a matter of fact, an approximately horizontal position of the bedding planes was surmised already several years ago by W. H. ZAGWIJN, who investigated the pollen content of the clay layers. More recently his interpretation seemed to be confirmed by the detailed study of the succession of lithology and structures (see correlation of strata).

COLOURS OF SAND AND OF MOLLUSC SHELLS

When sieving samples from the sea floor off the Dutch coast one is struck by the difference in colour of the successive grain size fractions. In general the coarse grades are rusty yellowish-brown, while the fine sand fractions have more greyish colours. As a rule the colours vary quite gradually in dependence of the grain sizes.

Microscopic examination reveals that the brown colour is caused by thin coatings on the surface of the sand grains, which apparently chiefly consist of iron hydroxides. These coatings are much more developed on large grains than on smaller ones.

The explanation of these differences in thickness of the coatings may be sought at least partly in the „antecedents" of the grains. Obviously, precipitates of iron hydroxides can only be formed in well oxygenated environment, i.e. they must have been formed when the sand grains were lying at or near the surface of the sediment, or while they were moving over it. In

[6]) It may be remarked that off IJmuiden low ridges are found in front of the coast (fig. 9), which may represent the remnants of formerly higher bars (fig. 24). During his sampling at sea off Scheveningen the author got the impression that similar ridges, but even lower than at IJmuiden, were present in this latter area. However, the place and depth observations were not accurate enough to allow more definite conclusions on this point. No relief is therefore shown on the profiles of the present sea floor in figs. 6, 15, 17, 23 and 26.

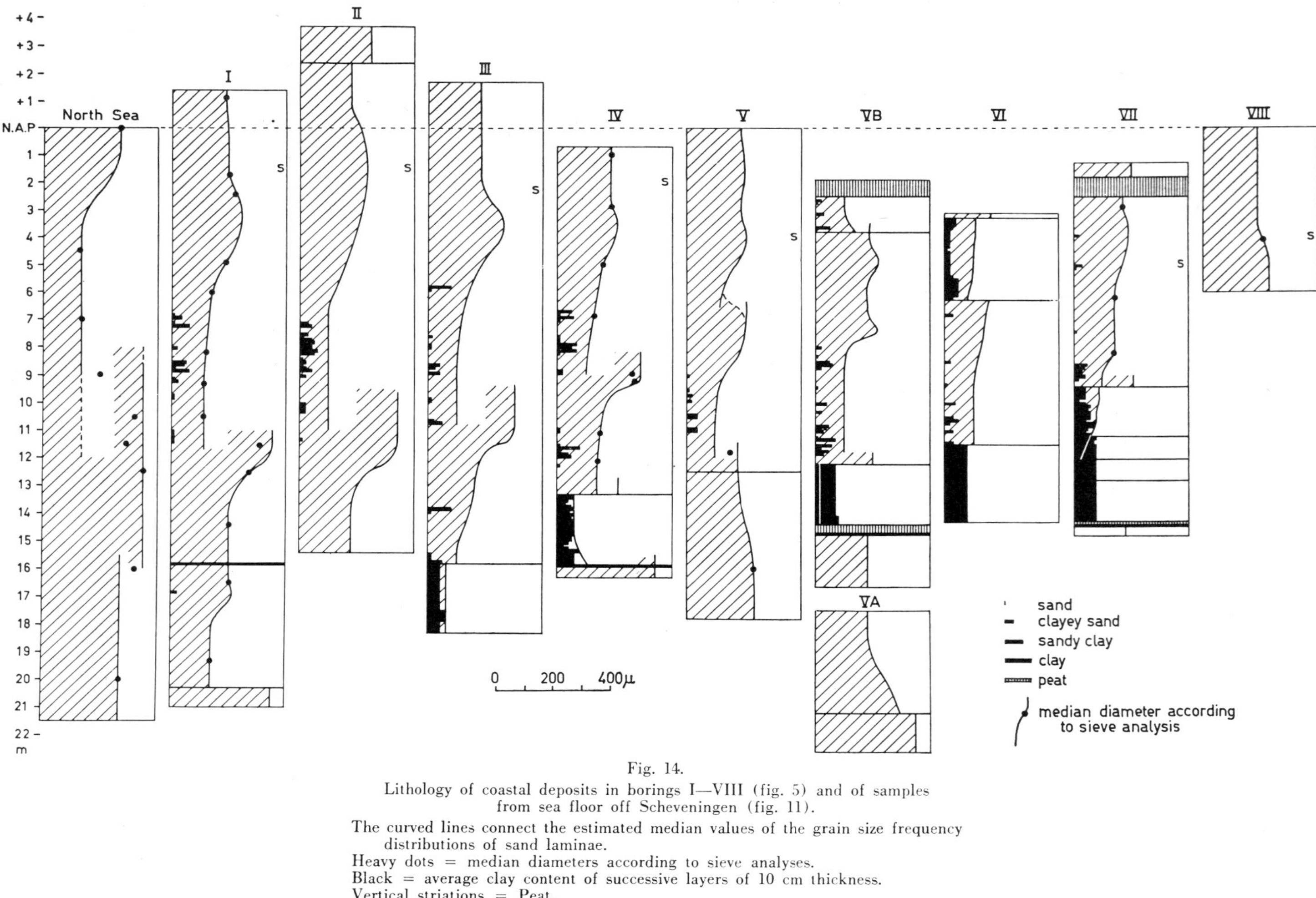

Fig. 14.

Lithology of coastal deposits in borings I—VIII (fig. 5) and of samples from sea floor off Scheveningen (fig. 11).

The curved lines connect the estimated median values of the grain size frequency distributions of sand laminae.

Heavy dots = median diameters according to sieve analyses.
Black = average clay content of successive layers of 10 cm thickness.
Vertical striations = Peat.
S = Highest shell layers.
Horizontal lines = Boundaries of formations and sedimentary units (see fig. 6).

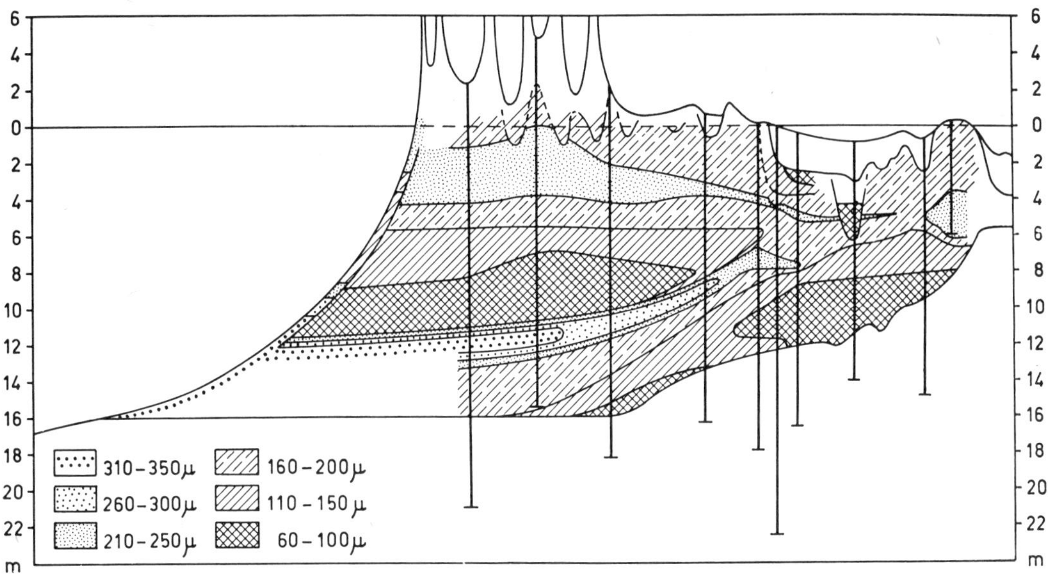

Fig. 15.
Inferred distribution of median diameters of sand laminae (see fig. 14) in profile from Scheveningen to Voorburg.

any case they cannot develop under anaerobic circumstances such as prevail at certain depths below the sediment surface. On the contrary in such conditions the coatings sooner or later disappear, whereby first black coloured, colloidal iron-monosulphides are produced, and then crystals or aggregates of crystals of iron bisulphide (ref. 21). The latter adhere much less strongly to the grains than the hydroxide films. When the sediment is eroded, and the grains are transported again through oxygen-rich environment, the precipitates at their surface wear off for a great part before they can be re-oxydized.

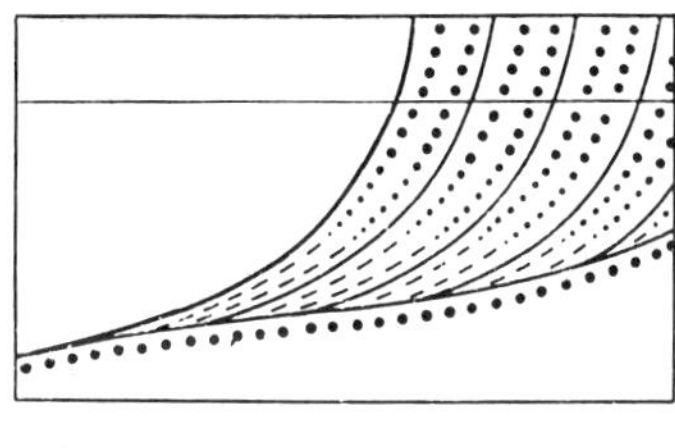

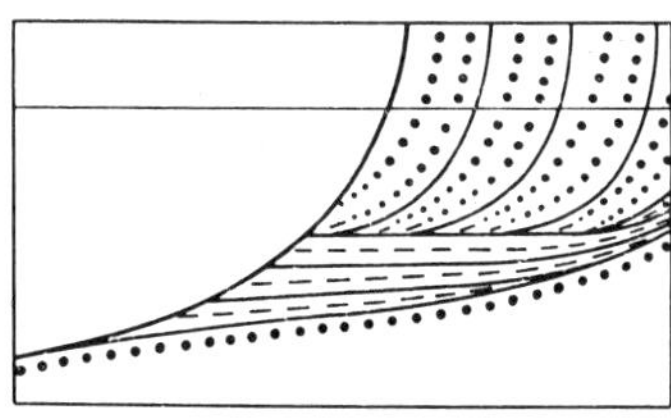

Fig. 16.
Two possibilities of formation of fine-grained deposits (rich in clay) on lower parts of subaqueous slope.
Above: Simultaneous deposition of fine-grained material in deeper water and of coarse sand on and along the beach;
Below: Deposition of fine-grained sediments completed before deposition of coarse sand on the upper parts of the slope.

Now, anaerobic conditions develop far more rapidly and to a much higher degree in fine-grained, muddy sediments than in coarse sands. This difference is not only due to the concentration of organic matter in fine-grained deposits (in consequence of its low settling velocities), but also to the different rate of ground water circulation. In muddy sediments lying on the sea floor the ground water must be practically stagnant. For these reasons it is to be expected that fine sand grains, which much of the time lie embedded in or between muddy deposits are more liable to lose their hydroxide coatings than coarse ones.

Comparison of sand samples from different localities proves that no fixed relation exists between grain size and colour. Equal size grades of different samples show a very unequal development of the iron-hydroxide coatings. Their average thickness is seen to vary with the mud content of the surrounding sediment. This is in agreement with the above given explanation.

The study of the sediment colours was extended to samples from the borings that were made through the barrier deposits between Scheveningen and Voorburg. One would perhaps expect that all these older sediments, as far as they lie below the ground water level,

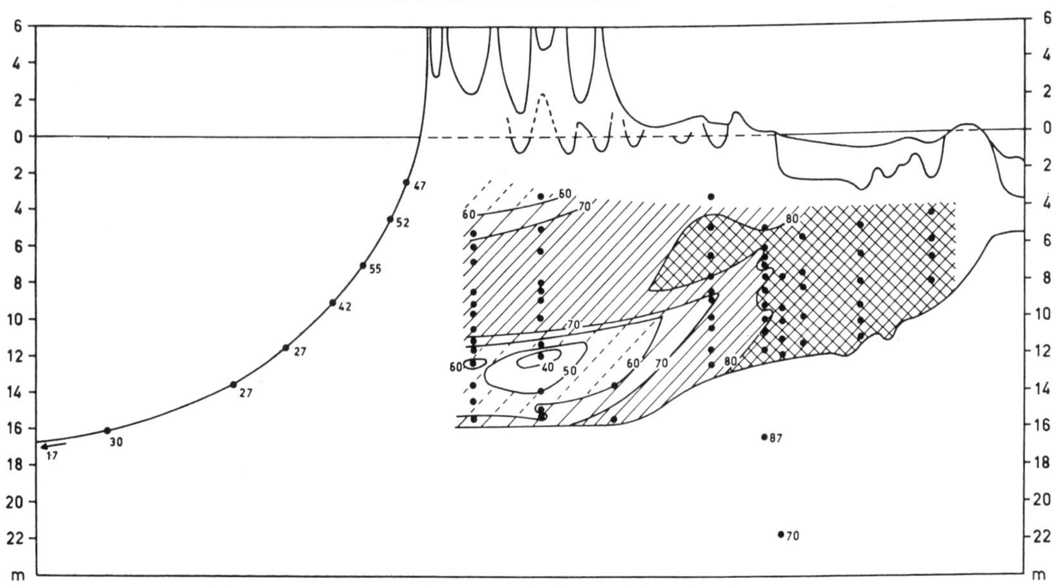

Fig. 17.
Distribution of indexes of greyness of coastal barrier sand between Scheveningen and Voorburg.

are of a uniform grey colour. However, though almost all samples are distinctly less brown than those from the sea floor, they show among each other considerable differences in the intensity of grey shades.

To investigate these relations more closely the following method was used. The sand was sieved into 4 fractions, of respectively 400—275 μ, 275—200 μ, 200—150 μ and 150—100 μ. Then equal fractions of all samples were arranged in order of decreasing brownness and increasing greyness. The four series were then each divided into 9 successive classes, and for each sediment sample the average of the class numbers of the 4 fractions was taken and multiplied by 10: the „index of greyness". In very fine-grained sediments only the fractions 200—150 μ and 150—100 μ could be used. Class numbers of the coarser grades were then obtained by extrapolation. The distribution of these indexes in the borings is shown in fig. 17. By comparing this distribution with that of the median diameters (fig. 15) it is seen that in the borings there is also a fairly close relation between greyness of the sand and general texture.

Regarding the origin of the coating substance, it seems probable that at least part of it was present on the sand grains before they were brought into the sea. In samples from the bed of the Rhine (taken near Vianen and Tienhoven) the sand grains appeared also to have coatings, though admittedly these are of a more purple-brown colour. Another part of the iron hydroxides may have been precipitated when the sand grains were lying on the sea floor. This latter conclusion is based on the examination of the colour of mollusc shells.

It was found that the shells on the sea floor off Scheveningen, sampled at depths from about 10 to 16 meters are dominantly brown coloured. In many cases the colouring substance has penetrated the whole shell. In other instances it is limited to certain layers. In addition, a precipitate of brown material is often present on the outer surface of the shells. Their surface then may have a fatty lustre. Chemical analyses [7]) of such shells prove that the brown precipitate indeed largely consists of iron compounds, see table III. Since it is most unlikely that the molluscs themselves have precipitated so much iron, it must be concluded that it was deposited later on. When such strong precipitation can take place on shells lying on the sea floor it is very probable that at least part of the coatings of the sand grains have originated under the same circumstances.

TABLE III

Iron contents of samples of mollusc shells, fractionated according to colour (North Sea floor)

Sample	Locality (fig. 11)	Depth	% Fe_2O_3
1. Fresh, white shells	12	13½ m	0.10
2. Fresh, white shells	13	15½ m	0.24
3. Old, yellowish shells	10	11½ m	0.28
4. Old, yellowish shells	12	13½ m	0.38
5. Old, brown shells	11	12½ m	2.43
6. Old, brown shells	12	13½ m	3.19
7. Old, blue shells	4	4½ m	0.27
8. Old, blue shells	5	7 m	0.20

[7]) These analyses were kindly carried out by Dr. A. J. DE GROOT (Groningen).

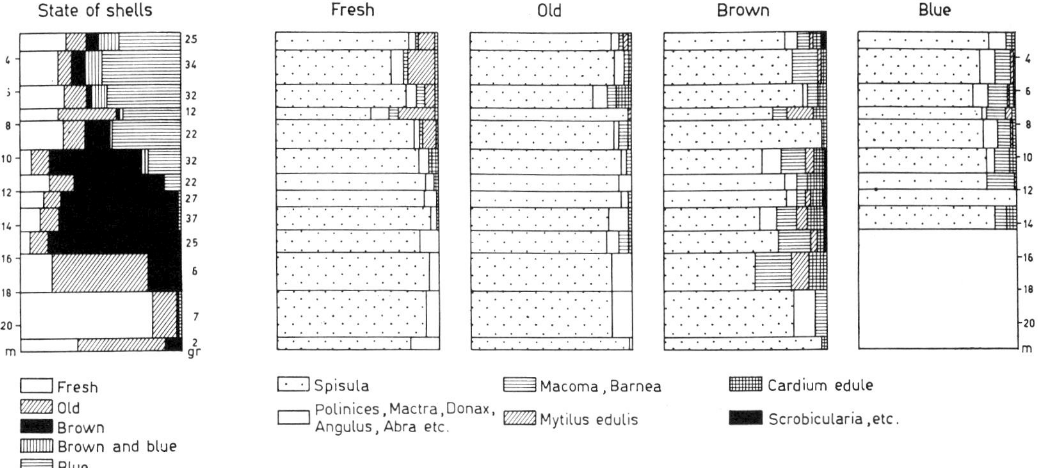

Fig. 18.

Colours of mollusc shells from sea floor off Scheveningen.
Left column: Weight percentages of colour fractions. On right side of column: total weight of samples.
Other columns: Group percentages (cf. figs. 7 and 8) of mollusc shell samples, fractionated according to colour. For numbers of specimens counted, see table IV.

TABLE IV

Station (fig. 11)	Depth	Fresh	Old	Brown	Blue
2	2½ m	188	51	19	180
4	4½ m	56	17	9	84
5	6¾ m	44	16	8	68
7	7 m	52	8	3	38
8	9 m	213	17	8	82
9	10½ m	34	21	42	19
10	11½ m	63	36	54	11
11	12½ m	72	32	80	4
12	13½ m	158	58	102	3½
14	15½ m	26	18	58	—
16	16 m	32	29	4	—
22	20 m	36	18	7	—
23	21½ m	33	20	3	—

The brown shell material mostly has an old, worn appearance. Much of it consists of shell fragments. Fresh-looking shells, on the other hand, are often white, cream-coloured, grey or pink. In general one may distinguish in the samples from the sea floor between: 1) fresh specimens with the original colours unchanged; 2) old-looking, more or less yellowish or very light brown shells; 3) intensely brown shell material; 4) blue coloured shells; and 5) shells that are partly blue and partly brown. The relative weights of these categories in a number of samples from varying depths are given in fig. 18.

The blue colour of shells is caused by infiltration of iron sulphides, and thus is produced only when the shells have stayed for a sufficiently long time in strongly reducing environment. The distribution of the blue shells should therefore show some relationship to the texture and mud content of the sediments. As a matter of fact, it follows from fig. 18 that the blue colours are almost exclusively found at depths smaller than about 12 metres. This approximately coincides with the outer limit of deposition of the fine grained and muddy deposits (fig. 14).

While therefore it seems warranted to conclude that at least some of the blue shells got their colouring under a cover of mud on the present sea floor, other shells must have been coloured in an earlier period. The lower limit of their distribution on the sea bottom not only coincides with the outer limit of deposition of fine-grained material at the present day, but it also corresponds to the base of the fine-grained zone in the older barrier deposits (fig. 6). It is known that in post-Roman times the coast has undergone a considerable recession. This retreat must have been accompanied by erosion of the shallow parts of the sea floor in front of it. During this erosive stage blue shells must have been washed out from the upper fine-grained zone of the older sediments.

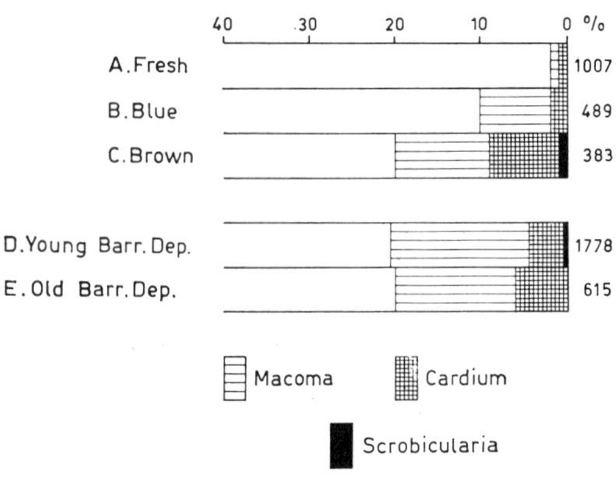

Fig. 19.

Average percentages (by number) of *Macoma balthica*, *Cardium edule* and *Scrobicularia plana* in colour fractions of samples from sea floor off Scheveningen and in old coastal barrier deposits.

A—C: Averages of all samples of fig. 18;
D: Average of all samples of fig. 8 above top of lower coarse grained zone (broken line);
E: Average of all samples of fig. 8 below top of lower coarse grained zone.
Total numbers of specimens counted are indicated at right.

Such a mixed origin of the blue shells is in agreement with the quantitative malacological composition of the blue shell material, see fig. 19. On the average this is approximately intermediate between the composition of the assemblages of fresh shells and that of the material from the upper series (zones I and II) of the older barrier deposits, notably as regards the percentages of *Macoma* and *Cardium*.

Although most of the blue shells acquired their colours in the zone of fine-grained and muddy deposits, they are also abundant in the beach- and „near-beach" sands. On the contrary, only quite subordinate percentages are found below the lower depth limit of the muddy deposits. This confirms the theory that, owing to the asymmetrical character of the wave motion (and its decreasing intensity towards deeper water), the coarse bottom particles are mainly transported up the subaqueous slope, in shoreward directions.

In contrast to the blue shells, which are found both in the sea and in the old barrier deposits, the brown ones are limited to the present sea floor (apart from a few exceptions in zone II of boring II). If they have originally been present in the old sediments, their brown colour must have disappeared or changed into blue under the influence of the reducing conditions. Since the sand grains in these sediments locally are still covered by remnants of brown coatings, it may perhaps be concluded that the colouring substance adheres more strongly to quartz grains than to carbonate material. Perhaps the iron is precipitated in a slightly different form.

As stated above, the brown shells have their maximum abundance on the sea floor at depths between 10 and 16 metres. This must be of significance with regard to the origin of the iron compounds. One possibility which has to be considered is that the iron is precipitated from the sea water. Then the abundance of the brown shells would have to be interpreted as the result of a very long time of exposure to the sea water. However, leaving aside the difficulty to explain why they should have been exposed so much longer than e.g. the shells that lie on the sea floor beyond this depth of 16 metres, it is very doubtful whether such great quantities of iron can be precipitated in the course of only a few thousand years. Sea water is extremely poor in dissolved iron, and moreover, no indications for such precipitation are found in other shelf seas at places where the shells have been exposed to sea water for more than 10,000 years.

The most likely explanation in the opinion of the writer is that the iron was deposited from ground water, emerging on the sea floor. It may be taken for granted that water has been emerging on the subaqueous slope of the barrier, because rains and other atmospheric precipitation caused a constant overpressure on the ground water body in the barrier formation, which was only partly compensated by subaerial drainage. One result of this outflow of ground water on the sea floor has been dealt with in earlier papers (e.g. ref. 23), viz. the development of thin slabs of recent sandstone and mudstone, which are cemented by calcium carbonate. Very probably the deposition of iron hydroxides is due to the same phenomenon. The iron which in the reducing environment of the barrier sediments was present in the ground water as easily soluble bivalent compounds, must have been precipitated when it became oxydized to ferric form during the emergence of the ground water on the sea bottom.

Presumably this ground water flow was concentrated in the coarser grained sediments, lying above and below the (upper) zone of fine sands and clays (fig. 15). The strongest flow may have passed through the sands immediately underlying this clayey series, because of their large median diameters and low contents of fine admixtures (fig. 12), accounting for a high permeability. This precisely corresponds to the zone of maximum abundance of brown shells. Most of these found at higher levels evidently have been derived by wave action from the area where the lower zone of coarse sands crops out on the sea floor.

The average quantitative composition of these brown shell assemblages (fig. 19) has a much greater similarity to that in the old barrier deposits than to the composition of the assemblages formed by the recent fauna. Apparently the latter have undergone little deposition of iron hydroxides. It is not known whether this is due to the slowness of iron precipitation, or to a strongly decreased ground water flow in the recent past. Another possibility is perhaps that the maximum precipitation of iron takes place at a certain depth below the sediment surface. Since the sea floor in these parts has probably not been raised by sedimentation of new sand layers, and rather seems to have been eroded, the precipitation of iron hydroxides would then necessarily be restricted to the shells of the older barrier deposits.

The shells that are partly brown and partly blue, may have been chiefly derived by wave action from the lower zone of coarse sands. Their two-coloured character then has resulted from partial reduction of the iron hydroxides when they were temporarily embedded in the muddy sediments of shallower depths.

MINOR STRUCTURES

The minor structures of the sediments were studied by means of lacquer peels, of partially dried samples in metal trays, and of thin sections of indurated material. The methods have been described in former publications (VAN STRAATEN, 1954, 1959). The lacquer peel method is of special advantage for sandy sediments, the metal tray and thin section methods for fine grained deposits, rich in clay. Examples of lacquer peels are given in Pl. 12-13, while a number of representative metal tray samples are shown on Pl. 14. The data are summarized in fig. 20.

The dune sand deposits of the coastal barrier formation are mainly characterized by cross bedded structures (Pl. 12, nr. 2). Here and there they contain intercalations of soils in which most or all of the original laminations have disappeared by homogenization (Pl. 12, nr. 1). Instead of the primary structures one may find in these soils vertical structures, made by plant roots, and patchy distributions of dark (humic) and lighter material. Where the laminations have been preserved, they usually show parallelism over several decimeters of core length, i.e. there are relatively few discordances, much fewer than in the subaqueously formed barrier sediments. The inclination of the laminae between successive discordances varies greatly. In many cases they are approximately horizontal (at least in the core sections).

In the beach sands directly below these dune sands almost all laminations are parallel and horizontal. Cross laminations are limited to rare layers of not more than a few centimetres thickness (Pl. 12, nr. 3). The deposits are underlain in most borings by wildly cross-laminated sands (Pl. 12, nr. 4), which apparently were formed under the influence of strong waves and currents, on the lower parts of the beach, or in shallow water below the low tide level.

Going further down in the borings one comes first into sandy sediments that show only a moderate degree of cross lamination (Pl. 12, nr. 5) and then to the zone with clay layers. The latter are usually more or less horizontally laminated (Pl. 14, nrs. 1-4), while the intervening sand layers may show minor cross laminations (Pl. 13, nrs. 1-15; Pl. 14, nrs. 1-4). These structures are rather similar to those of the channel floor deposits of tidal environments (ref. 21, 25). Presumably they were formed under comparable circumstances, the sandy laminae during stages of relatively strong tidal currents, and the clayey laminae around the times of slack water (especially after storms). A difference must nevertheless exist in the effect of waves. While waves hardly ever influence the bottom of channels in tidal flat areas (because of the small length of fetch), they may be of considerable importance for the movements of sand on the sea floor (at least when they are of sufficient size).

In the lower zone of coarse sands one often finds cross laminations of distinctly larger dimensions than those in the overlying fine grained sediments (Pl. 13, nrs. 16-18). As far as these cross laminations are due to rippling of the sand, the difference in scale may be explained by its coarser texture. It is known that the minimum size of ripple marks in coarse sand is always greater than in fine sands. The structures in the fine grained deposits in the lowest parts of the barrier formation are similar to those in the upper zone of fine sands and clays mentioned above.

In general the open sea deposits in the coastal barrier system are characterized by abundance of thinly laminated structures and by great scarcity of animal burrows. A few small burrows are noticed here and there, but their total quantities are quite negligible when compared to those in other formations, e.g. of tidal flats or of deeper shelf areas in non-tidal seas. No layers were found where the activity of bottom animals had resulted in a more or less complete obliteration of the original laminations.

This sparsity of organic structures is not caused by poverty of the bottom fauna. The great numbers of washed shells of endobenthonic molluscs and the abundant remains of other animals that live in the sediment (e.g. *Pectinaria belgica, Echinocardium cordatum*) prove that in general the environmental conditions were favourable for bottom life.

The lack of burrows therefore must mean that in some manner only a very small part of the original burrows have been preserved. In all probability this was the result of the alternation of stages of rapid deposition and of erosion, either locally or more regionally. Under such conditions few traces of the burrowing fauna can be left. Just when the animals have succeeded in disturbing the laminations, the upper layers of the sediment are eroded again, and replaced by a thick mass of new sediment with fresh laminations (see fig. 21). The rapid reworking of the upper sediment layers in the coastal barrier environment, of course, was due to the strength of the tidal currents and the waves.

In connection with the textures of the separate laminae in the coastal barrier deposits it may be remarked that there is a distinct lower grain size limit of laminae that are free or almost free of clay. When, at the turn of the tide, the water motion decreased so far that grains smaller than 40 μ could be deposited, a

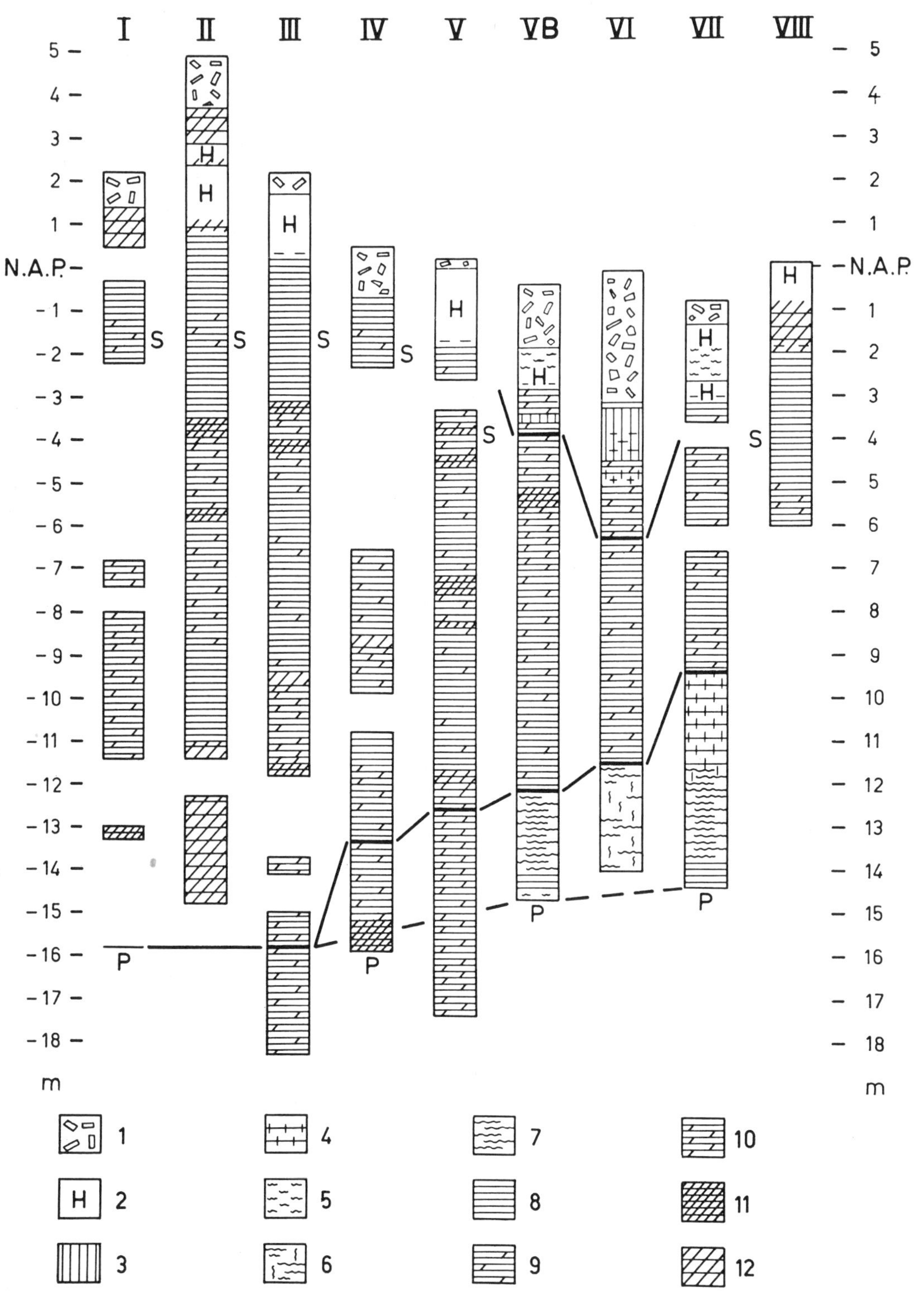
I
II
III
IV
V
VB
VI
VII
VIII
5
4
3
2
1
N.A.P.
-1
-2
-3
-4
-5
-6
-7
-8
-9
-10
-11
-12
-13
-14
-15
-16
-17
-18
m
H
S
P
1
2
3
4
5
6
7
8
9
10
11
12

◀

Fig. 20.
Distribution of structures in borings between Scheveningen and Voorburg.

1. Structures disturbed by human activities.
2. Soils, more or less homogenized by plant roots and burrowing land animals, see Pl. 12, nr. 1.
3. Laminations disturbed by burrowing marine animals, see Pl. 14, nr. 5.
4. Laminations partly disturbed by burrowing marine animals, see Pl. 14, nr. 6.
5. Highly uneven laminations of peat deposits.
6. Irregular lenticles and striations with vague burrows and structures of plant roots, in lake sediments.
7. Somewhat irregular, thin laminations of lacustrine clays, see Pl. 14, nr. 9.
8. Thin, parallel laminations of open sea-, beach- and dune deposits of barrier formation, see Pl. 12, nr. 3 and Pl. 13, nrs. 1, 2 and 3.
9. As 8, but with occasional intercalations of cross-laminated layers, see Pl. 12, nr. 5.
10. As 8, but with many intercalations of cross-laminated layers.
11. Wildly cross laminated deposits, see Pl. 12, nr. 4.
12. Cross-bedded deposits, see Pl. 13, nr. 16, 17 and 18.

S = Highest shell layers.
P = Top of Pleistocene (connected by broken line).
Drawn lines between borings = Lower and upper boundaries of open sea-, beach- and dune deposits of barrier formation.

great part of the suspended mud apparently was deposited at the same time. Thin sections show that the clay (or mud) laminae indeed contain high percentages of grains of 10 to 40 μ diameter, at least in their lower parts. The upper parts of the clayey laminae may be more purely lutaceous. These relationships are also found in the sediments of the Wadden Sea (ref. 21, 25, 27). It means that much of the clay in these environments is transported as floccules that have the same settling velocity as sand grains of about 40 μ. Part of these floccules may themselves contain minor amounts of silt [8]).

Where such floccules are drifted into brackish water environments they disintegrate to a certain extent, by peptization, and the resulting deposits then show laminae of pure silt, with grain sizes of e.g. 10 to 20 μ, but without clay. Such laminae are found in the „Hydrobia-clay" and in the brackish lake deposits underlying the coastal barrier formation in boring VII.

CORRELATION OF STRATA

Although there can be little doubt about the lateral continuity of the major lithologic zones, encountered in the borings between Scheveningen and Voorburg, it is mostly impossible to correlate separate strata from one boring to the next. Owing to short-comings of the boring technique it often could not be avoided that smaller or larger gaps were left open between the successive cores of each boring. Moreover, in some cores the sediment appeared to have been considerably disturbed.

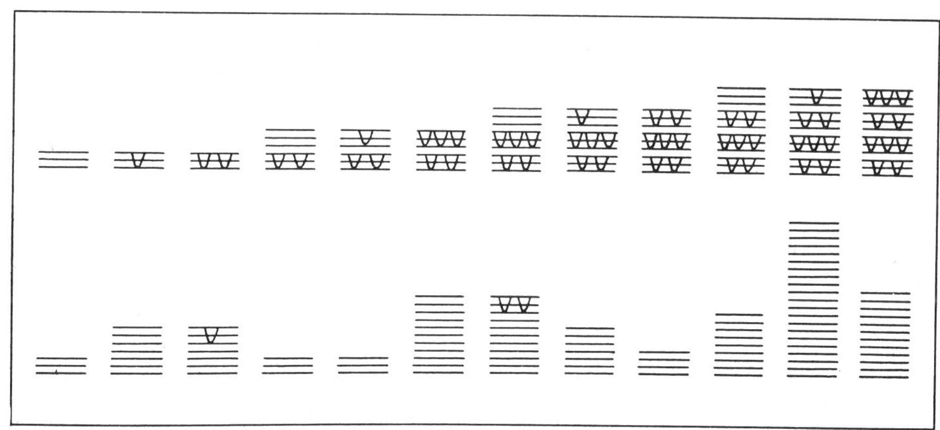

Fig. 21.
Different effects of burrowing animals on sediment structures.
Above: in environments of continuous, slow deposition.
Below: in environments of rapid deposition, alternating with stages of erosion.

[8]) The large size of the floccules, deposited along the Dutch North Sea coast certainly is not typical for marine conditions in general. It is probably restricted to the relatively muddy waters of nearshore environments.

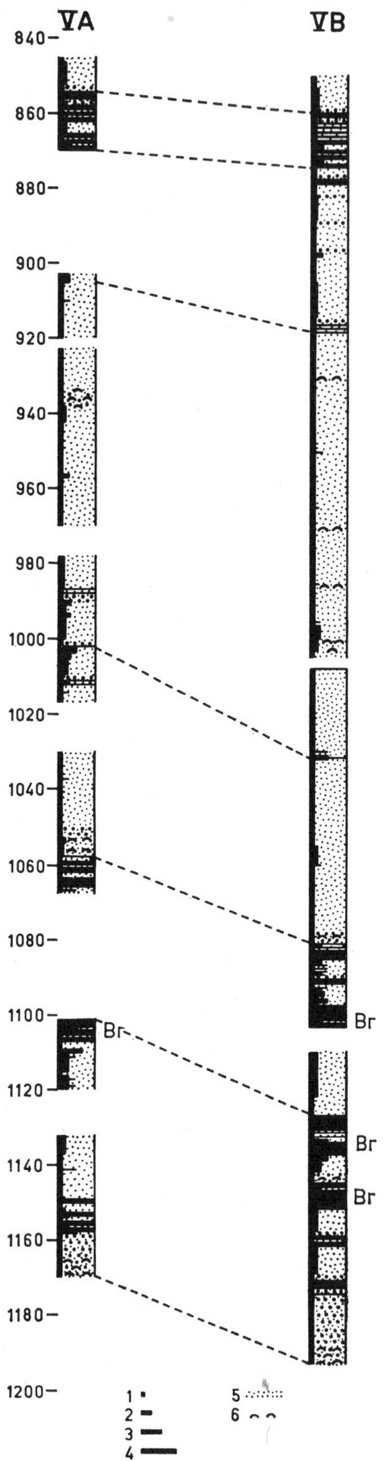

But even if the series of cores had been absolutely faultless, it is most improbable that a much better correlation would have been possible. For, it follows from the character of the barrier deposits (see e.g. the foregoing section) that during their formation the conditions of sedimentation and erosion must have strongly varied from place to place. Thus it cannot be expected that there are many strata with a horizontal continuity of more than a few hundred metres. Nevertheless a few cases were encountered where neighbouring borings show certain similarities in the succession of lithological and structural properties.

The thickest series of strata that seem to be correlatable was found in borings V A and V B, which are separated from each other by a distance of only 330 metres (figs. 22 and 23). The correlation of the layers at 11.00 m in boring V A and 11.20 m in boring V B is confirmed by pollen analysis (ZAGWIJN, 1965). They mark the first appearance of *Fagus* pollen. The boundary surface between sediments with and without *Fagus* pollen could further be traced through boring VI (10.60 m) to boring VII (5.80 m). Between borings VI and VII it therefore rises steeply towards the land, on the average about 5 m over a horizontal distance of 1 km. Yet this slope is still only half as steep as that of the present sea floor at Scheveningen between the beach and the 5 m depth contour (5 m over 500 metres horizontally). On the contrary, from boring V A to boring V B the layers slope down a little in coastward direction.

The other correlations that can perhaps be made refer to layers, or rather sequences of layers in the fine-grained zone of borings I to IV (see fig. 23). Photographs of lacquer peels of these sequences are shown on Pl. 13. Admittedly their correlation is based on very meagre evidence. Yet it seems to be corroborated by pollen analysis of the clayey sediments of this zone, at least as regards the supposed continuity of one particular layer. This latter corresponds in all four borings to (among other things) a maximum of the percentages of the *Alnus* pollen (see ZAGWIJN, 1965). The palynological argument in itself is perhaps not very strong either, but its accurate agreement with the result based on lithology and structures certainly renders the given correlation more acceptable.

The implications following from this correlation with respect to the manner of supply of sand to the upper parts of the barrier have been indicated already on p. 57. It would further follow that towards the end of the deposition of the clayey layers the sea has been not deeper than some 4 to 5 metres. Clearly the conditions must then have been completely different from those of the present day.

Fig. 22.
Correlation of sequences in borings V A and V B (see fig. 5).

1. Sand
2. Clayey sand
3. Sandy clay
4. Clay
5. Coarse sand
6. Shells

Br. Brown colour (owing to admixture of peat detritus)

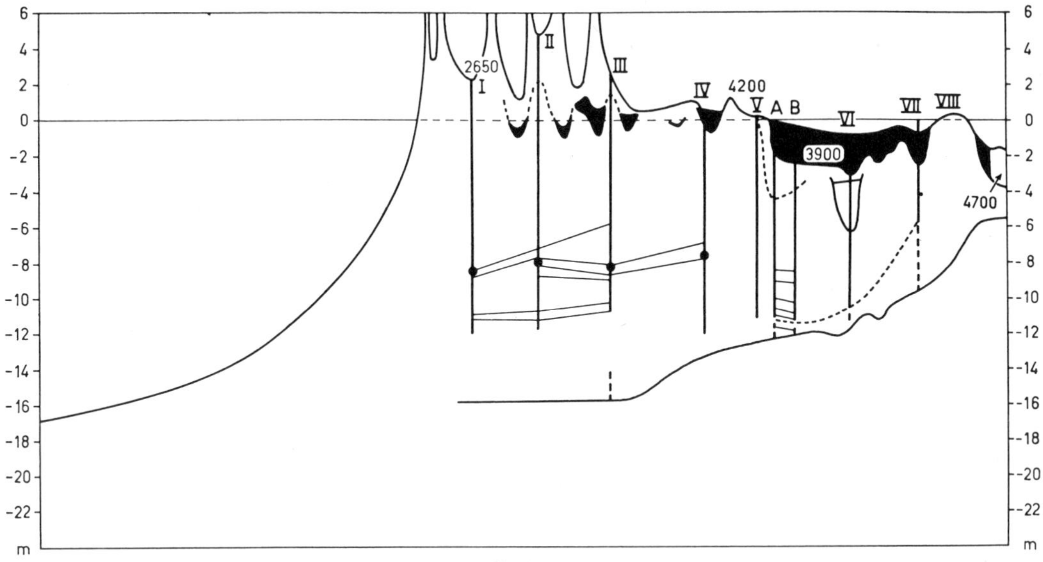

Fig. 23.
Age of deposits and correlation of strata.
Age of peat layers (black) by C 14-method. Age of barrier ridge by archeological dating.
For correlation of strata see fig. 22 and Pl. 13, nrs. 1—6.
Heavy dots: Maximum percentages of *Alnus* pollen in clay layers (after ZAGWIJN, 1965).
Drawn vertical lines: Sediments (in borings) deposited after first appearance of *Fagus* pollen (ref. 30).
Broken vertical lines: Sediments (in borings) formed before first appearance of *Fagus* pollen (ref. 30).

It was stated in the section on lithology and grain sizes that fine grained sediments may probably be formed in such shallow depths, if the deposition takes place in the shelter of a large subaqueous bar. Such a bar could have developed under influence of wave action transverse to the coast.

At present no high subaqueous ridges of this kind exist along the coast of South and North Holland. Nor is it probable that they have been there in the recent past. The ridges shown on old sea charts off Katwijk and IJmuiden (see e.g. fig. 24) may have been a little higher than those found at present in the area west and northwest of IJmuiden (fig. 9), but it does not seem likely that they rose to within 4 or 5 metres below the sea surface. Moreover, these former ridges stood obliquely to the shore. If the fine grained deposits of the coastal barrier have indeed been formed on the landward side of a subaqueous bar (or bars), then it is more probable, as will be shown later on, that the latter ran approximately parallel to the coast.

It could perhaps be objected that the accumulation of a large bar in front of the coast, though effectively protecting the depression on its landward side from heavy waves, may in some cases lead to an increase of the strength of the tidal currents passing through it. VEENSTRA (1964) has found that the bottom sediments in the depressions between the subaqueous bars in the Hinderbank area, about 35 kilometres off the Belgian coast, are in general very coarse grained, in consequence of the removal of fine material by tidal currents. Yet, it seems possible that the force of the tides on the shoreward side of a supposed former bar off Scheveningen was reduced, owing to the immediate vicinity of an outer tidal delta, viz. the one in front of the inlet of Den Haag (fig. 2). One could even think of a decrease in strength under influence of subaqueous sand accumulations, or of a foreland, at the mouth of the Rhine, which was not more than some 10 kilometres distant. On the other, northeastern side the outer delta of the Old Rhine pass may have had a protective effect.

While the above given correlation (if correct) concerns layers that probably correspond to time levels, this is not necessarily true of the major lithologic zones. It is obvious, for example, that the beach sand deposits of the coastal barrier formation, although showing approximately the same properties of structure and texture at all places, have been formed one after another, during a period of much more than a thousand years. Their age decreases from landward to seaward, and the internal bedding planes make distinct angles with their upper and lower boundaries.

But also the lower coarse grained zone may not be synchronous in all its parts. In this case it is more likely that the age diminishes in the opposite direction, from the seaward side towards the land. At least, this is suggested by the landward decrease of the grain sizes in the coarse grained zone (fig. 15), which seems

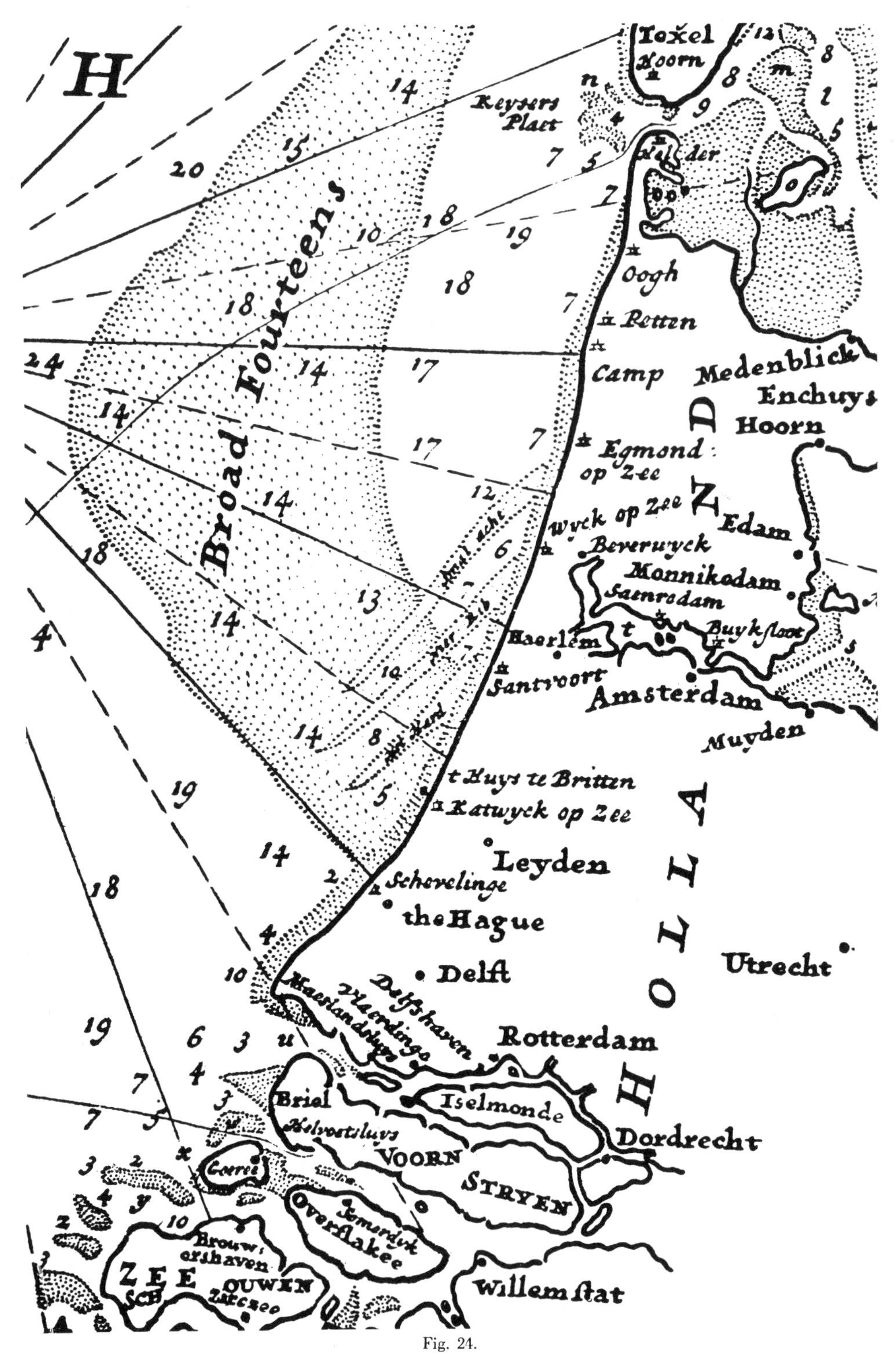

Fig. 24.
Part of chart by H. MOLL (circa 1720) of Southern North Sea, with location of the three submarine bars „Smal acht", „Uiter Rib" and „Het Hard".

to point to a gradual extension towards shallower water, by deposition of sand that has come from its more seaward parts. The bedding planes in this zone should then be inclined towards the land, more or less parallel to the downward sloping surface of its terminal portion in borings V to V B. Quite possibly the coarse zone represents the successive stages of a (Subboreal) subaqueous bar, which, under the influence of wave action, slowly migrated towards the coast. Then the muddy character of the sediments between 8 and 12 metres depth in borings V to VII could be at least partly due to the sheltering effect of this ridge, while their slight inclination in borings V A and V B could mean that they were deposited on the lower parts of its landward slope.

By supposing that the formation of the coastal barrier sediments took place in this manner, alternatively under protection of a subaqueous bar and without such protection, one can better understand the curious succession of coarse-grained and fine-grained zones. The following picture then evolves. The lower coarse-grained zone has formed on the seaward slope of such a ridge, while the sediments of the lower fine-grained zone were deposited on its landward side. Coastward migration of the ridge resulted in the superposition of its sediments on those of the lower fine-grained zone. At the same time the coast accreted in seaward direction, which eventually put an end to the landward migration of the subaqueous bar.

Around that time a new ridge had developed much further out in the sea, well beyond the shore-line of the present day. In its lee the sediments of the upper fine-grained zone were formed, which owing to continued coastal advance were subsequently covered by shallow water- and beach sands of the upper coarse-grained zone. Thus the alternation of coarse- and fine-grained zones would have been produced by the interplay of coastward migration of subaqueous bars and of seaward migration of the coast.

ACE OF THE DEPOSITS

No Carbon-14 datings have been made of shells from the borings between Scheveningen and Voorburg. The quantities of bivalve material were insufficient for such determinations, and washed shells were considered to be of too little value. The age of the barrier deposits therefore could only be inferred by indirect methods.

In the first place, three Carbon-14 datings are available from the base of peat layers in the depressions between barrier ridges in the vicinity. They are indicated (in years B.P.) on fig. 23. The age of the ridge of boring V (4200 years), which is added in this figure, is based on an archeological find.

The peat layers, of course, could not form before the depressions were closed off from the sea by a well developed barrier ridge, with at least a beginning of dunes on it. Therefore, their ages must be distinctly smaller than those of the ridges which lie directly seaward from them. If the difference in all cases is put at 200 years, it would follow that the coast reached its most landward position at Voorburg approximately 4900 years ago, and that the outermost parts, which are preserved along the present shore, date from about 2850 years B.P.

A still more indirect method is based on the facies of the sediments and their relation to sea level at the time of their formation. Where indications for the position of sea level can be obtained in this way, an estimate of the age of the sediments can be made on account of what is known by other methods about the course of the sea level rise. A series of C-14 age determinations of the basal layers of peat formations in the Netherlands has been published by JELGERSMA (1961). Later corrections of these datings were given by the same author in 1963 (ref. 17).

In fig. 25 a number of these data are plotted against depths of the samples. They refer exclusively to peat layers formed on old inland dunes in the vicinity of Dordrecht, South-Holland (ref. 2, 14). Accordingly, their depths cannot have been influenced to any important extent by later compaction of the underlying deposits. As far as samples younger than about 6000 years are concerned, the depths are seen to lie on a very smooth curve. Now it may be assumed that the peat started to grow at, or slightly below the local high water level in the adjoining estuaries. On the average this may have been 1¼ metres above mean sea level along the coast. In fig. 25 the inferred mean sea level curve is therefore drawn at 1 metre below the depths of the peat samples.

The facies indications for former sea levels as provided by the cores of the coastal barrier deposits are unfortunately rather vague. Moreover, they are influenced to an unknown degree by compaction of lower lying clay layers. The indications consists of the depths of:

a) the base of the cross-bedded dune sands, formed well above mean high tide level;
b) the highest shell layers, deposited at or above mean high tide level;
c) the top of the strongly cross-laminated sands of the „zone of heavy wave action", probably roughly corresponding to mean low tide level;
d) the burrowed tidal flat sediments in boring VI, the top layers of which were apparently formed below the original high water level, their base lying in the vicinity of the original low tide level.

In borings I to III the beach deposits must have been lowered by compaction of the clayey layers of the upper fine-grained zone. Considering the comparatively high silt contents and the small total thickness of these layers, their influence may not have exceeded some 20 cm. Subsidence of the tidal flat deposits in boring VI took place by compaction of both the muddy tidal channel sediments directly below them, and of the clayey deposits in the basal parts of the barrier formation. It is estimated here at 40 cm. The beach sand

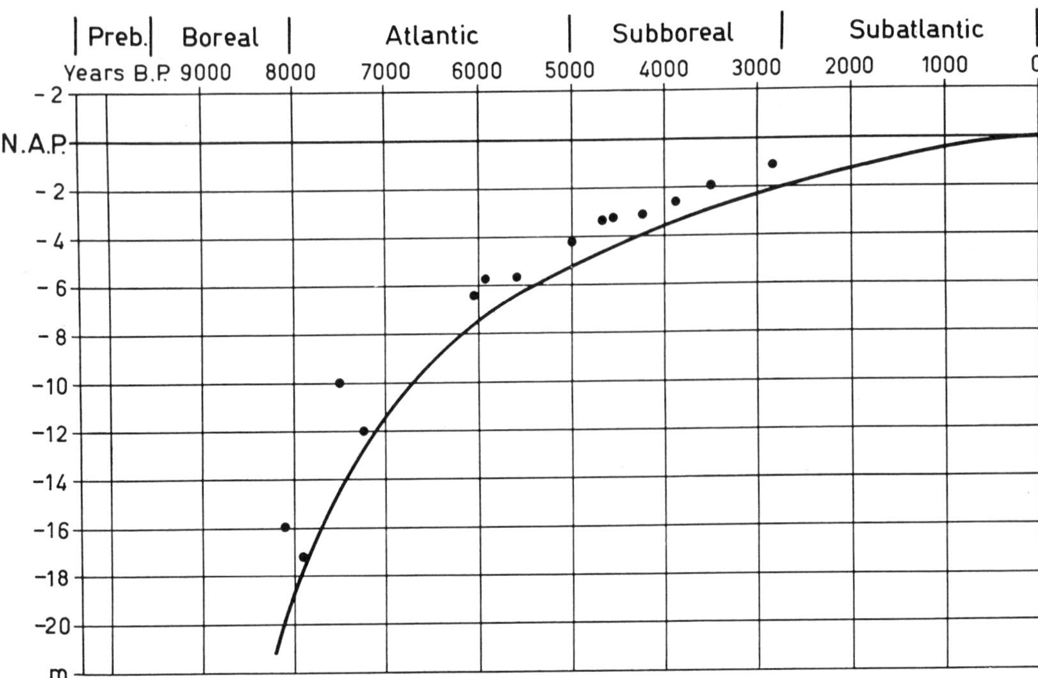

Fig. 25:
Relative sea level rise in South-Holland, following from C 14 age determinations of basal peat layers on old inland dunes (after JELGERSMA, ref. 9, 10).

TABLE V

Indications for relative position of sea level, corrected for (estimated) influence of compaction

	Borings I—III	Boring VI	Boring VIII
Base of cross-bedded dune sands	+ 0.90 m	—	— 1.60 m
Highest shell layers	— 1.50 m	—	— 3.60 m
Top of cross-laminated wave zone deposits	— 3.10 m	—	— 5.60 m ?
Top of burrowed tidal flat sediments	—	— 2.80 m	—
Base of burrowed tidal flat sediments	—	— 4.40 m	—
Estimated position of mean sea level	— 2.30 m	— 3.50 m	— 4.80 m
Corresponding age (according to curve of fig. 25)	2900 yrs	3900 yrs	4750 yrs

accumulation of the innermost ridge (boring VIII), finally, must have been lowered by later compaction of the underlying intracoastal deposits. It is also taken at about 40 cm. The data are summarized in table V (cf. fig. 20).

These conclusions would agree rather well with the C-14 age determinations of the basal peat layers in the depressions behind the barrier ridges and with the archeological dating of the ridge of boring V (Huis ten Bosch). It would follow from them that:

1) the peat on the landward side of the innermost ridge started to grow about half a century after the formation of this ridge;
2) the tidal flat deposits of boring VI were formed approximately three centuries after the ridge of boring V had developed, and directly before the area became covered by peat;
3) the peat in the depression corresponding to boring I began to grow some 250 years after the formation of the outer parts of the barrier at Scheveningen;
4) the accretion of the coast, from Voorburg to the position of the present shoreline, took place in the course of only about 1850 years.

RECONSTRUCTION OF COASTAL EVOLUTION BETWEEN SCHEVENINGEN AND VOORBURG

By combination of all data derived from the borings between Scheveningen and Voorburg, one arrives at the following reconstruction of the coastal evolution in this area (fig. 26):

The Holocene sedimentation begins with the formation of a peat layer: the lower peat. Owing to the rapid rise of the sea the upgrowth of the peat cannot keep pace with the elevation of the ground water table. It is drowned, and a lake is formed in the area:

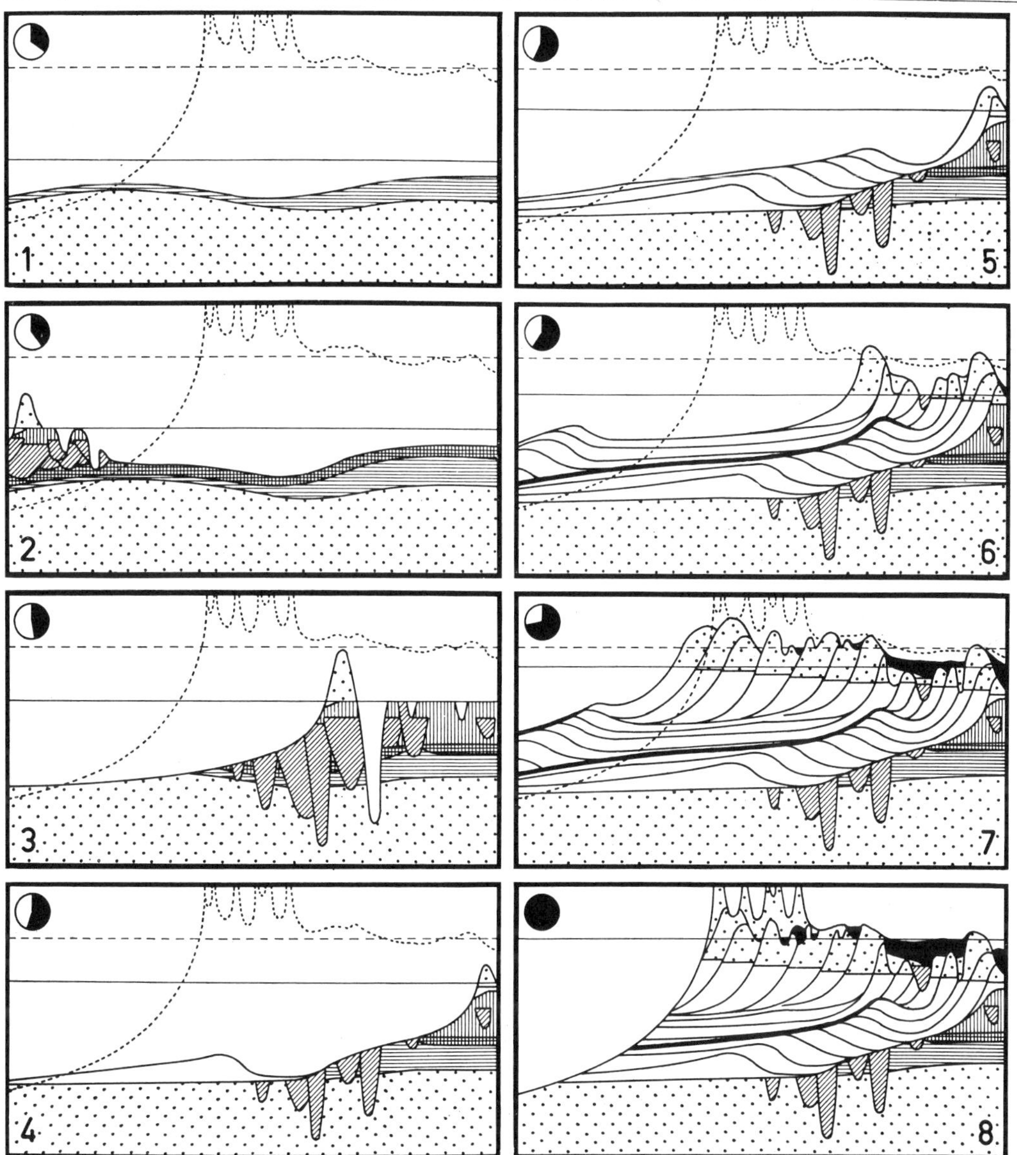

Fig. 26. Reconstruction of coastal evolution.

1. Sea level circa 10 m below N.A.P.; Time circa 4750 B.C.; Lake stage.
2. Sea level circa 8 m below N.A.P.; Time circa 4200 B.C.; Lagoon stage.
3. Sea level circa 6 m below N.A.P.; Time circa 3400 B.C.; Tidal flats stage.
4. Sea level circa 4.8 m below N.A.P.; Time circa 2800 B.C.; Completion of innermost barrier ridge.
5. Sea level circa 4.3 m below N.A.P.; Time circa 2500 B.C.; Formation of wide depression between barrier ridges.
6. Sea level circa 3.8 m below N.A.P.; Time circa 2200 B.C.; Completion of second major barrier ridge.
7. Sea level circa 2.2 m below N.A.P.; Time circa 850 B.C.; Maximum of coastal advance.
8. Sea level approximately at N.A.P.; Present situation.

Vertical dimensions of diagrams: 29 m; Horizontal dimensions: 14 km.

Explanation of symbols:

Punctation (basal deposits): Pleistocene.
„ (upper parts of formation): Dune and beach deposits formed above mean sea level.
Black: Peat.

Vertical striations: Tidal flat depostis.
Oblique „ : Channel deposits.
Cross „ : Lagoon clays.
Horizontal „ : Lake clays.
White, with stratification: Open sea deposits.

Stage 1. Time: circa 4750 B.C. (6700 years B.P.) (Middle Atlantic). Sea level (according to curve of fig. 25): 10 m below N.A.P. In the lake a clay layer of 2 m thickness or more is deposited. Shells of fresh-water gastropods (a.o. *Planorbis*) were found in this clay in boring V A (at 13.00 and 13.50 m depth). Continuing rise of the sea leads to inward migration of the coast. The lake is transformed into a lagoon:

Stage 2. Time: circa 4200 B.C. (Middle Atlantic). Sea level: 8 m below N.A.P. In this lagoon the „Hydrobia-clay" (Pl. 14, nr. 8) is deposited. Nea the inlets tidal deltas have developed which partly rise above low tide level, forming tidal flats.

Stage 3. Time: circa 3400 B.C. (Late Atlantic). Sea level: 6 m below N.A.P. The coast has been cut back further eastward, and now lies within the investigated area. On its landward side a tidal flat landscape has come into existence, by gradual extension of the inner deltas of the lagoon of the foregoing stage. Deep channels are present between the tidal flats, many of which have cut down into the Pleistocene. A considerable portion of the original lacustrine and lagoonal clays has been eroded by these channels and partially or wholly replaced by more sandy channel floor deposits.

The coastal barrier consists of a thin wedge of beach sand, and of a narrow belt of dunes, partly covering the intracoastal deposits on the landward side of the beach. On the subaqueous slope in front of the beach the older sediments of tidal flat environments, and the last remnants of lagoon and lake deposits are eroded by the waves (and currents) of the sea. Most of the eroded material is removed from the area by longshore transport. Part of it disappears via tidal inlets further down the coast into tidal flat areas, where it may be deposited.

Stage 4. Time: circa 2800 B.C. (Earliest Subboreal). Sea level: 4.8 m below N.A.P. The coast has reached its innermost position. Relatively little remains of the older intracoastal deposits that had become exposed on its seaward side. The former existence of tidal flat environments in this area is proved by the deposits of the deepest parts of the tidal channels, which, owing to their depth escaped erosion. Here and there thin outwash residues are left on the sea bottom. They are composed of relatively coarse grained elements.

At a distance of some 5 kilometres from the coast of that time a subaqueous sand ridge has formed. It partly consists of the residues of the marine erosion, which have gradually accumulated under the influence of waves approaching the coast. Another part of the sand may have been freshly eroded (after removal of the former erosion residue) by wave action further offshore. A third portion, finally, may have been supplied by longshore drift, e.g. from the submarine outer delta in front of the inlet south of Den Haag, or from the mouth of the Rhine. Fine-grained, muddy sediments are deposited in the lee side of this bar.

Stage 5. Time: circa 2500 B.C. (Early Subboreal). Sea level: 4.3 m below N.A.P. The barrier ridge has increased in height by deposition of dune sand. The coast has accreted a little, mainly owing to longshore supply of sand. Peat has just started to grow on the landward side of the dunes.

The volume of the subaqueous ridge has considerably augmented, as the result of further transverse and longshore sand transport in deeper water. Its crest has migrated towards the coast, whereby the earlier, fine-grained deposits became buried below the coarse sand of the bar. On the lee side of the bar the deposition of fine-grained material continues. The lowermost layers contain admixtures of coarse sand (fig. 14): the reworked outwash residue of the preceding (Atlantic) stage of marine erosion.

Stage 6. Time: circa 2200 B. C. (Middle Subboreal). Sea level: 3.8 m below N.A.P. Since stage 5 the submarine ridge has further increased in height. Its crest came to lie at a depth of not more than some $2\frac{1}{2}$ metres below mean sea level, so that only relatively small waves could pass over it. Owing to the absence of strong wave action along the shore this latter remained of low elevation. No high ridges of sand could be thrown up on the backshore during storms. During the later stages of coastal accretion a backshore was hardly developed at all. The littoral environment then consisted of a broad plain of low ridges and swales. Part of the latter may have contained water even at low tide. They were gradually filled up with washed beach sand, alternating with thin layers of mud: the beach plain deposits between 2 and 4 metres depth in borings V A and V B, and between 3 and 5 metres depth in boring VII.

While the coast accreted in seaward direction, the submarine bar migrated further towards the shore. At a given moment its crest became buried under the advancing subaqueous coastal slope. The bar then finished to exist as a morphological feature. The beach became again exposed to the full force of the waves, and the backshore grew up to greater height. With increasing elevation above high tide level better conditions were created for dune vegetation, which further accelerated the upgrowth of the coastal area.

In this stage the accretion of the coast could take place under influence of both longshore and transverse supply of sand. The transverse supply was the result of wave erosion of the seaward parts of the original ridge surface. However, this erosion soon stopped again, owing to the appearance of a new ridge, further from the coast, which once more prevented the passage of large waves. Fine grained sediments, rich in mud were deposited in the lee of this second ridge. In the investigated area they have an approximately horizontal position. At the time of stage 6 the formation of this muddy series has been almost completed.

Stage 7. Time: circa 850 B.C. (Late Subboreal). Sea level: 2.2 m below N.A.P. Since stage 6 the coast has advanced over a distance of more than 4 kilometres, in the course of little more than 1300 years. This rapid accretion was possible owing to the shallowness of the sea floor, which was not deeper than some 4 metres. At first the accretion was mostly (or completely) the result of sand supply by longshore drift. Later on transverse supply of sand from the subaqueous ridge may have added its share. This latter transport must then have taken place over the top of the muddy series, when the depression on the leeside of the ridge had been filled up to roughly the level of its crest. It is not impossible that the crest itself, after the earlier stages of vertical upgrowth, has again been lowered, whereby the transverse sand transport must have increased in intensity. But, since the whole ridge has disappeared by later erosion (stage 8), any reconstruction of the evolution of the ridge would be mere guess-work.

Shortly after the formation of the barrier ridge of Huis ten Bosch (boring V) a tidal flat landscape developed in the low area on the landward side of this ridge (the tidal flat series between 3.30 and 6.30 m in boring VI). The area subsequently became covered by peat swamps.

The accretion of the coast from the ridge of Huis ten Bosch to its outermost position still proceeded by successive formation of ridges and elongate depressions. With the help of vegetation the ridges soon grew up to chains of low dunes. The depressions later on became the sites of peat growth, when the local ground water table had been raised sufficiently to keep their bottom saturated with water.

Stage 8. Present situation. Great changes have taken place since the end of the Subboreal. The coast has retreated again, and the sea floor has been eroded to a depth of at least 16 metres. No high submarine ridges are present anymore, and the beach has become exposed to large waves. Moreover, huge masses of wind blown sand have been deposited on top of the old, subboreal dune ridges.

The main cause of the retreat of the coast was probably the gradual exhaustion of the sources of the sand. Its effect may have been strengthened by the continued rise of sea level relatively to the land, and by a change of the climatic conditions.

SOURCES AND TRANSPORT OF SAND

Since silt and clay form only minor constituents of the barrier formation, the phenomena of coastal accretion or erosion are primarily a matter of sand economy. Whether in a given area the supply or the removal of sand predominates depends on a great many factors. In studying their influence it is convenient to deal separately with the longitudinal components of sand movement, parallel to the coast, and with the transverse components, at right angles to it.

Transverse supply of sand to the shore takes almost exclusively place under influence of waves approaching from the sea over the subaqueous coastal slope. The asymmetrical oscillations of the water particles cause an upslope movement of sand, at least if the inclination of the bottom is not too steep. Coarse sand thereby tends to be concentrated on the beach and in shallow depths. Deposition of fine sand is limited to more quiet water, below the zone of most vigorous wave action.

Where accumulation of sand derived from the sea bottom leads to accretion of the shore, the subaqueous profile gradually steepens, both by the deposition on its higher parts and by the erosion of its lower parts. The accumulative effects then become more and more counteracted by the (indirect) effects of gravity, and eventually the transverse supply of sand to the shore stops altogether.

At the end of the Atlantic the slope of the sea floor in front of the coast was very gentle. From 0 to 10 metres depth the average inclination at Voorburg was only 1 : 450 (against 1 : 150 along the present coast at Scheveningen). Beyond the depth of 10 metres it was still much smaller, the (original) slope corresponding approximately to the general inclination of the pre-Holocene terrestrial surface. Probably these low angles of declivity were at least partly due to the preceding quick ascent of the sea. The intensity of sand transport under influence of the waves, diminishes strongly, of course, with increase of depth. As long as sea level, during the Atlantic, was rising rapidly, the erosion of the shore must therefore have been continuously ahead of the erosion of the intermediate and lower parts of the subaqueous slope. Hence, no steep profiles could develop, and the profile remained out of equilibrium with the waves.

When the rise of the sea subsequently slowed down, the equilibrium could be gradually restored. Great quantities of material were available for submarine erosion, and a persistent supply of sand towards the coast was the result. Yet, in the investigated area, only part of this sand actually reached the shore. Much of it was intercepted by the submarine bars. In this way the bars increased considerably in volume. Since, moreover, they promoted the deposition of fine grained sediment on their lee-side, the depths of the near-shore zone of the sea were quickly reduced. This brought about that coastal accretion by longshore supply could take place at a much higher rate than would have been possible in the absence of the bars.

It is clear that longshore transport of sand has indeed contributed to the formation of the coastal barriers. At certain times, e.g. shortly after the completion of the innermost barrier ridge, the coast advanced by deposition of relatively coarse sand, which could hardly have been supplied by transverse wave action, since the sea floor in front of the beach was effectively protected from wave motion by the first submarine bar. The fine-grained material that was deposited in this sheltered zone was probably also largely brought along by longshore transport.

The main source of the longshore supply must presumably be sought in southwesterly directions. It may be supposed that along the coast southwest of Hoek van Holland, and in front of it, great masses of sand were present in the Subboreal, consisting for a great part of reworked delta deposits of the rivers Rhine and Meuse. Some of the sand may even have been directly supplied by these rivers (ref. 27). A smaller quantity has perhaps come from the mouth or the delta of the Old Rhine distributary.

On the other hand, it does not seem likely that during the Subboreal a strong longshore transport has occurred from north (northeast) to south (southwest). Among other things the heavy-mineral composition of the present beach sand is not in favour of such a hypothesis [9]). As is known, the composition of the sands south of Bergen differs markedly from that north of Petten (ref. 7). In the latter area it is strongly influenced by reworking of Pleistocene, glacial formations. In the former the sands correspond mineralogically to those of the Rhine (cf. ref. 27). It is true that the composition of the present beach sands south of Bergen may differ a little from that of the Subboreal barrier deposits, but the differences cannot be great, since much of the present shore material must have been reworked from the older barrier. If a major portion of the older barrier material had been supplied from the North one would surely expect a distinct northern influence on the mineral composition of the beach sands south of Bergen.

Regarding the relative effects of transverse and longitudinal sediment transport for the accumulation of the Subboreal coastal barrier deposits, it was mentioned already on p. 57 that the author has come back from his original view that transverse supply would have been a factor of minor importance. It now seems that such a supply accounts at any rate for an essential part of the lower coarse grained zone of the barrier complex, which probably represents the deposits of a large submarine bar.

By accepting the bar hypothesis one cannot only better understand the distribution of sediment types in the cross section at Scheveningen, but also the areal relationships of the barrier ridges. It the latter had primarily originated by longshore supply, in the manner of spits, it is very difficult to account for the continuity of the broad depression that separates the innermost ridge from the next one. The depression is not just a local feature at Voorburg, but it is found with approximately the same width elsewhere along the coast: between Rijswijk and Leiden, between Leiden and Hillegom, and between Uitgeest and Alkmaar. This uniformity of the ridge pattern is hard to reconcile with the normal shape of spits, but it is easily explained as the result of effects of transverse sediment supply. It presumably is connected at all places with the evolution of the first submarine bar.

As yet there are insufficient data available for quantitative estimates of the relative shares of transverse and longitudinal transport in the formation of the barrier complex. It is certain, though, that both have been of great intensity. From the age determinations it follows that the whole complex has formed in a period of not more (probably less) than 2000 years. Since the area of the cross section of the barrier formation at Scheveningen at the end of the Subboreal may be taken at a minimum of 150.000 m^2 (see figs. 6 and 26, nr. 4), this means an average annual accretion of 75 m^2, or an average hourly accretion of 86 cm^2.

It half of this accretion has been due to transverse supply of sediment, and if the latter has taken place in the shape of asymmetrical wave ripples, with an average thickness of 1 cm, it would follow that these ripples have moved from the source areas with an average velocity of 43 cm/hour. If it be assumed (rather arbitrarily) that sand was moved from these areas by waves having a minimum height of $\frac{1}{2}$ m and an average period of 6 seconds, and that such waves occur in $\frac{3}{4}$ of the time, then the average advance of each ripple per wave would be a little less than 1 mm. This seems a rather high value, and since the frequency of the waves that are able to move the ripples can hardly have been understimated, it would follow that transverse sand transport has probably accounted for less than half of the total supply. This leaves more than half to be explained by longshore transport.

For the sake of simplicity it will be supposed that all sediment supplied by longshore drift originally came from the areas southwest of Scheveningen and from the mouth of the Old Rhine. Whereas between Scheveningen and Leiden the sediment theoretically may have been brought down from the southwest and from the northeast, the barrier deposits north of this place must then have a purely southwesterly origin. If the average cross section of this part of the Subboreal barrier system is taken at only 100.000 m^2 and its length at 55 km, it should have a total volume of 5.5×10^9 m^3. If half of this volume has been supplied by longshore transport, in the course of 2000 years, an annual average of at least $2\frac{3}{4}$. 10^6 m^3 must have passed the northern side of the Old Rhine mouth. In reality the quantities must have been considerably greater, because part of the material carried along the coast has disappeared through the inlets near Uitgeest and north of Alkmaar (III and IV on fig. 2) into intracoastal areas.

The estimates that are given for the intensity of longshore drift along open sea coasts of the present time are usually not larger than 1.10^6 m^3 per year. The same value has been suggested for the drift along the present Dutch barrier island Vlieland. From the above it seems to follow that the longshore transport during the formation of the Subboreal barrier complex was at least 3 times stronger.

A sediment supply of $2\frac{3}{4}$. 10^6 m^3 per year corresponds to an average of 88.000 cm^3/sec. Taking the average width of the zone of longitudinal transport during the Subboreal at 3 km, and the thickness of the layer corresponding to the total quantity of moving sand at 1 cm, an average velocity of $^1/_3$ cm per second results. This is far more than what is known by observations about the rates of ripple migration under influence of tidal currents. Moreover, the directions of

[9]) At present no sufficient data are yet available on the mineralogy of the older barrier deposits.

these currents are reversed twice per tidal cycle, so that only the difference in transporting power of the ebb and the flood currents can lead to a residual displacement of the bottom sand. Secondary currents, generated by wave motion of the water (undertow etc.) can add only little to this residual displacement. It must therefore be concluded that a great part of the sediment that was carried parallel to the coast, has moved in suspension or by saltation. Much of this latter transport probably took place in the breaker zone.

Furthermore, the possibility has to be considered that wind action on the beach sand has contributed to the longitudinal transport. One would then have to assume that in the Subboreal the dominant and prevailing winds had more southerly to southwesterly directions than those of the present time, which come from the westsouthwest and west, and tend to carry the beach sand towards the interior.

It is known, on paleobotanical grounds, that the Subboreal climate has differed slightly from the present climate. Accordingly, it may be taken for granted that the wind pattern has also been slightly different. An indication that the effective directions of the winds during the Subboreal were more parallel to the coast than nowadays may consist in the distribution of the old dunes. Whereas the topography of the young, postroman dune landscape is relatively independent of the underlying relief of ridges and swales (figs. 2 and 6), the Subboreal dunes seem to be more restricted to the original ridges.

REFERENCES

1. BAAK, J. A. (1936): Regional petrology of the southern North Sea. — Thesis Leiden. Veenman en Zonen, Wageningen, 128 pp.
2. BENNEMA, J. en L. J. PONS (1952): Donken, fluviatiel laagterras en Eemafzettingen in het westelijke gebied van de grote rivieren. — Boor en Spade, V, pp. 126-137.
3. BENNEMA, J. en L. J. PONS (1957): The Holocene deposits in the surroundings of Velsen and their relations to those in the excavation. In: The excavation at Velsen. — Verhand. Kon. Nederl. Geol. Mijnb. Gen., Geol. Ser., XVII, 2, pp. 199-218.
4. BENTHEM JUTTING, T. VAN (1943): Mollusca (I), C. Lamellibranchia. — Sijthoff, Leiden, 477 pp.
5. BURCK, P. DU (1957): Een bodemkartering van het tuinbouwdistrict Geestmerambacht. — Versl. Landbouwk. Onderz., 63, 3, 159 pp.
6. DOEGLAS, D. J. (1950): De interpretatie van korrelgrootteanalysen. — Verhand. Kon. Nederl. Geol. Mijnb. Gen., Geol. Ser., XV, pp. 247-328.
7. EDELMAN, C. H. (1933): Petrologische provincies in het Nederlandsche Kwartair. — Thesis Amsterdam. Centen Uitg. Mij., Amsterdam, 104 pp.
8. HAANS, J. C. F. M. (1954): De bodemgesteldheid van de Haarlemmermeer. — Versl. Landbouwk. Onderz., 60, 7, 154 pp.
9. JELGERSMA, S. (1961): Holocene sea level changes in the Netherlands. — Meded. Geol. Stichting, Ser. C-VI-7, 100 pp.
10. JELGERSMA, S., J. DE JONG en W. H. ZAGWIJN (1963): Toelichting bij de excursie naar het duingebied tussen Wijk aan Zee en Zandvoort op 26 sept. 1963. — Palynologenconferentie 1963.
11. MEER, K. VAN DER (1952): De bloembollenstreek. — Versl. Landbouw. Onderz., 58, 2, 155 pp.
12. MOLL, H. (circa 1720): A chart of part of the coast, sands and banks of England and Holland. — Bowles and Son, London.
13. PANNEKOEK, A. J. et al. (1956): Geological history of the Netherlands. — Staatsdrukkerij, Den Haag, 154 pp.
14. PONS, L. J. en J. BENNEMA (1958): De morfologie van het pleistocene oppervlak in westelijk midden-Nederland, voorzover gelegen beneden gemiddeld zeeniveau (N.A.P.). — Tijdschr. Kon. Ned. Aardr. Gen., 75, pp. 120-139.
15. PONS, L. J. en A. J. WIGGERS (1958): De morfologie van het pleistocene oppervlak in Noord-Holland en het Zuiderzeegebied, voor zover gelegen beneden gemiddeld zeeniveau (N.A.P.). — Tijdschr. Kon. Ned. Aardr. Gen., 75, pp. 140-153.
16. PONS, L. J. en A. J. WIGGERS (1959-60): De Holocene wordingsgeschiedenis van Noord-Holland en het Zuiderzeegebied. — Tijdschr. Kon. Ned. Aardr. Gen., 76, pp. 104-152 and 77, pp. 1-57.
17. PONS, L. J., S. JELGERSMA, A. J. WIGGERS and J. D. DE JONG (1963): Evolution of the Netherlands coastal area during the Holocene. — Verhand. Kon. Nederl. Geol. Mijnb. Gen., Geol. Ser., 21, 2, pp. 197-208.
18. ROO, H. C. DE (1953): De bodemkartering van Noord-Kennemerland. — Versl. Landbouwk. Onderz., 59, 3, 202 pp.
19. STRAATEN, L. M. J. U. VAN (1951): Longitudinal ripple marks in mud and sand. — Journ. Sedim. Petrol., 21, 1, pp. 47-54.
20. STRAATEN, L. M. J. U. VAN (1953): Rhythmic patterns on Dutch North Sea beaches. — Geol. en Mijnb., 15, pp. 31-43.
21. STRAATEN, L. M. J. U. VAN (1954): Composition and structure of recent marine sediments in the Netherlands. — Leidse Geol. Meded., XIX, pp. 1-110.
22. STRAATEN, L. M. J. U. VAN (1956): Composition of shell beds formed in tidal flat environment in the Netherlands and in the bay of Arcachon (France). — Geol. en Mijnb., 18, pp. 209-226.
23. STRAATEN, L. M. J. U. VAN (1957 a): Recent sandstones of the coasts of the Netherlands and of the Rhône delta. — Geol. en Mijnb., 19, pp. 196-213.
24. STRAATEN, L. M. J. U. VAN (1957 b): The Holocene deposits. In: The excavation at Velsen. — Verhand. Kon. Nederl. Geol. Mijnb. Gen., Geol. Ser., XVII, 2, pp. 158-183.
25. STRAATEN, L. M. J. U. VAN (1959): Minor structures of some recent littoral and neritic sediments. — Geol. en Mijnb., 21, pp. 197-216.
26. STRAATEN, L. M. J. U. VAN (1961): Directional effects of winds, waves and currents along the Dutch North Sea coast. — Geol. en Mijnb., 23, pp. 333-346 and pp. 363-391.
27. STRAATEN, L. M. J. U. VAN (1963): Aspects of Holocene sedimentation in the Netherlands. — Verh. Kon. Nederl. Geol. Mijnb. Gen., Geol. Ser., 21, 1, pp. 149-172.
28. VEEN, J. VAN (1936): Onderzoekingen in de Hoofden. — Landsdrukkerij, Den Haag, 252 pp.
29. VEENSTRA, H. J. (1964): Geology of the Hinder Banks, Southern North Sea. — Hydrographic Newsletter, Vol. 1, pp. 72-80.
30. ZAGWIJN, W. H. (1965): Pollen-analytic correlations in the coastal barrier deposits near The Hague (The Netherlands). — Meded. Geol. Stichting, N.S. 17, pp. 83-88.

PLATE 10 A.

Excavation at IJmuiden.

West-east section. The barrier deposits dip towards the West at an average angle of circa 2½°.

PLATE 10 B.

Excavation at IJmuiden.

Clay layers on rippled sand surface in west-east section. Depth of lower clay layer approximately 5.2 m — N.A.P. Height of section on photograph circa 1 m.

PLATE 11.

Excavation at IJmuiden.

Asymmetric wave ripples (steeper sides pointing eastward) and *Spisula* shells in lower foreshore- or breaker zone deposits. Length of knife circa 20 cm.

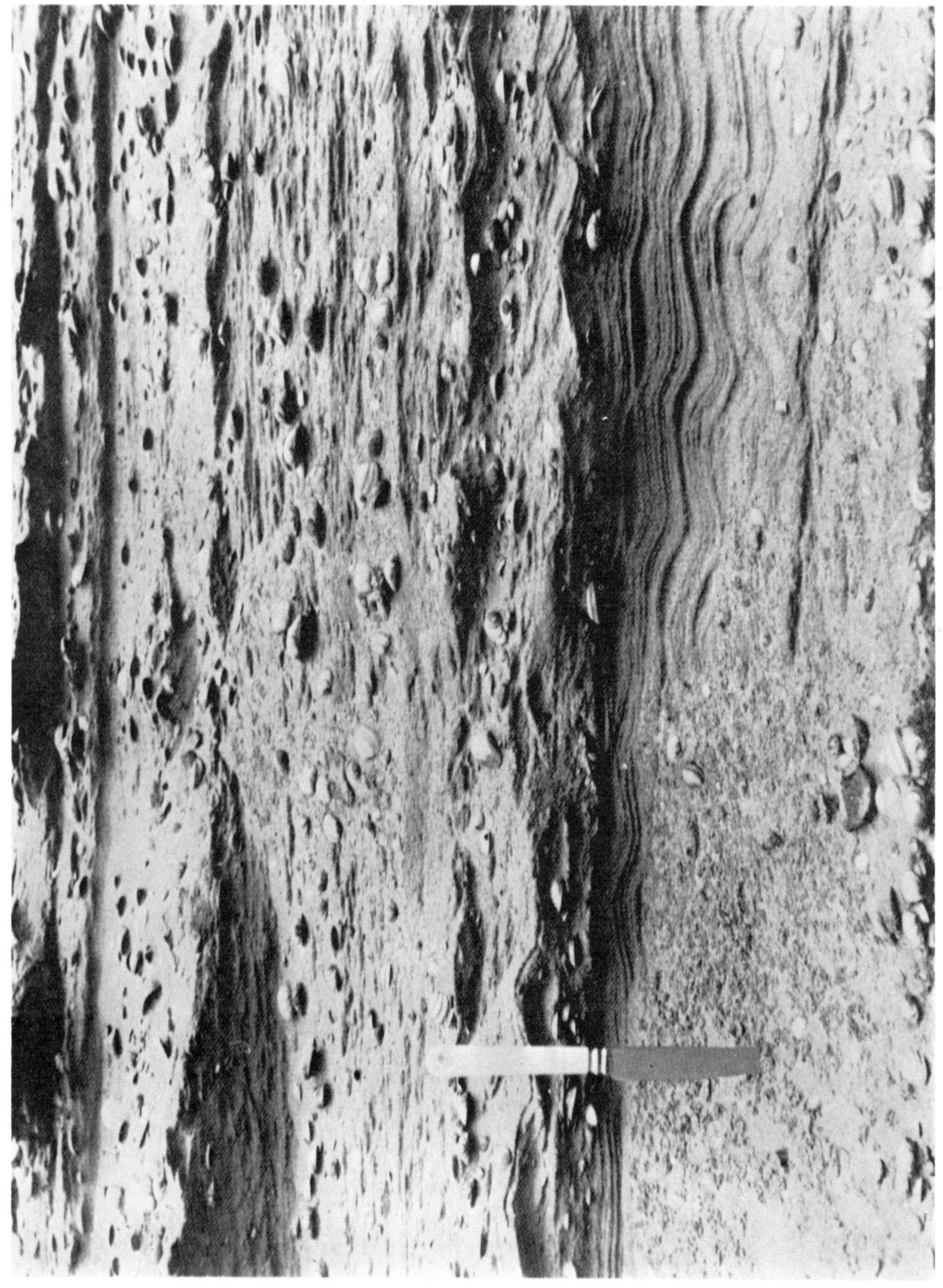

PLATE 12.

Lacquer peels of cores.

Depths in m below N.A.P. (+ = above N.A.P.).

1. Dune soil, sand. Laminations obliterated by homogenization. Boring III. + 1.29—+ 1.04.
2. Cross bedded dune sand. Boring VIII, 0.43—0.68.
3. Sand of backshore with thin, parallel laminations. Boring VIII, 3.45—3.70.
4. Sand of foreshore, or breaker zone, intensely cross laminated. Boring III, 4.00—4.25.
5. Sand of submarine slope of barrier. Thin, parallel, laminations with subordinate cross laminations. Boring V, 6.17—6.42.
6. Clayey sand of tidal flat, with bivalve of *Scrobicularia* in position of life. Laminations disturbed by burrowing animals. Boring VI. 3.57—3.82.
7. Sandy clay of shallow tidal channel. Predominantly horizontal laminations, with cross laminations in sand layers and with rare burrows. Boring VI, 6.00—6.25.
8. Sandy clay of deep tidal channel. Predominantly horizontal laminations with occasional cross laminated sand layers. Boring III. 17.58—17.83.
9. Sand of deep tidal channel. Rather strongly cross laminated structures. Boring V. 12.57—12.82.
10. Lake clay, with plant roots and with irregular lenses of silt. Boring VI, 12.45—12.70.

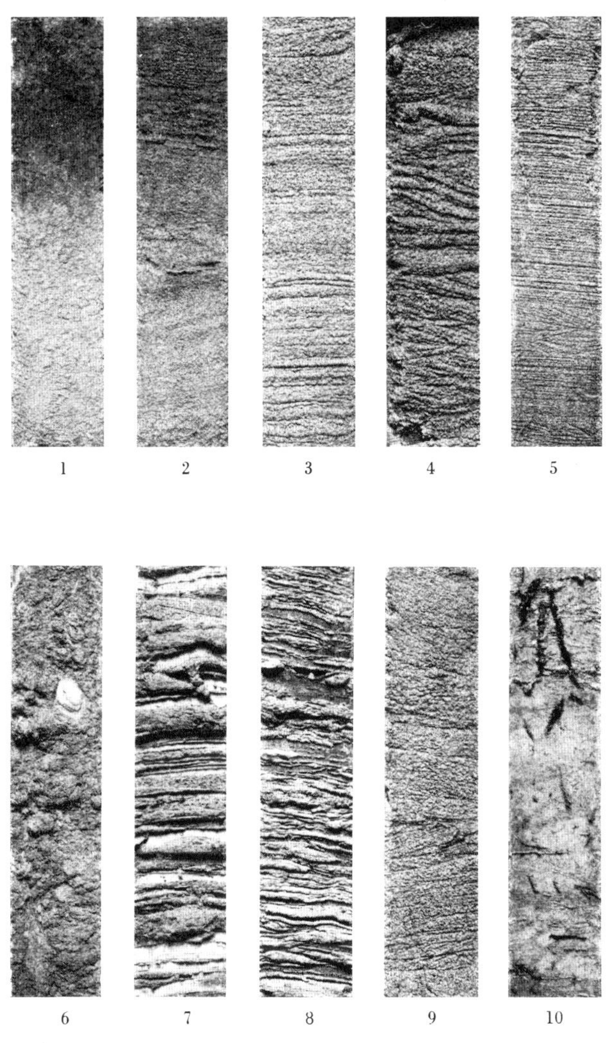

PLATE 13.

Lacquer peels of open sea deposits (cores).

Correlation of strata in borings I—IV (see fig. 23).

Depths in m below N.A.P.

1.	Boring	I	8.32—8.57	7.	Boring	IV	6.75—7.00	13.	Boring	I	10.70—10.95
2.	Boring	II	6.95—7.20	8.	Boring	II	8.02—8.27	14.	Boring	II	10.60—10.85
3.	Boring	III	5.64—5.89	9.	Boring	III	8.47—8.72	15.	Boring	III	10.19—10.44
4.	Boring	I	8.65—8.90	10.	Boring	IV	7.56—7.81	16.	Boring	I	11.10—11.35
5.	Boring	II	7.53—7.78	11.	Boring	II	8.61—8.86	17.	Boring	II	11.07—11.32
6.	Boring	III	7.97—8.22	12.	Boring	III	8.83—9.08	18.	Boring	III	10.55—10.80

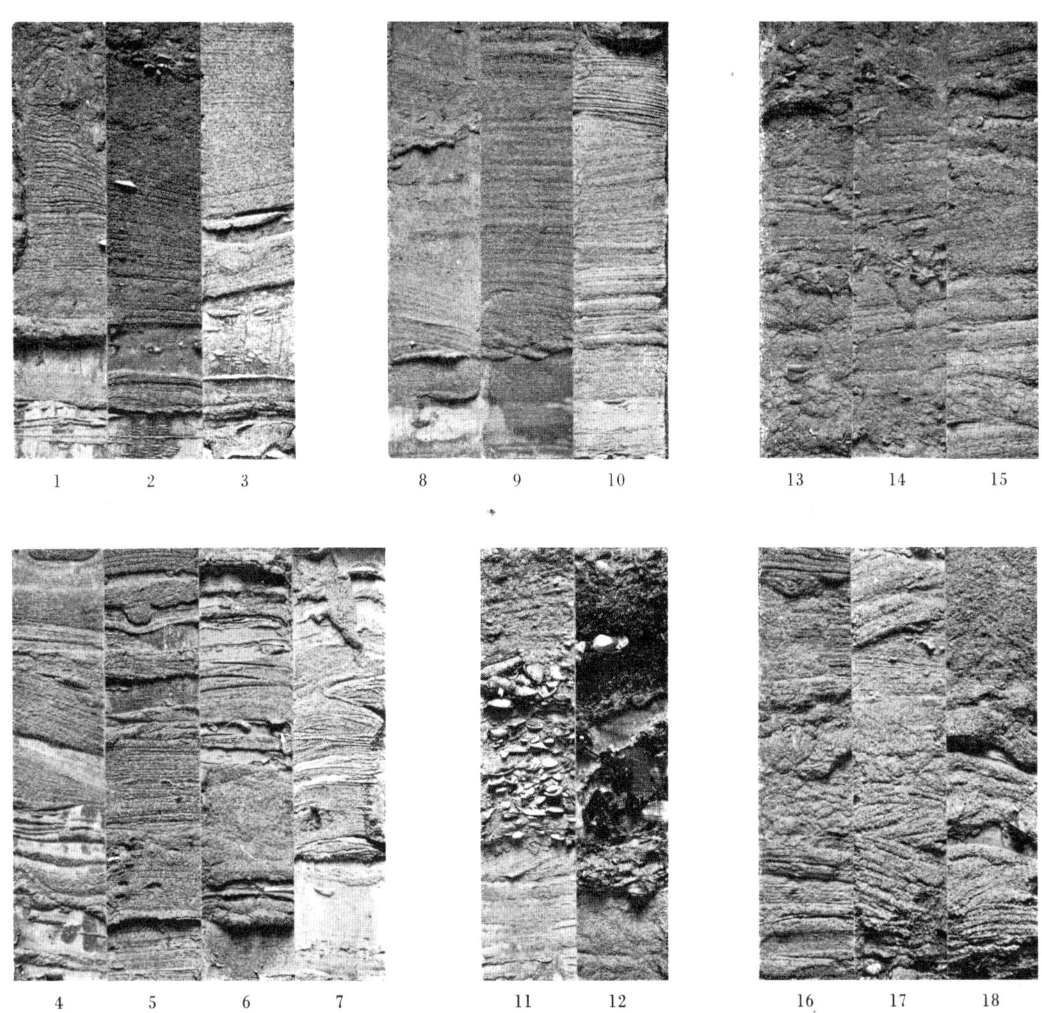
1
2
3
8
9
10
13
14
15
4
5
6
7
11
12
16
17
18

PLATE 14.

Metal tray samples.

Vertical sections, natural size. Depths in m below N.A.P.

1. Open sea deposit. Sandy clay, with horizontal laminations. Boring II, 7.16—7.23.
2. Open sea deposit. Clayey sand, with horizontal and cross laminations. Boring II, 8.10—8.17.
3. Open sea deposit. Sandy clay, with horizontal and cross laminations. Boring V B, 8.61—8.68.
4. Open sea deposit. Clayey sand, with horizontal and cross laminations. Boring V B, 8.69—8.76.
5. Tidal flat deposit. Clayey sand, rich in burrowing structures. Boring VII, 10.57—10.64.
6. Tidal flat deposit. Sandy clay with horizontal laminations partly disturbed by burrowing animals. Boring VII, 10.89—10.96.
7. Tidal flat deposit. Clayey sand, primary structures almost completely disturbed by burrowing animals. Boring VII, 10.89—10.96.
8. Lagoon clay. Primary laminations almost completely disturbed by burrowing animals. Many cross sections of *Hydrobia* shells. Boring VII, 11.76—11.83.
9. Lake clay, with irregular silt lenses and plant roots. Boring VII, 12.62—12.69.

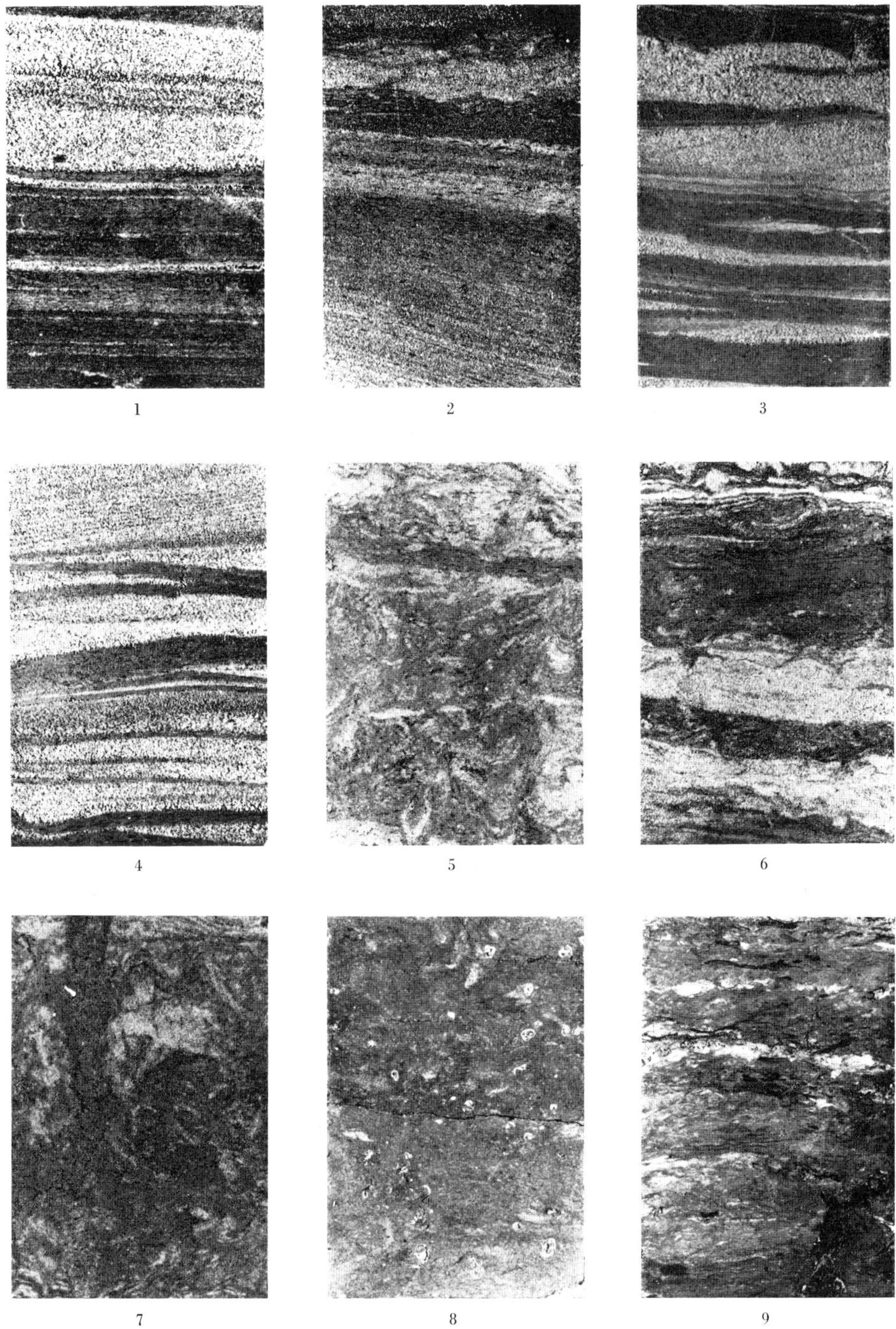

Editor's Comments on Paper 12

12 **Gill:** *Evolution of Shoreline Barriers*

The ephemeral nature of barriers that is stressed throughout this volume is voiced here by E. D. Gill in a paper reprinted from the *Victorian Naturalist.* Gill briefly traces scientific thought concerning sea level from constant change, to permanency, then back to change. If sea level is not stationary, it would follow that unconsolidated shore features are temporal forms. Thus barriers of all sorts must move up and down, and back and forth, with the changing shoreline. He takes as his example Australia's famous Ninety Mile Beach in Gippsland, eastern Victoria. This region will again be discussed, by E. C. F. Bird, in one of the last papers in this volume.

Edmund D. Gill was born in Auckland, New Zealand, in 1908, and received his education at the University of Melbourne. Honorary Secretary of the ANZAAS Quaternary Shoreline Committee from 1952 to 1972, Honorary Secretary, and later President, of the Royal Society of Victoria, Gill is a Life Member Honoris Causa of the society and was awarded the Research Medal in 1967. In 1966 Mr. Gill was Visiting Professor at the California Institute of Technology. He is a member of the INQUA Quaternary Shorelines Commission and President of the Pacific and Indian Oceans Subcommission. He is also presently serving as the Deputy Director of the National Museum of Victoria.

Reprinted from *Vict. Nat.*, **84,** 282–283 (1967)

Evolution of Shoreline Barriers

12

By Edmund D. Gill*

A young Greek named Anaximander stood up in the market place in Athens in the 5th century B.C. and stated "Change is the nature of things, all is flux, nothing stays the same, all things are in alteration, change is the nature of things". This view of the world was lost in the Middle Ages, when people thought of all things as static. The "everlasting hills" were thought always to have been there. The level of the sea was thought to have been ever constant. Human history was thought of as a pond, rather than as a river. In our own century we have come to expect change, we have come to look for progress, we agree with Anaximander that change is the nature of things. The process of change-over from a static view of things to a dynamic view of things has gradually extended into all areas of thinking, including the origin of the rocks and the history of the sea.

We now know that sea level is constantly fluctuating. It is dynamic and not static. It is always changing, but the disturbance of its level by tides and by meteorological conditions makes it hard for us to observe these changes. However, the evidence of shoreline morphology, the evidence from the analysis time tables, and the understanding of the variation in sea level as a function of change in the ice régime of the earth has brought us to see that sea level is dynamic. Now if sea level is oscillating, shoreline structures cannot be static, but must also be a function of dynamic processes. Shoreline sand dunes (or "barriers" as they are now generally called) are constantly changing, new ones being built and old ones modified or even destroyed. Morphologically, the only stable barriers are the fossil stranded dunes that have been lithified into rock, or anchored by vegetation. Although shoreline barriers (like other shoreline features) are liable to change, there is evidence that some barriers are old. Some in Victoria have existed for 5,000 years or more as is shown by radiocarbon dating. Also unconsolidated stillwater marine shell beds of this age, although readily eroded, have survived on the open coast for this period of time (e.g. at Seaspray in Gippsland), and so must have been protected by barriers that have existed for at least that long. This does not mean that the barrier has necessarily always been the same, nor does it infer that the barrier has always been in exactly the same place. It is suggested that the barriers that have protected these beds have themselves been mobile, accommodating themselves to changing meteorological conditions and changing levels of the sea. What is the rapprochement between evidence for longevity of barriers and evidence for their instability? It is suggested that they are constantly evolving, being destroyed in part and rebuilt—hence the lack of consolidation, and lack of secondary carbonates. It is suggested that they migrate to and fro with changes of sea level and hence barriers are found sometimes overlying shell beds, sometimes overlying soils, sometimes overlying former shore

* National Museum of Victoria.

platforms, and so on. This concept matches the mobility of sea level with a comparable mobility of shoreline structures. It provides a dynamic and evolutionary view of both sea level and shore lines.

Let us take an example. The Ninety Mile Beach in Gippsland, eastern Victoria, is an open-coast high-energy sandy beach. The tidal range varies from about six to ten feet; storms beat on the coast so that there is often high wind-energy as well as high wave-energy. All along this coast there is a barrier behind the beach, and there is an ample sand supply to maintain this barrier. Let us consider a cross section of the coast where it is simple in structure. At Seaspray, about a mile inland, there is an ancient coast which is approximately parallel to the present one. From the fossil cliff, swampy flats extend towards the sea, then comes the sand barrier, then a sandy beach, and the ocean. A large drain has been dug for two miles on the inland side of the barrier, from the village of Seaspray eastwards to Lake Reeve—a lake inside the barrier. The drain provides an excellent cross section and shows that below the swampy beds there are still-water marine shell beds which are stratified. Such beds must have been laid down below low water level. The top of these beds is now approximately six feet above low water level on the coast, so the sea must have been at least six feet higher relative to the sea than it is at present. Where tested, the shell beds pass underneath the sand barrier. Pieces of cemented shell bed are found on the ocean beach, and some distance .off shore the fishermen report a reef, the nature of which has yet to be determined. It could be the site of an ancient barrier in that the beds are of stillwater marine facies. They must have been laid down behind an ocean barrier, because the geographic context in an open ocean sandy beach. There must have been protection for the waters in which those shell beds were laid down. In that the shell beds pass underneath the barrier, it can be concluded that the position of the barrier has changed since the shell beds were deposited. It was noted after a storm that cliffs up to twelve feet high occurred on the ocean side of the barrier. Where the sea had cut away the barrier, the sand was uncompacted. It can therefore be inferred that as a result of sea attack, the barrier is cut and then filled again according to the prevailing conditions. In other places there are juvenile soils in the barrier indicating that parts of it are relatively old. With this and other evidence it is possible to construct a concept of the barrier changing position with altering sea level and altering meteorological conditions—a dynamic and not a static concept of the barrier. With rising sea level the barrier would gradually migrate inland, and conversely with falling sea level it would migrate seawards. However, its continuity since mid-Holocene times (the date provided by radiocarbon assay for the shell beds) is assured, because otherwise these beds that can so readily be eroded would have been destroyed. As the site is an open high-energy coast, these beds would certainly have been rapidly removed if at any time the barrier had ceased to exist. There has therefore been a continuity of barrier, but with change of position and nature according to the prevailing conditions. Detailed boring, dating of the top and bottom of each shell bed, and dating of the various elements in the structure of the barrier could elucidate this story in some detail. For the present, attention is drawn to the dynamic nature of sand barriers.

Mobility of sea level is matched by mobility in shoreline structures.

Editor's Comments on Paper 13

13 Mackenzie: *Environments of Deposition on an Off Shore Barrier Sand Bar, Moriches Inlet, Long Island, New York*

In this paper M. G. Mackenzie is mainly concerned with the sedimentary environments to be found on barrier islands in the vicinity of an inlet. Samples from the barrier profiles on each side of Moriches Inlet were analyzed for mean diameter and standard deviation. Heavy-mineral analyses were made as well.

Mackenzie defines two regimes of sedimentation in the barrier island inlet area: (1) the backbar (bay and dunes) and (2) the forebar (berm and ocean beach). The backbar sands are fine grained and better sorted than the forebar sands. He concludes that there is an increase in heavy-mineral content by weight with smaller mean diameter and better sorting. Recognition of these various parameters should aid in the identification of fossil barrier islands.

In 1960 Michael G. Mackenzie earned his B.S. in geology from Yale University and during the same year did field work in the Cretaceous in Wyoming. He was awarded an M.S. in geology at New York University in 1962; joined Mobil Oil Company, as a petroleum geologist, in 1963; and is presently a consulting petroleum geologist in New Orleans. He anticipates receiving his Ph.D. from Tulane University as this volume goes to press.

Reprinted from *Tulane Studies Geol.*, **5**, 67–80 (1967)

ENVIRONMENTS OF DEPOSITION ON AN OFFSHORE BARRIER SAND BAR, MORICHES INLET, LONG ISLAND, NEW YORK

13

MICHAEL G. MACKENZIE
MOBIL OIL CORPORATION
NEW ORLEANS, LOUISIANA

CONTENTS

I. ABSTRACT

Environments of deposition associated with an offshore barrier sand bar system at Moriches Inlet, Long Island, New York, were studied by mechanical analysis and heavy mineral analysis. Samples collected from six traverses normal to the barrier trend were statistically defined by measurement of mean diameter ($M\phi$) and standard deviation ($\sigma\phi$). Variations in heavy mineral content in different parts of the sand bar are related to the concept of hydraulic equivalent size in sedimentation. By relating threshold velocity (V_t) to grain characteristics, the concept of hydraulic equivalent size, developed for water-transported sands, can be effectively extended to wind-blown particles.

Two distinct sedimentary regimes are defined by the methods used in this study, namely, a forebar and a backbar. The small dimensions of the environments studied preclude further subdivision by these methods.

Results are discussed with reference to fossil shoestring sand bodies found in the geologic record. It is concluded that lateral mineralogical and textural variations should be combined with gross geometric properties in studies involving the genesis of shoestring sands.

II. INTRODUCTION

A. Purpose and Scope

The purposes of this study are to delineate environments of sedimentation associated with an offshore barrier sand bar by the study of grain size distribution, degree of sorting, and heavy mineral variations. The study was designed to test the applicability of these parameters in the delineation of sedimentary environments in the sand bar system.

The Moriches Inlet area, Long Island, New York, was chosen for this study because it includes several recognizably distinct environments of deposition in proximity. In addition, it is clear that a single source accounts for detrital materials found within this system of environments. It is probable that if only one suite of materials was introduced into the environmental system, then any variation found within that suite would indicate local environmental change only, and would not reflect influences of materials from different sources.

Rittenhouse's (1943) concept of hydraulic equivalent size is used to explain the variation in heavy mineral content. If this concept can be extended to explain variations of heavy minerals in beach and dune sands, then such explanations as "selective sorting", or, "lag concentrate" would be modified by a more quantitatively useful interpretation.

The writer contends that if studies of heavy mineral concentrations in sedimentary materials lend more emphasis to the concept of hydraulic equivalent size, then heavy minerals may assume greater importance as sensitive and reliable indicators of depositional conditions.

B. Location, Extent, and Description of Area

The Moriches Inlet area is part of an offshore barrier sand bar system on the south shore of Long Island, New York. The barrier island system is separated from the mainland by approximately one mile at Moriches Inlet. The area of study (Figure 1) is bounded on the north by Moriches Bay and on the south by the Atlantic Ocean. The area is bisected by the 800-900 foot wide Moriches Inlet and extends 3000 feet west and 3000 feet east from the shores of the inlet. The barrier is between 1000 and 2000 feet wide in the study locality. Moriches Inlet truncates Fire Island to the west and Cupsoque Beach to the east, and lies within Brookhaven Township, Suffolk County, New York.

Sand dunes average 15-20 feet in elevation and form the most conspicuous features of the barrier system. The dunes consist of two or three main ridges trending parallel to the barrier. Along the traverses across the barrier, the first dune ridge was encountered at distances varying between 360 and 480 feet across the berm from the ocean swash zone. The dunes are in most places well anchored by grasses and bushy plants. The only other notable topographic feature is the ocean beach face (ocean beach swash zone), which has a gradient of between five and ten percent.

Moriches Bay is from one to three feet deep at mean sea level, at distances from 2000 to 3000 feet into the Bay from the barrier margin. Several small, marshy islands are present in the Bay, and inlet delta accumulations are situated in the Bay immediately north of Moriches Inlet. During average tidal fluctuations considerable portions of these delta accumulations remain subaerial.

During the three day sampling period, winds were from the south and southwest at an average velocity of about 10 mph. Tidal fluctuation was between 1.07 feet and 1.29 feet, as measured on the north shore of Moriches Bay.

C. Geologic Setting and Regional Patterns of Sedimentation

The south shore of Long Island forms a transition zone between Pleistocene glacial deposits of the Island to the north and the continental shelf to the south. Geologic formations available for erosion by either drainage or wave action consist entirely of glacially-derived sediments. A geologic map is presented in Figure 2.

Regional patterns of sedimentation on the south shore of Long Island have been determined in several previous studies. Colony's (1932) study of littoral materials of the south shore shows a net westward drift of sand. Beach erosion and engineering studies show accretion and westward migration of the east sides of inlets present on the southern coastline of Long Island. The Beach Erosion Board, Corps of Engineers, U.S. Army (1961) estimates that the annual littoral transportation rate is of the order of 450,000 cubic yards per year along most of the coast, and that 300,000 cubic yards per year (822 cubic yards per day) westward drift occurs in the vicinity of Moriches Inlet. The headlands physiographic province (Figure 2) is the chief source of clastic materials on the south shore of Long Island.

Lucke (1934) and Nichols (1964) show that the contribution of sediments by streams to the barrier beaches is negligible. Accumulations of fine clastic materials are found in the estuaries of the small streams that enter Moriches Bay from the north. However, the bulk of sediment in the south shore bays is derived from the barriers and inlets.

Although the presence of some sand in the barrier system may result from the near shore bottom drag of incoming waves, the amount of attrition due to this process is difficult to ascertain.

The drainage basin associated with streams

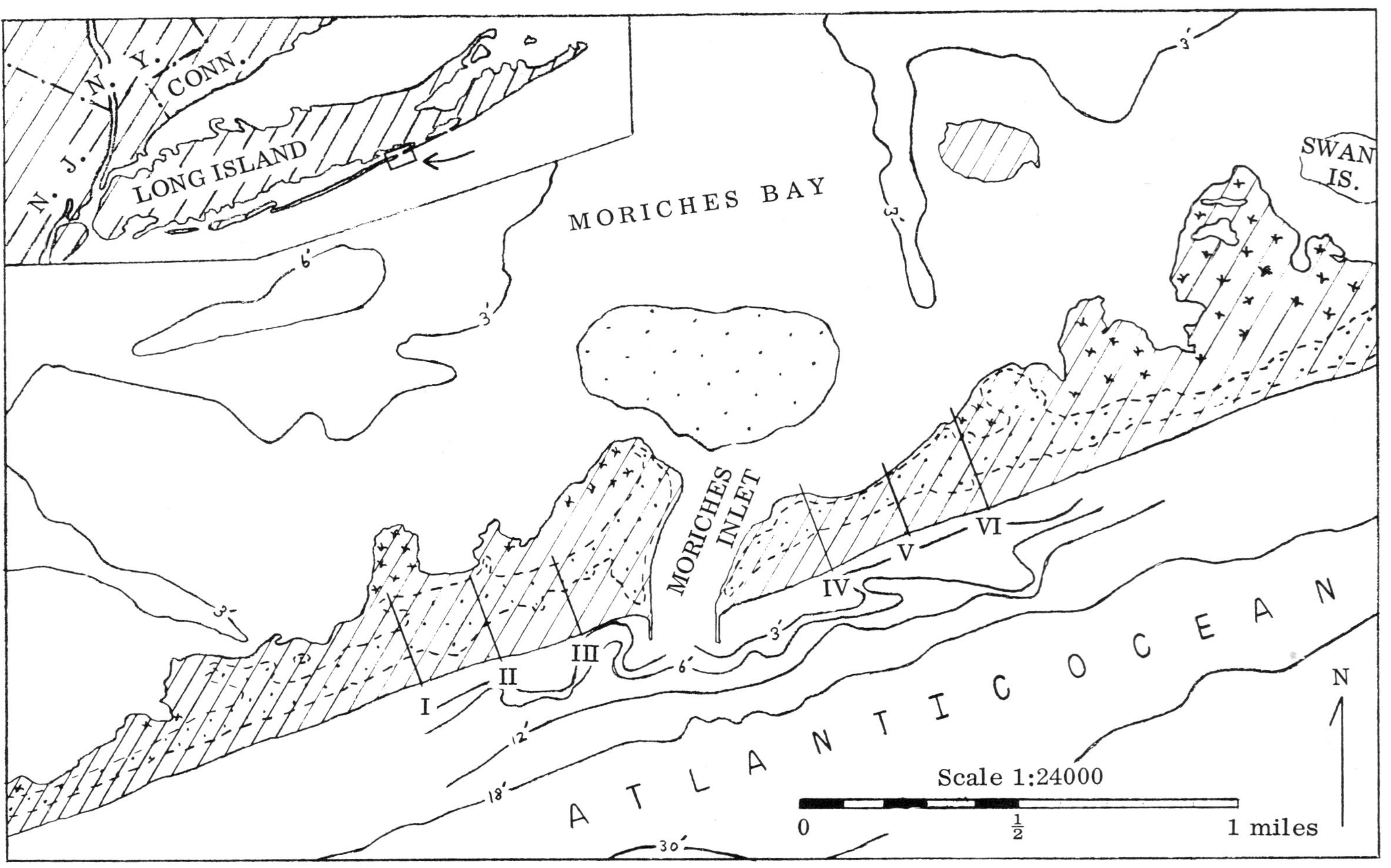

Figure 1. Map and Inset Map (1 inch = 37 miles) of Moriches Inlet area and sampling traverses. Dotted area = sand dunes; "planted" area = marsh; irregular dotted area = subaerial sand deposit. (after USGS, 1957)

that flow into Moriches Bay is underlain by the Manhasset Formation and by the Ronkonkoma Moraine and sediments derived as outwash from the Ronkonkoma Moraine. Clastic materials derived by littoral drift from the shoreline east of Moriches Inlet are also eroded from the Manhasset and Ronkonkoma units. Compositions of these source formations are shown in the legend description of Figure 2.

III. Acknowledgments

Acknowledgment is expressed for helpful criticism by the faculty of the Department of Geology at New York University. Thanks are extended to Dr. A. W. McCrone, thesis advisor, and to Dr. L. E. Spock for aid in mineral identification. Dr. Garrett Briggs, Dr. J. P. McDowell, and Dr. H. C. Skinner, faculty members in the Department of Geology at Tulane University, are acknowledged for their constructive criticism. Miss Jeanne Danker contributed x-ray analyses towards the identification of the finer-sized clastic particles. Messrs. D. A. Ravenhall of Brookhaven National Laboratory and S. B. Cross, Department of Public Works, County of Suffolk, New York, supplied wind and tide data. Mr. W. H. Hillicker acted as field assistant.

IV. Methods of Study

On September 23, 24, and 25, 1961, 30 surface samples were collected to a depth of 2 cm along six traverses normal to the long axis of the barrier. Three traverses were located on each side of the inlet at 1000 foot intervals. Sampling on all six traverses includes samples from the following environments (traverse locations and the sampling pattern are shown in Figures 1 and 4):

a. Intertidal ocean beach swash zone
b. Midpoint of berm (between swash zone and foredune)
c. Foredune slope
d. Backdune (fronting bay) slope

The three traverses east of the inlet include samples from two additional environments:

e. Intertidal bay (protected) beach swash zone
f. Bay bottom samples (approximately 100 feet from barrier)

LEGEND

Beach and Dune Deposits

Swamp and Marsh Deposits

Thin Outwash from Harbor Hill Moraine, with outcrops of Manhasset Formation: very thin deposits, forming sandy plains

Ronkonkoma Moraine: structureless gravels, locally stratified; widespread sands

Outwash from Ice along the Ronkonkoma Moraine: sloping sandy plains

Thin Outwash from Ice along the Ronkonkoma Moraine, with outcrops of Manhasset Formation: very thin deposits, forming sandy plains

Montauk Till Member of the Manhasset Formation: faintly banded clay, sand, with many Triassic igneous and sedimentary pebbles and boulders

Thin Till: bouldery till, mostly a mixture of red clay, sand, and boulders

Figure 2—Legend. Description of surface deposits and key to lithologic symbols used on the geologic map of eastern Long Island (page 71).

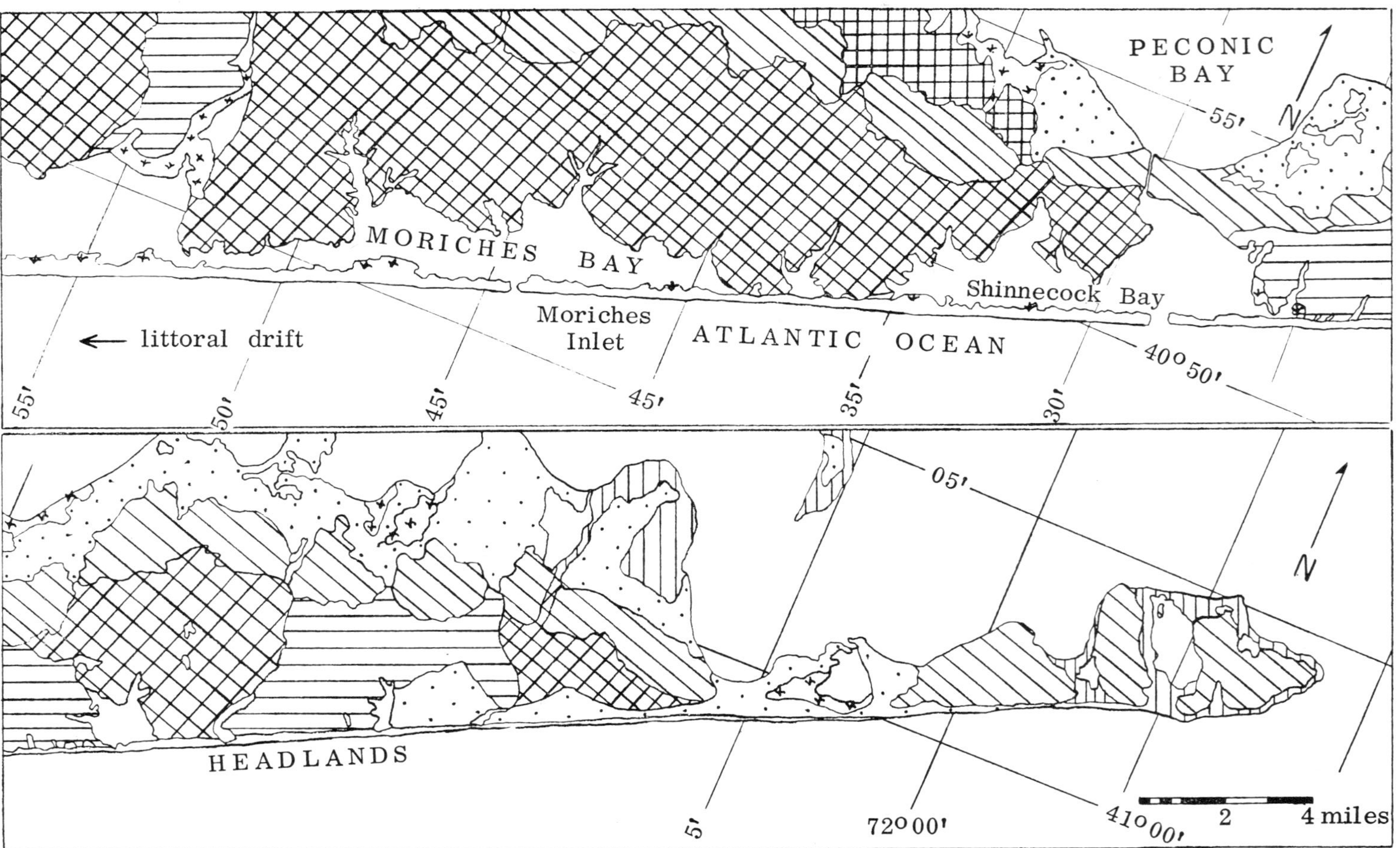

Figure 2. Geologic map of eastern Long Island (after Beach Erosion Board, "Geomorphology", 1961).

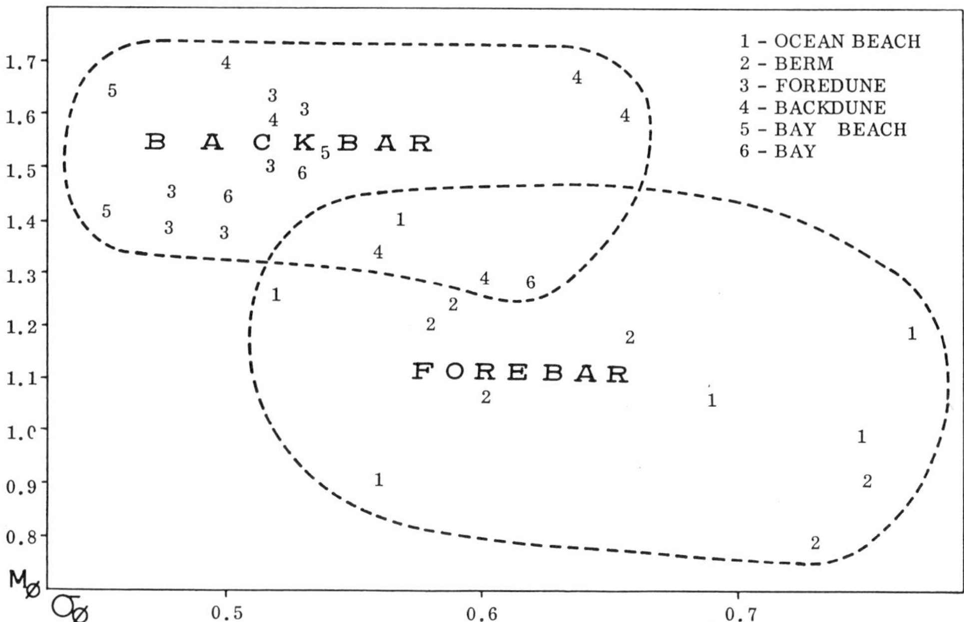

Figure 3. Diagram showing mean diameter ($M\phi$) plotted against standard deviation ($\sigma\phi$). Both parameters are in phi units. Resulting separation into forebar and backbar environments is shown with dashed lines.

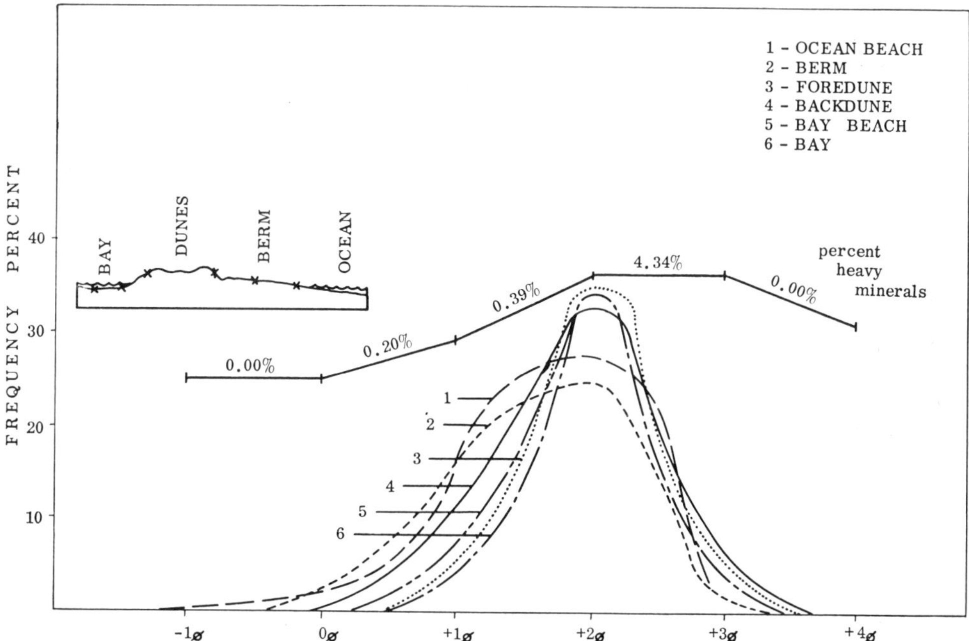

Figure 4. Diagram showing size frequency distribution of samples collected along six traverses in the Moriches Inlet area. Inset cross section shows localities (X) sampled along the traverses. The average heavy mineral percent by weight of each size grade is shown above curves.

Except for those taken from the bay, all samples were collected to a depth of 2 cm with a small, flat-bottomed plastic shovel with centimeter marks on the inner walls. The bay samples were collected by hand to depths estimated at 2 cm.

Each sample was split by means of a Jones sample-splitter to obtain representative 40-50 gram samples. The material was then warmed or boiled gently in dilute (3.75% HCl A.C.S. Standard) HCl for 20 minutes to remove shell fragments. Percent carbonate by weight varied between 0.00 and 4.50 percent. Almost all samples contained less than one percent carbonate by weight.

After acid treatment the material was reweighed and sieved in a mechanical sieve shaker for fifteen minutes. Material retained on each sieve was weighed. Size frequency distributions are shown in Figure 4.

Heavy mineral separations were made on the sieve fractions. Bromoform ($CHBr_3$; S.G. - 2.89 at 20° C) was the heavy liquid used. For each separation approximately 150 ml of bromoform were placed in 300 ml separatory funnels. Results of these separations show that size grades between —1ϕ and 1ϕ (2 mm - 0.5 mm) contain negligible amounts of heavy minerals. Size grades >3ϕ (<1/8 mm) were either absent or present in insignificant amounts. Consequently the 1ø - 2ø and 2ø - 3ø size grades were the only ones used for heavy mineral analysis.

A mineralogical count was made of 563 grains from the heavy separates of samples from the ocean swash zone, dune, and protected swash zone. The mineralogical content of the light separates was cursorily examined.

The raw data for mechanical analysis are considered in terms of the phi scale ($\phi = -\log_2 d$, where d=diameter in mm) in order to allow use of basic statistical methods and to facilitate graphic representation. Because of similarity in the central tendencies of the size distributions of most of the samples, moment measures are used instead of quartile measures. For computing logarithmic means ($M\phi$) and logarithmic standard deviation ($\sigma\phi$), the following formulae were used:

$$M\phi = \Sigma fm/100$$

(f =frequency by weight)
(m=midpoint of phi class)

$$\sigma\phi = \sqrt{n_2 - n_1^2}$$

(n_1 and n_2 are moment measures)

Mean diameter ($M\phi$) is plotted against standard deviation ($\sigma\phi$) in Figure 3.

V. Discussion of Results

Results of the analysis of 30 sand samples are considered below. Three samples (I-2, V-3, V-5) are rejected for heavy mineral distribution analysis because they contain aberrant heavy mineral contents more than five times in excess of the average for their respective environments. Although the relatively small number of samples taken from each environment precludes rigorous statistical confidence analysis, the three samples rejected are well outside of the standard 5% confidence interval.

The writer realizes that the data might have been more accurately representative of deposition within the Moriches Inlet area if the samples had included material from a larger vertical interval than the 2 cm obtained. Also, a complete identification of all heavy minerals counted, as well as a 0.5ϕ interval heavy mineral frequency distribution, would have been important additions. However, the depositional trends found in this study indicate that these elaborations of procedure would only serve better to define the results herein presented.

A. Heavy Minerals

Table 1 shows the results of grain counts of the heavy mineral fraction of three samples collected from different environments. The heavy mineral suite found in the Moriches Inlet area is garnet-rich (average 50.9%). Staurolite (average 11.0%) and opaque minerals (magnetite and/or ilmenite, average 9.7%) are second and third in abundance.

The profile in Figure 5 shows the average percent by weight of heavy minerals found in the environments studied. The ocean beach swash zone and the berm show lowest values. The dune environment shows highest values whereas bay beach samples contain inconsistent percentages. Bottom samples from the bay contain heavy mineral concentrations intermediate between those in

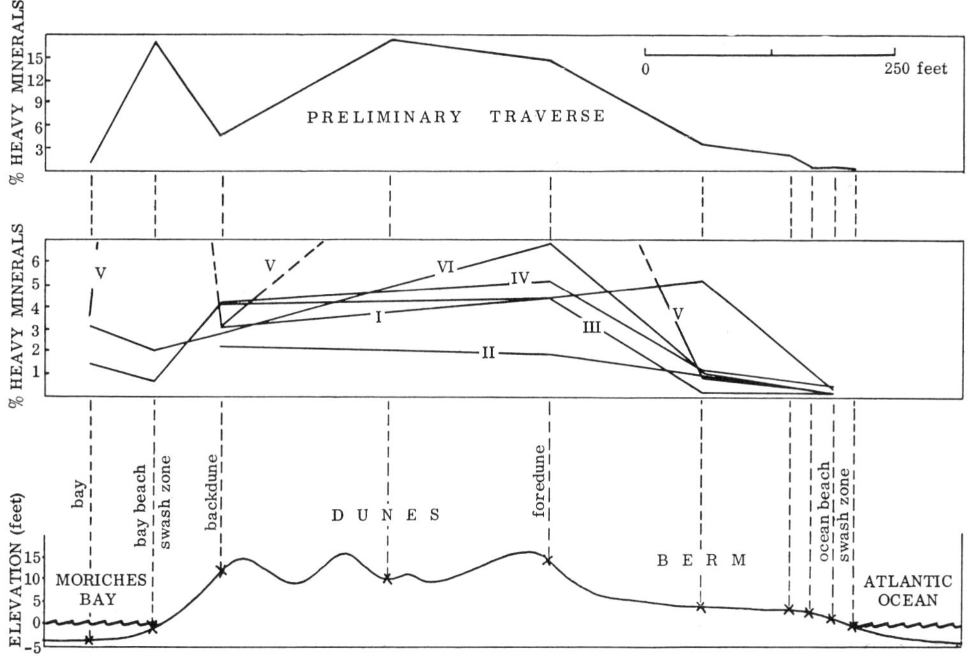

Figure 5. Schematic cross section of a typical traverse in the Moriches-Inlet area. Diagrams above section show heavy minerals as percent of total weight of each sample. Locations of sampling on preliminary traverse are shown on section (X).

the dune sands and those in berm and ocean beach sands.

Figures 6a and 6b indicate that depositional regimes can be subdivided into a forebar (ocean beach and berm) and a backbar (dunes, bay sediments) when heavy mineral content by weight is compared with $M\phi$ and $\sigma\phi$. The best separation occurs when heavy mineral percentages are plotted against $M\phi$, in Figure 6a. The overlap of fields in Figure 6b may be partly explained by the fact that whereas degree of sorting is quite similar in both beach and dune sands, $M\phi$ shows a more marked contrast between the two environments.

Figure 7a shows heavy mineral content

TABLE 1

MINERALOGICAL GRAIN COUNTS, INDICATING NUMBER OF GRAINS AND PERCENT BY NUMBER OF TOTAL HEAVY MINERALS COUNTED

Sample Number	Garnet	Magnetite Ilmenite	Zircon	Kyanite	Rock Frag's	Epidote	Staurolite	Other
I-3	90 45.0%	21 10.5%	2 1.0%	1 0.5%	3 1.5%	6 3.0%	28 14.0%	49 24.5%
V-5	107 53.0%	24 11.8%	9 4.5%	2 1.0%	4 2.0%	3 1.5%	17 8.4%	36 17.8%
7	87 54.7%	11 6.9%	3 1.9%	2 1.3%	0 0.0%	5 3.1%	17 10.7%	34 21.4%
Average Totals	94.7 50.9%	18.7 9.7%	4.7 2.5%	1.7 0.9%	2.3 1.2%	4.7 2.5%	20.7 11.0%	39.7 21.2%

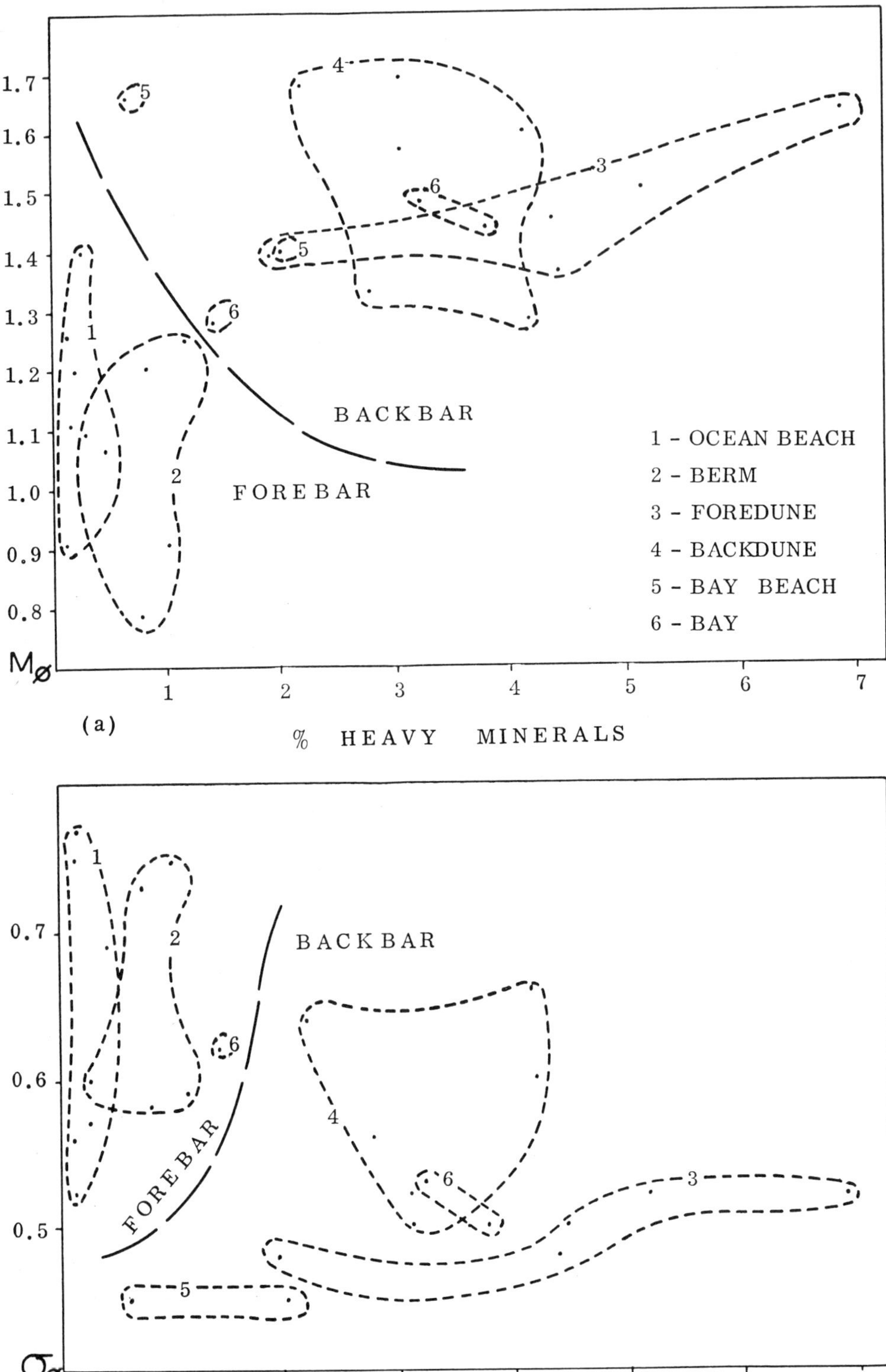

Figure 6. Diagrams showing heavy mineral percentage by weight plotted against mean diameter (a) and sorting (b).

compared with the percent by weight of the samples that occur in the size interval 1ø - 2ø (1/2 mm - 1/4 mm). Figure 7b is a similar plot utilizing the interval 2ø - 3ø. Figure 7a shows far better separation into a forebar-backbar system than does Figure 7b. The environments containing consistently higher percentages of heavy minerals (dune and bay) show, in Figure 7a, a much smaller degree of overlap with the environments containing lower percentages of heavy minerals (ocean beach, berm). There is no tendency for heavy mineral content to vary with the 2ø - 3ø size content of the samples, even though this size range contains a much greater amount of heavies per unit weight than any other size range (see Figure 4). There is a distinct tendency for heavy mineral content to vary with the sample content in the 1ø - 2ø size range; that is, one phi size lower (one Wentworth size larger) than the range wherein most of the heavies are concentrated.

The concept of hydraulic equivalence among particles in a fluid medium permits a quantitative expression of factors governing the distribution of heavy minerals in the Moriches Inlet area.

In 1943, Rittenhouse published a comprehensive review of factors controlling heavy mineral transportation and deposition. He expanded the idea of equivalency, and considered the relationships between light and heavy minerals from the standpoint of "hydraulic equivalent size" and "hydraulic ratio". In his discussion of variations found in sediments collected along a stream traverse, Rittenhouse defines three factors that affect size distributions in the same sample or in different samples taken from the traverse (Rittenhouse, 1943, p. 1743):

1. The hydraulic conditions which vary with time and position and are a composite of many interacting. conditions
2. The hydraulic equivalent size which is also the net effect of several factors
3. The relative availability for deposit of the different sizes of each mineral

Rittenhouse also stated that the size distributions of light and heavy minerals in a deposit will reflect the net effect of temporal changes in hydraulic conditions. He indicates that (referring to the above factors):

"At any instant, however, all kinds and sizes of mineral grains that are accumulating will be subject to the same hydraulic conditions, whatever they may be. At other instants, this will also be true.

"Consequently, the differences in size distribution of different minerals in the same sample will be due to the second and third factors."

Within the Moriches Inlet area it is axiomatic that hydraulic conditions must change with time at any given point. While this is also indicated in Rittenhouse's study (for deposits found along a stream traverse), a consideration of the relatively small dimensions sampled enabled Rittenhouse to assume little variation between samples in the relative availability for deposit of each size grade of each mineral.

Many factors must be considered in attempting to explain sediment variations found in the Moriches Inlet area. It has been stated above that several environments of deposition are recognized in the offshore barrier system. It is known that sediment found in all the local environments is directly and indirectly derived from materials brought into the area by ocean wave action. Also, because these materials undergo net migration from the ocean beach to the bay, each environment must have a distinctive sediment assemblage introduced into it that will in part be deposited within that environment and in part be deposited elsewhere. If this were not true, no selectivity would exist, and no environments could be differentiated. It was thought that these environments would be recognizable by means of methods employed in this study.

Because it was found that the concept of hydraulic equivalent size can be used to explain the distribution of heavy minerals at Moriches Inlet, the obvious differences between a stream traverse (Rittenhouse's study) and a barrier sand bar must be noted. Moreover, the sand dunes are wind-blown. It seemed that differences in effect on sediments between the two fluid transporting media might be accounted for by considering a theoretical relationship established by Bagnold (1941). Bagnold found empirical indications that the threshold velocity (V_t= velocity in a turbulent flow of air over a rough surface needed to start a sand grain moving from rest) in air varies as the square root of the grain diameter:

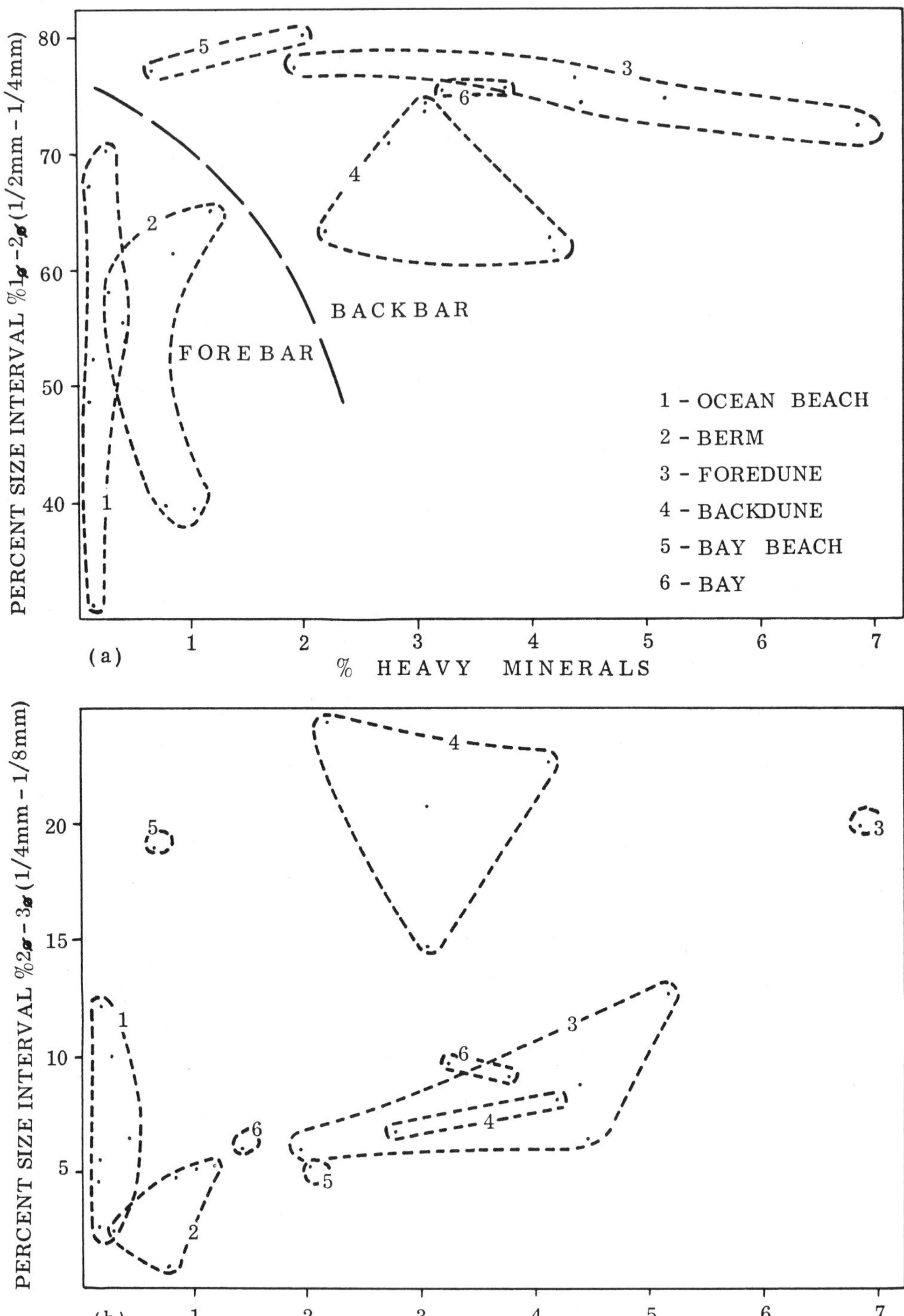

Figure 7. Total percent by weight of heavy minerals in sample plotted against percent of sample in the given size ranges.

$$V_t = A \sqrt{\frac{\sigma - p}{p} Gd}$$

where:

V_t=threshold velocity

A =a constant (for grain sizes considered here)

σ =grain density

p =fluid density

d =grain diameter

G =gravitational constant

The expression is especially interesting from the viewpoint of the present study as the grain density σ is included. To apply the threshold velocity relationship to this study, the following is noted:

$$V_t = A \sqrt{\frac{\sigma - p}{p} Gd} \quad (1)$$

but because $p = 1.22 \times 10^{-3}$, then, effectively,

$$V_t = A \sqrt{\frac{\sigma}{p} Gd}$$

but A, p. and G are constants, and

$$V_t = \sqrt{\sigma d} \quad (2)$$

If the principles of hydraulic equivalent size apply, then there should be some threshold velocity that is the same for a given heavy mineral particle as it is for a hydraulically equivalent light mineral so that

$$V_t = \sqrt{\sigma_1 d_1} = \sqrt{\sigma_2 d_2}$$

or,

$$\sigma_1 d_1 = \sigma_2 d_2 \quad (3)$$

Hence, if $\sigma_1 > \sigma_2$, then $d_1 < d_2$, and vice versa.

The relationship may seem oversimplified, but it is adapted as a working formula. The concept of hydraulic equivalence may be tested, as Rittenhouse points out, without knowledge of rounding, sphericity, grain surface features, or other such parameters. If a distinct grain size relationship is found between heavy minerals and light minerals in sedimentation processes, then the concept is valid, regardless of the various factors inherent in producing that relationship.

Using quartz on the one hand and heavy minerals on the other hand, expression (3) has been calculated for the most common heavy minerals that were found in the sands of the Moriches Inlet area.

Table 2 shows the computed hydraulic equivalent sizes corresponding to the given sizes of the identified heavy minerals as calculated for wind-blown particles. The distribution of the heavy minerals in the dunes is significant when related to the computed hydraulic equivalent sizes which predict the relationship shown in Figure 7, namely, the variation of heavy mineral content with the content of light minerals of relatively larger diameters (hydraulic equivalent sizes). These data indicate that the concept of hydraulic equivalent size is applicable to wind-blown sands, as well as to water-transported sands, and explains the distribution of heavy minerals found in the Moriches Inlet area.

B. Mechanical Analysis

Sands in the dunes and along the barrier margin of the bay show better sorting and finer texture than sands in the berm and ocean beach environments. Although there is approximately a 0.5ϕ difference in the Mϕ of samples between the forebar and backbar areas, all samples but three taken from the ocean beach are fine-grained sands (2ϕ - 3ϕ). Sands from the ocean beach and berm environments yield a $\sigma\phi$ only 0.12ϕ

TABLE 2

EQUIVALENT DIAMETERS (CALCULATED IN MM) OF WIND-BLOWN PARTICLES. VALUES AT LEFT ARE THE DIAMETERS OF THE HEAVY MINERALS FOR WHICH THE HYDRAULIC EQUIVALENT DIAMETERS HAVE BEEN CALCULATED

	Garnet	Magnetite Ilmenite	Zircon	Kyanite	Epidote	Staurolite
1.5ϕ (0.375 mm)	0.542	0.699	0.662	0.511	0.481	0.524
2.0ϕ (0.250 mm)	0.361	0.466	0.442	0.341	0.321	0.349
2.5ϕ (0.188 mm)	0.272	0.350	0.332	0.256	0.241	0.262

units larger than do sands found in the backbar area. When $M\phi$ is plotted against $\sigma\phi$, a good separation becomes apparent between the forebar sediments and the sediments associated with the bay and with the dunes (see Figure 3) The size frequency curves in Figure 4 also show good separation.

Little difference can be expected between the physical properties of samples collected from the bay and samples collected from the bay beach. Moreover, only three samples were collected from both the bay and the bay beach. It is probable that the similarity of the bay samples to the dune samples is caused by the effective sorting action of the constantly fluctuating shallow tidal currents that are especially strong along the interior margin of the bar. Similarity between bay beach and dune materials is expected because part of the dunes is undergoing active erosion by the bay waters.

Ocean beach and berm materials are almost identical mechanically. This uniformity is probably a result of the migration of sand across the relatively narrow berm (average width 420 feet). The berm is probably not broad enough to enable sediments derived from the ocean beach to acquire a texture distinct from that of the local provenance, or, ocean beach.

The ocean beach and berm samples show negative skewness. This may in part be due to the fact that offshore bottom samples in this area may be negatively skewed (Beach Erosion Board, Corps of Engineers, 1961). The dune sands yield normal frequency curves whereas the curves for bay and bay beach samples are slightly negatively skewed. Because only the forebar and the backbar sedimentary regimes can be delineated by mechanical analysis, inter-environmental inheritance of textural characteristics must be a strong factor in sedimentary patterns in the Moriches Inlet area.

C. Shoe-String Sand Bodies

The origin of fossil shoe-string sand bodies is of long-standing interest to petroleum geologists. Lenticular sand bodies have been explained as fossil stream channels or ancient sand bars. The problem of origin has been pursued by stratigraphers, but there are few accounts of lateral sedimentary variation within the sand bodies. Bass (1936) and Bass, Leatherock, Dillard, and Kennedy (1937) give data on texture and mineralogy, but these data are ascribed to the sand body as a whole. There has been no attempt to describe local lateral textural changes.

Shoe-string sands that represent ancient barrier bars may contain textural and mineralogical variations similar to those found in this study. Because similar heavy mineral species are present throughout a local barrier section, diagenetic solution of less stable species would not mask primary differences due to sedimentation. Dune sands will remain relatively enriched in the residual stable species.

VI. Conclusions

This study leads to the following conclusions concerning sediments in the Moriches Inlet area:

A. Because of the relatively small geographic dimensions (normal to the barrier trend) of the environments studied, only two regimes of sedimentation, the forebar and the backbar, can effectively be differentiated by the methods used in this study. Subdivisions of these regimes must be delineated by other means

B. Backbar sands (bay and dunes) are better sorted and finer-grained than sands in the forebar regime (berm, ocean beach)

C. The concept of hydraulic equivalent size is applicable to wind-blown sands as well as to water-transported sands

D. Heavy mineral content by weight increases with better sorting and smaller mean diameter

E. Studies of lateral changes in mineralogy and mechanical properties of sands will aid in the interpretation of the history of formation of fossil shoestring sand bodies.

VII. Selected References

Bagnold, R. A., 1941. The physics of blown sand and desert dunes: Methuen & Co., Ltd., London, 265 pp.

Bass, N. W., 1936. Origin of the shoestring sands of Greenwood and Butler Counties, Kansas: Amer. Assoc. Petrol. Geol., Bull., v. 18, pp. 1313-1345.

Bass, N. W., D. Leatherock, W. R. Dillard, and L. E. Kennedy, 1937. Origin and distribution of Bartlesville and Burbank shoestring oil sands in parts of Oklahoma and Kansas: Amer. Assoc. Petrol. Geol., Bull., v. 21, pp. 30-66.

BEACH EROSION BOARD, CORPS OF ENGINEERS, U. S. ARMY, 1961. Geomorphology of the south shore of Long Island, New York: Tech. Mem. No. 128, 50 pp.

BEACH EROSION BOARD, CORPS OF ENGINEERS, U. S. ARMY, 1961. Littoral materials of the south shore of Long Island, New York: Tech. Mem. No. 129, 59 pp.

COLONY, R. J., 1932. Source of the sands of the south shore of Long Island and on the coast of New Jersey: Jour. Sed. Petrology, v. 2, pp. 150-159.

LUCKE, J. B., 1934. A theory of evolution of lagoon deposits on shore lines of emergence: Jour. Geology, v. 42, no. 6, pp. 561-584.

NICHOLS, M. M., 1964. Characteristics of sedimentary environments in Moriches Bay *in* Papers in marine geology, Shepard Commemorative Volume: Macmillan Company, New York, pp. 363-383, illus., table.

RITTENHOUSE, GORDON, 1943. Transportation and deposition of heavy minerals: Geol. Soc. America, Bull., v. 54, pp. 1725-1780.

July 31, 1967

Editor's Comments on Paper 14

14 Hoyt and Henry: *Influence of Island Migration on Barrier-Island Sedimentation*

The barrier inlet area investigated by Mackenzie in the preceding paper is the site of barrier island migration reported on here by J. H. Hoyt and V. J. Henry, Jr. Migrating inlets rework and deposit sediment to a greater depth than that of other portions of the barrier, thus preserving for later investigation what may well be the only remaining evidence of the barrier's existence.

Hoyt and Henry have found that channel sediments, modified by migration, interfinger landward with lagoonal salt-marsh sediments and seaward with shallow neritic deposits. As the inlet migrates, the end of the island in the migrating direction is eroded, and deposition occurs on the margin of the channel away from the direction of migration. Stratification dip at the forward, or growing, end of the barrier island has a general trend in the direction of migration.

John Harger Hoyt was born in 1928 in Pontiac, Michigan. He received his baccalaureate degree in geology at the University of Michigan in 1951, and M.S. there in 1952. Hoyt spent the following six years in Colorado, Wyoming, Texas, and Louisiana as a subsurface geologist for the Shell Oil Company. He then entered the University of Colorado, where he received his Ph.D. in 1960. In the same year he joined the University of Georgia as a research assistant at the Marine Institute on Sapelo Island and an assistant professor of geology at the university's Athens, Georgia, campus. His research, resulting in 70 articles and abstracts, concerned modern nearshore and coastal processes and development of Pleistocene eustatic shorelines and related features. Hoyt's tragic death in a soaring-plane accident in 1970 terminated his short, but brilliant, career. John H. Hoyt was, without a doubt, one of the most important barrier island investigators of our time. It is with deep sorrow that we dedicate this book to him.

Vernon J. Henry, Jr., was born in Port Arthur, Texas, in 1931. He received the B.S. degree in geology at Lamar State College of Technology in 1953, then attended Texas A & M University, where he obtained his M.S. and Ph.D. in oceanography in

1955 and 1961, respectively. Henry became an assistant professor in 1961 and an associate professor in 1966. He was Director of the University of Georgia Marine Institute from 1964 to 1971 and is presently professor at both the Skidaway Institute of Oceanography and the University of Georgia. Henry's current research interests are the geology of coasts and continental shelves; in this, he is continuing the work on which he collaborated with John Hoyt.

14

Reprinted from the *Geol. Soc. Amer. Bull.*, **78,** 77–86 (1967)

JOHN H. HOYT
VERNON J. HENRY, JR.
Marine Inst. and Dept. Geology, University of Georgia, Sapelo Island, Georgia

Influence of Island Migration on Barrier-Island Sedimentation

Abstract: Barrier islands migrate along some coastal areas in the direction of dominant sediment transport. At the forward end of the island deposition occurs on the margin of the channel where the environment strongly influences the characteristics of barrier-island deposits. The depth of the channel, for example, exceeds two-to-three times that of other environments associated with barrier deposits. Moving along the coast, the channel erodes and reworks the deposits of other environments. The reworked area extends landward and seaward of the inlet and is several miles wide. The depth of reworking and the subsequent deposition preclude further modification by other agencies of the barrier-island environment. The erosion that accompanies transgression and regression may remove the upper level of barrier deposits, leaving modified channel sediments for interpretation and identification.

Sedimentary modifications produced by island migration include textural changes, gross shape of the deposit, and steepening and reorientation of stratification. Recognition of the reorientation of stratification is particularly important in paleocurrent analysis. Modified channel sediments interfinger seaward with shallow neritic deposits and landward with lagoonal salt-marsh sediments.

Although the duration of the Holocene high stand of the sea was too short to permit major migration-modification of Holocene islands, there was probably enough time for the extensive reworking of many ancient deposits. Studies on channel sediments and on the extent of island migration can provide information on the environment during the deposition of coastal sediments.

CONTENTS

INTRODUCTION

Barrier islands, represented by large volumes of sediments, are important depositional features of coastal plains. Although modern barrier islands are easily recognized, their ancient counterparts are often obscured by the reworking, modification, and incompleteness of the preserved sediments. Knowledge of their characteristics is important, however, because these islands can provide a wealth of information on the paleogeography, strand-line movements, sedimentation, and geologic history of an area.

The characteristics of barrier sediments are

Geological Society of America Bulletin, v. 78, p. 77–86, 6 figs., January 1967

ted by several factors; along coasts with a inant longshore transport, however, the sitional environment of the tidal inlets that rate the islands is particularly important. interpretation of influences of this tidal-environment on present-day sediments establishment of criteria for the recog- f similar influences on ancient sediments objectives of this paper.

portant studies have been made on the re- of inlet sedimentation to modern barrier . For example, Shepard and Moore) comment as follows:

apparent that inlet deposits are not well under- and their recognition in older sediments d be subject to some doubt. This is unfortu- because many of the samples obtained under arrier islands appear to be more closely related ose of the inlet deposits than to other environ- ents."

Johnson (1919) mentions the importance to shoreline development of inlets and their shifts in position, but does not consider specific problems of sedimentology. Price (1952) and Bruun and Gerritsen (1960) amply document the lack of stability of many inlets.

ACKNOWLEDGMENTS

This research represents part of the investigations supported by the National Science Foundation (Grants NSF-G16426 and NSF-GP-1380). Information on the vertical distribution of sediments was obtained with a portable core rig lent by the Esso Production Research Company of Houston, Texas (formerly Jersey Production Research Company of Tulsa, Oklahoma). Radiocarbon analyses were made by the Exploration Department and the Geochemical Laboratory of Humble Oil and Refining Company, Houston, Texas; The Marine Laboratory, Institute of Marine Science of the University of Miami, Miami, Florida; and the Radiocarbon Dating Laboratory, Department of Geology, Florida State University, Tallahassee, Florida. The efforts of these organizations are sincerely appreciated. We are indebted also to Drs. Robert J. Weimer and David G. Darby for assistance in the field and for stimulating discussions on many aspects of barrier-island sedimentation and wish to thank our colleagues at the Marine Institute for their help and encouragement.

DESCRIPTION OF AREA

The area of study (Fig. 1) is the central part of the Georgia coast and includes Sapelo Island, other smaller barrier islands, and several inlets. For the first 10 miles offshore the ocean floor slopes seaward at approximately 4 feet per mile. Further seaward the slope decreases to about 1.6 feet per mile at the edge of the continental shelf. The tidal range averages about 7 feet, with spring tides exceeding 9 feet.

Sapelo Island is largely Pleistocene, having fine-grained sand deposits in littoral and shallow neritic environments overlain by accumulations of dune sands. Separated from the Pleistocene area of the island by a narrow salt marsh is a strip of Holocene beach and dune sediments, which borders the Atlantic Ocean. The salt-marsh sediments are predominantly silts and clays with lesser amounts of fine sand.

The Georgia coast is bordered by short, relatively wide, barrier islands of Pleistocene and Holocene age, separated from the mainland by 4–6 miles of salt marsh (Hoyt, Weimer, and Henry, 1964). They are 7–18 miles long, 2–4 miles wide, and have one major inlet about every 10 miles.

The inlets, 1–4 miles wide and 40–80 feet deep, are in some places as deep as 100 feet. Offshore the inlet channels are flanked by low bars that are emergent at low tide and constrict the flow of water, causing strong currents. Channel depths of 20–30 feet below mean low water 3–4 miles offshore are common and exceed ocean depths at comparable offshore distances (Fig. 2). At the distal ends of these channels are tidal deltas or crescentic bars. Water depths at low tide are commonly less than 10 feet over the shallow parts of these bars.

ISLAND MIGRATION

Sediment movement along a coast results in gradual shifts of barrier islands and of the inlets separating them. In the study area, the geomorphology of coastal features indicates a dominant sediment transport from north to south along the front of the barrier islands. Bars and spits, formed near the mouth of small creeks flowing into the ocean, are directed to the south. Most noticeable is the truncation of dune ridges on Blackbeard Island at the north end of the Sapelo Island complex (Fig. 1), which indicates considerable erosion. On the south end, deposition has caused an approximately three-quarter-mile southward advance of Sapelo Island during the present high stand of the sea. Along the island front, sand is transported southward by littoral and longshore currents and deposited on the northern margin of the Doboy Sound inlet.

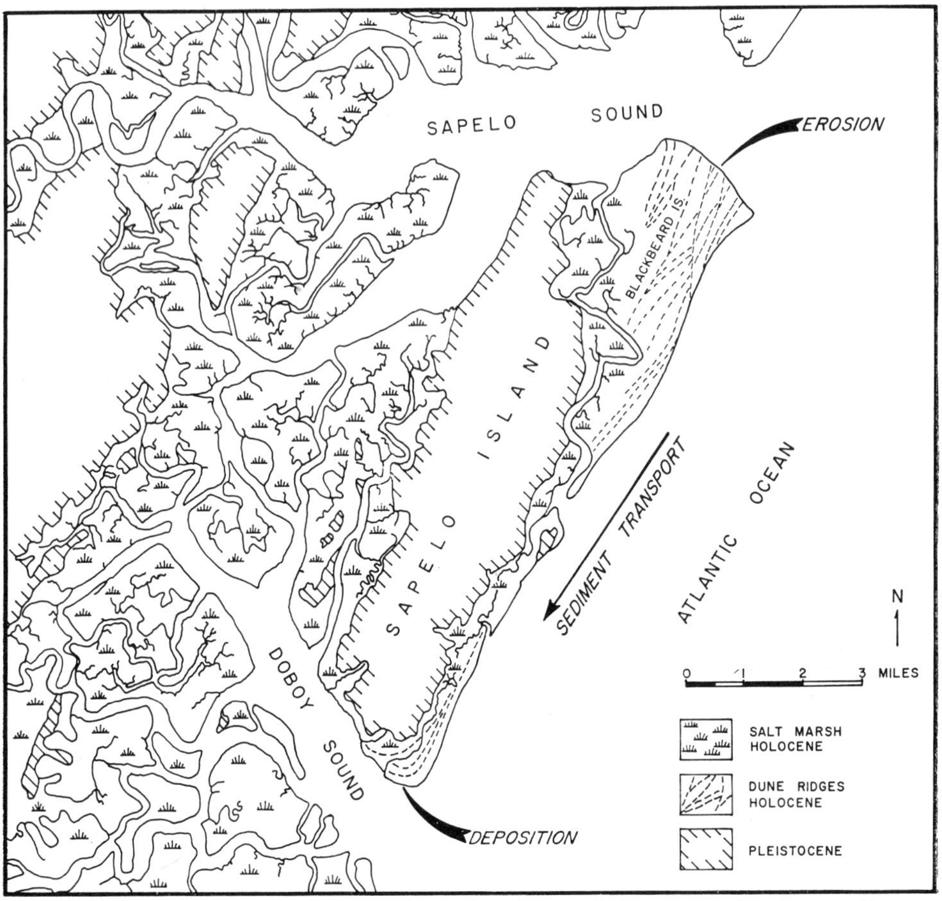

Figure 1. Map of Sapelo Island, Georgia, showing distribution of Pleistocene and Holocene sediments, dune ridges on Blackbeard and Sapelo islands, and dominant longshore drift

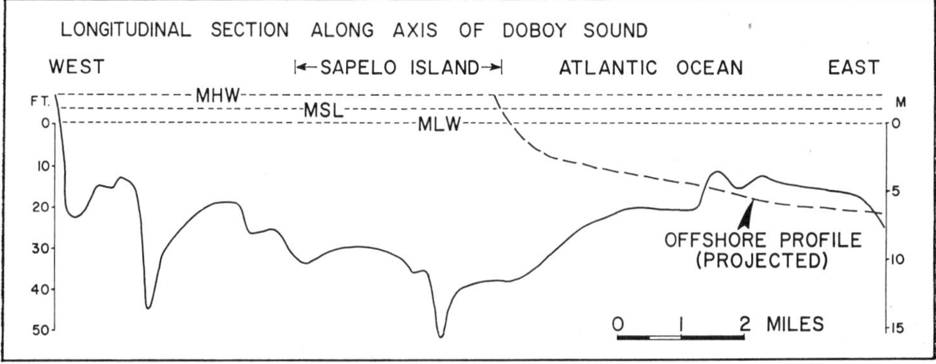

Figure 2. Longitudinal section along maximum depth of Doboy Sound, Georgia; offshore section away from influence of channel (dashed line). Symbols: MHW = mean high water; MLW = mean low water; MSL = mean sea level.

Assuming that a relatively constant volume of water moves through the inlets and that the cross-sectional area of the inlet is maintained, a southward shift of one island must be accompanied by erosion and reworking of the next island to the south. The southward migration of islands and inlets therefore is due to the accumulation of sediments on the margins of these islands, and the depositional environment of the inlets (Fig. 3) shapes the characteristics of the sediments. Confirming the southward migration of the Doboy Sound inlet during the Holocene high stand of the sea are the results of radiocarbon analyses of shell material (Table 1), taken from sediments in the depositional area on the margin of the migrating island.

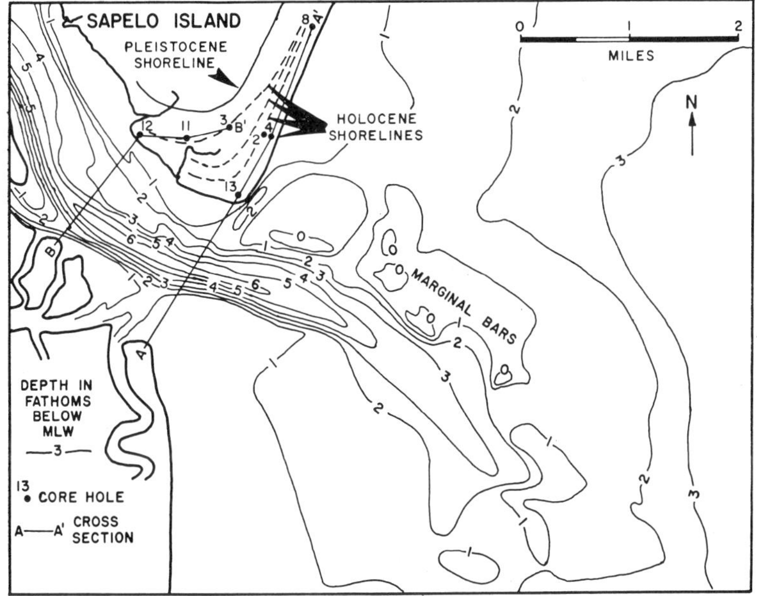

Figure 3. Map of Doboy Sound inlet, Georgia, showing Pleistocene shoreline, Holocene shorelines, marginal channel bars, hydrography, and location of cross sections. MLW = mean low water.

The stratigraphic positions of sediment samples, obtained with a portable core rig, are shown in the locations of the core holes (Fig. 3) and the cross sections (Figs. 4 and 5). The youngest material (1065 years B.P. from core hole no. 13) comes from the southernmost tip of the island 20 feet below shells dated as next youngest (1475 years B.P. from core hole no. 4) and obtained from a location more than half a mile to the north.

An anomalous date (8195 years B.P.) for sample 4-28 implies probable contamination at its depth of −22 feet by underlying Pleistocene deposits. This is supported by a stratigraphic judgment that the sample lay only a few feet above the Pleistocene-Holocene contact. Moreover, curves for that period, presented by McFarlan (1961), Curray (1961), and Shepard (1963), indicate that the sea level was considerably below −22 feet. On the other hand, dates from samples 8-26 (25,475 years B.P.) and 12-43 (27,720 and 29,925 years B.P.) show that these samples are from a previous high stand of the sea and add to the evidence on late-Pleistocene, sea-level fluctuations (Hoyt, Weimer, and Henry, 1965).

DEPOSITIONAL ENVIRONMENTS

Littoral

Holocene littoral sediments in the area of study are characterized by fine-grained, well-sorted, angular sand with minor thin lenses of clay. Shell layers are common. Stratification is even and continuous with common seaward dips of less than 4 degrees (Hoyt and Weimer,

1963). A notable exception to this is the steep, landward slope of small bars that develop on the lower foreshore (Hoyt, 1962). These bars move up the beach forming a deposit having inclinations as high as 30 degrees.

Nearshore-Neritic

Sediments of the nearshore-neritic environment are similar to the littoral, but are less well sorted and contain larger quantities of silt and clay. The turbid waters of the study area obscure visual observations of the ocean bottom and the sandy-sediment cores are of little value. Fathometer records show little relief over much of the nearshore area. Scuba-diving observation of the bottom indicates that much of the area slopes less than 2 degrees and rarely as high as 6 degrees. At the seaward end of the inlet channels, however, are deposits that slope 10–20 degrees, and as much as 20–30 degrees, in 30–40 feet of water.

Limited observation of the ocean floor shows that it bears symmetrical, low-amplitude ripples with wave lengths of 4–6 inches. Judged by the attitude of the depositional interface and by a few large-diameter-oriented cores, much of the stratification in the nearshore-neritic environment is probably low-angled, continuous, and even.

Salt-Marsh

The lagoon commonly associated with barrier islands is replaced in this area by an extensive salt marsh. The dominant flora, *Spartina alterna-flora*, grows luxuriantly from mean sea level to high-tide level. A small area of exposed sediments between mean sea level and low tide consists mainly of sloping channel banks, mud flats, and bars (Ragotzkie and Bryson, 1955). In general, the salt-marsh sediments are silts, clays, fine sand, and organic detritus. Fines settle out at slack water and are either incorporated in the accumulating sediments or picked up again as the currents increase with the changing tide. With time, however, the marsh sediments increase, but at a rate dependent on sediment supply and current energies.

Migrating-Inlet

Of primary importance in this study is the migrating-inlet environment. Its main depositional area lies where longshore and littoral currents are supplying sediments. In the case of Doboy Sound, this area is the north margin because the dominant transport is to the south. Part of the sediment supplied to the inlet mouth remains on the channel margin. Other parts are carried either landward by flood-tide currents or seaward by the ebb. Sand-size sediment carried landward usually joins the channel deposits, whereas the finer material is divided between the channel and the marsh. The sand carried seaward is incorporated in bar deposits flanking the offshore channel or is added to the delta forming at the offshore mouth of the channel.

Surprisingly, in this high-energy environment, layers of silt and clay are found. Thin layers are often present in the troughs of large shifting ripples which bury the fine material. Also, the configuration of the inlet may permit areas of low-velocity currents in which silt and clay can settle. The strong (2–3 knots) currents near the channel inlet, however, remove most of the silt and clay-size sediment from the depositional area, leaving the sand-size sediment which contains abundant shell material, much of it small pieces.

The deposit formed at the channel inlet is shaped by the amount of migration that occurs. The deposit that formed at the mouth of Doboy

TABLE 1. RADIOCARBON DATES OF SEDIMENTS FROM MARGIN OF MIGRATING ISLAND, DOBOY SOUND INLET

Sample no. (core-hole and depth)	Lab. no.	Approximate elevation of sample (datum mean sea level)	Age (years before present)
13-27 to 30	ML-115	−21 to 24 feet	1065 ± 65
2-15 to 17½	0-1876	−9 to 11½ feet	1625 ± 105
4-7 to 10	0-1874	−1 to 3 feet	1475 ± 105
4-25	0-1873	−19 feet	3375 ± 115
4-28	ML-116	−22 feet	8195 ± 110
8-26	0-1875	−20 feet	25,475 ± 1150
12-43	FSU-8	−37 feet	27,720 ± 760
12-43	0-1869	−37 feet	29,925 ± 2000

Sound during the present high stand of the sea, for example, is wedge-shaped. Processes working longer without major sea-level changes, however, produce an elongate lens which thins landward and seaward and extends in the direction of migration. The length of the lens depends on the migration rate and on the time available between shore relocations due to relative sea-level changes. The width and thickness deposition and on the composition of the sediment being reworked. The sediment transported to the Doboy Sound inlet is chiefly fine-grained sand. Finer sediments are carried in suspension and are generally not deposited in the inlet or along the offshore channel. Pleistocene sediments eroded by the migrating inlet contain medium- and coarse-grained sands which are incorporated in the deposit.

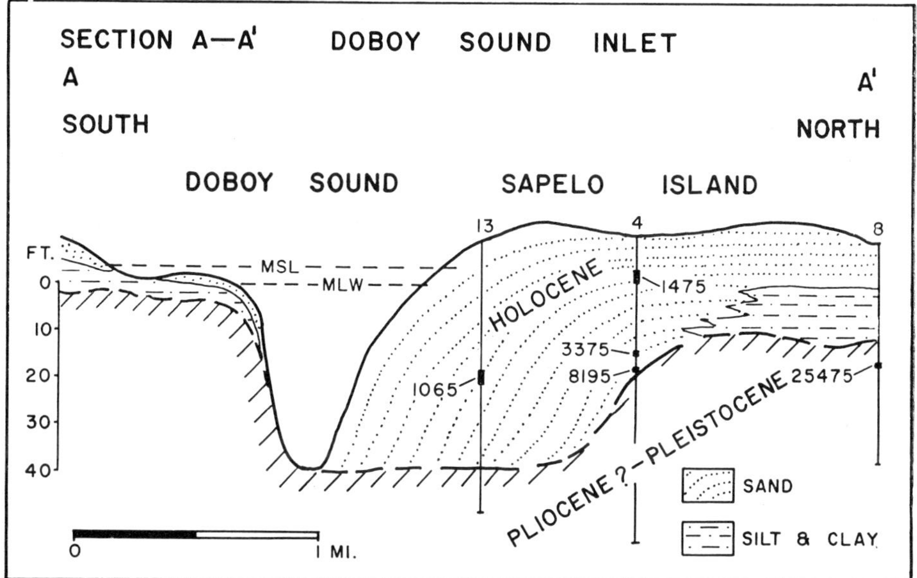

Figure 4. Cross section at Doboy Sound inlet, Georgia; vertical exaggeration 100 to 1. *See* Figure 3 for location and Figure 2 for explanation of symbols.

of the lens vary with the depth of the migrating channel and the morphology of the sea bottom.

In the Sapelo Island area (Figs. 4 and 5), the width of the lens being formed is approximately 6–8 miles and its length on the island's southern tip, about three-quarters of a mile. The relatively brief duration of the modern high stand of the sea precludes more extensive migration. During the geologic past, however, sufficient time was probably available for migration of considerable distance.

EFFECTS OF MIGRATION ON SEDIMENTARY PROPERTIES

The reworking of sediments which accompanies migration of the channel inlet alters the nature of the sediments in important ways. The texture of the sediment may change, for example, depending on the material available for

The main effect of migration on sediment properties is the modification of stratification. The littoral and nearshore-neritic sediments deposited along the barrier-island fronts of the Georgia coast have low depositional slopes, rarely over 6 degrees. In the littoral sediments the stratification commonly dips seaward; in the shallow neritic is a variety of slope directions. In contrast to these gentle slopes, the sediments of the migrating channel commonly have depositional slopes with moderate (10–20 degrees) and high (20–30 degrees) angles. On these slopes the sediments are cross-stratified in sets of various dimensions, including sweeping planes up to tens of feet long in the slope direction.

There are three major areas of medium- to high-slope development in the migrating channel environment. These include (1) the depositional margin of the tidal channel inlets, (2) the steep face of asymmetrical megaripples developed by tidal currents in the channel, and (3) the steep face of sand waves (Hoyt and Henry, 1964). Slopes as steep as 30 degrees, but commonly less, are found in each of the three areas. The megaripples have amplitudes up to 3 feet and wave lengths of 20–40 feet, and the sand along the offshore channel dips in the direction of island migration, *i.e.*, almost parallel to the regional shore trend. Where megaripples and sand waves are developed on the channel margin, the depositional slopes are directed either up or down the channel. Although the attitude of numerous megaripples is difficult to determine in the turbid waters, the majority of depo-

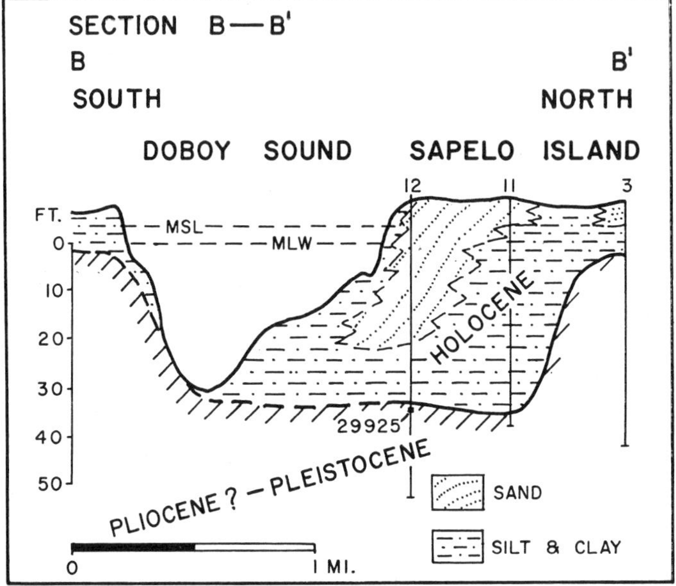

Figure 5. Cross section of Doboy Sound, Georgia, 1 mile landward of inlet; vertical exaggeration 100 to 1. *See* Figure 3 for location and Figure 2 for explanation of symbols.

waves, amplitudes up to 12 feet and wave lengths up to 300 feet.

Because of the curving nature of the inlet-channel margin, the dip direction of stratification varies in response to both the orientation of the channel margin and the configuration of the ripples and sand waves. The orientation of the channel, for example, is nearly parallel to the regional shoreline where the inlet channel joins the island front. At this site where much of the sediments transported by littoral and sublittoral currents are deposited, the channel margin slopes seaward, and depositional slopes of megaripples and sand waves dip parallel to the shore. In general, however, the trend of the inlet channel is almost perpendicular to the trend of the regional shoreline, and stratification developed on the channel-inlet margin and sitional surfaces appear to slope up the channel away from the ocean. The sand waves also are oriented with the steep slopes away from the ocean. Under slightly different current conditions, however, many of the steep depositional slopes may be directed down the channel toward the ocean.

Thus, stratification dips influenced by migration in this area have three common dip directions—one (in response to the configuration of the channel margin) dipping in the direction of migration and the other two (in response to the configuration of ripples and sand waves) dipping at right angles to the regional shore trend, either toward or away from the ocean. Because the last two dip directions develop on the first one (*i.e.*, the ripples and sand waves form on the channel margin), their stratifications have a component in the direction of migration produced by the particular slope of the channel margin. On the other hand, a small number of stratification dips, formed where the inlet joins the island front, slope in the direction opposed to migration.

PRESERVATION AND INTERPRETATION OF BARRIER-ISLAND SEDIMENTS

The geologic record of littoral and nearshore-neritic sediments is dependent on their preservation by a burial adequate for protection against erosion and modification. Emergence exposes barrier-island sediments to subaerial erosion, whereas submergence subjects some deposits to marine reworking depending, of course, on the amount of sea-level change, wave energies, and supply of sediments. In either case, part of the deposit is reworked and redeposited. If the original deposit is thin or if the reworking is extensive, much of the barrier islands is altered or removed.

Compared to the littoral and nearshore sediments, those near the bottom of a migrating channel are buried deeply and thus are not likely to be subsequently eroded or reworked. The sediments, added on top of the migrating-inlet sediments, may be reworked by the next migrating inlet that sweeps along the coast. Although upper sediments may be reworked following a relative change in sea level, the burial depth of the migrating-inlet sediments favors their preservation over that of the overlying littoral and nearshore deposits.

Although the inclination of stratification helps to determine paleocurrents and paleogeography (Potter and Pettijohn, 1963), the curving nature of inlets greatly complicates the analysis of cross-stratification dip directions. Measurements should be restricted to single depositional units, and data should be compiled from small areas. Where the inlet joins the island front, stratification produced by megaripples and sand waves parallels the shore and dips in the direction either of migration or opposed to migration. Elsewhere along the inlet much of the steeply dipping stratification slopes either landward or seaward and has a component in the direction of migration because of the slope of the channel margin. As most of the inlet channel is oriented approximately perpendicular to the shoreline, landward and seaward dips predominate. This knowledge obtained locally can be applied on a regional scale to the entire shoreline trend and the paleogeography of the basin.

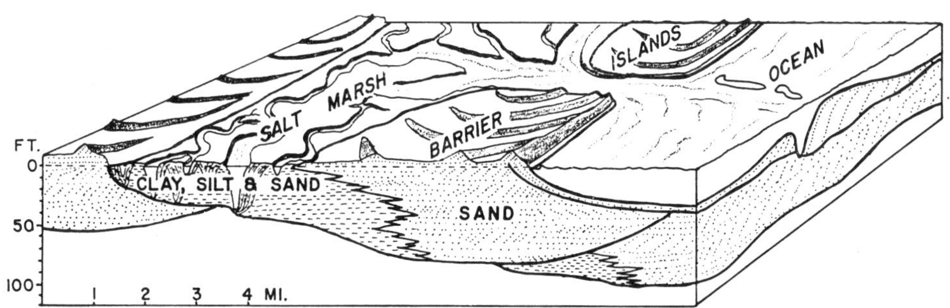

Figure 6. Idealized diagram of a migrating barrier island

Ideally, barrier islands can be recognized by their associated sediments. Lagoonal or salt-marsh sediments are present on the landward side of the islands; littoral and nearshore sediments, in the vicinity of the islands; and neritic sediments, in increasing water depths in the offshore direction. Often, however, this idealized picture is so altered by both erosion and reworking that sections of the sediment are missing.

Further disrupting the idealized picture, even though associated landward and seaward elements are still present, are the migration influences on the barrier system. On the whole, the basal contact of the sediments affected by migration is undulatory in the direction parallel to the shore trend and lens-shaped in the direction perpendicular to the shore (Fig. 6). The migration deposits are thickest in the vicinity of the shoreline. Landward they interfinger with marsh or lagoonal deposits and seaward are commonly flanked by tidal delta deposits. The tidal delta deposits, in turn, interfinger with shallow neritic deposits having low-angle dips.

Such idealized conditions are modified by erosion and reworking as a result of relative changes in sea level. The effect of these changes on the sedimentary deposits depends on several factors, including the amount the sea level rises or falls, the rate of sea-level change, the slopes of sea bottom and land surface, the wave and current energies, and the nature of the sediments involved. All the changes are related to sediment supply and to energies available for sediment distribution. Many of these factors, discussed also by others, are not unique to migrating islands.

The lagoonal area landward of the barrier chain is a geologically transient feature. The rate at which the lagoon fills is governed by such conditions as sediment supply, tidal range, spacing of inlets, and fauna and flora. Under most conditions, however, sedimentation occurs during slack water causing a reduction in inlet-water volume. Reduction in tidal flow permits a reduction in the cross-sectional area of the inlet. This process ends with the complete filling of the lagoon, the closing of the inlets, and the terminating of the barrier-island chain.

CONCLUSIONS

Along barrier-island coasts with a dominant longshore current, a part of the sediment is deposited in a migrating-inlet environment. These sediments, unlike those of the littoral and shallow neritic environments, have a great abundance of stratification dips in the medium- (10–20 degrees) or high-angle (20–30 degrees) range. The distribution of dip directions is fan-shaped with a minimum of inclinations in the direction opposed to migration. Coarse sediment, if available, is concentrated near the inlet of the migrating channel.

On a large scale, the shape of the sediment body affected by the migrating inlet is elongate in the direction parallel to the shoreline and lens-shaped perpendicular to the shoreline. The width of the sediment lens formed by the shifting inlet is as much as 6–8 miles. The length is a function of the rate of migration and the time available. Relative changes in sea level and varying rates of sediment accumulation produce a variety of sediment-body shapes from shoestring to blanket deposits, depending on the amount of reworking. Because of the depth of sediment deposition in the inlet and the inability of other processes to rework the inlet deposits, these sediments may be the ones commonly preserved in the geologic record.

REFERENCES CITED

Bruun, P., and Gerritsen, F., 1960, Stability of coastal inlets: Amsterdam, North Holland Publishing Co., 123 p.

Curray, J. R., 1961, Late Quaternary sea level; a discussion: Geol. Soc. America Bull., v. 72, p. 1707–1712

Hoyt, J. H., 1962, High-angle beach stratification, Sapelo Island, Georgia: Jour. Sed. Petrology, v. 32, p. 309–311

Hoyt, J. H., and Henry, V. J., Jr., 1964, Formation of high-angle marine stratification, central Georgia coast, p. 246 *in* The Geological Society of America, Abstracts for 1963: Geol. Soc. America Special Paper 76, 341 p.

Hoyt, J. H., and Weimer, R. J., 1963, Comparison of modern and ancient beaches, central Georgia coast: Am. Assoc. Petroleum Geologists Bull., v. 47, p. 529–531

Hoyt, J. H., Weimer, R. J., and Henry, V. J., Jr., 1964, Late Pleistocene and Recent sedimentation, central Georgia coast, U.S.A., p. 170–176 *in* Van Straaten, L. M. J. U., *Editor*, Deltaic and shallow marine deposits: Amsterdam, Elsevier Publishing Co., v. 1, 464 p.

—— 1965, Age of Late Pleistocene shoreline deposits, coastal Georgia (Abstract): Abs. Internat. Assoc. for Quaternary Research, VII Internat. Cong., p. 228

Johnson, D. W., 1919, Shore processes and shoreline development: New York, John Wiley & Sons, Inc., 584 p.

McFarlan, E., Jr., 1961, Radiocarbon dating of Late Quaternary deposits, south Louisiana: Geol. Soc. America Bull., v. 72, no. 1, p. 129–158

Potter, P. E., and Pettijohn, F. J., 1963, Paleocurrents and basin analysis: New York, Academic Press, Inc., 296 p.

Price, W. A., 1952, Reduction of maintenance by proper orientation of ship channels through tidal inlets, p. 243–255 *in* Johnson, J. W., *Editor*, Coastal Eng.: Proc. 2nd Conf., 393 p.

Ragotzkie, R. A., and Bryson, R. A., 1955, Hydrography of the Duplin River, Sapelo Island, Georgia: Marine Sci. of the Gulf and Caribbean Bull., v. 5, p. 297–314

Shepard, F. P., 1963, Thirty-five thousand years of sea level, p. 1–10 *in* Clements, T., *Editor*, Essays in marine geology in honor of K. O. Emery: Los Angeles, Calif., Univ. of Southern Calif. Press, 201 p.

Shepard, F. P., and Moore, D. G., 1955, Central Texas coast sedimentation: characteristics of sedimentary environment, recent history, and diagenesis: Am. Assoc. Petroleum Geologist Bull., v. 39, p. 1463–1593

Manuscript Received by the Society January 13, 1966
Marine Institute of the University of Georgia Contribution No. 116

Editor's Comments on Paper 15

15 **Hoyt:** *Barrier Island Formation*

With this paper, J. H. Hoyt launched the second phase of the long-lived barrier island controversy. In it he provides a fine review of the barrier island literature to 1967 and puts forth his thesis that barrier islands are developed essentially by the partial submergence of preexisting coastal ridges.

Hoyt points to the absence of neritic sediments and open ocean beach landward of barrier islands as evidence that they have not developed from offshore bars. Further, he rejects continuous development through submergence (because original formation of the barrier is not explained) and formation from emergence (because of insufficient evidence of a higher-than-present Holocene sea level). He does accept, although relegated to a minor role, barrier islands formed by some bars or breached spits. In the latter cases, open marine characteristics would be found on the landward side of the barrier island. Holding that partial submergence of dune and beach ridges, flooding the area landward of them thus forming lagoons and islands, was the major cause of barrier island development, Hoyt foresees further reworking of the barrier under subsequent conditions.

Reprinted from the *Geol. Soc. Amer. Bull.*, **78**, 1125–1135 (1967)

JOHN H. HOYT *Marine Institute and Department of Geology, University of Georgia, Sapelo Island, Georgia*

Barrier Island Formation

15

Abstract: Empirical data fail to substantiate classical theories of barrier island formation from offshore bars. Specifically, the absence of open ocean beach and neritic sediments landward of barrier islands suggests that barriers have not developed from offshore bars. Formation of barrier islands from emergent bars is also rejected, because evidence from many areas of the world does not support a sea level higher than present during the Holocene. Also unacceptable is the hypothesis of continuous barrier development throughout the Holocene submergence because it does not explain the original formation. Barrier islands which form from barrier spits or, in some instances, from bars are accepted, but these methods are not regarded as the general mechanism of barrier island formation.

The hypothesis proposed here maintains that a barrier island is initiated by the building of a ridge immediately landward of the shoreline from wind- or water-deposited sediments. Slow submergence, as during the late Holocene, floods the area landward of the ridge forming a barrier and a lagoon. Once formed, the island may migrate parallel, or normal, to the coast or may remain stationary depending on sediment supply, local hydrodynamic conditions, and land-sea stability. The width of the lagoon depends on the slope of the mainland surface, amount of submergence, sediment infilling, and erosion. Slow submergence or negligible sedimentation is necessary to maintain the lagoon. Emergence in excess of lagoonal depth terminates the barrier system.

CONTENTS

INTRODUCTION AND ACKNOWLEDGMENTS

Barrier islands are ubiquitous along the gently sloping coastal plains of the world, and large volumes of sediments are deposited in the associated environments. Although the facility for recognizing ancient barrier sediments is imperfect, studies of shoreline deposits indicate that barrier islands have been significant features during former periods. In spite of the importance of barriers, their mode of formation is poorly understood, and traditional theories are not supported by recent data.

Barrier island sediments and those of neighboring environments have been the subject of considerable investigation. Recent studies have been made by Shepard and Moore (1955), Fisk (1959), LeBlanc and Hodgson (1959), Rusnak (1960), McIntire and Morgan (1962), Zenkovitch (1962), Hails (1964), Hoyt and others (1964), and Thom (1965). This partial list will lead to earlier studies and will define the properties of barriers. Theories of barrier island formation have been presented by deBeaumont (1845) and Gilbert (1885) and have been reviewed by Johnson (1919), whose conclusions have been widely accepted and quoted in discussions of barrier island formation.

The support of National Science Foundation by Grants NSF-G16426, NSF-GP1380, and NSF-GA704 is gratefully acknowledged.

Geological Society of America Bulletin, v. 78, p. 1125–1136, 8 figs., September 1967

DISCUSSION OF BARRIER ISLANDS

Barrier islands are sometimes called offshore bars, but, as pointed out by Price (1951) and Shepard (1952), this terminology is to be avoided. Bars are water-covered at high tide, whereas islands are not. Barrier islands, barrier beaches, and spits are commonly referred to as barriers.

The barrier islands discussed in this paper are constructional features formed of detrital sediments and thus are distinguished from organic formations like coral reefs. Barrier islands may consist of organic debris; however, their common materials are sand and gravel.

Generally, barriers are elongate islands, parallel to the shore and separated from the mainland by a bay, lagoon, or marsh area. Individual islands range from a few miles to over 100 miles long. Width is commonly only a few miles. The islands along a coast may form a barrier chain or a series of islands separated by channels or sounds. The islands consist of one or more, and sometimes numerous, ridges of beach and dune sediments which mark successive shoreline positions during progradation. These ridges, which paralleled the shore at the time of formation, may be low, barely exceeding high-tide level, or high, reaching over 100 feet in altitude. The lagoons vary considerably in width from a few miles up to several tens of miles. Lagoonal sediments tend to be finer than the sediments of the barriers and contain larger percentages of silt and clay. Grain size also decreases offshore in the area beyond the surf zone; however, in many areas the sediment in greater than 50 feet of water is relict from an earlier depositional cycle. In part, this is due to the brief duration of the Holocene, which precludes the deposition of thick sedimentary sections over much of the shelf area.

THEORIES OF BARRIER ISLAND FORMATION

If a survey of geologic textbooks can be used to indicate the accepted means of barrier formation, it shows that the favorite theory is the upward building of offshore bars to become islands (Fig. 1). This theory was originally proposed by deBeaumont (1845), and it was strongly supported by Johnson (1919), who probably influenced later writers. DeBeaumont theorized that waves approaching the shore stir up sea-floor sediment and that where they break, losing much of their energy, the sediment would accumulate, forming a bar and subsequently an island.

As an alternate theory, Gilbert proposed in 1885 that material for the bar is transported along the shore rather than coming from erosion of the sea floor. A spit forms which is converted to a barrier island by subsequent breaching. This theory was unfavorably reviewed by Johnson (1919) because the offshore profiles of a sample of barriers did not conform to the interpretation of the theorical shape.

Price (1963) has seen small barriers form a short distance from shore during periods of high water associated with storms. A bar develops in front of the beach and builds vertically almost to the level of the temporarily raised sea level. Subsequently, when sea level returns to normal level, the bar remains as a low barrier.

Zenkovitch (1962) briefly mentions two methods of barrier formation. One involves the sinking of a wave-built terrace, the other the submergence of an alluvial plain. The latter hypothesis is similar to that, independently derived, presented in this paper.

Recently Leontyev and Nikiforov (1966) suggested that barriers formed from offshore bars exposed during a general lowering of sea level in late Holocene time. The proposed high stand of the sea is correlated with the climatic optimum about 6000 years ago. In other cases they suggest uplift has brought bars above sea level to form islands.

DISCUSSION OF BARRIER ISLAND FORMATION THEORIES

Several difficulties arise in applying the theory of barrier island formation from offshore bars: (1) Studies by Evans (1943) and Leontyev and Nikiforov (1966) suggest that, although offshore bars form under certain wave conditions, the upward development of the bar stops as the water level is approached. Apparently wave wash over the top of the bar prevents accretion above water level. In wave tank experiments conducted by McKee and Sterrett (1961), some bars attained a height equal to the maximum wave height (approximately 3 inches), but in most cases the bars did not exceed the still-water level. The applicability of these studies to natural situations is unknown. (2) If barriers develop directly from bars, such genesis should be observable somewhere in the world. The absence of abundant examples of various stages of barrier development argues that barriers are not forming in this manner.

This does not deny that minor barriers may form from bars, but the observed cases have been small, short-lived features formed close to the shoreline and can hardly be compared to the major barriers under discussion. (3) Most damaging to the theory of barrier formation from bars is the absence of beach and shallow neritic deposits landward of the barriers. If the barriers developed from bars, then open-ocean conditions should have prevailed landward of the bar during early stages of bar formation. The contact of Holocene salt-marsh deposits and Pleistocene sediments has been examined at many locations along the Georgia coast, and Holocene beach deposits have not been found landward of the salt marsh (Fig. 2). Detailed coring and study of Laguna Madre, Texas, by Fisk (1959) (Fig. 3) and Rusnak (1960) indicates that open-marine conditions did not prevail in the lagoon during the development of Padre Island. In investigation of Galveston Island and West Galveston Bay, Texas (barrier and lagoon), LeBlanc and Hodgson (1959) and Bernard and others (1959) found no open marine beach deposits landward of the barrier island. Borings were made in San Antonio Bay, Texas, landward of Matagorda Island by Shepard and Moore (1960), who commented: "There is no indication from the sediments, however, that during deposition this was an open-gulf environment." Studies of the barriers and lagoons along the Dutch coast by Straaten (1965) indicate that beach and open marine sediments are not found landward of the barriers. Similar findings are reported by Shepard and Moore (1955), Shepard (1960), Zenkovitch (1962), Hoyt and others (1964), and Hails (1965) from several areas of the world. Study of Pleistocene barriers associated with former high stands of the sea along the Georgia coast has also failed to find open marine sediments which could be interpreted as accumulating landward of an offshore bar later developing into a barrier (Hoyt and Hails, 1967).

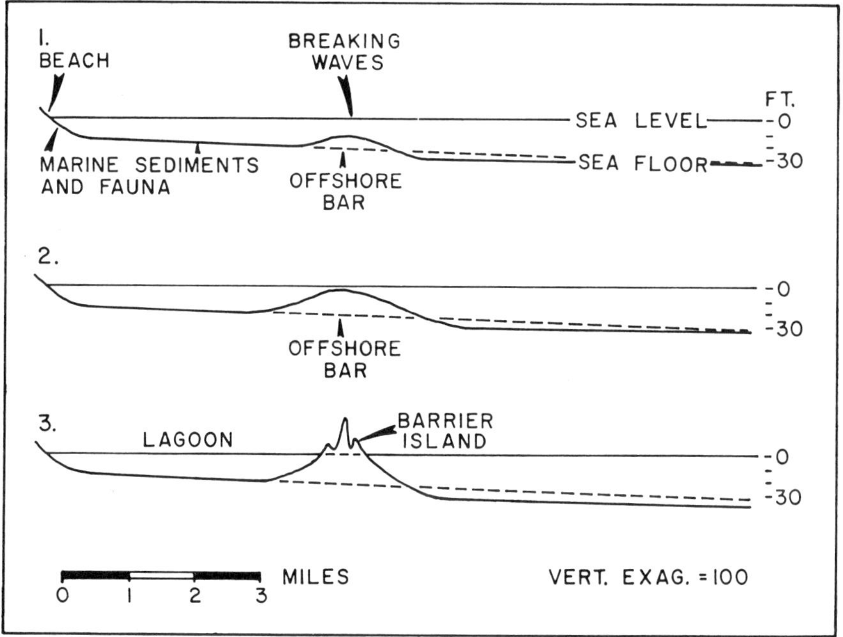

Figure 1. Idealized cross sections showing barrier island formation from an offshore bar. 1, Waves agitate sea floor and deposit sediment to form bar in area of energy loss. 2, Sediment accumulates to near sea level. 3, Bar is converted to island with lagoon on landward side.

Taken together, the evidence presented by

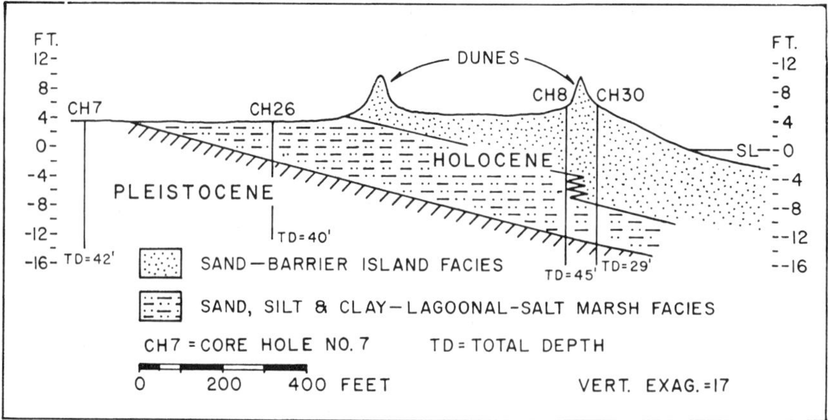

Figure 2. Cross section of narrow Holocene barrier and salt marsh along front of Sapelo Island, Georgia. SL = sea level.

detailed study of barriers strongly suggests that they have not developed from offshore bars. There may be minor exceptions, but it seems apparent that the major barrier features that rim the coastal plains of the world have formed in some other manner.

As pointed out by Shepard (1963a), Gilbert's suggestion (1885) that barriers form by the accretion of sediments transported along the shore by littoral and longshore currents should not be dismissed in spite of Johnson's objections (1919) that modern barrier profiles do not fit his theoretical profile. Part of the problem may lie in a misinterpretation of Gilbert's observations, for there can be little argument that barriers do form from spits by the breaching

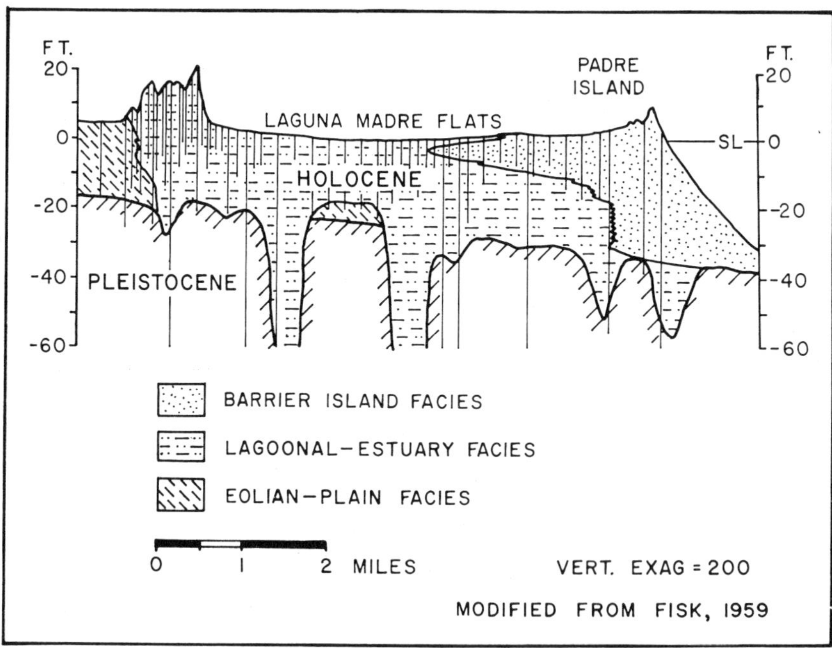

Figure 3. Cross section of Laguna Madre and Padre Island, Texas. SL = sea level.

of these features during storms or at other times (Fig. 4). The formation and growth of spits is observable, as is their breaching to form barrier islands. In evaluating profiles across barrier islands, Johnson apparently did not consider the effects of submergence, but assumed that the barrier sediments accumulated on pre-existing sea floor. In spite of the acceptance of barrier formation on a limited scale from spits, this mechanism does not seem adequate to account for the major barrier island systems, and probably it is limited to small segments of the coast where an abundance of sediment is available for littoral and longshore transport. This is substantiated by studies suggesting that the sediments of the barriers have been derived from offshore, rather than contributed directly by rivers (Shepard, 1960; Pevear and Pilkey, 1966).

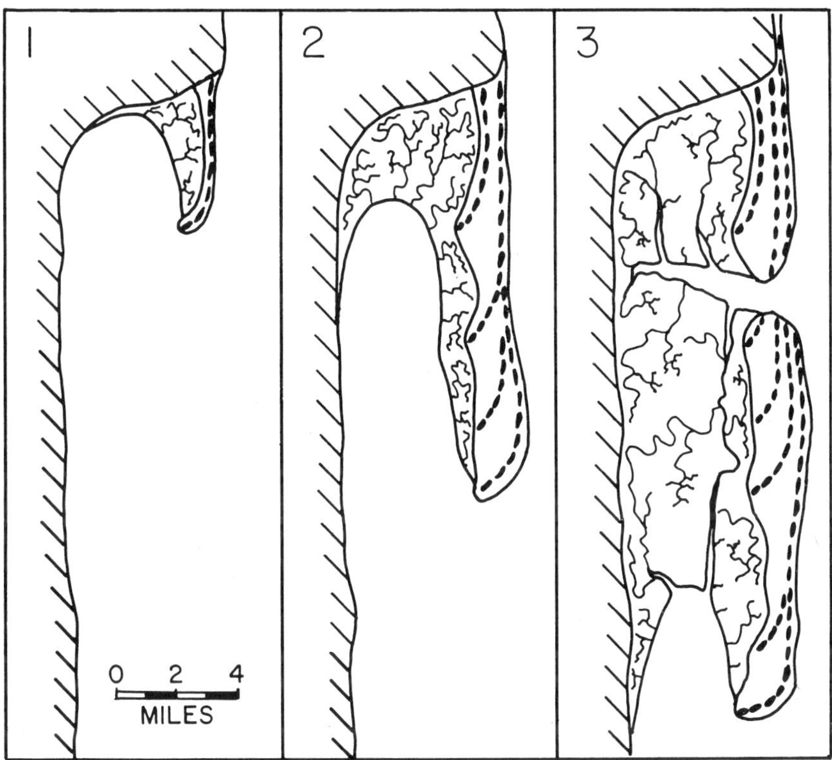

Figure 4. Idealized diagram showing barrier island formation from spit. 1 and 2, Spit develops in direction of longshore sediment transport. 3, Spit breached to form barrier island

The explanation of barrier islands proposed by Leontyev and Nikiforov (1966) is dependent on a high stand of the sea during the Holocene. While this may be true for some coasts (Fairbridge, 1961), it apparently does not hold for other coasts (Jelgersma, 1961; Shepard, 1963b; Hoyt and others, 1964; Scholl, 1964; Hails, 1965; Straaten, 1965), and thus cannot be accepted as a general theory. In addition, the absence of open-marine sediment landward of the barrier poses the same objections to this as to the offshore-bar theory. If areas are found which have experienced late Holocene fluctuations in sea level for any of a number of causes, then, in these areas, barriers may have developed from bars. Again, the appropriate sediments should be encountered landward of the barrier. Tectonic movements may bring a part of the sea floor above water level to produce barrier islands; however, this is a local phenomenon, not applicable on a world scale.

An hypothesis of unknown origin suggesting that barriers developed at a low stand of the sea and have transgressed the continental shelf during the Holocene submergence does not solve the problem of the original formation. Although data are not available, it is uncertain whether barriers could be maintained during the rapid rise in sea level. It is probable that barriers formed on the continental shelf, but they have not been positively identified or related to sea-level fluctuations.

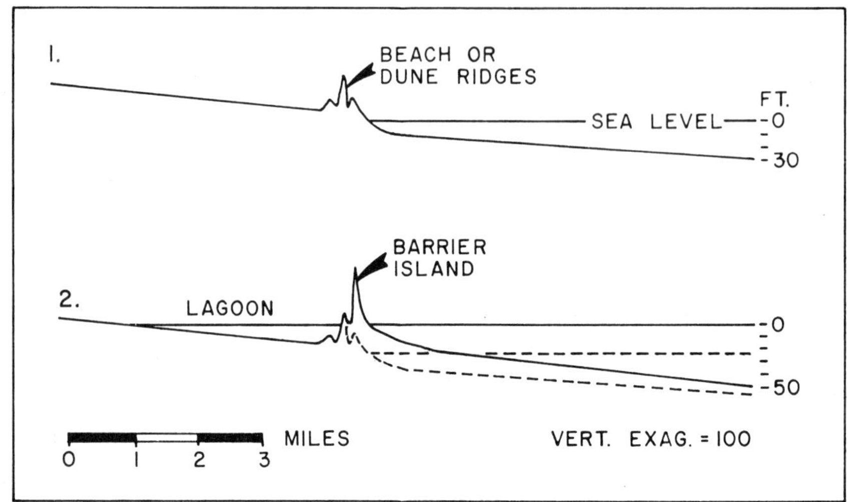

Figure 5. Formation of barrier islands by submergence. 1, Beach or dune ridge forms adjacent to shoreline. 2, Submergence floods area landward of ridge to form barrier island and lagoon.

SUBMERGENCE HYPOTHESIS OF BARRIER ISLAND FORMATION

Conditions that must be incorporated in an hypothesis of barrier island formation are: (1) the absence of open marine beach or shallow neritic sediments and fauna landward of the barrier; (2) the ability of barrier island systems to reform after they have been terminated by an emergence; (3) the absence of a worldwide, higher than present sea level during the Holocene; and (4) development and maintenance of a barrier system during a slow rise in sea level. An important aspect of the Holocene epoch is the rapid submergence which began approximately 18,000 years ago (Shepard, 1963b; Curray, 1965). The rate of this submergence apparently slowed considerably 3000 to 4000 years ago (Shepard, 1963b; Scholl, 1964). Tide-guage measurements for the past several decades indicate that a very slow submergence is still occurring in some areas (Donn and Shaw, 1963).

The hypothesis of barrier island formation presented here considers these conditions. At any level the sea must meet the land at the shoreline. Along sand beaches the wind will form dunes immediately landward of the shoreline, which may sometimes be over 100 feet high. If the beach consists of very coarse sand, gravel, or shells, the swash of waves forms a beach ridge which may be as much as 20 feet above high-water level. Thus, by either wind or wave action or by a combination of both, a topographic ridge is formed along the upper edge of the beach. Such ridges are ubiquitous along sediment shorelines of the world. If sediment is abundant, a series of ridges may form along a prograding shoreline and over a period of time become a sizable accumulation.

If, during or following the formation of the ridge, there is a relative submergence, the area landward of the ridge will be flooded to form a lagoon. The ridge then becomes a barrier island (Fig. 5). The width of the island depends on the amount of progradation which took place prior to or during submergence. Progradation, in turn, is related to sediment supply, hydrodynamic factors, and so forth. The length of the island depends on the continuity of the ridge. Areas of low elevation become inlets; however, the original configuration is modified by sedimentational processes, and some inlets may be closed and others opened.

The depth of the lagoon depends on the amount of submergence and the original altitude of the area; the depth, however, may be decreased by sedimentation and increased by scour. The width of the lagoon is also dependent on the amount of submergence and

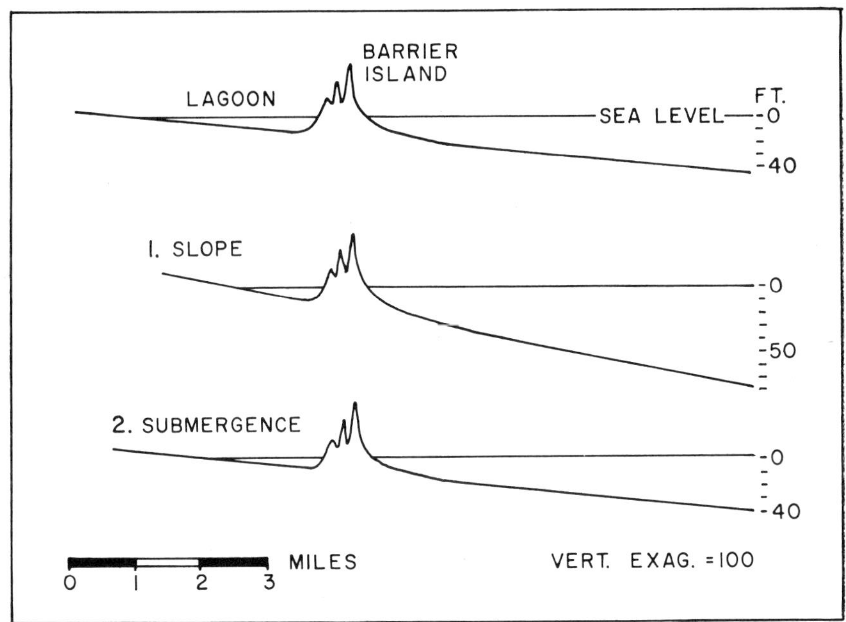

Figure 6. Lagoonal width and depth are functions of 1, slope and 2, submergence. Sediment deposition and erosion modify dimensions of lagoon.

on the slope of the land surface (Fig. 6). With a steep slope, for instance, the lagoon may be very narrow. The width of the lagoon may be increased by erosion and narrowed by sedimentation. In many areas the lagoon is almost completely filled with sediment and converted to a salt marsh.

Once formed, the barriers have a normal history for barrier islands and are maintained as long as there is a balance of sediment supply, rate of submergence, and hydrodynamic factors (Fig. 7). With abundant sediment the island progrades seaward, but, if the sediment is reduced, the maintenance of the dune-beach ridge system may be impossible. Submergence may result in a landward advance of the shoreline in excess of the amount predicted from the island morphology (Bruun, 1962; Schwartz, 1965). Too fast or too slow a submergence rate is detrimental to continued barrier development. A fast rate without sufficient sediment allows the sea to overwhelm the barrier, whereas a slow rate allows the lagoon to become filled.

The submergence, which is a vital part of this theory, was provided during the Holocene by the melting of the continental glaciers and the consequent increase in ocean volume. Glacier melting has produced repeated submergences during the Pleistocene, and along the southeastern United States coast there are at least six sequences of barriers (Hoyt and Hails, 1967). The periodic emergences and submergences in response to the waxing and waning of continental ice sheets is superimposed on a general emergence of undetermined cause, perhaps related to increased storage of water in the Greenland and Antarctic ice caps or to tectonic instability of some oceanic basin. A result of the general emergence is that the shoreline of the oldest exposed Pleistocene barrier is the highest topographically and that the younger shorelines stabilized at successively lower altitudes.

Because the barriers have not been transgressed, they retain the general topography and morphology characteristic of barriers. Submergences during more ancient geologic times than the Pleistocene, in general, must have been related to tectonic movements and may have occurred at a slower rate than during the Pleistocene and Holocene. Although a slower rate of submergence provides a greater length of time for the filling of the lagoon, the essential elements for a barrier system remain.

Cretaceous sediments in the Western Interior provide excellent examples of barrier-facies sequences (Sears and others, 1941; Weimer, 1961). The sequence from land to sea consists of fluvitile deposits; carbonaceous shale with fresh- to brackish-water fauna and flora; dune, backshore, and littoral sandstone; nearshore neritic sandstone; and marine shale. In places coal may be present with the carbonaceous shale. The sequence of facies may shift landward or seaward, and thus submergence may be accompanied by transgression or regression of the shoreline, depending on sedimentological factors.

Eolian deposits should be expected within the barrier island depositional unit because the dunes may not be completely reworked during submergence, particularly if abundant sediment supply results in a seaward shift in the position of the island. The apparent rarity of eolian deposits within the barrier facies suggests that barrier island migration (Hoyt and Henry, 1967) or some other mechanism has reworked the barrier deposits. Clean sand, interpreted as dune deposits, was found at 8 to 15 feet below sea level in Holocene sediments at St. George Island, Florida (Schnable, 1966). Radiocarbon dates from associated deposits suggest that the modern barrier formed prior to 4500 years ago when sea level was 12 to 15 feet below present level and that the dunes were inundated 4000 to 4500 years B. P.

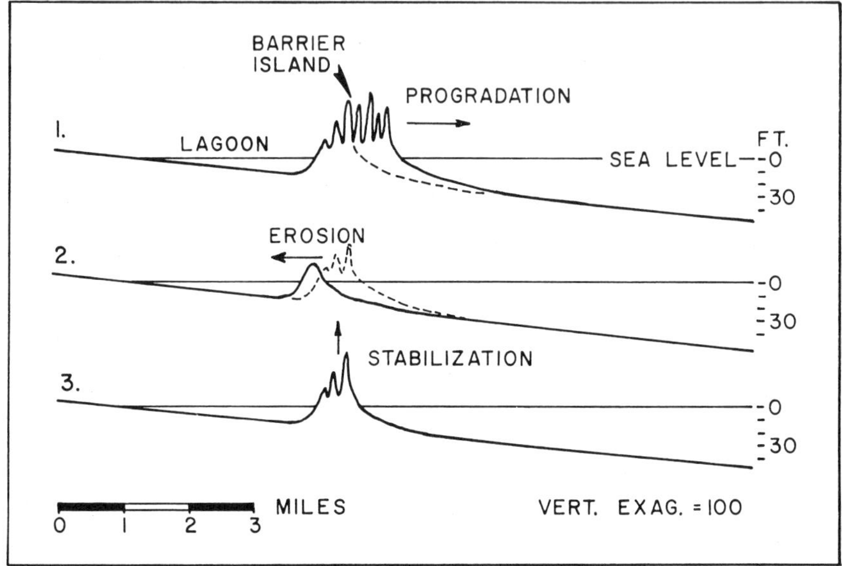

Figure 7. A barrier island may 1, prograde; 2, erode; or 3, remain in place, depending on sediment supply, rate of submergence, and hydrodynamic factors.

A relative emergence in excess of the lagoonal depth terminates the barrier system (Fig. 8). New barriers may form at a relatively lower elevation if the conditions outlined for the submergence hypothesis are met. In addition, topographic highs on the sea floor (such as bars) may form the nucleus for island development. In general, emergent sequences do not contain extensive lagoon sediments. General emergence with small intervening submergences tends to produce thin, discontinuous barrier and lagoonal sediments in contrast to the thick sequences that may accumulate during submergence.

It is not expected that the submergence hypothesis of barrier formation covers all types of barrier islands. Barrier island formation from spits is an observable phenomenon and explains islands in several areas. Emergent bars may also produce barriers where emergence is active, such as areas of tectonic uplift or isostatic rebound along glaciated coasts. On a worldwide scale, however, barriers formed in these ways appear to be of minor importance.

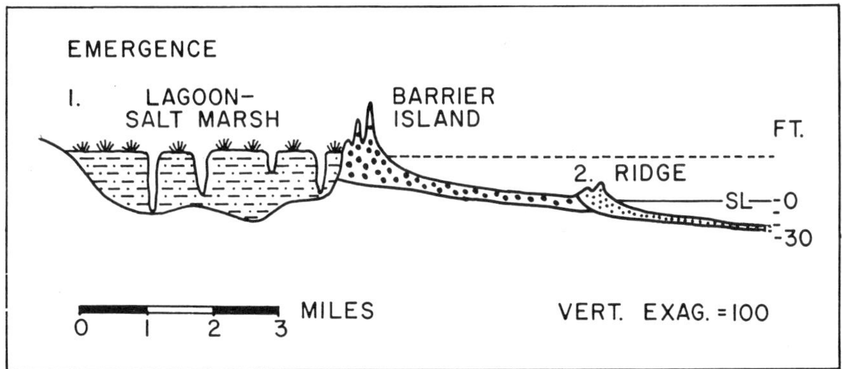

Figure 8. 1, Barrier island-lagoon system terminated by emergence. 2, Ridge forms adjacent to shoreline of lower sea level. SL = sea level.

CONCLUSIONS

The theory of barrier island development from offshore bars as the general mechanism of barrier formation is believed invalid because: (1) field observations and wave-tank experiments suggest that bars do not build appreciably above high-water level; (2) there is an absence of examples of the conversion of bars to barriers at various stages of development; and (3) open-marine sediments and organisms commonly are not found landward of barrier islands, as would be required if the barriers developed from bars. The formation of barrier islands from emergent offshore bars is also rejected in the absence of a worldwide, higher than present stand of the sea during Holocene time. This explanation would also require open marine sediments landward of the barrier.

The proposed hypothesis of barrier island formation is: (1) the sea intersects the land along the shoreline; (2) either, or both, dune and beach ridges form adjacent to the shoreline; and (3) submergence (as during the late Holocene) floods the area landward of the dune-beach ridges, forming lagoons and islands. The islands may shift landward, seaward, or remain stationary, depending on sediment supply, rate of submergence, hydrodynamic factors, and so forth. Emergence in excess of lagoonal depth terminates the barrier system.

Barrier islands may, in some instances, form from spits and from offshore bars. In these cases marine sediments and organisms will be present landward of the barriers.

With sufficient sediment supply, barriers may be maintained during slow submergence and lagoonal, barrier, and associated facies accumulated in transgressive or regressive sequences. Emergence may also produce sheet deposits; however, the lagoonal sediments will be largely missing.

REFERENCES CITED

Bernard, H. A., Major, C. F., Jr., and Parrott, B. S., 1959, The Galveston barrier island and environs: a model for predicting reservoir occurrence and trend: Gulf Coast Assoc. Geol. Soc. Trans., v. 9, p. 221–224

Bruun, P., 1962, Sea level rise as a cause of shore erosion: Jour. Waterways and Harbors Div., Am. Soc. Civ. Engineers Proc., v. 88, p. 117–130

Curray, J. R., 1965, Late Quaternary history, continental shelves of the United States, p. 723–735 *in* Wright, H. E., Jr., and Frey, D. G., *Editors*, The Quaternary of the United States: Princeton, N. J., Princeton Univ. Press, 922 p.

deBeaumont, E., 1845, Leçons de géologie pratique: Paris, p. 223–252

Donn, W. L., and Shaw, D. M., 1963, Sea level and climate of the past century: Science, v. 142, p. 1166–1167

Evans, O. F., 1943, The origin of spits, bars, and related structures: Jour. Geology, v. 20, p. 846–865

Fairbridge, R. F., 1961, Eustatic changes in sea level, p. 99–185 *in* Ahrens, L. H., and others, *Editors*, Physics and chemistry of the earth, Volume 4: New York, Pergamon Press, 317 p.

Fisk, H. N., 1959, Padre Island and the Laguna Madre flats, coastal south Texas: 2nd Coastal Geography Conf., Louisiana State Univ., Baton Rouge, La., p. 103–151

Gilbert, G. K., 1885, The topographic feature of lake shores: U. S. Geol. Survey 5th Ann. Rept., p. 69–123

Hails, J. R., 1964, The coastal depositional features of southeastern Queensland: Australian Geographer, v. 9, p. 207–217

—— 1965, A critical review of sea-level changes in eastern Australia since the last glacial: Jour. Inst. Australian Geographers, v. 3, p. 63–78

Hoyt, J. H., and Hails, J. R., 1967, Deposition and modification of Pleistocene shoreline sediments in coastal Georgia: Science, v. 155, p. 1541–1543

Hoyt, J. H., and Henry, V. J., Jr., 1967, Influence of island migration on barrier island sedimentation: Geol. Soc. America Bull., v. 78, p. 77–86

Hoyt, J. H., Weimer, R. J., and Henry, V. J., Jr., 1964, Late Pleistocene and Recent sedimentation, central Georgia coast, U.S.A., p. 170–176 *in* Straaten, L. M. J. U. Van, *Editor*, Deltaic and shallow marine deposits: Amsterdam, Elsevier Pub. Co., v. 1, 464 p.

Jelgersma, S., 1961, Holocene sea level changes in the Netherlands: Meded. Geol. Stichting, Ser. C-VI-7, 100 p.

Johnson, D. W., 1919, Shore processes and shoreline development: New York, John Wiley and Sons, Inc., 584 p.

LeBlanc, R. F., and Hodgson, W. D., 1959, Origin and development of the Texas shoreline: Gulf Coast Assoc. Geol. Soc. Trans., v. 9, p. 197–220

Leontyev, O. K., and Nikiforov, L. G., 1966, An approach to the problem of the origin of barrier bars: Intern. Oceanographic Cong., 2nd, Abs. of Papers, p. 221–222

McIntire, W. G., and Morgan, J. P., 1962, Recent geomorphic history of Plum Island, Massachusetts and adjacent coasts: Louisiana State Univ., Coastal Studies Inst., Tech. Rept. No. 19, 44 p.

McKee, E. D., and Sterrett, T. S., 1961, Laboratory experiments on form and structure of longshore bars and beaches, p. 13–28 *in* Peterson, J. A., and Osmond, J. C., *Editors*, Geometry of sandstone bodies: Tulsa, Okla., Am. Assoc. Petroleum Geologists, 240 p.

Pevear, D. R., and Pilkey, O. H., 1966, Phosphorite in Georgia continental shelf sediments: Geol. Soc. America Bull., v. 77, p. 849–858

Price, W. A., 1951, Barrier island, not "offshore bar": Science, v. 113, p. 487–488

—— 1963, Origin of barrier chain and beach ridge, p. 219 *in* The Geological Society of America, Abstracts for 1962: Geol. Soc. America Special Paper 73, 355 p.

Rusnak, G. A., 1960, Sediments of Laguna Madre, Texas, p. 153–196 *in* Shepard, F. P., and others, *Editors*, Recent sediments, northwest Gulf of Mexico: Tulsa, Okla., Am. Assoc. Petroleum Geologists, 394 p.

Schnable, J. E., 1966, The evolution and development of part of the northwest Florida coast: Sedimentological Research Laboratory, Fla. State Univ., Tallahassee, Fla., Contr. No. 12, 231 p.

Scholl, D. W., 1964, Recent sedimentary record in mangrove swamps and rise in sea level over the southwestern coast of Florida, Part 1: Marine Geology, v. 1, p. 344–366

Schwartz, M., 1965, Laboratory study of sea-level rise as a cause of shore erosion: Jour. Geology, v. 73, p. 528–534

Sears, J. D., Hunt, C. B., and Hendricks, T. A., 1941, Transgressive and regressive Cretaceous deposits in southern San Juan Basin, New Mexico: U. S. Geol. Survey Prof. Paper 193-F, p. 110–119

Shepard, F. P., 1952, Revised nomenclature for depositional coastal features: Am. Assoc. Petroleum Geologists Bull., v. 36, p. 1902–1912

—— 1960, Gulf Coast barriers, p. 338–344 *in* Shepard, F. P., and others, *Editors*, Recent sediments, northwest Gulf of Mexico: Tulsa, Okla., Am. Assoc. Petroleum Geologists, 394 p.

—— 1963a, Submarine geology: New York, Harper and Row, 557 p.

—— 1963b, Thirty-five thousand years of sea level, p. 1–10 *in* Clements, T., *Editor*, Essays in marine geology in honor of K. O. Emery: Los Angeles, Univ. Southern California Press, 201 p.

Shepard, F. P., and Moore, D. G., 1955, Central Texas coast sedimentation: characteristics of sedimentary environment, recent history, and diagenesis: Am. Assoc. Petroleum Geologists Bull., v. 39, p. 1463–1593

—— 1960, Bays of central Texas coast, p. 117–152 *in* Shepard, F. P. and others, *Editors*, Recent sediments, northwest Gulf of Mexico: Tulsa, Okla., Am. Assoc. Petroleum Geologists, 394 p.

Thom, B. G., 1965, Late Quaternary coastal morphology of the Port Stephens–Myall Lakes area, N.S.W.: Royal Soc. New South Wales Jour. and Proc., v. 96, p. 23–36

Van Straaten, L. M. J. U., 1965, Coastal barrier deposits in South- and North-Holland: Meded. Geol. Stichting, new ser., no. 17, p. 41–87

Weimer, R. J., 1961, Spatial dimensions of Upper Cretaceous sandstones, Rocky Mountain area, p. 82–97 *in* Peterson, J. A. and Osmond, J. C., *Editors*, Geometry of sandstone bodies: Tulsa, Okla., Am. Assoc. Petroleum Geologists, 240 p.

Zenkovitch, V. P., 1962, Some new exploration results about sand shores development during the sea transgression: De Ingenieur, no. 17, Bouw en Waterbouwkunde, 9, p. 113–121

Manuscript Received by the Society February 20, 1967
Marine Institute of the University of Georgia Contribution No. 128

Editor's Comments on Paper 16

16 Cooke: *Barrier Island Formation: Discussion*

Of the many commentaries published following the preceding barrier island paper by Hoyt, this short note by C. W. Cooke was the first. He maintains that storm waves, breaking upon shoals, pile up sand to form barrier islands and sees the development and landward migration of the islands as occurring without a change in sea level. Cape Hatteras and Padre and Sapelo islands are cited as examples. In this, Cooke belongs to the de Beaumont school of origin in place by wave action.

Cooke also raises another possibility to add to the accumulating hypotheses about barrier island genesis. He points to the Florida Keys as coral reefs and shoals of a former early Pleistocene (Pamlico) sea level, higher than the present stand of the sea which thus exposes them today. Whether or not these may be called barrier islands is a matter for discussion.

Following graduation from Johns Hopkins University, where he received his A.B. in 1908 and Ph.D. in 1912, Charles Wythe Cooke had a long and distinguished career with the U.S. Geological Survey. He joined the survey in 1910 as a junior geologist and subsequently became assistant geologist (1913), paleontologist (1917), associate geologist (1919), geologist (1920), scientist (1928), senior scientist (1941), and stratigrapher and paleontologist (1952). In 1956 Cooke moved to the Smithsonian Institution as Research Associate. His research interests during his remarkable career included coastal plain stratigraphy, fossil echinoids, geomorphology, coastal terraces, and Carolina Bays. Cooke was born in Baltimore, Maryland in 1887 and died in Daytona Beach, Florida, in 1971, having devoted over 50 years of his life to geology.

Reprinted from the *Geol. Soc. Amer. Bull.*, **79,** 945–946 (1968)

C. WYTHE COOKE

16

Barrier Island Formation: Discussion

In a recent paper Hoyt (1967) ". . . maintains that a barrier island is initiated by the building of a ridge immediately landward of the shore line from wind- or water-deposited sediments. Slow submergence, as during the late Holocene, floods the area landward of the ridge forming a barrier and a lagoon." Partial submergence of a beach ridge certainly might produce a barrier island and probably has done so in the remote past, but whether or not there has been sufficient submergence in the late Holocene to produce a recognizable barrier island is a moot question, which will not be argued here.

Barrier islands may be formed without a change of sea level. Consider the Hatteras Banks, which are as typical a barrier as one might wish for. They separate the shallow water of the North Carolina sounds from the deeper water of the Atlantic Ocean. Storm waves, rolling shoreward from the deep Atlantic, break against the shoals and pile up sand on them. Sand is blown across the dunes and falls into the sounds behind them, causing the islands to migrate landward. That this migration continues without noticeable change of sea level is shown by tree trunks that reappear as the dunes march across them.

Figure 2 of Hoyt (1967, p. 1128), a cross section of Sapelo Island, could be redrawn to be like his Figure 3 of Padre Island, which shows a sand-barrier facies resting directly on the Pleistocene at the edge of deep water, but this frontal strip of Sapelo Island appears to have been completely eroded away, leaving the barrier sand resting on salt-marsh deposits. This interpretation seems the more probable, since salt marshes do not usually face a windward shore. I infer that Sapelo Island and Padre Island were formed in the same way as the Hatteras Banks, without noticeable change of sea level.

Barrier islands may also be produced by emergence: consider the Florida Keys. Like the Hatteras Banks, the keys separate the deep water of the Straits of Florida from the shallow water of Florida Bay, but the keys originated in a manner entirely different from that of the Hatteras Banks. The Florida Keys ". . . are of two types. The eastern keys, which terminate at Loggerhead Key, are long, narrow islands composed of limestone (Key Largo limestone) containing large heads of corals in place, just as they grew. They evidently were formed as coral reefs that grew at the edge of deep water in the Pamlico sea, to whose surface they did not quite reach. The western keys, which lie behind the eastern keys and extend beyond them to Key West, were merely a shoal in the Pamlico sea. They are similar in origin to the rim of the Everglades and are composed of the same kind of oölitic limestone" (Cooke, 1939, p. 57–58). The Florida Keys were recently described by Hoffmeister and others (1967).

Although the Florida Keys were formed below sea level, they now stand a few feet above it because the net post-Pamlico emergence exceeded the pre-Recent emergence by about 25 feet. The keys are islands, because Florida Bay, which separates them from the mainland, was deeper than 25 feet in Pamlico time.

References Cited

Cooke, C. W., 1939, Scenery of Florida interpreted by a geologist: State of Florida Dept. Conservation Geol. Bull. no. 17, 118 p.

Geological Society of America Bulletin, v. 79, p. 945, July 1968

Hoffmeister, J. E., Stockman, K. W., and Multer, H. G., 1967, Miami Limestone of Florida and its Recent Bahamian counterpart: Geol. Soc. America Bull., v. 78, p. 175–190.

Hoyt, J. H., 1967, Barrier island formation: Geol. Soc. America Bull., v. 78, p. 1125–1136.

Manuscript Received by the Society January 5, 1968
Honorary Research Associate, Smithsonian Institution, Washington, D.C.
Post Office Box 5368, Daytona Beach, Florida 32020

Editor's Comments on Paper 17

17 Hoyt: *Barrier Island Formation: Reply*

In answer to Cooke, Hoyt first documents an appreciable rise in sea level during the Holocene to support his submergence hypothesis. Hoyt doubts the formation of Cape Hatteras from offshore bars or shoals, as proposed by Cooke, because evidence of an open marine environment is missing along the inner shore of Pamlico Sound. This argument will surface again and again in the debate that follows. He cites similar reasoning for Padre Island, Texas, and Sapelo Island, Georgia. Finally, the Florida Keys are dismissed as irrelevant to the discussion on two counts: they are relict Pleistocene features, and they formed from coral reefs.

Reprinted from the *Geol. Soc. Amer. Bull.*, **79**, 947 (1968)

JOHN H. HOYT

17

Barrier Island Formation: Reply

The comments of C. W. Cooke indicate that slight elaboration of a hypothesis of barrier island formation (Hoyt, 1967) is desirable. Cooke raises, but does not discuss, the question of "sufficient" submergence in the late Holocene. Part of the problem may result from differences in interpretations as to when the Holocene began. A discussion of this problem was given by Curray (1965, p. 732–733). According to one definition, the Holocene encompasses the past 18,000 years, during which sea level has risen about 320 feet (Shepard, 1963; Curray, 1965; Hoyt and others, 1965). It is evident that, on this basis, the late Holocene rise in sea level could have been as much as 150 feet. The point is that, whatever limit of the Holocene is used, there has been a significant submergence during the past several thousand years. It is during the latter part of this submergence that the major barrier islands are believed to have formed.

Cooke describes the present modification and migration of the barrier islands in the Cape Hatteras, North Carolina, area. Whereas the modern processes are observable and easily understood, it is unlikely that these barriers formed originally from offshore bars or shoals while the sea was at this level. If this were the case, open marine sediments, fauna, and shoreline morphology would have developed along the mainland shore of Pamlico Sound. The fact that this mainland shore does not show evidence of fronting on the ocean indicates that the Hatteras barrier islands formed before the sea attained its present altitude.

Once barrier islands have formed, they may be modified in several ways. The cross sections of Padre Island, Texas, and Sapelo Island, Georgia (Hoyt, 1967, Figs. 2 and 3), show two types of modification: accretion and progradation at Padre Island, and erosion and shore retreat at Sapelo Island. Again, the original formation of the barriers occurred at a lower sea level, as evidenced by the lack of open marine sediments, fauna, and shoreline morphology landward of the barriers.

As suggested on page 1132 (Hoyt, 1967), topographic prominences on the sea floor could, in some instances, form a nucleus for barrier island development during emergence. The Florida Keys, however, are not relevant to the present discussion because they formed as coral reefs, and, in addition, they are relict Pleistocene features.

References Cited

Curray, J. R., 1965, Late Quaternary history, continental shelves of the United States, p. 723–735, *in* Wright, H. E., Jr., and Frey, D. G., *Editors*, The Quaternary of the United States: Princeton, N. J., Princeton Univ. Press, 922 p.

Hoyt, J. H., 1967, Barrier island formation: Geol. Soc. America Bull., v. 78, p. 1125–1136.

Hoyt, J. H., Smith, D. D., and Oostdam, B. L., 1965, Pleistocene low sea level stands on the southwest African continental shelf: Inter. Assoc. Quaternary Research, Abstracts VII Intern. Cong., p. 227.

Shepard, F. P., 1963, Thirty-five thousand years of sea level, p. 1–10, *in* Clements, T., *Editor*, Essays in marine geology, in honor of K. O. Emery: Berkeley and Los Angeles, Univ. California Press, 201 p.

Manuscript Received by the Society February 14, 1968

Marine Institute and Department of Geology, University of Georgia, Sapelo Island, Georgia

Marine Institute of the University of Georgia Contribution No. 147

Geological Society of America Bulletin, v. 79, p. 947, July 1968

Editor's Comments on Paper 18

18 **Weidie:** *Bar and Barrier Island Sands*

Alfred Edward Weidie was born in New Orleans in 1931. He received his B.A. from Vanderbilt University in 1953 and after two years in the U.S. Marine Corps, went on to Louisiana State University, where he received his M.S. in 1958 and Ph.D. in 1961. Weidie became an assistant professor at Louisiana State in 1961, an associate professor in 1966, and Department of Geology Chairman there in 1967. His research interests are concerned with structural geology and physical stratigraphy.

Weidie is concerned here with the sedimentary characteristics of depositional strike sands as exhibited in bars and barriers. He describes these as elongate, parallel, or subparallel to sedimentary strike. In the case of barrier island sands the barrier shoreface and inner shelf deposits form in a marine environment located on the downdip side, while lagoonal, or alluvial, deposits are found on the updip and landward side. Barrier and bar sands increase in grain size upward in their structure. Barrier sands tend to be coarser, better sorted, and have a higher grain-to-matrix ratio than bar sands. While determination of these parameters is useful for purposes of petroleum exploration, they are advantageous, too, in tracing the ancestry of barrier islands.

Reprinted from the *Gulf Coast Assn. Geol. Soc. Trans.*, **18**, 405–415 (1968)

18

BAR AND BARRIER ISLAND SANDS

A. E. WEIDIE[1]
Louisiana State University
New Orleans, Louisiana

ABSTRACT

Elongate sand bodies may be classified as depositional dip sands (alluvial, channel, etc.) or depositional strike sands (bar, barrier, etc.). Bar and barrier sand bodies usually exhibit a number of characteristic properties; among these are shape, relations to surrounding strata, mineral content, grain size and sorting, size and sorting trends, and carbonate and heavy mineral content. These properties define the sand body and permit its identification. Depositional strike sands are excellent potential reservoirs for hydrocarbons. They are deposited in close proximity to source beds, the marine shales, and in many cases their plano-convex geometry makes them "natural" traps.

INTRODUCTION

It has long been recognized that there are marked differences in the size, shape, and petrology of quartzose sand bodies. Size and shape are most often considered to be intimately related to the genesis of the body or to reflect its development in a specific environment of deposition. The petrology and mineralogy of a sand body are reflections of source area, sediment transport, and the sum of physical and chemical conditions prevailing at the locus of deposition. Excellent summaries of various types of sand bodies have been given by Potter (1967) and Shelton (1967).

In recent years a great deal of information has accumulated on elongate sand bodies. These sands would be classified as prisms or shoestrings under Krynine's (1948) classification which uses the ratio of width to thickness. Elongate sand bodies may be divided into two groups on the basis of their position with respect to regional depositional strike. Depositional dip sands are elongate perpendicular or nearly perpendicular to sedimentary strike. Among the types of depositional dip sands which have been recognized are alluvial, channel, valley-fill, shoestring, and bar finger sands. Depositional strike sands are elongate parallel or sub-parallel to sedimentary strike; beach, bar, barrier island, and chenier sand bodies are of this type. Depositional strike sands are usually straight or gently curving in contrast to depositional dip sands which are often anastomosing and branching.

Both types of sand bodies have often served as reservoir rocks for petroleum or natural gas. Future exploration for these types of reservoirs will doubtlessly increase. Increasing emphasis must be placed on basic stratigraphic methods and techniques for the discovery of these reservoirs.

DISCUSSION AND DEFINITION

Bar and barrier island sands are depositional strike sands which form at some distance from the shoreline. Bar sands are defined as submerged features in contrast to the emergent barrier island. These definitions are in accord with those advocated by Price (1951) and Shepard (1952,1960). The most commonly accepted origin of bar sands is an accumulation of material at the point of wave break. Laboratory experiments by McKee and Sterrett (1961) have produced this type of deposit. In some cases a number of parallel "break-point bars" may be formed. Hoyt (1967) has discussed the origin of barrier island sands and presented strong evidence that they owe their origin to slow submergence. This slow submergence may be caused by eustatic rise in sea-level or along a slowly subsiding coastline.

DEPOSITIONAL ENVIRONMENTS

Ideally bar and barrier island sand bodies can be differentiated by studying the properties of the sand body as well as the adjacent strata. Barrier island sands are usually flanked by lagoonal or alluvial deposits updip and normal marine deposits downdip. Bar deposits will most often be flanked by marine deposits on both sides. Both bar and barrier sands show upward increases in grain size; but barrier sands exhibit a general tendency to an overall coarser size, as well as a higher grain to matrix ratio. Barrier sands will generally exhibit better sorting than bar sands.

Bar sands will commonly exhibit a variation in depositional environments. As stated above, the normal condition is that the bar is flanked by marine deposits on both sides. This is true in the case of offshore bars caused by breaking waves, the so-called "break-point bars" (see Fig. 1A).

River mouth bars are most often flanked by fresh to brackish waters deposits. Seasonal salinity variations related to stream discharge may be quite large (Fig. 1B). These bar sands have comparatively short longitudinal dimensions; e. g., the North Pass bar of the Mississippi River is between one and two miles long. In most ancient deposits these bars would not be recognizable because they would appear as the distal end of a channel or "bar finger" sand.

Bay, estuary, or embayment bars (Fig. 1C) have larger dimensions than river mouth bars and are distinguishable in ancient sediments. Brackish to normal marine environments flank these bars. Salinity conditions adjacent to the bar are influenced by a number of variables such as wave, current, and tide intensity; bar or sill depth; and the amount of fresh water influx into the bay, estuary, or embayment. A detailed description of the petrology, paleogeography, and depositional environment of an Upper Cretaceous bar, the Monterrey bar sand, is given below.

Barrier island environments (Fig. 2) are generally more uniform than those of bars. Hayes and Scott (1964) give an

[1] I would like to acknowledge the assistance of a number of students at LSUNO. John Long assisted in the field work near Monterrey, Mexico. Long, C. P. Cameron, and E. J. Ritchie assisted at Petit Bois Island and performed the mechanical analysis of these sediments. Long and Cameron assisted in the work on the New Orleans Barrier Island sand. The work on the Monterrey bar sand was funded by National Science Foundation Grant GA-1536.

excellent summary of environments of deposition on and adjacent to barrier islands. The barrier shoreface and inner shelf deposits form in a normal marine environment. The lagoon, bay, or sound environment landward of the barrier islands exhibits greater variability. Hypersaline to nearly fresh water conditions can exist and a high degree of short term variability is common (see Hayes and Scott, 1964; Shepard and Moore, 1960). The physical conditions in the lagoon or sound and quantity of terrigenous influx cause various subfacies to exist. In most cases the lagoon-sound environment exhibits a characteristic biofacies (see Hayes and Scott, 1964; Shepard and Moore, 1960; Phleger, 1960; Parker, 1960; and Walton, 1960).

EXAMPLES OF BAR AND BARRIER ISLAND SANDS

Monterrey Bar Sand

The Monterrey bar sand is located 15 miles west of Monterrey, Nuevo Léon, Mexico on the north side of the highway to Saltillo. The sand is within strata assigned to the Upper Cretaceous-Lower Tertiary Difunta Group. The Rancho Carbajal is located at the southern base of the ridge whose top is formed by the bar sand. Figure 3 shows a photograph of the bar sand and its plano-convex geometry is apparent in this photo.

The bar sand trends north-south for a distance of 3½ miles; its northern and southern extremities have been removed by erosion. The sand was deposited at the mouth of the Parras Basin and separated the basin from the ancestral Gulf of Mexico to the east. Studies of the basin indicate the embayment mouth was 25 to 30 miles wide, and the bar sand probably extended over much of this distance. The maximum thickness of the sand is 520 feet; a negligible amount has been removed from the top by erosion. The sand thins rapidly to the east and west. West of the sand are the coarse clastics of the Difunta Group which were deposited in the Parras Basin; to the east is the finer Mendez Shale of the Gulf Coastal Plain. The bar acted as an effective restriction to the eastward transport of the coarser terrigenous materials.

Figure 4 is a sketch of the Monterrey bar sand. It can be seen that there are two smaller bar buildups below the major sand. The entire sequence is a complex of multistroy sand bodies. The mean grain size diameters and grain to matrix ratios at various sample points are shown on the figure. The four samples at the base of the larger bar sand show smaller mean diameters than the three uppermost samples. The basal samples also show a lower grain to matrix ratio than the upper samples. Both basal and upper samples show larger grain sizes in a westerly direction verifying the restrictive nature of the bar. Average grain size diameters are also shown for the lower two bar sands and the intervening shaly siltstones.

The composition of the grains and matrix does not differ significantly. In all samples the matrix is a mixture of clay and micro-crystalline calcite (micrite) cement. The grain composition was grouped into three components: (1) the silica component (quartz and chert), (2) the rock fragment component (exclusive of carbonate rock fragments), and (3) the carbonate component of carbonate rock fragments, fossil fragments, and carbonate grains of indeterminate origin such as pellets, lumps, etc. Table I summarizes grain composition for the samples shown on Figure 4.

TABLE I

Sample	SiO^2 Component	R.F. Component	CO3 Component
1.	80%	11%	9%
2.	79%	16%	5%
3.	65%	20%	15%
4.	68%	23%	9%
5.	79%	21%	—
6.	67%	13%	20%
7.	81%	12%	7%
8.	56%	26%	18%
9.	57%	15%	28%
10.	75%	25%	—
11.	65%	35%	—

The major variation in samples is in the rock fragment and carbonate components. These vary horizontally as well as laterally in the section. Thin beds, pods, and pockets locally contain moderate large quantities of fossil fragments, and clay and shale clasts. As can be observed from Table I, some horizons are totally devoid of carbonate fragments.

The small grain size and high percentage of matrix attest to a low-energy environment. The bar complex at no time approached sea level. Water depths were not great, however, because some small-scale, poorly-developed cross—bedding is present. *Exogyra costata* is common throughout the section indicating shallow to moderate water depths.

Water depths immediately preceding the development of the bar complex were considerably shallower as evidenced by a large sand stratigraphically below the lowest bar and a few miles to the west which shows characteristics of a higher energy environment. Figure 5 shows this sand with its well-developed cross-bedding and zones of scour and erosion. Alternating inflow and outflow of water near the embayment mouth is clearly shown by the dips of the cross-beds. As water depth at the embayment mouth increased these opposed currents partially negated each other and permitted the buildup of the bar complex.

Gulf Coast Barriers - Petit Bois Island

Petit Bois Island is one of a series of barrier islands about 8 miles offshore from the coastlines of Mississippi and Alabama (Fig. 6). Petit Bois is about 6 mile long and has a maximum width of about 3/ mile. The island differs from other Gulf Coast barriers in having virtually no forest cover. Scrub grass is the dominant vegetation. The dunes of Petit Bois island are smaller than the other islands, most being 5-8 feet above mean sea level, and the largest only 10-12 feet above M.S.L. Low dune height and absence of forest cover are attributable to frequent wash-over during hurricanes and severe storms. Figure 2 illustrates the subenvironments of the barrier island. As can be observed in the figure there is a basic physiographic symmetry across the island.

A detailed sampling program was undertaken at the western end of the island. Numerous samples were taken from each of the sub-environments and analyzed. Figure 7 shows the western end of Petit Bois island and the sample stations on the island and in the adjacent waters. No samples were taken immediately west of the island in Horn Island Pass because this area is frequently dredged to maintain the channel depth. Dredging is not believed to have affected any sample stations since they are east of the dredge area and the longshore drift in this area is from east to west.

Mean grain size diameters at each sample location are shown in Figure 7. Water depths at sample points on Traverses A, B, and C are also shown. Traverses R1-R2 and R2-R4 are along the littoral zone with samples taken at approximately 300' intervals from the zone between high and low tide levels. Traverses R1 through R4 are through the fore-dune, back-dune, and mid-island subenvironments with samples taken from dune crests and troughs.

Dune and mid-island samples show little variance in grain size. Mean diameters vary from .30 mm. to .38 mm. (medium sand), have sorting coefficients (Trask) of 1.10+ 0.10 and standard deviations (S or b) of 1.05+ 0.15. Sorting values (Trask and b) show a slight tendency to better sorting in a gulfward direction. The dunes are irregular in shape but show a poorly developed north-south alignment. This is probably attributable to the prevailing southerly winds and the occasional strong "northers."

Beach samples from the sound littoral (Traverse R1-R2) and gulf littoral (Traverse R2-R4) show much larger grain size variation. Two major trends may be noted: (1) sound beaches are coarser grained than gulf beaches, and (2) sound beaches show a decrease in grain size in a westward direction. The cause(s) of these trends are unknown, but a number of contributing factors may be predicated:

1) The "northers" blow more finer material to the gulf littoral zone and adjacent shore face.

2) Storm and hurricane washover may carry coarser material to the sound littoral area and sound shoreface.

3) Westward longshore drift and tidal currents through Horn Island Pass may more effectively winnow the beach sediments on the northwest shore.

Both gulf and sound beaches are well-sorted. Heavy mineral bands and concentrates are frequent on gulf beaches and rare to absent on sound beaches. Calcium carbonate (dominantly fossil fragments) content is higher on gulf beaches than sound beaches.

Gulf and sound samples (Traverses A, B, & C) show similar variability exists on the barrier and sound shore faces as exists on the gulf and sound littoral regions. Mean grain sizes are larger in the sound than the gulf and all samples show decreases in grain size as water depth increases. A few random samples north of the end lines of Traverses A and B in water depths greater than 20 feet are muds or muddy silts. There is a rapid transition from medium and fine sands to mud and silt throughout the area of study in the sound paralleling the minus 20' contour. This rapid transition does not occur on the gulf side. Samples south of Traverse C taken in greater water depths show grain size decreases from .30+ .05 mm. (Traverse C) rather uniformly to .125 mm. (depth 20') and then remain fairly constant to a depth of 35 feet where the last sample was taken. Both gulf and sound samples are moderately to well-sorted and sorting becomes poorer with increased water depth. Heavy mineral content of sound samples is negligible (1-2%) but higher (3-5%) in gulf samples to depths of 30-35 feet. $CaCO_3$ content is low in the sound, generally less than 5%, and higher (4-10%) in the gulf.

Pine Island (New Orleans) Barrier

Treadwell (1955) was the first to publish any information on the Pine Island Barrier although the existence of a large sand body beneath the surface had long been known to engineering and contracting firms of the New Orleans area. Additional data has subsequently been presented by Kolb and Van Lopik (1958), Corbeille (1962), and Saucier (1963).

The Pine Island Barrier trends east-west through Orleans and Jefferson Parishes, Louisiana It is 35 miles long and has a maximum width of about 4 miles. The maximum thickness of the barrier sand is between 35 and 40 feet and occurs along the southern flank of the barrier. The barrier came into existence about 5000 years B.P. (Saucier, 1963) as a result of the sea level rise that began about 18,500 B.P. (McFarlan, 1961). Saucier (1963) presented a detailed map and cross-sections of the Pine Island Barrier. Louisiana State University students have surveyed much of the island using a resistivity meter and a portable seimograph, but have not changed the basic map by Saucier which is shown as Figure 8.

Borings for foundation tests have penetrated the barrier island under the LSUNO campus. Samples of these borings were given to the author by the Eustis Engineering Company of New Orleans. Ten samples have been analyzed and show mean grain sizes ranging from .136 mm, to .181 mm. with all samples averaging .155 mm. Corbeille (1962) published analyses of 13 samples from the barrier island whose means ranged from .122 mm. to .210 mm. The average of these 13 means was .159 mm. Our data corroborates Corbeille's in that there is a general increase in grain size upward. On the whole, the sand of the barrier ranges from very fine to coarse sand size, is usually well-sorted, and in places is apparently of dune origin near the crest (Saucier, 1963).

The cross-sections of Figure 8 show the facies relations beneath and adjacent to the island. Bay-sound deposits flank the barrier on the north and nearshore Gulf deposits on the south. Marine faunas typical of open gulf waters have been noted from two localities on the south flank of the barrier (Rowett, 1957; Corbeille, 1962).

Frio Barrier Bar System (South Texas)

Boyd and Dyer (1964) have documented the existence of a barrier bar system with prolific hydrocahbon production in South Texas. This system is a complex of multistory sand bodies more similar to the Monterrey Bar Sand than the Petit Bois or Pine Island Barriers. The Frio Barrier Bar System is at least 120 miles long, 25-40 miles wide, and 4000-6000 feet thick. The barrier bar system trends nearly parallel to the present Texas coastline (Fig. 9) and thickens to the southwest in the Rio Grande embayment where it grades into deltaic sediments.

Boyd and Dyer (1964, p. 312) describe the sediments as,

> "The Frio barrier bar consists of coarse to fine grained, well sorted, porous quartzose sands which grade updip into lagoonal shales and downdip into inner continental shelf shales. Sand sized material from the area of the

Frio depocenter to the southwest was transported laterally by longshore currents and deposited on a gently sloping continental shelf due to energy loss. Small rivers and streams doubtless influenced sedimentation in local areas. These bar deposits were reworked by eolian action, oscillatory wave motion, and tidal currents. Storms resulted in washover fan deposits in the lagoonal area behind the bar."

It is probable that during its long and complex history parts of the system were at times true barrier islands and at other times submerged bars along the main trend of the system (Fig. 10).

The barrier bar system is widest to the southwest where deposition was greatest. Here the ancestral Rio Grande deltaic deposits were prograding across the continental shelf and the barrier bar shifted gulfward on the NE flank of the prograding delta. Were it not for the unusual sea-level fluctuations of Pleistocene and Recent times, a similar depositional sequence of great thickness could have built up adjacent to the Mississippi Delta.

SUMMARY AND CONCLUSIONS

Bar and barrier island sand bodies usually exhibit marked physical characteristics which enable their recognition in ancient deposits. Recent bar and barrier sands serve as excellent "presents" which "may be the Key to the past."

Bar and barrier sands are excellent reservoirs for hydrocarbons. The sands are usually fairly clean and well-sorted and are deposited in close proximity to probable source beds, the downdip marine shales.

Bar and barrier sands are common at present and should be frequent throughout the stratigraphic section of the Gulf Coast. The following conditions are regarded as optimal for bar and barrier sand development:

1) A broad low-relief coastal plain and continental shelf.

2) Slow regional subsidence.

3) A locus of maximum deposition (an ancient deltaic complex or depocenter). This serves both as a sediment source and as a cause of local subsidence (delta flank depressions) of an area in which bars and barriers may develop.

4) A longshore current or drift pattern which aids in the formation of bars and barriers from spits and increases their size.

Bar and barrier sand bodies are worth more exploration effort, for in addition to their ideal reservoir properties and source bed relationships, their characteristic geometry makes them excellent "natural" traps.

BIBLIOGRAPHY

Boyd. D. and B. F. Dyer, 1964, Frio Barrier Bar System of South Texas; Trans. Gulf Coast Assoc. Geol. Soc., Vol. XIV, pp. 309-322:

Corbeille, R. L., 1962, New Orleans Barrier-Island; Trans. Gulf Coast Assoc. Geol. Soc., Vol. 12, pp. 223-230.

Hayes, M. O., and A. J. Scott, 1964, Environmental Complexes, South Texas Coast; Trans. Gulf Coast Assoc. Geol. Soc., Vol. 14, pp. 237-240.

Hoyt, J. H., 1967, Barrier Island Formation; Bull. Geol. Soc. Amer., Vol. 78, No. 9, pp. 1125-1136.

Kolb, C. R. and J. R. Van Lopik, 1958, Geology of the Mississippi River Deltaic Plain, Southeastern Louisiana; U. S. Corps of Engineers Waterways Experiment Station Technical Reports 3-483 and 3-484, 2 volumes.

Krynine, P. D., 1948, The Megascopic Study and Field Classsification of Sedimentary Rocks; Jour. Geol., Vol. 56, No. 2, pp. 130-165.

McFarlan, E., Jr., 1961, Radiocarbon Dating of Late Quaternary Deposits, South Louisiana; Bull. Geol. Soc. Amer., LXXII, pp. 129-158.

McKee, E. D., and T. S. Sterrett, 1961, Laboratory Experiments on Form and Structure of Longshore Bars and Beaches, pp. 13-29 in Peterson, J. A. and Osmond, J. C., Editors, *Geometry of Sandstone Bodies;* Tulsa, Oklahoma, Amer. Assoc. of Petrol. Geol., 240 p.

Parker, R. H., 1960, Ecology and Distributional Patterns of Marine Macro-Invertebrates, Northern Gulf of Mexico, in Recent Sediments, Northwes Gulf of Mexico: Amer. Assoc. Petrol. Geol., pp. 302-327.

Phleger, F. B., 1960, Foraminiferal Populations in Laguna Madre, Texas; Science Reports Tohoku Univ. 2d Ser. (Geology), Special Vol. No. 4 (Hanzawa Memorial Volume), pp. 83-91.

Potter, P., 1967, Sand Bodies and Sedimentary Environments: A Review; Bull. Amer. Assoc. Petrol. Geol., Vol. 51, No: 3, pp. 337-365.

Price, W. A., 1951, Barrier Island, Not 'Offshore Bar'; Science, Vol. 113, No. 2939, pp. 487-488.

Rowett C. L., 1957, A Quaternary Molluscan Assemblage from Orlaens Parish, Louisiana: Trans. Gulf Coast Assoc. Geol. Soc., Vol. VII, pp. 153-164.

Rusnak, G. A., 1960, Sediments of Laguna Madre, Texas; pp. 153-196 in Shepard, F. P., and others, Editors, Recent Sediments, Northwest Gulf of Mexico: Tulsa, Oklahoma, Amer. Assoc. Petrol. Geol., 394 pp.

Saucier, Roger T., 1963, *Recent Geomorphic History of the Pontchartrain Basin;* Louisiana State Univ. Press, Baton Rouge, 114 p.

Shelton, John W., 1967, Stratigraphic Models and General Criteria for Recognition of Alluvial, Barrier-Bar, and Turbidity-Current Sand Deposits; Bull. Amer. Assoc. Petrol. Geol., Vol. 51, No. 12, pp. 2441-2461.

Shepard, Francis P., 1952, Revised Nomenclature for Depositional Coastal Features; Bull. Amer. Assoc. Petrol. Geol., Vol. 36, No. 10, pp. 1902-1912.

Shepard, Francis P., 1960, Gulf Coast Barriers; in Recent Sediments, Northwest Gulf of Mexico: Amer. Assoc. Petrol. Geol., pp. 197-220.

Shepard, F. P., and D. G. Moore, 1960, Bays of Central Texas Coast; pp. 117-152 in Shepard, F. P. and others, Editors, Recent Sediments Northwest Gulf of Mexico: Tulsa, Oklahoma, Amer. Assoc. Petrol. Geol., 394 p.

Treadwell, R. C., 1955, Sedimentology and Eucology of Southeast Coastal Louisiana; Louisiana State Univ., Baton Rouge, Technical Report 6, Project No. N7, ONR, pp. 1-78.

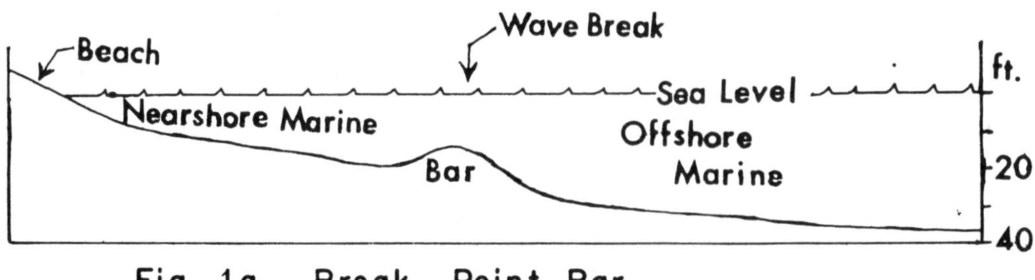

Fig. 1a. Break Point Bar

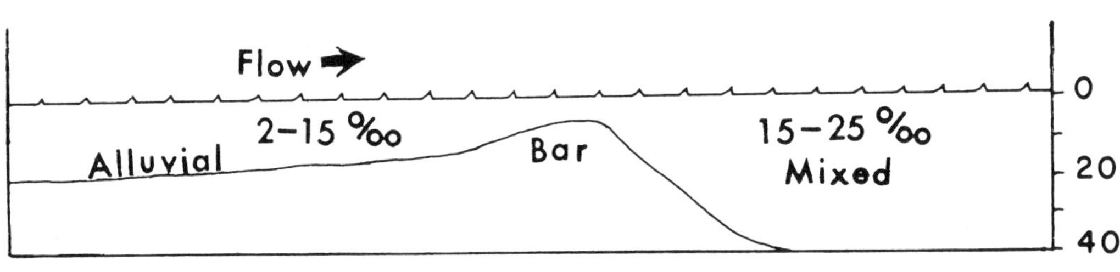

Fig. 1b. River Mouth Bar

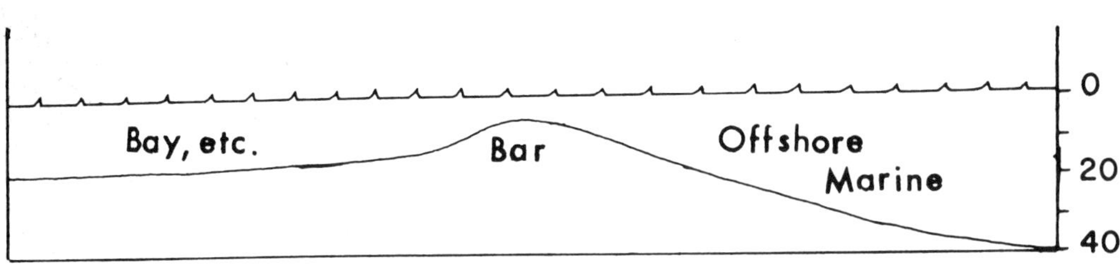

Fig. 1c. Bay, Estuary or Embayment Bar

Figure 1. Types of Bar Sands

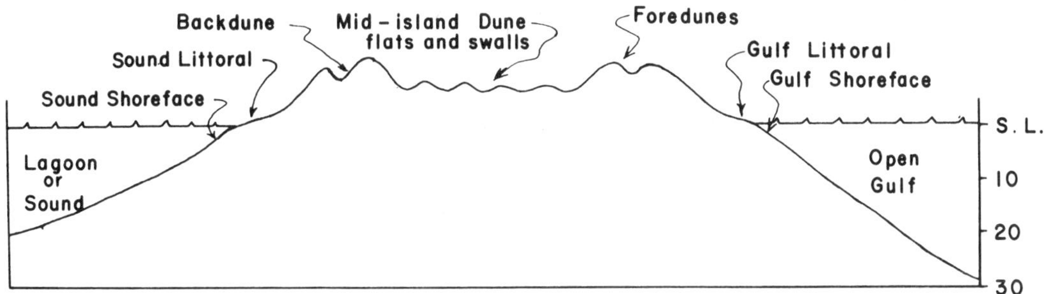

Figure 2. Barrier Island Environments

Fig 3. View to North of Monterrey Bar Sand on Monterrey - Saltillo Highway.

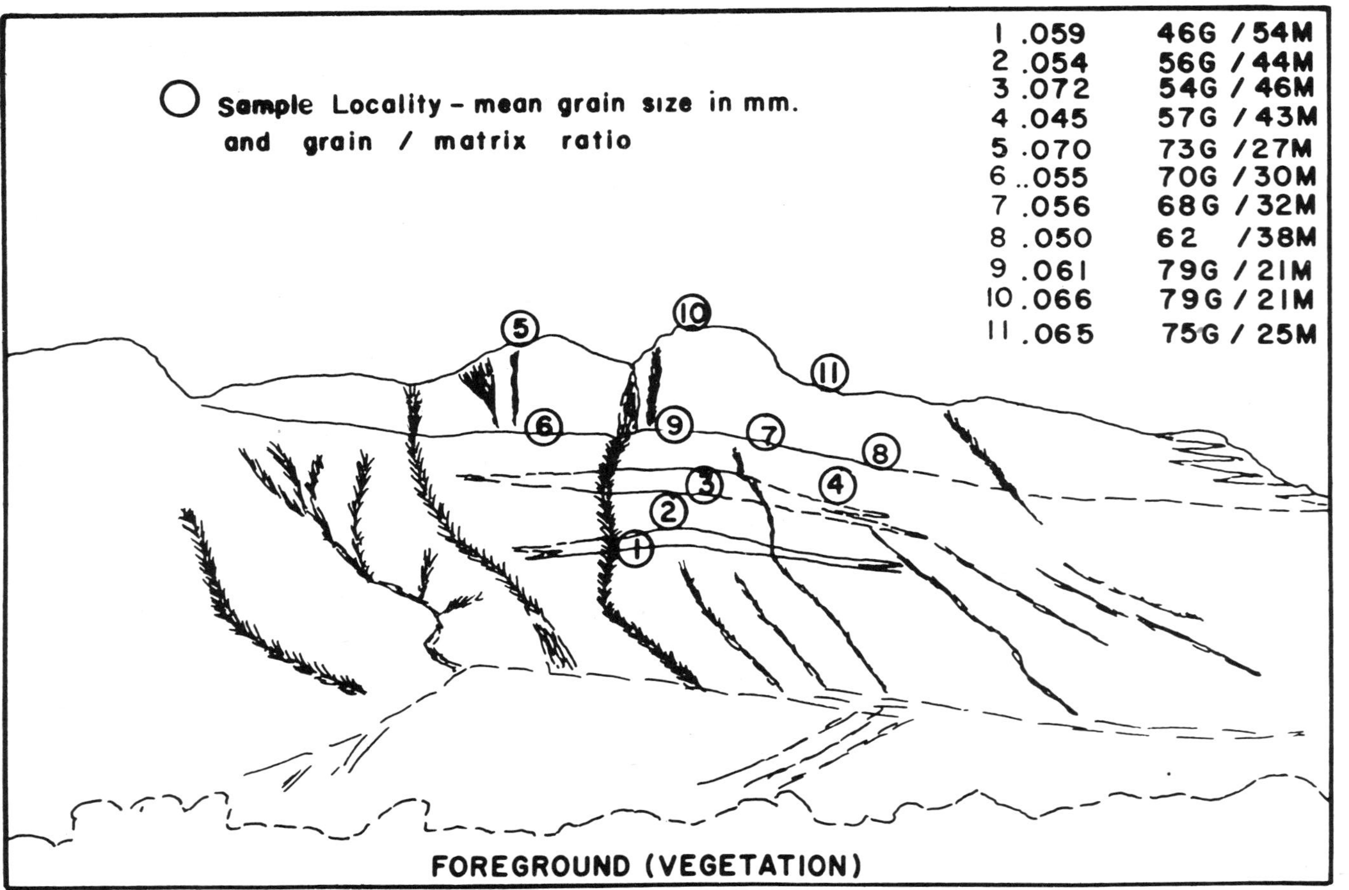

Fig. 4. Sketch of Monterrey Bar Sand

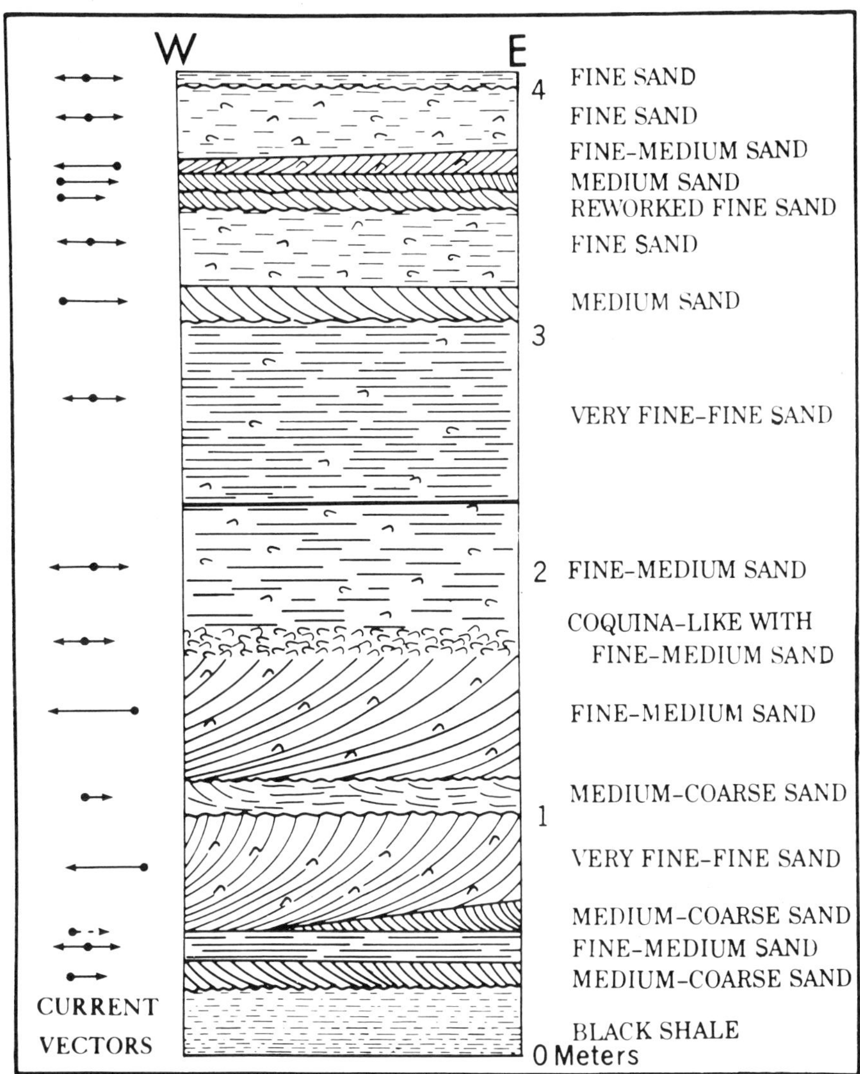

FIGURE 5. Cross bedded sandstone stratigraphically below and to the west of Monterrey

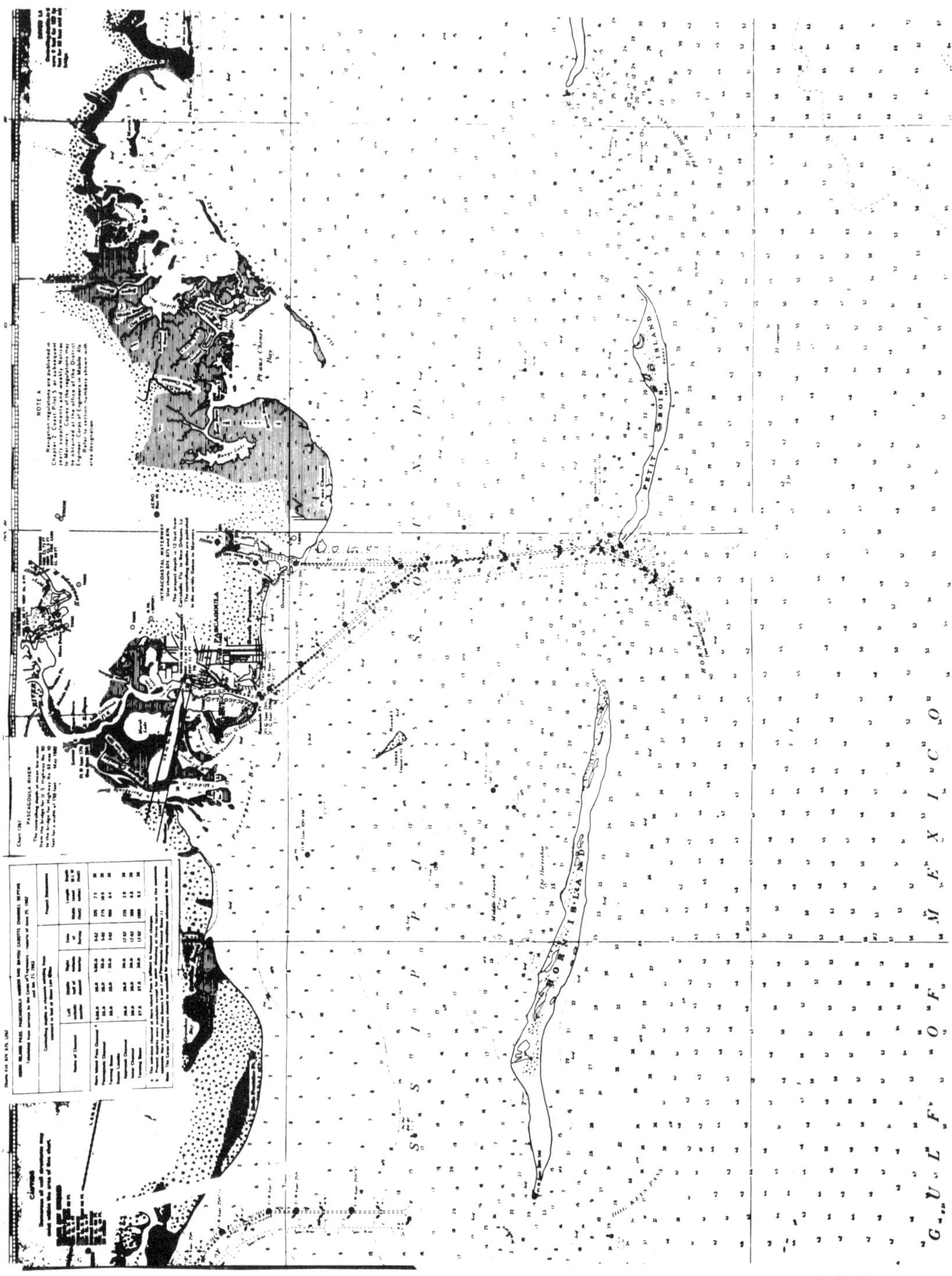

Fig 6. Gulf Coast Barrier Island, Mississippi and Alabama.

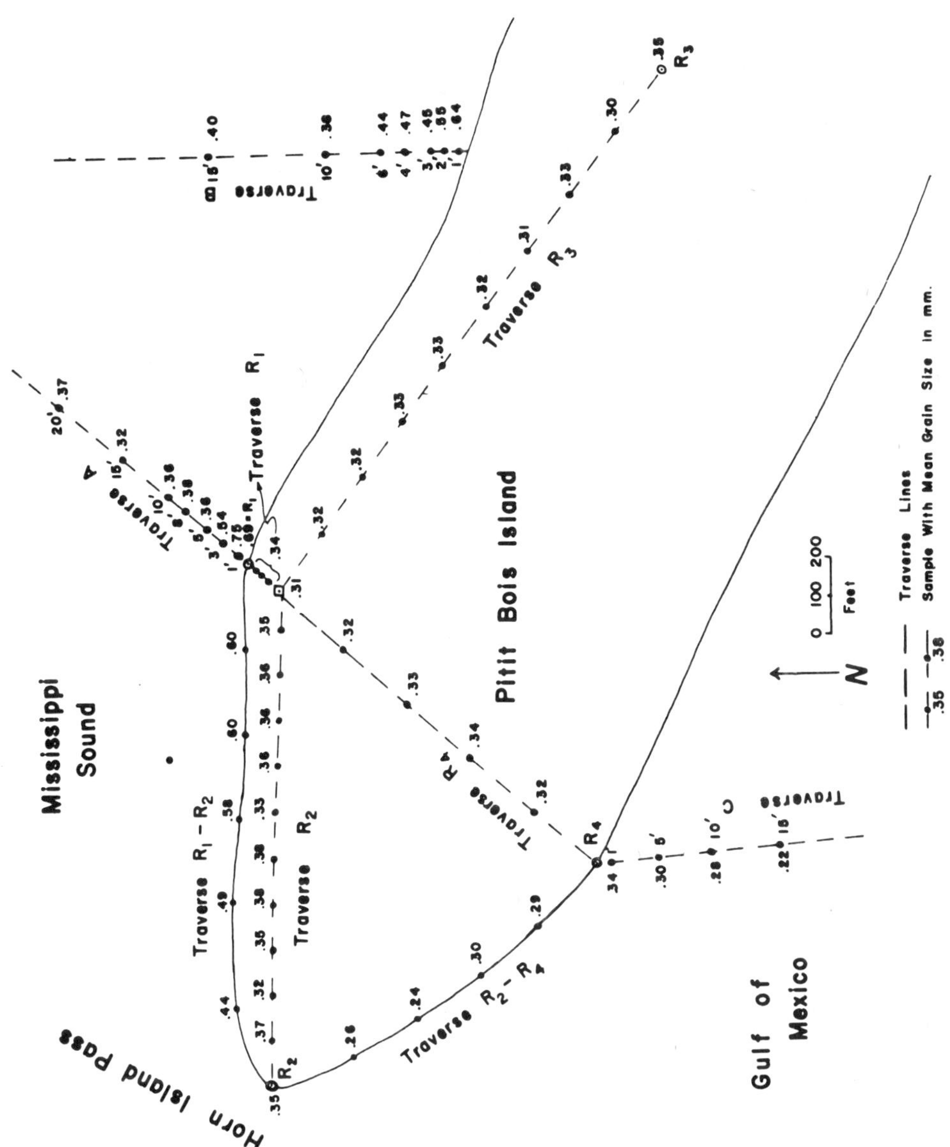

Fig. 7 Western Petit Bois Island

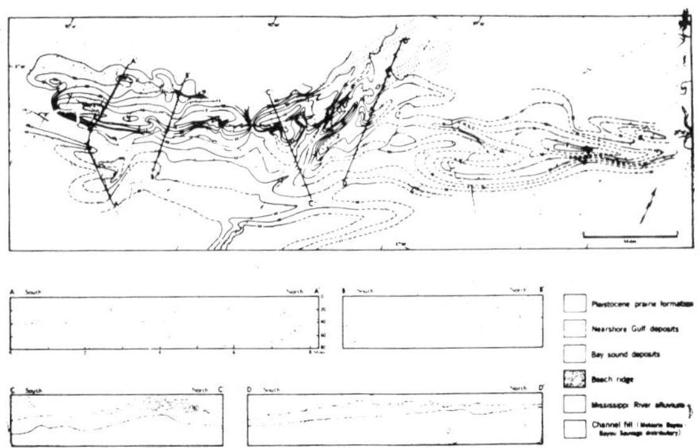

Fig 8. PINE ISLAND BARRIER; Structure contours and cross-sections (after Saucier, 1963)

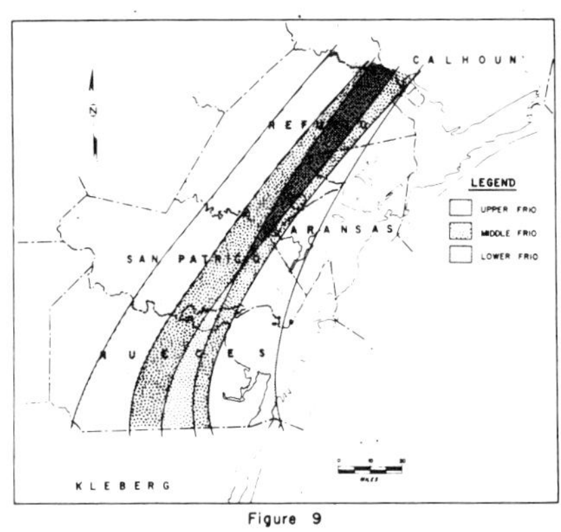

Figure 9

MAXIMUM SAND DEPOSITION AND TREND OF FRIO BARRIER BAR SYSTEM (after Boyd and Dyer, 1964)

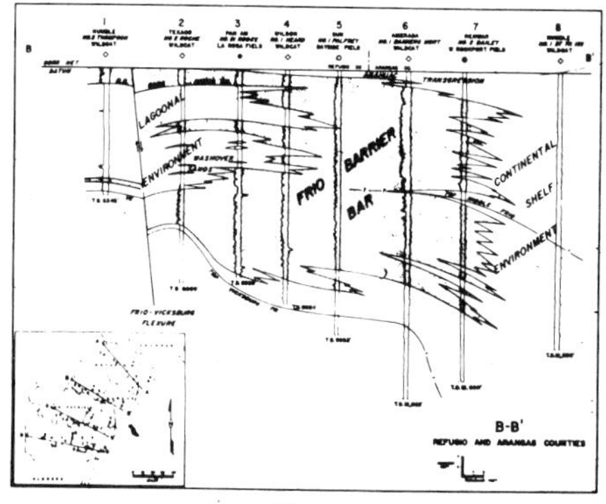

Figure 10

FRIO BARRIER BAR STRATIGRAPHIC DIP SECTION (after Boyd and Dyer, 1964)

Editor's Comments on Paper 19

Opposed to Hoyt's ridge engulfment hypothesis, J. J. Fisher maintains, in the fashion of Gilbert, that most barrier islands originate through the breaching of spits. Fisher does, however, agree with Hoyt that formation from submarine bars during emergence is unproved.

Fisher's field evidence reveals, as did Hoyt's, the absence of an open marine coast behind barriers, but Fisher finds the sublagoonal stratigraphy to be more like that expected behind a spit than the coastal forest that should prevail behind Hoyt's beach or dune ridge. Fisher also questions the ubiquity of large, long, and straight ridges along a submerging coast. He also holds that the known time sequence in barrier islands is indicative of spit progradation seaward rather than ridge growth upward.

At both the outset and closing of this very effective paper, Fisher points out that there may be, descriptively and genetically, more than one type of barrier island shoreline. This is a point well worth considering, as various divergent hypotheses are proposed, attacked, and defended.

At the time that his discussion paper was published in the *G.S.A. Bulletin,* John J. Fisher was 36 years old. He had received his B.A. at Rutgers University in 1958 and his M.S. and Ph.D. at the University of North Carolina in 1962 and 1967, respectively. From 1962 to 1964 Fisher studied the Cape Hatteras and Cape Lookout National Seashores under a grant from the National Parks and Monument Association and has been doing similar work at the Cape Cod National Seashore since 1969. He has also carried out investigations in coastal processes for the U.S. Department of the Interior, Water Resources Program, and for the Army Corps of Engineers, Coastal Engineering Research Center. In 1971 and 1972 Dr. Fisher studied coastal engineering aspects of Venice, Italy, and Amsterdam, The Netherlands.

19

Reprinted from the *Geol. Soc. Amer. Bull.*, **79**, 1421–1425 (1968)

JOHN J. FISHER *Department of Geology, University of Rhode Island, Kingston, Rhode Island*

Barrier Island Formation: Discussion

The recent theory advanced by Hoyt (1967) pertaining to a possible hypothesis for the origin and development of barrier islands implies certain conditions that the author questions in view of known field evidence. The author agrees with Hoyt (1967, p. 1126) and others that the classical Johnsonian theory which relates the origin of barrier islands to shorelines of emergence has never been satisfactorily demonstrated, either in the laboratory or in the field. The author also agrees with Hoyt that recent evidence suggests that it may actually be coastal submergence rather than emergence that is responsible for the origin of barrier islands. The author feels, however, that the basic mechanisms producing the islands during submergence are far different from those proposed by Hoyt. He proposes that barrier islands develop initially as subaerial dunes which are then submerged to form barrier islands. The author, in contrast, has recently proposed that barrier islands are formed as complex spits on the shoreline of submergence (Fisher, 1967). It is therefore to this point of the formation of barrier islands in Hoyt's theory rather than the question of submergence that the following points are directed.

It should be pointed out, however, at the outset of this discussion that genetically there appears to be more than one type of barrier island and that to neglect these differences might lead to conflict between certain theories of origin and the field evidence upon which these theories are built. All geomorphic features referred to as barrier islands by some coastal workers are obviously not the same feature (Fig. 1). Barrier-island chains along the Middle Atlantic states, such as Long Island, New Jersey, Delmarva, and North Carolina (Fig. 1A), as well as the eastern coast of Florida (Fig. 1C) and the Gulf Coast (Fig. 1D) appear to be barrier islands in the classical sense of Davis and Johnson; however, there are significant differences in their basic morphology. Fenneman (1938) and more recently Thornbury (1965) classified the coast of the Middle Atlantic states as an "embayed" shoreline, recognizing the large "bays" or estuaries (for example, Chesapeake Bay) found along this coast. Although not stated by either of these authors, it should be pointed out that numerous other "bays" or, more correctly, drowned river valleys are found behind the barrier-island chains as an intrinsic part of the lagoonal system (*compare* Albermarle Sound, North Carolina). In contrast, along the eastern coast of Florida there is a line of barrier islands behind which there are long, extensive lagoonal "tracts" or "lagoonal rivers" parallel to the coast (Thornbury, 1965). The Gulf Coast barrier islands, unlike those of the other coasts, are associated with extensive deltas of various river systems. In addition, these barrier islands are composed of both relict beach ridges and cheniers. Cheniers are not simply the Gulf Coast "equivalent" of relict beach ridges of other areas, as is commonly implied. To the author, they are basically two different geomorphic relict coastal features and hence reflect different origins, a view shared by R. J. Russell (1967, oral commun.). Finally, the Sea Islands of South Carolina and Georgia (Fig. 1B) appear to resemble least of all the classical description of a barrier island. Fenneman (1938) considered the Sea Islands to have formed by erosion of a lower coastal plain terrace during coastal submergence. In contrast, Zeigler more recently (1959) suggested that the Sea Islands are actually composed of three types of islands: "eroded remnant islands," "marsh islands," and "beach ridge islands" (true barrier islands?). In general, the discussion that follows will consider the implications of Hoyt's theory in relation to the origin and development of the classical barrier islands (those of the Middle Atlantic states and the Gulf Coast).

First, there is the unresolved problem in Hoyt's theory about the origin of the dunes and beach ridges along the shoreline before submergence. There is no evidence that dunes and ridges to heights of 20 and 30 feet would develop along a stable coast. The shore must be stable, in Hoyt's theory, in order to allow time for the dunes to form. Hoyt states in sup-

Geological Society of America Bulletin, v. 79, p. 1421–1426, 4 figs., October 1968

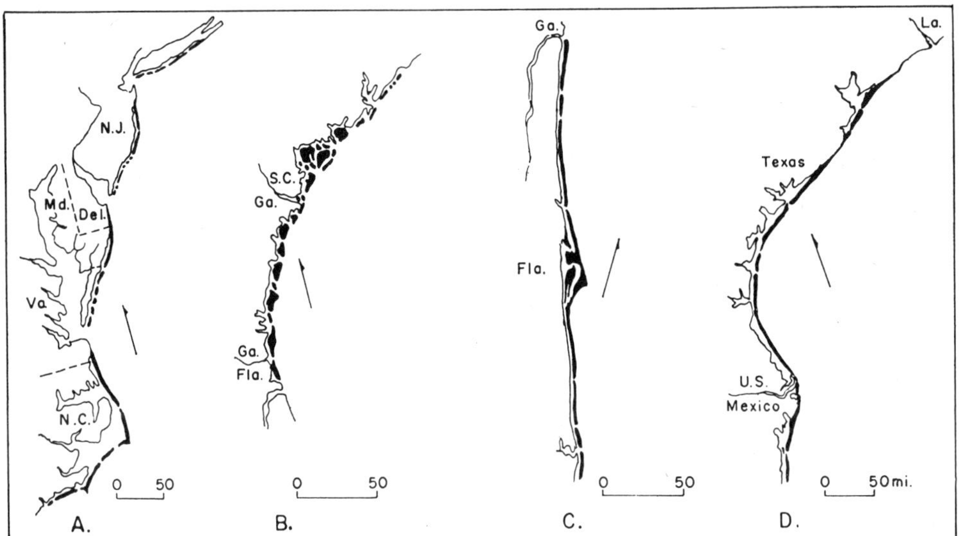

Figure 1. Barrier-island chain shorelines of the eastern United States. A. Middle Atlantic states. B. Sea Islands of South Carolina and Georgia. C. Eastern Coast of Florida. D. Western Gulf of Mexico.

port of dune building that "such ridges are ubiquitous along sediment shorelines of the world" (1967, p. 1130). This comparison with the present-day barrier shorelines is not valid because these shorelines, by the assumption of Hoyt's theory, are second-cycle shorelines of submergence rather than stable shorelines. In addition, the author believes that most beach ridges and barrier dunes along present-day barrier islands are actually relict features, having formed in the past as beach ridges under submerging coastal conditions.

The second questionable point in Hoyt's proposed theory is the implied straightness of the initial presubmerged shoreline. The sequence of development presented by Hoyt (1967, p. 1130) is that "the sea must meet the land at the shoreline, . . . dunes will form immediately landward of the shoreline . . . and with submergence . . . the ridge then becomes a barrier island." Considering that present-day barrier-island shorelines tend to be linear (*compare* Figs. 1, A, C, D), Hoyt's theory that the islands are beach ridges or dunes formed along a submerging coast would require submergence of a flat, featureless plain. Submergence of almost any coastal plain with its streams, valleys, and uplands would produce an irregular shoreline and hence, by Hoyt's theory, an irregular barrier chain. In Johnson's theory, the barrier shoreline was straight because he claimed that it developed on a hypothetically smooth plain, the emerged offshore sea bottom. Thus, when analyzed in map view, the sequence of shoreline development suggested by Hoyt does not seem to be valid. At least, it is not valid for the regular linear barrier islands of the Atlantic and Gulf Coasts.

A third problem in Hoyt's theory is his interpretation of the subsurface sediments beneath barrier islands and their associated lagoons. The author agrees with Hoyt (1967, p. 1127) that studies of subsurface sediments along the Gulf and Georgia barriers indicate the lack of open-sea marine sediments in the lower lagoonal sediments. Such marine deposits should be present if the barrier developed as a result of emergence of the sea floor in the classical Johnsonian theory. However, applying Hoyt's theory, if barrier islands do develop from a core of subaerial shoreline dunes, then the coast directly behind these dunes should be vegetated to the extent that not only shrubs but also trees would be present (Fig. 2A). The author has mapped pine-oak forests located only 500 feet behind the protective beach and dune ridges in the relatively unspoiled Cape Hatteras National Seashore Park. Burke (1962) and Brown (1959) have reported similar forest communities along the North Carolina coast. As to the possibility of salt-spray injury keeping the coast free of trees, Brown (1959, p. 66) reported that the effective zone of salt-spray toxicity was only about 0.1 to 0.3 mile inland

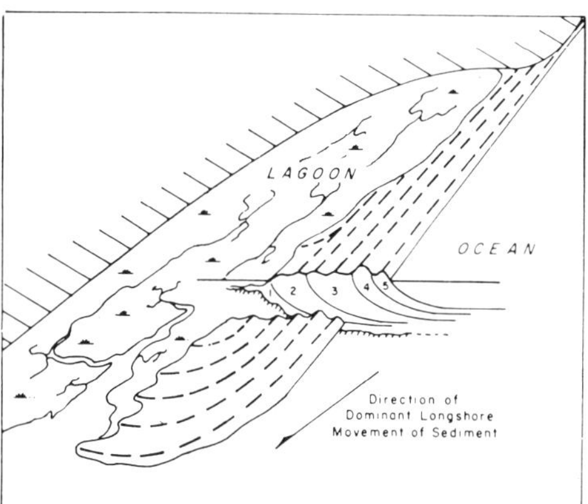

Figure 4. Possible development of barrier islands as complex spits, under action of longshore movement and related to known time sequences of barrier islands of Gulf Coast (*compare* Fig. 3B).

dunes become the barrier islands. This aspect of the theory is also open to question. Most, if not all, inlets open and close during a period of time that is short when compared to the longer geologic history of a barrier chain. On the Outer Banks barrier-chain shoreline of North Carolina and Virginia, some 30 inlets have opened and closed during the 350 years of recorded history in this area (Fisher, 1962). During that time, 6 others have opened and remained open. This coastal history therefore suggests that a continuing hydraulic process is responsible for the opening (and closing) of inlets rather than an origin as "areas of low elevation" before submergence. The presence of inlets in types of shorelines other than barrier chains also suggests that a hydraulic mechanism originates and localizes all inlets, not the unique combination of events suggested by Hoyt's theory.

Regarding other possible effects of coastal submergence, Hoyt points out that Bruun (1962) and Schwartz (1965) have suggested that recent submergence has resulted in shoreline erosion ("landward advance of the shoreline in excess of the amount predicted from the island morphology" [Hoyt, 1967, p. 1131]). If Hoyt accepts this premise, and it appears that he does, then the submergence of the initial dune-covered shoreline of his theory should cause *erosion* of that shoreline, not the building up of it to a barrier island (*compare* Hoyt, p. 1130, Fig. 5). To explain the development of certain shorelines by arguing the need for a "fast rate" or "slow rate," "without sufficient sediment" or "with abundant sediment" only begs the question as to the validity of the basic theory proposed in explaining the origin of known barrier-island chains.

The author feels that the present-day evidence, both surface and subsurface, suggests a theory of origin for barrier-island shorelines as complex spit chains. Hoyt (1967, p. 1129) feels, however, that a theory of development of barrier-island chains as spits requires that the material building the spit be supplied from rivers. However, spits do grow along many shorelines where there are no neighboring river systems. An offshore source for barrier-island sediments, as suggested by Hoyt, does not preclude this material's being deposited as spits by longshore movement. In fact, it was Gilbert (1885) who first suggested that barriers formed by accretion of sediments transported along the shore by various longshore currents. Interestingly enough, Hoyt does not consider longshore transport anywhere in his theory as an agent of either deposition or erosion. Longshore transport, together with the recent changes in sea level, are perhaps the two most important coastal processes operating along unconsolidated sediment shorelines.

Johnson, in his classical theory, attempted to explain the origin of barrier islands as the result of certain unproven processes that were to take place during coastal emergence. Hoyt, in the theory he proposed, called upon a series of fortuitous events to take place during coastal submergence. However, both Johnson and Hoyt, in their respective theories, ignored the demonstrated significant effects of longshore movement in the origin and development of barrier islands. Hoyt, in his proposed theory, however, did attempt to relate the increasing evidence that the origin of barrier islands is related to coastal submergence rather than emergence. The question remains, however, as to whether interpretation of field evidence from the Gulf and Atlantic Coasts actually supports Hoyt's theory. Thus, from the author's point of view, three problems yet remain in the study of barrier-chain shorelines. First, are they definitely features of submergence? Second, are they primary features (complex spits) or second-cycle features (drowned subaerial dunes or emerged offshore bars)? Third, is there genetically, as well as descriptively, more than one type of barrier-island shoreline? Hoyt, in his proposed theory, has advanced a solution to the first problem, but the latter two problems remain.

except at one place where it was 1.35 miles inland. The presence of barrier dunes tends to protect the forest even further from salt-spray injury. Thus, if barriers formed as proposed by Hoyt, the subsurface base of the lagoonal deposits should be a veritable buried forest (*compare* Fig. 2B). This is definitely not the situation. Salt marsh and root fragments have been found in some deposits (Shepard and Moore, 1955, 1960), but nothing indicative of subaerial tree and shrub deposits. So, while the evidence of subsurface lagoonal deposits does not indicate open marine sediments as required by the Johnsonian classical theory, neither does the evidence seem to indicate the subaerial sediments implied by Hoyt's theory. Furthermore, the sediment types actually found at the base of the lagoons along the Gulf Coast are described as a lagoonal or "bay" facies (Shepard and Moore, 1960, p. 133) similar to present-day lagoonal deposits. If this is the case, then the base of the lagoon could never have been above sea level as a subaerial region behind coastal dunes as Hoyt's theory would imply. Shepard and Moore further conclude (1960, p. 134) that the subsurface sediments indicate that during the development of the barrier islands it seems "likely that there was a barrier across most of the mouth of the bay" (lagoon). From the author's point of view, this barrier could have been a spit.

Figure 3A has been constructed by the author to show another inconsistency in Hoyt's theory, that of the dating of subsurface barrier sediments and the ages implied in his theory. Figure 3A is an exact copy of Hoyt's Figure 5 (1967, p. 1130) with numbers added by the author to indicate the probable sequence of development. First, according to Hoyt's theory, a prograding-dune or beach-ridge stage develops with the sequence of ridges becoming progressively younger seaward (ridges 1 through 3). Later, with submergence, the dune ridge becomes a barrier island and grows upward and seaward by developing a second set of prograding ridges (ridges 4 and 5). However, the known time sequence of barrier-island development along the Gulf Coast based on radioactive carbon dating (Bernard and others, 1962) indicates that the pattern is actually only that of a seaward prograding with no substantial vertical growth (Fig. 2B). However, in contrast to Hoyt's theory, this known time sequence of barrier-island development is adequately explained by spit development. This feature, Hoyt feels, does not adequately "account for the major barrier island systems" (Hoyt, 1967, p. 1129). However, Figure 4, based upon Figure 4 in Hoyt's article, illustrates that the known time sequence of the development of barrier islands (*compare* Fig. 3B) does indeed fit the development of barrier islands from spits. In fact, the author has recently suggested (Fisher, 1967) that the well-known barrier islands of the Middle Atlantic states from Long Island to North Carolina have developed in just such a manner. This spit development has taken place primarily during the most recent hypsithermal submergence.

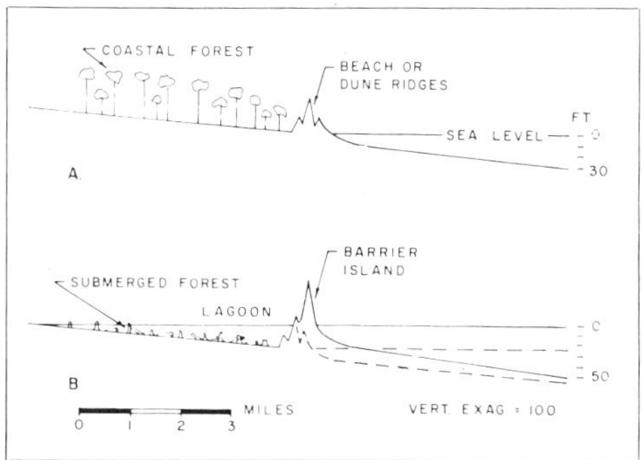

Figure 2. Subsurface lagoonal sediment characteristics after Hoyt's theory. A. Initial dune ridge with associated subaerial vegetation. B. Barrier island formed by submergence with buried vegetation incorporated into lagoonal sediments.

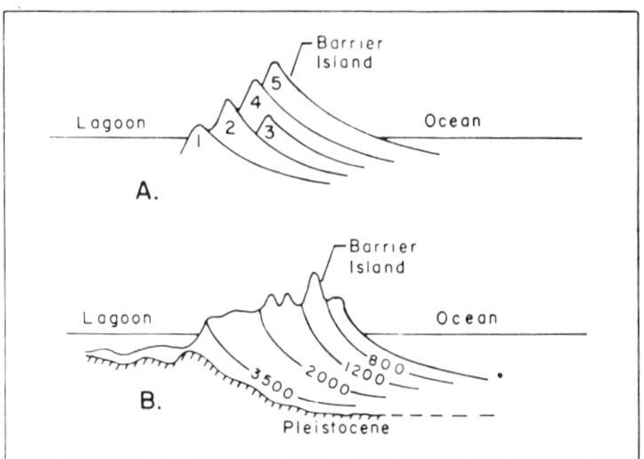

Figure 3. Sequence of development of barrier islands. A. After Hoyt's proposed theory. B. After known radioactive carbon dating of Gulf Coast barriers by Bernard and others (1962).

Hoyt also suggested (1967, 1130) as part of his theory that the origin of most inlets in a barrier chain is related to "areas of low elevation" among the initial dunes on the shoreline. Upon submergence by the rise of sea level, these "low areas" become inlets while the

References Cited

Bernard, H. A., LeBlanc, R. J., and Major, C. F., 1962, Recent and Pleistocene geology of southeast Texas, p. 175–224 *in* Geology of the Gulf Coast and central Texas and guidebook of excursions: Houston Geol. Soc.

Brown, C. A., 1959, Vegetation of the Outer Banks of North Carolina: Louisiana State Univ. Coastal Studies Series, no. 4, 179 p.

Bruun, P., 1962, Sea level rise as a cause of shore erosion: Am. Soc. Civil Engineers Proc., Jour. Waterways and Harbors Div., v. 88, p. 117–130.

Burke, C. J., 1962, The North Carolina Outer Banks: a floristic interpretation: Jour. Elisha Mitchell Scientific Soc., v. 78, p. 21–28.

Fenneman, N. M., 1938, Physiography of Eastern United States: New York, McGraw-Hill, 714 p.

Fisher, J. J., 1962, Historical geography of North Carolina inlets—a geomorphic re-evaluation: Assoc. Am. Geographers, S. E. Div., Memorandum Folio, v. 14.

—— 1967, Origin of barrier island chain shorelines: Middle Atlantic states (Abstract): Geol. Soc. America Ann. Meeting Program, p. 66–67.

Gilbert, G. K., 1885, The topographic features of lake shores: U. S. Geol. Survey 5th Ann. Rept., p. 69–123.

Hoyt, J. H., 1967, Barrier island formation: Geol. Soc. America Bull., v. 78, p. 1125–1136.

Schwartz, M., 1965, Laboratory study of sea-level rise as a cause of shore erosion: Jour. Geology, v. 73, p. 528–534.

Shepard, F. P., and Moore, D. G., 1955, Central Texas coast sedimentation: characteristics of sedimentary environment, Recent history and diagenesis: Am. Assoc. Petroleum Geologists Bull., v. 39, p. 1463–1593.

—— 1960, Bays of central Texas coast, p. 117–152 *in* Shepard, F. P., and others, *Editors*, Recent sediments, northwest Gulf of Mexico: Am. Assoc. Petroleum Geologists.

Thornbury, W. D., 1965, Regional geomorphology of the United States: New York, Wiley, 609 p.

Zeigler, J. M., 1959, Origin of the sea islands of southeastern United States: Geog. Rev., v. 49, p. 22–237.

Manuscript Received by The Society April 10, 1968

Editor's Comments on Paper 20

20 Hoyt: *Barrier Island Formation: Reply*

In this, his reply to Fisher's discussion paper, Hoyt concurs with Fisher on the possible multiplicity of means of formation of barrier islands, yet maintains that Gulf Coast and Atlantic Coast barriers are not significantly different.

Answering Fisher's criticisms, Hoyt points out that (1) dune and beach ridges develop quickly along stable or unstable shorelines and may be maintained for considerable periods of time; (2) the present straightness of barrier island shorelines is the result of reworking by coastal processes, not the shape of the initial ridges; and (3) forest remnants over the whole of the exposed continental shelf are removed by unknown processes. On this last point, Hoyt steadfastly maintains that bay or lagoonal facies from the base of the lagoons indicate that lagoons formed behind a submerged ridge. He argues that an open marine coast must precede the growth of a protective spit.

Other details are discussed, and the possibility of some breached-spit barrier islands is accepted, but Hoyt holds to his original ridge submergence hypothesis.

20

Reprinted from the *Geol. Soc. Amer. Bull.*, **79**, 1427–1431 (1968)

JOHN H. HOYT *Marine Institute and Department of Geology, University of Georgia, Sapelo Island, Georgia*

Barrier Island Formation: Reply

Many of the ideas expressed by Fisher are similar to those considered in developing a hypothesis of barrier-island formation (Hoyt, 1967a). In presenting a succinct and useful scheme it was not possible to include a detailed history of rejected mechanisms. Now, however, it is worthwhile and necessary to consider the comments of one knowledgeable in this field.

There is apparent agreement that there may be more than one means of formation of barrier islands. This was noted in my paper (Hoyt, 1967a, p. 1133). In discussing the morphology of barrier islands, however, Fisher limits his comments to the bays, drowned river valleys, and coast-parallel lagoonal tracts and fails to indicate significant differences in the islands that suggest diverse origins. It is unfortunate that in considering the barrier islands of the Georgia and South Carolina coast Fisher ignores a number of recent papers which describe this area in considerable detail (Hoyt and Weimer, 1963; Hoyt and others, 1964; Colquhoun, 1965; Hoyt and others, 1966; Hoyt and Hails, 1967; Hoyt and Henry, 1967; Thom, 1967; Henry and Hoyt, 1968; Logan and Henry, 1968). These papers clearly indicate that the Georgia and South Carolina coasts are bordered by barrier islands which differ in no genetically significant way from those to the north and south. The Georgia and South Carolina islands are constructional barriers similar to those of the Gulf Coast (Fisk, 1959; Shepard, 1960; Bernard and others, 1962) and the Atlantic Coast (Fischer, 1961; Smith, 1961; McIntire and Morgan, 1962).

Except for the Mississippi River Delta, the deltas of Gulf Coast rivers had little influence on the origin of the major Gulf Coast barriers. In general, the barriers pre-date the deltas, and at the time of barrier formation the rivers were still filling their lower valleys (LeBlanc and Hodgson, 1959; Bernard and others, 1962). Thus, it is difficult to see how the origin of Gulf Coast barriers would be significantly different from those of the Atlantic Coast. The use of the term "chenier" in conjunction with Gulf Coast barriers should be carefully considered. Cheniers form in an entirely different manner and sequence from barriers and in a different sedimentological environment. The two features should be carefully distinguished (Hoyt, 1967b; 1968, in press). In general, cheniers and barriers are mutually exclusive.

Fisher criticizes three aspects of my paper: (1) the origin of dune and beach ridges along the shoreline, (2) the implied straightness of the initial presubmergence shoreline, and (3) the lack of buried forests and shrubs beneath barrier islands and lagoons.

(1) In criticizing the origin of dune and beach ridges, Fisher makes unwarranted assumptions concerning the stability of the coast and the time required for the ridges to form. There is no need for the shore to be "stable" to "allow time" for the ridges to form. Dunes and a dune ridge 15 to 20 feet high have developed on Sapelo Island during the past 8 years, and a major ridge with dunes more than 30 feet high, now several hundred yards from the beach, has formed during the past 200 years (Land, 1964). Beach ridges may form even more rapidly and be deposited during a single storm (Davies, 1957; Shepard, 1960; Thom, 1964). Such rapid formation suggests that dune and beach ridges could develop at all stages of the Holocene transgression. It is difficult to believe that there is something unique about modern conditions not present during the past that favors dune- and beach-ridge development. Further, the dune and beach ridges do not necessarily form before submergence, but rather can form during submergence. On the other hand, there is abundant evidence that dune ridges have developed along stable shorelines during the past 3000 to 4000 years. The majority of Holocene dune and beach ridges conspicuous on barrier islands have formed during this time. A well-documented example of a dune ridge developed during the past 3500 years is provided by the studies of Galveston Island, Texas (Bernard and others, 1962). Fisher's belief that dune and beach

Geological Society of America Bulletin, v. 79, p. 1427–1432, 1 fig., October 1968

ridges along barrier shorelines are relict features is questionable, as even a brief observation shows that they are forming under present conditions (Shepard, 1960; Land, 1964; Hayes, 1967). Studies of Pleistocene barrier islands along the southeastern coast of the United States indicate that dune ridges were common features of the islands that developed along the interglacial shorelines when sea level was higher than present (Oaks and Coch, 1963; Hoyt and others, 1964; Colquhoun, 1965; Hoyt and Hails, 1967; Thom, 1967). In summary, dune and beach ridges have formed during the past and are forming today along stable and unstable coasts, and there is no reason to believe that this has not always been the case.

(2) In considering the straightness of modern barrier-island shorelines, Fisher makes a serious error in confusing present linearity with the morphology at the time of original formation. The modifications that occur along the shoreline, the erosion and progradation, have resulted in the present coastal configuration, and the modern islands may bear little resemblance to their shape and extent at the time of origin. A flat, featureless plain is not required, nor would the original barriers be continuous along the coastline. Thus, the argument presented by Fisher has no bearing on the hypothesis of barrier formation.

(3) Fisher modifies my Figure 7 by the addition of a forest landward of the dune ridges. Although eloquently done, the modification is incomplete. Forests and fallen trees, at times of lower sea level, covered the continental shelf and have been reported from several areas (Lyell, 1850; Shumway and others, 1961; Emery, 1968). Trees also covered the exposed coastal plain, and by Fisher's reasoning many areas should have extensive accumulations of fallen logs. There must be some mechanism which removes the vegetable matter soon after death and it can be assumed that the processes of weathering, oxidation, insect destruction, fire, and so forth were operable during the submergence of the lagoonal area just as they are in terrestrial environments today. Considering that the submergence rate during the past 3000 to 4000 years was less than 0.6 feet per century (Bloom and Stuiver, 1963; Scholl, 1964), there was ample time for the destruction of most vegetable material. Fisher's discussion of salt-spray toxicity is irrelevant because in the lagoonal area it is the rise of the water table and the encroachment of brackish waters that results in the replacement of forest communities with a succession, first of fresh-water marsh grasses and finally of halophytes. In addition, in many areas erosion and reworking by tidal channel and inlets result in the dissemination and destruction of vegetable matter at the base of the lagoon (Hoyt and Henry, 1967). Thus, Fisher's contention that trees and shrubs must be encountered in lagoonal areas is not valid. Although not mentioned by Fisher, the study of peat formation and deposition in coastal areas offers a more reasonable method for preservation of vegetable material (Chapman, 1960; Bloom, 1964; Scholl, 1964; Redfield, 1965). The peat record supports submergence during the past several thousand years, and the protected environment in which it is found indicates the conditions necessary for its formation and preservation.

The recovery of lagoonal or bay facies from the base of lagoons along the Gulf Coast (Shepard and Moore, 1960) was an important criterion in developing the submergence hypothesis. These are the sediments that would accumulate if the lagoons were protected by a barrier island formed from a submerged dune ridge. Note that this is not the case if the barrier formed as a spit, for at an early stage of spit formation open marine conditions must prevail in the lagoonal area. Fisher's view that the base of the lagoon could never have been above sea level is not true, as a lowering of sea level of more than 300 feet during the Pleistocene is well established (Shepard, 1963; Curray, 1965; Hoyt and others, 1965). The absence of subaerial sediments, in some areas, suggests the removal of these sediments and the contained vegetable material by some process such as reworking by inlet and tidal channels.

In criticizing my Figure 5, Fisher must realize that this illustration is idealized and is not capable of duplicating the infinite variety of barrier islands. Although a minor point, his Figure 3A is hardly an exact copy of my Figure 5. Again Fisher fails to recognize the modification through erosion, reworking, progradation, and inlet migration that alters the barrier accumulations. Galveston Island is an excellent example, for the record of deposition is well known for the past 3500 years (Bernard and others, 1962). But is it not strange that the history of this island began 3500 years ago? Where are the earlier deposits? It should be evident that they were seaward of the 3500-year B.P. shoreline and were reworked during the submergence and transgression that accompanied the accumulation of the 3500-year

B.P. deposits. Fisher fails to relate the concepts of Bruun (1962) and Schwartz (1965) to their effect on barrier-island history. The development of Galveston Island demonstrates the importance of these ideas. Since 3500 years ago there has been relative stability of land and sea in the Galveston area and the record has been one of progradation.

Although it is difficult to discern from his comments, Fisher apparently supports my ideas on the history of inlets, "the original configuration [of inlets] is modified by sedimentational processes, and some inlets may be closed and others opened" (Hoyt, 1967a, p. 1131). The original inlets, before modification by coastal processes, occupied the natural low areas between ridges. That water would flow in low areas and in existing channels hardly seems a "unique combination of events."

Fisher is incorrect in indicating that the author suggests that the sediments for building spits are derived from rivers, as a careful reading of page 1129 reveals. He is also mistaken in saying that the author does not consider longshore movement of sediment important to the modification of barriers and sediment movement. This was, in fact, the subject of another paper (Hoyt and Henry, 1967) which detailed the effects of longshore sediment transport on the morphology of barrier islands and the characteristics of the resulting deposits. The subject was not reiterated in the paper on barrier-island formation; however, reference was made to the earlier paper for those who wished to pursue this aspect of the problem.

As an alternative to the dune-ridge submergence hypothesis, Fisher reproposes the breached-spit theory of Gilbert (1885). This theory was reviewed in my paper and accepted for special situations. Field evidence, however, indicates that it must be rejected as a mechanism for the major barrier-island coasts. This is demonstrated by Figure 4 of Fisher's paper, which is here redrawn, with corrections (Fig. 1). Although it was emphasized in my paper, it is repeated that, if barrier islands developed from spits, the mainland coast of the lagoon must have been, at an early stage of spit development, an open marine environment with appropriate sediments, fauna, and morphology. If it is contended that these conditions were met at a lower sea level, then their presence must be demonstrated by subsurface investigation. This has not been done, and the detailed studies cited and illustrated in my paper indicated that it is not the case (Shepard and Moore, 1955, 1960; Fisk, 1959; LeBlanc and Hodgson, 1959; Bernard and others, 1962; Hoyt and others, 1964; Hails, 1965; Hoyt and Hails, 1967) and the breached-spit theory must be rejected.

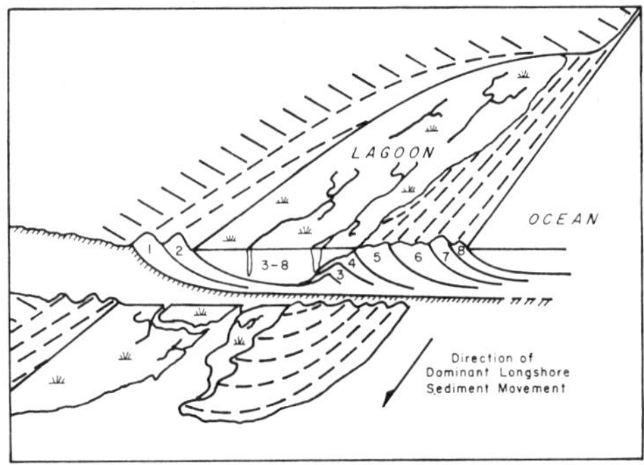

Figure 1. Idealized diagram showing spit development. Redrawn from Fisher's Figure 4 with the addition of dune, littoral, and shallow neritic sediments (units 1 and 2) along mainland shoreline. Open marine conditions must exist along the mainland prior to spit development.

In summary, the submergence hypothesis is based upon conditions which are believed to be generally acceptable: (1) submergence has accompanied the melting of the continental glaciers, and (2) dune and beach ridges, ubiquitous along modern shorelines, form sufficiently rapidly to have been common during the Holocene transgression. Further, the decrease in the rate of submergence during the late Holocene (approximately 0 to 4000 years B.P.) allowed, in many areas, the progradation of the shoreline and successive dune- and beach-ridge development or both. The barriers are continually modified in response to changing hydrodynamic conditions, sediment supply, and particularly the influence of inlet migration. The result is a great variety of shapes, sizes, thicknesses, and sediment relations which, except for minor specialized cases, are believed to have originated by a single process and have, since formation, undergone complex histories of modification and development.

References Cited

Bernard, H. A., LeBlanc, R. J., and Major, C. F., 1962, Recent and Pleistocene geology of southeast Texas, p. 175–224 *in* Rainwater, E. H., and Zingula, R. P., *Editors*, Geology of the Gulf Coast and central Texas and guidebook of excursions: Houston Geol. Soc., 392 p.

Bloom, A. L., 1964, Peat accumulation and compaction in a Connecticut coastal marsh: Jour. Sed. Petrology, v. 34, 599–603.

Bloom, A. L., and Stuiver, M., 1963, Submergence of the Connecticut coast: Science, v. 139, p. 332–334.

Bruun, P., 1962, Sea level rise as a cause of shore erosion: Am. Soc. Civil Engineers Proc., Jour. Waterways and Harbors Div., v. 88, p. 117–130.

Chapman, V. J., 1960, Salt marshes and salt deserts of the world: New York, Interscience, 392 p.

Colquhoun, D. J., 1965, Terrace sediment complexes in central South Carolina: Atlantic Coastal Plain Geol. Assoc. Conference 1965, 62 p.

Curray, J. R., 1965, Late Quaternary history, continental shelves of the United States, p. 723–735 *in* Wright, H. E., Jr., and Frey, D. G., *Editors*, The Quaternary of the United States: Princeton, N. J., Princeton Univ. Press, 922 p.

Davies, J. L., 1957, The importance of cut and fill in the development of sand beach ridges: Australian Jour. Sci., v. 20, p. 105–111.

Emery, K. O., 1968, Relict sediments on continental shelves of world: Am. Assoc. Petroleum Geologists Bull., v. 52, p. 445–464.

Fischer, A. G., 1961, Stratigraphic record of transgressing seas in light of sedimentation on Atlantic coast of New Jersey: Am. Assoc. Petroleum Geologists Bull., v. 45, p. 1656–1665.

Fisk, H. N., 1959, Padre Island and the Laguna Madre Flats, coastal south Texas: Baton Rouge, Louisiana, Louisiana State Univ., 2nd Coastal Geography Conf., p. 103–151.

Gilbert, G. K., 1885, The topographic features of lake shores: U. S. Geol. Survey 5th Ann. Rept., p. 69–123.

Hails, J. R., 1965, A critical review of sea-level changes in eastern Australia since the last glacial: Jour. Inst. Australian Geographers, v. 3, p. 63–78.

Hayes, M. O., 1967, Hurricanes as geological agents: Case studies of hurricanes Carla, 1961, and Cindy, 1963: Texas Univ. Bur. Econ. Geology, Rept. Inv., no. 61, 54 p.

Henry, V. J., Jr., and Hoyt, J. H , 1968, Quaternary paralic and shelf sediments of Georgia (Abstract): Geol. Soc. America Southeastern Section Program, p. 44.

Hoyt, J. H., 1967a, Barrier island formation: Geol. Soc. America Bull., v. 78, p. 1125–1136.

—— 1967b, Chenier versus barrier, genetic and stratigraphic distinction (Abstract): Am. Assoc. Petroleum Geologists Bull., v. 51, p. 471.

—— 1968, Chenier versus barrier, genetic and stratigraphic distinction: Am. Assoc. Petroleum Geologists (in press).

Hoyt, J. H., and Hails, J. R., 1967, Pleistocene shoreline sediments in coastal Georgia: Deposition and modification: Science, v. 155, p. 1541–1543.

Hoyt, J. H., and Henry, V. J., Jr., 1967, Influence of island migration on barrier-island sedimentation: Geol. Soc. America Bull., v. 78, p. 77–86.

Hoyt, J. H., and Weimer, R. J., 1963, Comparison of modern and ancient beaches, central Georgia coast: Am. Assoc. Petroleum Geologists Bull, v. 47, p. 529–531.

Hoyt, J. H., Henry, V. J., Jr., and Howard, J. D., 1966, Pleistocene and Holocene sediments, Sapelo Island, Georgia and vicinity: Geol. Soc. America, Southeastern Section, Field Trip no. 1, 158 p.

Hoyt, J. H., Smith, D. D., and Oostdam, B. L., 1965, Pleistocene low sea level stands on the southwest African continental shelf (Abstract): Abs. Internat. Assoc. for Quaternary Research, VII Internat. Cong., p. 227.

Hoyt, J. H., Weimer, R. J., and Henry, V. J., Jr., 1964, Late Pleistocene and Recent sedimentation, central Georgia coast, U. S. A., p. 170–176 *in* Van Straaten, L. M. J. U., *Editor*, Deltaic and shallow marine deposits: Amsterdam, Elsevier, v. 1, 464 p.

Land, L. S., 1964, Eolian cross-bedding in the beach dune environment, Sapelo Island, Georgia: Jour. Sed. Petrology, v. 34, p. 389–394.

LeBlanc, R. F., and Hodgson, W. D., 1959, Origin and development of the Texas shoreline: Gulf Coast Assoc. Geol. Socs. Trans., v. 9, p. 197–220.

Logan, T. F., and Henry, V. J., Jr., 1968, Subsurface Pleistocene sediments and stratigraphy of the central Georgia coast (Abstract): Geol. Soc. America, Southeastern Section Program, p. 53.

Lyell, C., 1850, Principles of geology: 8th ed., London, John Murray, 811 p.

McIntire, W. G., and Morgan, J. P., 1962, Recent geomorphic history of Plum Island, Massachusetts and adjacent coasts: Louisiana State Univ. Coastal Studies Inst. Tech Rept. no. 19, 44 p.

Oaks, R. Q., and Coch, N. K., 1963, Pleistocene sea levels, southeastern Virginia: Science, v. 140, p. 979–983.

Redfield, A. C., 1965, Ontogeny of a salt marsh estuary: Science, v. 147, p. 50–55.

Scholl, D. W., 1964, Recent sedimentary record in mangrove swamps and rise in sea level over the southwestern coast of Florida, Part 1: Marine Geology, v. 1, p. 344–366.

Schwartz, M., 1965, Laboratory study of sea level rise as a cause of shore erosion: Jour. Geology, v. 73, p. 528–534.

Shepard, F. P., 1960, Gulf Coast barriers, p. 338–344 *in* Shepard, F. P., and others, *Editors*, Recent sediments, northwest Gulf of Mexico: Tulsa, Oklahoma, Am. Assoc. Petroleum Geologists, 394 p.

—— 1963, Thirty-five thousand years of sea level, p. 1–10 *in* Clements, T., *Editor*, Essays in marine geology in honor of K. O. Emery: Los Angeles, Univ. Southern California Press, 201 p.

Shepard, F. P., and Moore, D. G., 1955, Central Texas coast sedimentation: characteristics of sedimentary environment, recent history, and diagenesis: Am. Assoc. Petroleum Geologists Bull., v. 39, p. 1463–1593.

—— 1960, Bays of central Texas coast, p. 117–152 *in* Shepard, F. P., and others, *Editors*, Recent sediments, northwest Gulf of Mexico: Tulsa, Oklahoma, Am. Assoc. Petroleum Geologists, 394 p.

Shumway, G., Dowling, G. B., Salsman, G., and Payne, R. H., 1961, Submerged forest of mid-Wisconsin age on the continental shelf off Panama City, Florida (Abstract): Geol. Soc. America, Ann. Meeting Program, p. 147.

Smith, D. D., 1961, Geomorphic and sedimentologic studies on the outer banks of North Carolina: 1st Nat. Coastal and Shallow Water Research Conference Proc., p. 459–461.

Thom, B. G., 1964, Origin of sand beach ridges: Australian Jour. Sci., v. 26, p. 351.

—— 1967, Coastal and fluvial landforms: Horry and Marion counties, South Carolina: Louisiana State Univ. Coastal Studies Inst. Tech Rept. no. 44, 75 p.

Manuscript Received by The Society May 13, 1968
Marine Institute of the University of Georgia Contribution No. 152

Editor's Comments on Paper 21

21 **Colquhoun, Pierce, and Schwartz:** *Field and Laboratory Observations on the Genesis of Barrier Islands*

A new aspect was introduced into the barrier island discussion by the presentation of the paper whose abstract is reproduced here. Colquhoun, Pierce, and Schwartz cite stratigraphic and experimental model evidence for the recognition of two types of barriers.

The two types are described as primary and secondary. The primary barrier forms at a higher shoreline, migrates landward during marine transgression, and is backed by lagoonal sediments overlying a former subaerial surface. During marine regression, this type progrades seaward only if the secondary barrier is absent. In comparison, the secondary barrier forms after the primary does during a stable, or slow fall of, sea level and overlies continental shelf sediments. What the authors are saying is that they see support for Hoyt's drowned ridge, behind which the lagoon overlies terrestrial sediment, as well as Fisher's spit, developing subsequently over the shelf bottom. Furthermore, an emergent bar *may,* under special conditions, develop in the secondary case during a slow fall in sea level, if a large "stable shelf" is present. Can there actually be two or three modes of barrier island genesis? The evidence proposed here would appear to support such a claim.

Donald John Colquhoun was born in Toronto, Canada, in 1932. He received a B.A. at the University of Toronto in 1953, an M.A. there in 1956, and a Ph.D. at the University of Illinois in 1960. Colquhoun became an assistant professor at the University of South Carolina in 1960, an associate professor in 1964, and a professor and Department of Geology Chairman in 1970. A Project Geologist for the South Carolina State Development Board since 1960, Colquhoun's interests include the application of sedimentological studies and techniques to geomorphology and stratigraphy, interpretation of coastal plain terraces, and regional stratigraphy.

Biographies of Pierce and Schwartz appear with other papers authored by them in this volume.

Reprinted from the *Geol. Soc. Amer. Ann. Meeting, Abstr.*, 59–60 (1968)

21

Field and Laboratory Observations on the Genesis of Barrier Islands

D. J. Colquhoun, J. W. Pierce, and M. L. Schwartz

Atlantic Coastal Plain Pleistocene sediments contain barrier island sand bodies from Delaware to Florida. Stratigraphically, two types of barriers are present. The primary type is formed at a higher shoreline, is bounded landward by back-barrier sediments overlying a former land surface, is often reworked and buried by encroaching deltas, and is stratigraphically thin and discontinuous in distribution. The secondary type is formed subsequent to the primary type, is underlain by continental-shelf sediments, is usually well expressed geomorphically and stratigraphically, and is thick and continuous in distribution.

Stratigraphic regional analysis indicates that the primary type forms during marine transgression and migrates landward with rise in sea level. The secondary type forms during sea-level stability or slow fall in sea level, after the primary type. Either type may prograde seaward with marine regression, but the primary type progrades only if the secondary type is not present.

Wave-tank experiments have duplicated the field observations for the development of the primary type with rise in water level. The secondary type has been shown to form subsequently, either with erosion of large headland areas during stability, or during slow regression if a large "stable shelf" has been allowed to develop. Either condition can be envisioned relative to the Atlantic Coastal Plain.

Editor's Comments on Paper 22

The parameters of present-day barrier islands as clues to ancient barriers, for the purpose of locating petroleum reserves, are the concern of this report, as they were in several of the earlier papers. Berryhill, Dickinson, and Holmes present a detailed compendium of the sedimentary characteristics of barrier islands. For specific factors, the reader can study the report itself. Essentially though, they see the alignment and geometry of the barrier islands, together with the relationship of the lagoonal muds to the barrier sands, as fundamental characteristics of the system. Secondary features are biota and inlet and washover stratigraphy.

The three authors of this most interesting paper are all presently employed by the U.S. Geological Survey.

Henry L. Berryhill, Jr., was born in 1921, and received his professional education in geology at the University of North Carolina: a B.S. in 1947 and an M.S. in 1949. He joined the Survey in 1947 and became a full-time geologist in 1948. From 1948 to 1965 Berryhill conducted geological investigations throughout the United States and in Puerto Rico. The rest of the 1960s was spent studying the eastern seaboard and Gulf Coast. Berryhill was transferred to Washington, D.C., where he became Deputy Assistant Chief Geologist, Marine Geology, in 1971. His current interests are ocean floor tectonics, sea floor sedimentation, and ocean mineral resources.

Kendall A. Dickinson was born in 1931. He received his B.A. in 1954, M.S. in 1959, and Ph.D. in 1962 from the University of Minnesota. In 1962 Dickinson joined the Survey, and until 1968 studied Upper Jurassic stratigraphy and sedimentology in Arkansas, Louisiana, and Texas. He is presently working on the sedimentology of the south Texas coast and the stratigraphic framework of the Gulf of Mexico continental shelf. Dickinson's research interests are coastal sedimentary processes and the resource potential of continental shelf stratigraphy.

Charles W. Holmes was born in 1937 in Detroit, Michigan. He attended St. Joseph's College in Indiana, where he received his B.S. in 1959, and Florida State

University, where he received his M.S. in 1962 and Ph.D. in 1965. Holmes was appointed to the faculty of Colgate University as an assistant professor in 1966. In 1967 he joined the Survey as a geologist and began a study of the geochemistry of the Gulf of Mexico. In 1969 he was party chief representing the survey in a cooperative study of the Gulf of Mexico with the U.S. Naval Oceanographic Office aboard the USNS Kane. Holmes is interested in recent sediments in shallow and deep marine environments, rates of sedimentation by radioelement disequilibrium, and geochemical characteristics of marine sediments.

Reprinted from the *Amer. Assn. Petroleum Geologists Bull.*, **53**, 706–707 (1969)

22

HENRY L. BERRYHILL, JR., KENDALL A. DICKINSON, and CHARLES W. HOLMES, U.S. Geol. Survey, Corpus Christi, Tex.

Criteria for Recognizing Ancient Barrier Coastlines[1]

Worldwide modern barrier coastlines constitute a minor part of the total coastlines of all the continents. The aggregate length of present barrier coastlines in the world is approximately 3,530 mi, distributed as follows: North America, 2,000 mi; Europe, 500 mi; South America, 350 mi; Africa, 300 mi; Australia, 200 mi; and Asia, 200 mi.

Barrier islands commonly border coastal plains adjacent to broad continental shelves. They form in areas of abundant sand accumulation where longshore currents are prominent. Sandstone lenses which represent ancient barrier islands would be expected in thick wedges of interfingering terrestrial and marine sandstone, siltstone, and mudstone. Barrier islands of Pleistocene age have been recognized inshore from present coastlines, and drowned Holocene barrier coastline features have been described on the continental shelves. Pre-Holocene linear sandstone bodies resembling barrier islands have been described in ancient rocks of Pennsylvanian, Cretaceous, and Tertiary ages.

Probable barrier island sandstone bodies in ancient rocks have been described by previous investigators on the basis of comparison with features of modern analogs: geometry, sedimentary structures within the sand lens, physical properties of the sand, and the nature of associated environments. Recognition criteria used in this report are based partly on previous work and partly on recent studies along the Texas and North Carolina coasts.

Barrier islands are linear, have a length to width ratio generally greater than 10:1 and commonly are less than 60 ft thick. Padre Island, Texas, consists of four morphological units that have characteristic sedimentary structures: beach, foredune, barrier flats, and wind tidal flats—though the development of the foredunes and wind tidal flats changes considerably from north to south. Along the North Carolina coast, wind tidal flats are absent, but accretionary beach ridges are locally prominent. Superimposed on the islands of both coasts are storm washovers of hurricane origin that breach the foredunes and channel inlets that cross the island and connect the sea with the lagoon behind the islands. Beaches contain laminae of different thicknesses that dip principally seaward; the sand is locally shelly and fine laminae of heavy minerals may be prominent. The foredunes are markedly cross-bedded in an oriented pattern that reflects strongly the predominant wind direction. Barrier flats are underlain by sand which ranges from structureless to highly laminated; vegetal remains are common. Wind tidal-flat sediments that border the lagoon are an interlayered mixture of sand beds containing some fine shell fragments, and laminae of clay and algal remains. Sand is fine grained throughout. However, shell fragments, locally abundant, exhibit greater variability in size, shape, and sorting. Sand which refills channel inlets ranges from horizontally bedded to structureless; this contrasts sharply with the cut-and-fill cross-bedded sand common in stream-channel deposits.

The associated lagoon sediments are organic and calcareous mud which interfingers with barrier-island sand; the fauna is less diverse than that of the open sea and unbroken shells are abundant. Tongues of sand —washover deltas and fans which are built by storm flood tides—are prominent local features of the lagoons. Marshes overlying peat are characteristic of the inshore side of the bays along the North Carolina coast.

The geometry and alignment of the barrier islands and the close association of the sand in the barrier island with the organic mud of the lagoon are the key factors for the recognition of a barrier coastline. Attendant washover deltas and fans, cross-cutting inlet fill, and associated biota are important supplementary aids.

[1] Publication authorized by the Director, U.S. Geol. Survey.

Editor's Comments on Paper 23

Once again, V. P. Zenkovich outlines the basic requirements for barrier island formation, this time with a few refinements. Quoting from page 35 of his paper: ". . . it may be established that any sea level oscillations are a favourable condition for barrier formation if the new surface, in the zone of wave action, is less inclined than that of the equilibrium profile for the given sediment coarseness." In the next sentence, a stable sea level is included as well. This of course, is the de Beaumont and Johnson emergent-submarine-bar concept, dependent on what Zenkovich referred to, in his earlier paper, as transverse debris drifting. As stated in that paper and this, submarine bars may build up through sea level in the offshore waters, or shore ridges (terraces) may form at the edge of the land. Further submergence of these ridges generates back-barrier lagoons, whose muds overlie former terrestrial surfaces, very much in the fashion outlined by Hoyt. In this, there may be some commonality between the two hypotheses.

It is interesting to note that in some cases of barrier island development cited (other than those supporting the above-mentioned types) Zenkovich provides for alternative modes of barrier origin. He describes the western Kamchatka peninsula barrier as the result of submergence of a series of ancient shore ridges. Initial development as barrier spits through longshore drift, with a large additional component provided by transverse debris drifting, is cited as the origin of the Sakhalin Island barrier system. While the Hoyt approach finds support in the former, there is a strong resemblance to the ideas of Gilbert and Fisher in the latter. There appear to be then, three ways in which barrier islands may be generated: bar emergence, spit breaching, and partial submergence of coastal ridges.

Reprinted from *Lagunas Costeras, Un Simposio, Mem. Simp. Intern. Lagunas Costeras,* UNAM–UNESCO, 27–37 (1969)

23

ORIGIN OF BARRIER BEACHES AND LAGOON COAST

V. P. Zenkovitch *

ABSTRACT

The report contains some information on the investigations carried out in the Soviet Union. Abundant sediment delivery from the sea bottom shoreward has been proved by the marine origin of the material (broken shells, oolithes) and by some geomorphological observations.

Prolonged subsidence of the coast evokes continual barrier beach formation at different levels. The lower ones may be submerged. As a whole they look like a staircase. It was evidenced by coring in the western Black Sea.

During relatively rapid subsidence beach barriers may consist of terrigenous material washed out and reworked at the margin of flat plains. Examples are known from the Chukchi peninsula, Kamchatka and other areas. In any case the longshore drift may complicate the structure of barrier beaches.

It is not clear until now barriers emerge from beneath the water, though new islands form in the regions of rapid shallow water sedimentation. It is suggested that the world-wide barrier formation has been caused either by the Holocene transgression as a whole or by Flandrian high sea level stand.

The sea level oscillations produce a series of barriers both in space and in the vertical section.

From the geologist's point of view any aquatory may be considered as a lagoon, if it has a limited or no connection with the sea. The usual salinity regimen is disturbed in such basins and this is a reason for the development of specific faunal complexes living in them. As the lagoonal waves are small and not active, the bottom is generally covered by a thick layer of fine mud.

In the geomorphologist's comprehension the lagoons are those shallow elongated bodies of water only, which are separated from the sea with a sand or shingle barrier and oriented along the general shore trend. There may be or may not be inlets through the barrier. The bays separated from the sea by the barrier spit, strictly speaking, should not be called lagoons because their mode of origin is quite different from that of true lagoons.

In the geological past the formation of many useful ores took place in lagoons. This is one of the main reasons for the great interest in the study of lagoons which recently arose.

The classic concept of geomorphologists (Davis, 1912, and Johnson 1919) is that lagoons are a common feature of shorelines of emergence. The formerly proposed theory of a lagoon's origin is well-known.

However, later on, it was established that the lagoons are not less frequently developed along the shores of submergence. This controversy stimulated several studies by the scientists of different countries which were conducted to confirm or modify the old

* Institute of Oceanology, USSR Academy of Sciences, Moscow, USSR.

theory. Investigations of this kind have been intensively conducted by Soviet scientists too. Nevertheless, the results obtained are almost unknown in the western side of the Atlantic Ocean. The main task of this report, therefore, is to explain the principal theoretical concepts obtained and the methods of reasoning in Soviet investigations devoted to the origin of lagoons.

First of all, there are many convincing examples showing that the sediment transported to the shore from the sea bottom is affected by an active process, which has being acting since previous sea level changes, as well as, in some cases, during long periods of vertical shore stability. For instance the broken and whole molluscs shells, which are certainly a thalassogenic product, are being transported shoreward and form a series of barriers along the Azov sea shore yearly. The calculations based on the biological data indicated that the productivity of *Cardium edule* is about 400-800 tons per 1 km shore-length. It is enough to build a shore ridge 100 km long and of 2 x 3 m in cross section (Samoilov, 1952).

The same process is going on at the Caspian sea shores. The admixture of recent ooliths exists along its eastern side. The Caspian sea level fell down to 2, 1 m during 1933-1941 owing to climatic reasons. This stimulated mighty aggrading of the beaches along all the shores and, at the same time, the rock benches were exposed on the bottom at the places previously covered by sand (Leontiev, 1961, 1965).

A row of shallow banks existed along the border between the deep middle part of Caspian (down to 700 m) and its northern shallow part (less than 5 m). These banks grew up by 3-4 m after the lowering of level and some of them turned out to be the flat ring-shaped islands (Leontiev, 1957). Around the Volga river avandelta arose a semicircular submarine flat bar with a relief of about 1 m.

Investigations carried out in the Caspian Sea indicated that the sediment shifting onshore is intensive when the bottom inclination is in the limits of tangent 0.01-0.001. Where the slope is less than this the wave influence on the bottom is perceptible farther seaward, beyond the limits of possible observations during storms. There may be also such a smooth bottom profile that the wave energy expends gradually without disturbing the sediments and reworking the bottom relief.

In some regions there are offshore barriers composed of boulders and pebbles. One of them, the isolated barrier-island Mehechkyn (figure 1), has been investigated at Anadyr Gulf in the Bering Sea (Zenkovitch, 1967). It was concluded that the source of stone material is on the sea bottom. The morainic cover is exposed over a large area (to a depths of about 30 m) and the bottom is subjected to washing out processes (Shcherbakow, 1961a.) The slope inclination here is more steep than in the Caspian Sea.

Experiments in the wave tank were undertaken by Leontiev and Nikiforov (1966). The authors lifted gradually a block with a flat sand covered surface up to the water level, imitating the growth of any tectonic structure. The waves washed out the margins of the structure and created a ringshaped barrier. This result may be considered as a supplementary evidence of the validity of the classic theory.

More frequently, however, the off-shore barriers are being formed either during or just after the land submergence and in this case their origin may be quite unlike the one described. Some interesting formations are particularly on the Black sea shores, which according to morphological and geodetical data, are everywhere subjected to recent submergence (Zhyvago, 1958). Many submarine relics of ancient offshore barriers have been discovered at a depth of 50 m (Nevessky, 1961; Zenkovitch, 1960,

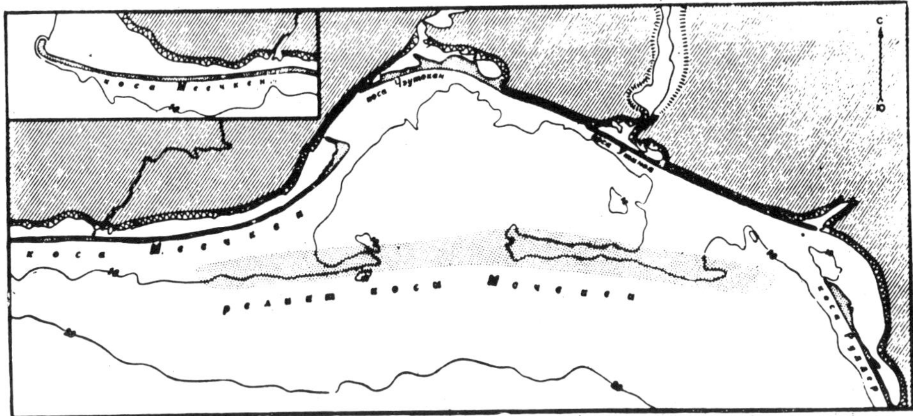

Figure 1. The plan of the Mehechkyn off-shore barrier.

1967). Some barriers appear as smooth banks on the bottom, but the majority of them are buried under the recent mud layer and were revealed using the vibro-piston corer (figure 2).

The lithological analyses proved that the top part of these submarine banks is composed of beach sediments of broken shells. Lagoonal muds have been discovered in the rear side of several barriers (figure 3).

Longitudinal geological sections of some gulfs such as Kalamitys, Karkinitys and Anapa (figure 4) look like a series of relict submarine barriers. They are similar to the land formations known from eastern side of the Caspian (Leontiev, 1961), the Atlantic coastal plain of the United States of America (Price, 1956) or those west of the Nile Delta (Shukri and Philip, 1956).

In other localities, however, it was observed that the barriers are not necessarily built with bottom materials. Some interesting phenomena have been discovered in the submerging Chukchi Peninsula and western Crimea coasts. In the first region a shingle offshore barrier extends for about 400 Km (Kaplin, 1957, 1964). The landward side of the lagoon borders the wide, gently sloping plain which is composed of fluvio glacial sediments enriched with broken stones.

The comparison of barrier shingle complexes and those of broken stones from the land proved their petrographical identity. Drillings of the lagoon bottom showed the peat layer on top of the permafrost fluvioglacial strata; no marine sediments were found in between. It means that during the Holocene transgression the waves washed out the land sediments of the plain's edge. As a result, the coarse fragments (shingle) have been thrown on the shore and fine particles (clay) carried away to the deep sea. So the lagoon aquatorium never has been a part of the sea and it developed as consequence of the submergence of the plain's edge below the sea level. This idea is supported by geomorphological evidences as the inner lagoon coast belongs to the liman type.

An almost similar sequence of strata was described in the lagoonal beds at Crimea peninsula, westward from Euphpatoria. The sequence from the top down is: *a*) salt, *b*) deluvium with broken stones of solid rocks and *c*) horizontally bedded Tertiary limestones. No traces of former marine sediments have been found in pits. The barrier, however, is composed mainly of shell fragments. The same material covers the sea bottom in front of the barrier for a distance of about 100 m. Farther seaward,

Figure 2. The vibro-piston corer lowering from board the ship.

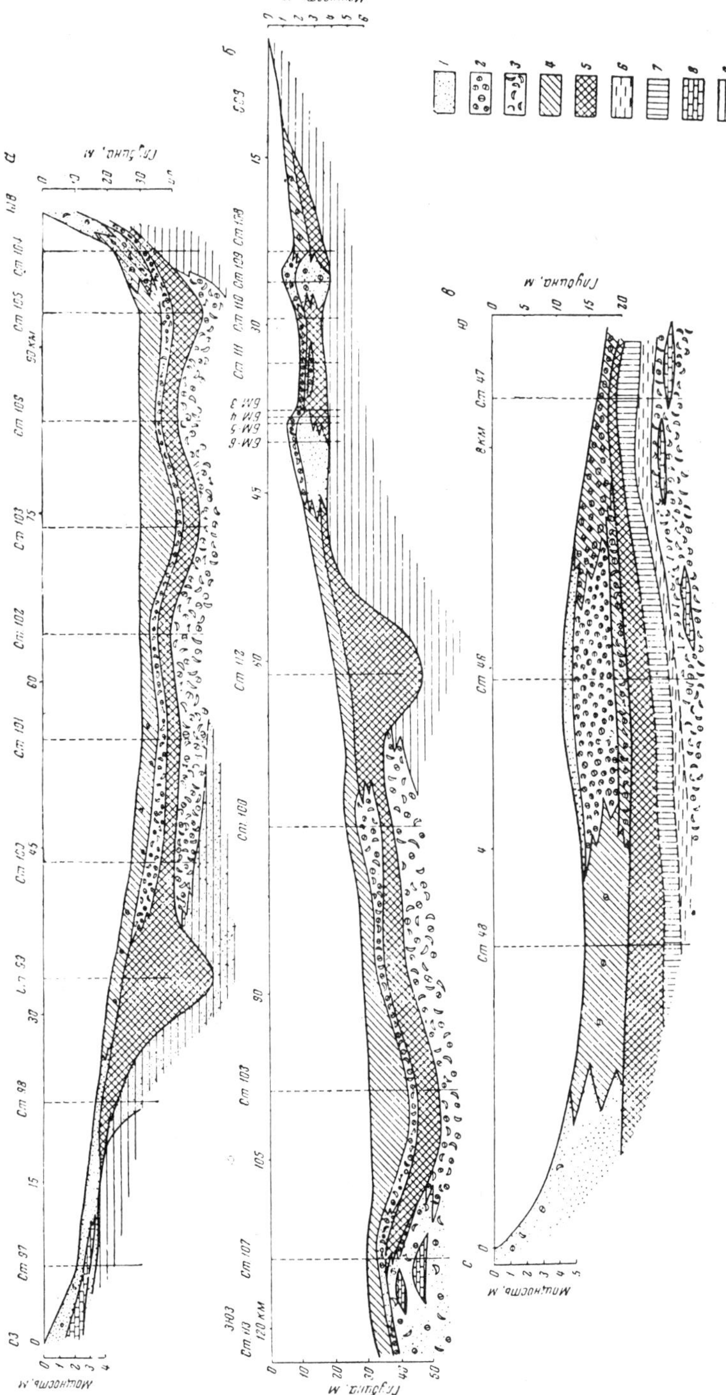

Figure 3. Bottom sections of some Black-sea gulfs (after E. Nevessky) 1. Sand; 2 and 3. Shells and shell sand; 4. Sea mud; 5. Lagoon mud; 6. other lagoonal sediments; 7. Clayley mud; 8. Lithyfised sediments; 9. Clay of terrestrial origin.

the bottom is bare for a distance of 6 km. At a depth of 20 m there is a layer of mud with shells of the same species as those present at the beach. In this case also the lagoons never have been a part of the sea aquatorium although the barriers consist of bottom material. The bottom sediment is being transported to the shore and the total sediment mass moves landward as the plain submerges (Zenkovitch, 1960).

The lithological composition of western Kamchatka peninsula barrier sediments does not permit a determination of their origin. But along the shore in some places it is observed that there exists a gradual transition between the lagoonal coast and the coast is bordered by a wide terrace containing a large series of ancient shore-ridges (figure 5). Recent submergence of the coast is proved by the peat layers revealed in borings 15 m below sealevel (Vladimirov, 1958, 1959; Zenkovitch, 1950).

The elevation of shore-ridges decreases landward and finally the landward side of the terrace is submerged under the water. If the rest of terrace which is not drowned is relatively wide (2-3 km), the coast strictly speaking does not correspond to a lagoon type. During the coastal retreat, however, some parts of the primary terrace were subjected to erosion. Their material has been reworked by the sea and has remained as a narrow barrier of usual appearance. The inner side of the lagoon is bordered by the "Tundra" flat composed of the fluvial-glacial sediments. There are no doubts concerning the origin of such a lagoon as it is proved by the shore-type transitions mentioned above.

The shape of the shore barriers in most cases is influenced by the longshore sediment drift. As a result of the repeated shift in opposite directions the barrier ends grow wider than elsewhere and are hooked landward. The same process is going on at the inlets. The sediments transported to them make the barrier wider and build the tidal deltas.

The waves of the lagoon itself may complicate the inner barrier side creating a series of either cuspate forelands or asymmetrical spits. Under favourable conditions there is a possibility that the lagoon-aquatorium can be divided forming a chain of rounded lakes (Zenkovitch, 1959).

The bays of liman type usually are being formed during the sea transgression into the lowlands dissected by smooth valleys. In young stages the sediments are mainly of abrasion origin and tend to fence off the bays from the sea by spits. The spits being fed by longshore drifting are bent landward and their outlines are irregular, with some preserved hooks evidencing the stages of growth. But at the same time there exists a supply of sediments from the bottom. The total sediment mass gradually increases and moves following the retreating land block.

If there are large scale sediment migrations or the so-called sediment-streams (net transport) in the region under consideration, the coast may undergo some peculiar transformations and finaly be converted into a lagoonal coast.

The best example of such a succession is known at the eastern side of Sakhalin Island (Vladimirov, 1961). The land here is composed of loose Quaternary and Pliocene sediments. Originally the coast was intensively embayed. After a period of erosion and submergence a smooth outline resulted with intermittent cliffed parts and long spits (figure 6). During this evolution the bulk of sediments was increasing due to sea bottom transport. There were several great sediment migrations, as a result of which some broad accumulative terraces have been created. The progressive submergence formed a lagoon at the rear part of the terrace. The tidal waters flowed in through the lows between the ancient shore-ridges.

The way by which the submarine bank

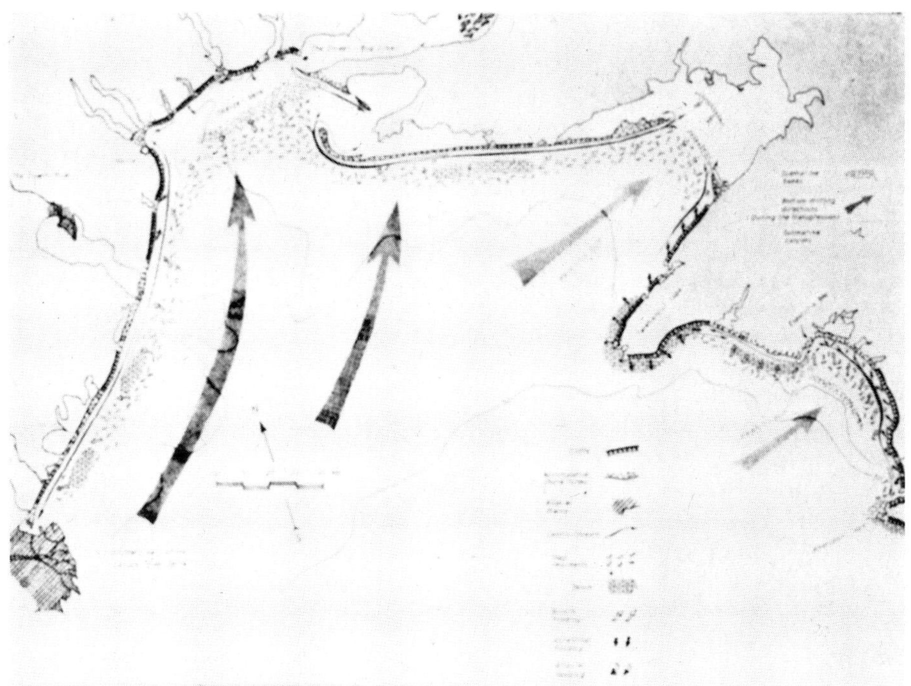

Figure 4. Dynamical scheme of the bottom and shore of North-western Black sea part. 1. Cliffs; 2. Aggradational forms; 3. Deltas; 4. Sediment streams; 5. Shell sediments; 6. Bench; 7. Material delivering from the bottom; 8. Abrasion sediment feeding; 9. Alluvial feeding.

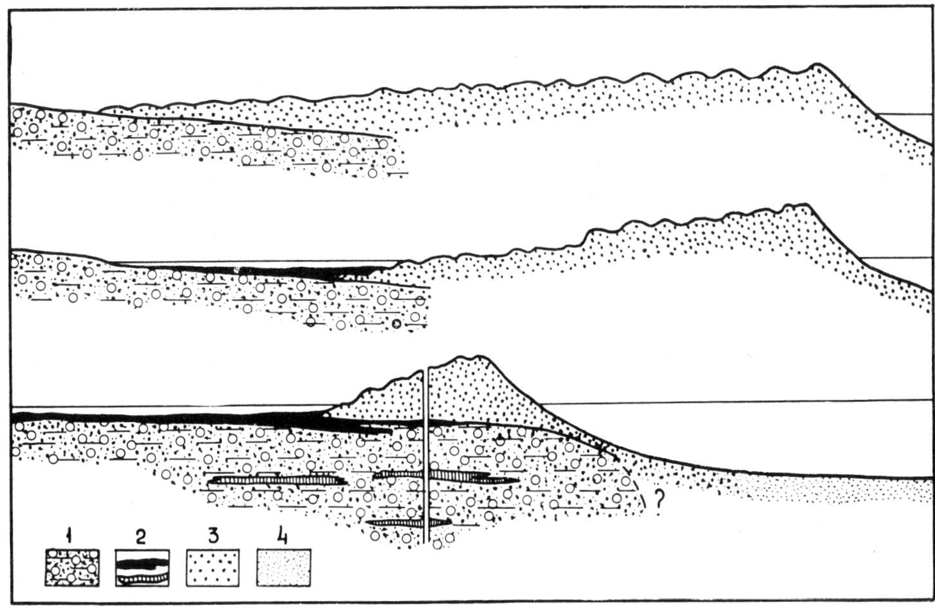

Figure 5. Profiles scheme of western Kamchatka coasts. 1. Fluvial-glacial sediments; 2. Lagoonal mud (above) and peat-beds; 3. Coarse beach sediments. 4. Marine silt.

becomes an exposed barrier is not clear until now. The small structures, as the submarine bars, have been observed many times to rise above the water during the spring ebbs or other temporary level lowering. They remain as relatively stable structures but only for a short time. On the other hand, the typical shore barriers are known to have been formed recently, for example Chandeleur islands near the Mississippi Delta, as well as very long Sakhalin Island in front of the Saint George outflow of the Danube Delta (Petrescu, 1957). Both regions are, however, characterized by abundant sand transported into the submarine slope and high molluskan productivity.

Usually there is no delivery of coarse sediment to the nearshore parts of the sea at the present time. Therefore, based on the widespread distribution of barriers all over the world (about 13% of the total shore length), some Soviet scientists (Leontiev and Nikiforov, 1965) consider that barrier formation, in general, was stimulated by the so-called Flandrian eustatic rise of ocean level about 3-4 thousand years before present (Fairbridge, 1961).

The author does not agree with this explanation as the only possibility. In many lagoonal regions there is no evidence of Flandrian rise of sea level higher that present. The submarine bars of different depth and width are nevertheless a common feature on almost every shallow part of the sea, sometimes far from the land. The development of small barriers has been observed many times above the sea level. But in favourable conditions such a feature becomes a center or "nucleus" of accumulation for new sediments. It may increase in dimensions by wind action; the sand is blown over the ridge and deposited beyond the new barrier.

If the sand transport from the bottom is fairly intensive as during the Holocene transgression, new bars would be added to

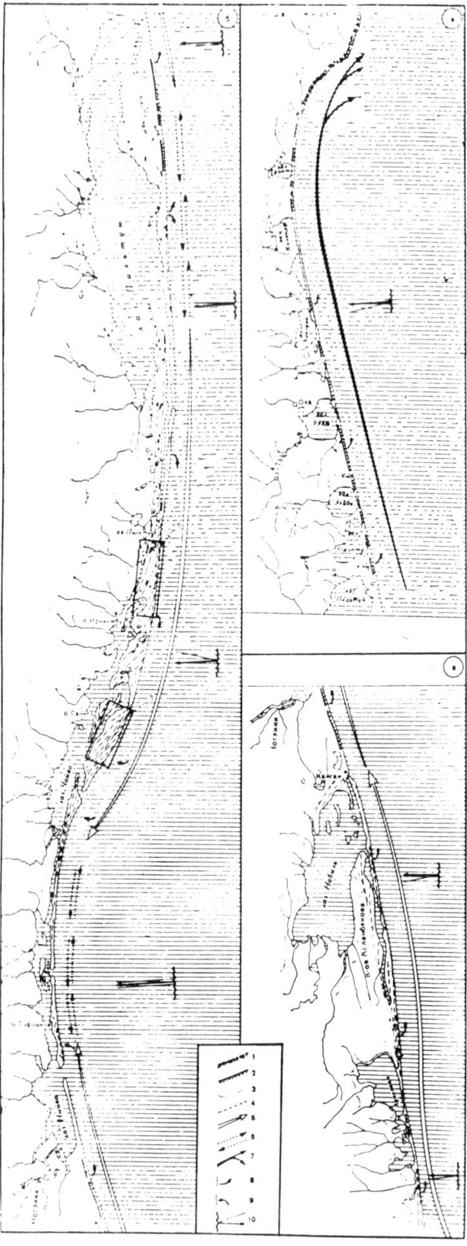

Figure 6. Dynamical scheme of the eastern Sakhalin island coast (after A. Vladimirov) 1-3 Cliffs; 4. Shore-ridges; 5-6. Sediment streams and migrations; 7-9. Feeding and loss of sediments; 10. The wave regime resultant and normal direction to the shoreline.

the first ridge fixed above the water. In the process of reworking the total sand mass will be shifted coastwards and the barrier may be increased in size.

A somewhat similar process may be acting on the sea bottom, as was pointed out above, at the boundary between the northern and the middle parts of Caspian Sea. This process may be occurring also at Ogurchinsky Island in the same sea. There is a submarine bank about 30 Km long, a continuation of the island's Southern end (Nikiforov, 1964). Recent soundings show that the bank grew up during the last decade higher than the former sea level lowering. As a consequence of this active process it is possible that new portions of the bank will reach the present sea level as a prolongation of the Ogurchinsky Island.

Based on the above data and on the general laws of waves transformations (Longuinov, 1963), it may be established that any sea level oscillations are a favourable condition for barrier formation if the new surface, in the zone of wave action, is less inclined than that of the equilibrium profile for the given sediment coarseness. The barrier formation of bottom material is occurring during sea-level oscillation, as well as perhaps a fairly long time after its stabilization. There are some places along USSR coasts where this process does not cease after the end of the Holocene transgression during the last 5000 years. The same process may be supposed to exist along the Pacific shores of North America (Cooper, 1958) and South America (Segerstrom, 1962). The recent rise of sea level, about 12 cm per 100 years, may add to the activity of the process.

There are some indications that if a sufficiently large barrier was formed on a gently inclined plain, it becomes a center of sediment accumulation during succesive sea

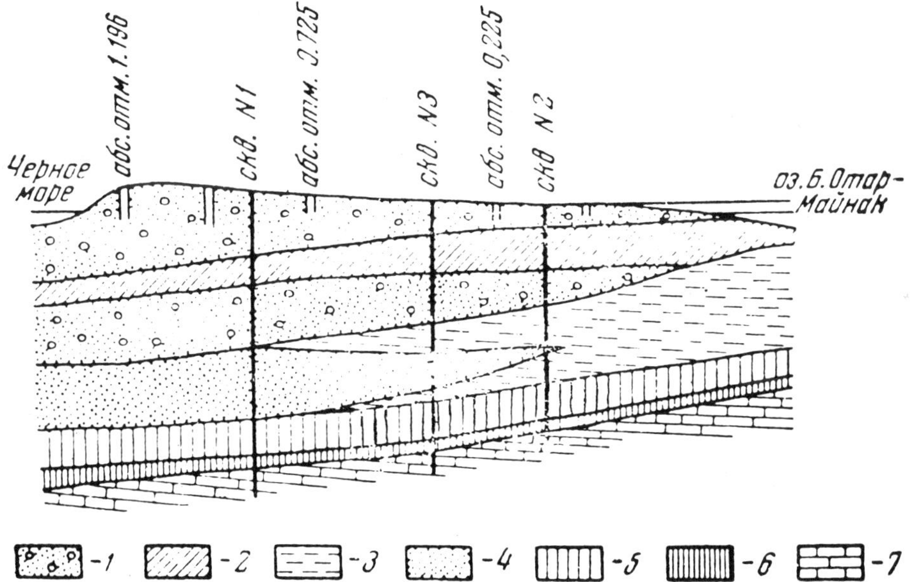

Figure 7. The inner structure of a barrier at west Crimea coast (after A. Dzens-Litovsky) 1. Beach sediments; 2-3. Different kinds of lagoonal mud; 4. Marine sand; 5. Deluvium; 6. Terrestrial clay; 7. Limestones.

level oscillations. When the new shore-line gets near the former barrier, a second one is formed, and so on. This statement is of great importance for geologists and it may be supported by many convincing examples. At western Crimea some composite barriers have been revealed by drillings (figure 7). Coarse sand barriers alternate with lagoonal mud and fine-grained marine silts (Dzens-Lytovsky, 1938).

Even more interesting results were obtained from the Arabat barrier (the Azov Sea) investigations. Drillings show the ancient barrier just beneath the recent one at a depth of 15-23 m. O. and V. Leontiev (1956) considered that this barrier is the same which was mentioned in the Strabo writings about 2000 years ago (Strabo, 1853). The calculated rate of submergence is therefore about 7-8 mm per year and this value nearly coincides with recent measurements (6 mm per year). Another barrier has been found under recent muddy sediments at a depth of 8 m in the open sea just in front of Arabat barrier (Shcherbakov, 1961b).

A barrier series of different age has been formed at the head of the Riga Gulf (Baltic Sea) since the ancient Ice-lake existed until recently (Ulst, 1957) when sea level sank and rose several times.

Similar examples of repeated barrier formation (Pleistocene and Holocene) are also known in the Gippslake region of Australia (Bird, 1965) as well as in the Western Gulf of Mexico (Bernard and Le Blanc, 1965).

If this interpretation of shore development (Leontiev *et al*, 1960) is confirmed by further investigations it may be a useful tool for geologists in searching for ores of lagoonal origin.

LITERATURE CITED

BERNARD, H.A. and LE BLANC, J. 1965. Résumé of the Quaternary Geology of the NW Gulf of Mexico Province. The Quatern. USA (to the VII INQUA Congr.).

BIRD, E.C.F. 1965. Geomorphological study of the Gippsland lakes. Australian Nat. Univ. Canberra.

COOPER, W. S. 1958. "Coastal sand-dunes of Oregon and Washington." *Geol. Soc. America, Mem.* 72.

DAVIS, W. M. 1912. *Die Erklärende Beschreibung der Landformen.* Leipzig.

DZENS-LYTOVSKY, A. I. 1936. "Geology of the Crimean salt lakes." *Acad. Sci.* USSR.

FAIRBRIDGE, R. W. 1961. Eustatic changes in sea level. Physics and Chemistry of the Earth, 4.

JOHNSON, D. W. 1919. *Shore Processes and Shoreline Development.* New York.

KAPLIN, P. A. 1957. "On the peculiar lagoon features of the NW coast of the USSR." *Trans. Oceanogr. Commn. Acad. Sci. USSR.* 2.

———. 1964. "Some regularities of lagoon development." *Oceanol.* (2.).

LEONTIEV, O. K. 1957. "On the origin of some islands of north Caspian Sea." *Trans. Oceanogr. Commn. Acad. Sci. USSR.* 2.

———. 1961. *Basis of sea shore geomorphology.* Moscow. Univ. Press, 417 p.

———. 1965. "On the geomorphological regions of the Caspian Sea coast." *Trans. Inst. Oceanol. Acad. Sci. USSR.* 76.

LEONTIEV, O. K. and LEONTIEV, V. K. 1956. *Trans. Acad. Sci., USSR Geogr. Ser.* (2)

LEONTIEV, O. K., MYAKOIN, V. S. and NIKIFOROV, L. G. 1960. "Inheritance of ancient shore processes on eastern Caspian coast for Quaternary period." *Trans. Compl. South. Geol. Exp. Acad. Sci. USSR.* (5.).

LEONTIEV, O. K. and NIKIFOROV, L. G. 1965. "On the reasons of world wide spread of shore barriers." *Oceanol.* (4.).

———. 1966. "Model investigations into the formation of shore barriers during the stage of regression." In: *Sea shores development under the earth crust oscillations.* Inst. Geol. Acad. Sci. Est. USSR.

LONGUINOV, V. V. 1963. "Dynamics of the

nontidal sea shore zone." *Publ. Acad. Sci. USSR*: 1-379.

NEVESSKY, E. N. 1961. "Some data on the post-glacial evolution of the Karkinitsky Gulf and the accumulation of bottom debris." *Trans. Inst. Oceanol. Acad. Sci. USSR*. 48.

NIKIFOROV, L. G. 1964. "Contributions to the shore barriers development." *Oceanol.* (4.).

PETRESCU, I. G. 1957. *Delta Dunarii.* Stiintifica (Ed), Bucuresti, 234 p.

PRICE, W. A. 1956. "Environment and history of identification of shoreline types." In: *Quaternaria*, 3.

SEGERSTROM, K. 1962. "Deflated marine terrace as a source of dune chains Atacama province, Chile." *United States. Geol. Surv. Prof. Papers*, N 450-C.

SAMOILOV, I. V. 1952. "The river mouths." *Geogr. Publ. House*, Moscow 1-526.

SHCHERBAKOW, PH. A. 1961a. "Some data on the post-glacial transgression at the Bering Sea." *Trans. Inst. Oceanol. Acad. Sci. USSR*. 48.

———. 1961b. "Contributions on the evolution of the northern and western coasts of the Azov Sea." *Trans. Oceanogr. Commn. Acad. Sci USSR*. (12.)

SHUKRI, N. M. and PHILIP, G. 1956. "The Geology of the Mediterranean coast between Rosetta and Bardia." *Bull. Inst. Egypte* 37 (2.).

STRABO 1853. Geographica Graece. Paris.

ULST, V. H. 1957. "The morphology and development history of the marine accumulations at the head of Riga Gulf." *Acad. Sci. Latvian SSR*: 1-178.

VLADIMIROV, A. T. 1958. "On the morphology and dynamics of the western Kamchatka coast." *Acad. Sci. USSR. Proc.; Geogr. Ser.* (2.).

———. 1959. "The Quaternary evolution of the western Kamchatka coast." *Trans. Oceanogr. Commn. Acad. Sci. USSR*. 4.

ZENKOVITCH, V. P. 1950. "On the lagoons formation." *Rep. (Doklady) Acad. Sci. USSR*, 75 (4.)

———. 1959. "On the genesis of the cuspate spits along lagoon shores." *J. Geol.* 67 (3.).

———. 1960. "Morphology and dynamics of the Black Sea soviet coast." *Ed. Acad. Sci. USSR*: 2:1-215.

———. 1967. *Processes of coastal development.* Oliver & Boyd, London.

ZHYVAGO, A.V. 1958. "Recent vertical movements of the USSR seas coasts." *Trans. Inst. Geodesy, Airphotosurv. Cartogr.* (128.)

Editor's Comments on Paper 24

24 Pierce: *Sediment Budget Along a Barrier Island Chain*

Jack Warren Pierce, coauthor of the earlier paper with Colquhoun and Schwartz, was born in 1927 in Springfield, Illinois. He was awarded the B.S. in 1949 and M.S. in 1950 at the University of Illinois and the Ph.D. at Kansas University in 1964. Pierce joined the Pure Oil Company in 1950 as a geologist and was promoted to district geologist in 1956. In 1963 he became an associate professor of geology at George Washington University, Washington, D.C. Since 1965 an adjunct professor at the university and Curator of Sedimentology at the Smithsonian Institution, Pierce's research interests are in sedimentology, stratigraphy, and marine geology.

Recently, Pierce calculated the net accretion along the barrier system between Hatteras Inlet and Cape Lookout to be 796,000 m^3 annually. A little less than half of this can be accounted for by longshore drift and biogenous sources. Where, then, does the balance come from? Pierce suggests relict sands or poorly consolidated sediments on the continental shelf, both moved shoreward initially by extremely long surface waves. As a contribution toward the building up of barrier islands, Pierce's view closely parallels that of Zenkovich's transverse debris drifting. If the offshore bottom is not the only source of barrier sediment, it appears that it may be at least an important factor.

Reprinted from *Sed. Geol.*, **3**, 5–16 (1969)

SEDIMENT BUDGET ALONG A BARRIER ISLAND CHAIN

24

J. W. PIERCE

Smithsonian Institution and *George Washington University, Washington, D.C. (U.S.A.)*

(Received April 22, 1968)
(Resubmitted September 4, 1968)

SUMMARY

The sediment budget for a stretch of coast along a barrier chain is calculated through use of historical records, covering a time span of nearly 100 years, mapped short-term changes, and estimates of volume changes caused by physical processes.

For the section studied along the southeastern United States, accretion has exceeded erosion. The sediment deficit requires an input at an average rate of 796,000 m^3 (1,041,000 cubic yards) annually. This deficit is partially filled by longshore drift from adjoining sections and by biogenous contributions. Currents are insufficient to carry material from the mainland across the lagoons.

Longshore drift and biogenous source cannot account for approximately 337,000 m^3 (441,000 cubic yards) of sediment annually. It is postulated that this material is being moved in from a reservoir on the continental shelf. This reservoir is either the unconsolidated relict sediments or outcrops of poorly consolidated Tertiary rocks which are only thinly veneered by Holocene sediments.

INTRODUCTION

Sediment budgets have been calculated for years along many coasts in support of engineering projects. Seldom do these types of studies appear in the geological literature with a sedimentologic, rather than an engineering, purpose.

Quantitative studies, such as these, are best accomplished along a stretch of the coast where the processes within a limited section are nearly independent of the processes in adjacent sections. Unfortunately, coasts with barrier islands seldom have stretches where the source areas, areas of loss, and processes can be isolated from those of the adjoining stretches.

This paper gives the sediment budget, both inputs and losses, to such a shoreline, using a portion of the southeastern United States as an example. This coast is considered typical of similar shorelines under like climatic conditions.

The section of the shoreline used lies between Hatteras Inlet, N.C., 35°10′N 75°40′W and Cape Lookout, 24°35′N 76°32′W, a distance of about 100 km (Fig.1). This stretch of coast lies between two giant cusps, Cape Hatteras and Cape

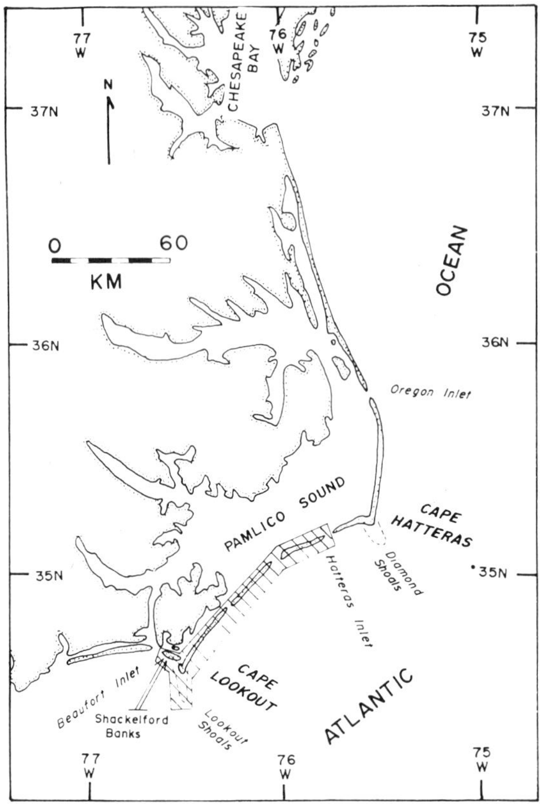

Fig.1. Map of part of the coast of southeastern United States with area of barrier island study cross-hatched.

Lookout, which in part isolate the section from adjoining stretches (LUTERNAUER and PILKEY, 1967). Data from adjoining areas were used to calculate the sediment input into the area.

METHODS OF STUDY

Quantitative changes in the barrier islands of this area have been measured by the U.S. ARMY CORPS OF ENGINEERS (1964). Their estimates of the volume of sediment eroded from and added to the shoreline have been used unchanged. For the coast north and south of the study area, volume estimates were obtained from mapped areal changes (U.S. CONGRESS, 1948).

Smooth sheets of U.S. COAST AND GEODETIC SURVEY (1866,1955) bottom surveys were used for measuring the changes in Lookout Shoals (Fig.1). Areas, enclosed by sea level, the 1.8-, 5.5-, 9.1-, and 12.8-m (1-, 3-, 5-, 7-fathom) isobaths, were determined by planimeter on the smooth sheets of these surveys (1866, scale 1:40,000; 1955, scale 1:20,000) and areas converted to volumes.

Results of an experimental dune-building project was used to estimate the loss of sand from the beach-longshore system by wind transport. Sand fences were installed, essentially parallel to the shoreline, in October and November 1960. Accumulation of sand was measured the following December, April, and June (SAVAGE, 1963).

Loss of sediment through tidal inlets was estimated for this area by using Corps of Engineers estimates for the New Jersey coast (CALDWELL, 1966). Although the New Jersey coast is somewhat removed from the study area, the wave and current regimen is similar for the two areas.

Detailed topographic surveys, over a short time span, were used to compare the amount of material eroded from the shore against the sediment build-up on the backshore. It was assumed that the sediment added to the backshore potentially could be washed over the barrier into the lagoon. This amount was used as a minimum estimate of the amount of sediment washed into the lagoon by storm waves.

QUANTITATIVE CHANGES

Barrier islands

Erosion is in no way dominant along the barrier islands on this stretch of coast. Of 165 equally-spaced sections, nearly one half indicated accretion and one half erosion (U.S. ARMY CORPS OF ENGINEERS, 1964). In gross overall aspect, the barrier islands have not been reduced in area to any great extent, although erosion has been extreme in places (Fig.2).

Volumetric calculations for the barrier islands are based on the empirical relationship that a gain (or loss) of 0.09 m^2 (1 sq. ft.) of area along the shoreline is equivalent to a volumetric gain (or loss) of 0.76 m^3 (1 cubic yard) of sediment (U.S. ARMY COASTAL ENGINEERING RESEARCH CENTER, 1966, p.216). This empirical relationship will give the order of magnitude of the volumetric change.

The long-term trend is one of accretion for the barrier islands. An integration of the changes that have occurred shows that 70,000 m^3 (91,000 cubic yards) have been added to the barrier complex between Hatteras Inlet and Cape Lookout.

The primary area of accretion has been on Cape Lookout at the south end of the study area. The total land area of Cape Lookout increased from 2.1 km^2 (530 acres) in 1866 to 4.4 km^2 (1,092 acres) in 1955. The changes are striking also in the orientation of the cape as shown on Fig.3 between 1864 and 1957. It should be noted that the area of accretion is on the east side of the cape while erosion has occurred on the west. This is opposite to the evolutionary trend occurring at Cape Hatteras (U.S. CONGRESS, 1948; RICHARDS, 1950; ATHEARN and RONNE, 1963).

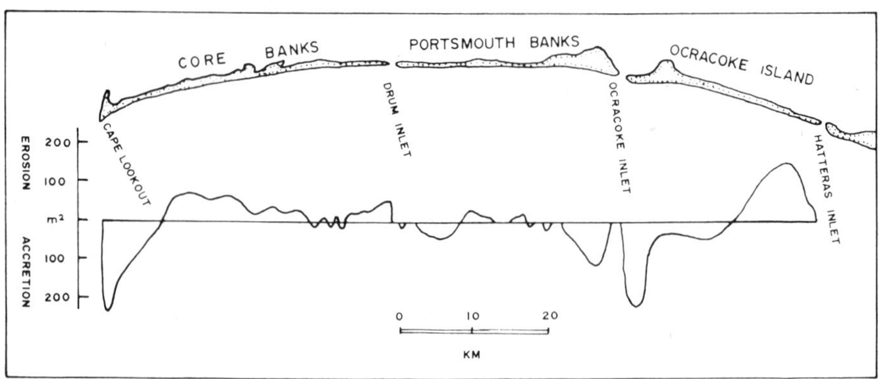

Fig.2. Schematic diagram showing long-term areal changes along barrier islands. Areas and amount of erosion or accretion are indicated by direction and amount of deflection from line of no change.

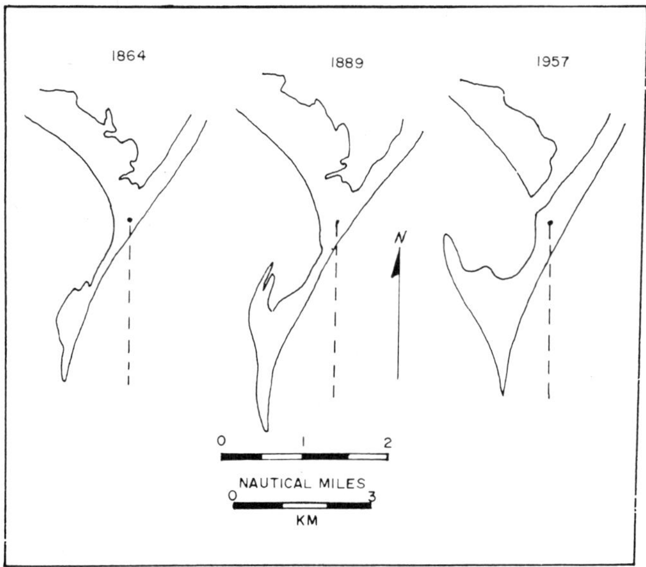

Fig.3. Changes in Cape Lookout, 1864–1957, as shown on U.S. Coast and Geodetic Survey charts.

Lookout Shoals

Lookout Shoals are a submarine continuation of the subaerial portion of the Outer Banks and Cape Lookout (Fig.1). The first major interruption in the 12.8-m (7-fathom) isobath is a deep channel approximately 22 km south of the tip of the cape, although the shoals continue to the south, at increasing depths, for an additional distance of 19 km.

Between 1866 and 1955, the volume of sediment making up Lookout Shoals increased by 17.6 million m^3 (23.02 million cubic yards) or at an annual rate of 198,000 m^3 (259,000 cubic yards).

It is believed that this value is a minimum estimate. Volumes were not determined south of the channel nor deeper than 12.8 m because of the lack of surveys. What limited survey data is available shows that sediments have been accreting outside of the measured area.

Loss through inlets

Some portion of the sediment involved in longshore drift will be lost through tidal inlets into the lagoon. A true quantitative estimate of the loss in this area is hard to evaluate because of lack of accurate periodic surveys in the lagoon although the size of the tidal deltas associated with the inlets imply a considerable mass of sediment.

Estimates of material loss through the inlets can be made by using the New Jersey coast as a model. Although about 650 km north of the study area, the New Jersey coast is exposed to nearly the same physical conditions as in the North Carolina coast. For the New Jersey coast, CALDWELL (1966) has estimated that 191,000 m^3 can be lost annually through each permanent inlet.

Three permanent inlets presently exist in the study area: Hatteras, Ocracoke, and Drum (Fig.1, 2). Other large inlets have existed in the past, some of which have been artificially closed. Until the advent of shoreline stabilization and dune building, many smaller inlets were also present.

Therefore, over the long term, it is estimated that 382,000 m^3 (500,000 cubic yards) is lost to the system through the inlets.

Loss across barriers

Sediment can be lost from beach-longshore system by transport across the barrier by overwash or by wind. Either effectively removes material from the system.

Overwash. Water washing over the barrier during storms will carry beach sediment into the lagoon. The amount of material lost by this process is difficult to evaluate because of lack of accurate periodic surveys. The size of the overwash fans suggests a significant amount of sediment could be lost in this manner.

A minimal estimate of this amount can be made by considering changes in barrier elevation over a short period of time. An increase in elevation of the backshore area represents loose sediment potentially available to being washed over the barrier into the lagoon.

Topographic surveys in 1960 and 1961 along the barrier show that everywhere erosion occurred there was a concurrent increase in the elevation of the backshore. Where accretion had occurred, no such change was present. The volume involved in the increased elevation is approximately 13% of the shoreline loss.

Those parts of the coast, between Hatteras Inlet and Cape Lookout undergoing erosion, had an average annual loss of 576,000 m^3 (753,000 cubic yards). If 13% of this material was used to build up the backshore and thus a potential source of overwash material, 75,000 m^3 (98,000 cubic yards) could be washed over the barrier into the lagoon annually.

Wind transport. Winds, if they have a velocity greater than the threshold velocity of approximately 402 cm/sec (9 miles/h) are capable of initiating transport of sand size sediment and moving it considerable distances (BAGNOLD, 1954, p.69; BELLY, 1964).

The amount of sand moved by surface creep and saltation can be estimated, for this area, by the amount of sand entrapped during an experimental dune-building project in 1960 and 1961 (SAVAGE, 1963).

Sand fencing, which could be supplied with sand only from the foreshore because of a heavily vegetated backshore, trapped 2.26 m^3 (2.96 cubic yard)/ft. (0.3 m) of beach during the 9-month project. The 9-month period occurred during the winter when stronger onshore winds prevail. Comparison of the wind-field for the 9 months to the wind-field over a 5-year period indicated that a proportionality factor of 0.73 was needed to smooth out the short-term fluctuations and to reduce the effect of strong on-shore winds during the shorter period.

Multiplying by this proportionality factor and extrapolating the 9-month transport rate to an annual basis reveals that 2.26 m^3 (2.89 cubic yard) of sand per 0.3 m (ft.) of beach can be moved, by wind, from the beach and trapped annually. This material is moved primarily by surface creep and saltation and may not be that which is permanently lost to the beach-longshore system.

It is doubtful if much of the sediment moved by surface creep or saltation would be permanently lost. Any sediment lying in an exposed position on the barrier islands would be moved back into the beach-longshore system by an offshore wind. Only that material, which was moved all the way across the barrier, would be permanently lost since the land area is relative devoid of vegetation and offshore winds predominate.

That sediment being moved in suspension is probably the only type that is moved all the way across the barrier during one period of onshore winds. Suspension is limited, except, under extremely strong wind conditions, to material with an equivalent diameter of less than 0.2 mm. Size analyses show that 14.7% of the sediment north of Ocracoke Inlet is less than 0.2 mm while to the south, 23.5% is of this size.

It was assumed, for the purpose of this problem, that the maximum amount of sand available for transport by suspension was 14.7% of the amount transported by surface creep and saltation north of Ocracoke Inlet and 23.5% of this amount south of Ocracoke Inlet. Also, it was assumed that only one half of the maximum

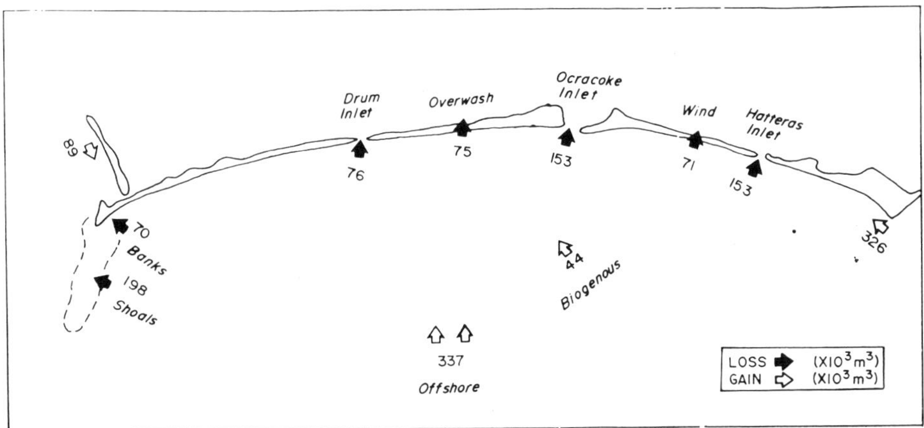

Fig.4. Map showing long-term volumetric changes along barrier islands. Solid arrows show estimated amount of losses to different areas; open arrows, source and amount at input necessary to balance losses.

available sand was in suspension long enough to be carried across the barrier into the lagoon.

With these assumptions and using 2.26 m^3, per 0.3 m (ft.) of beach, for the volume of surface creep and saltation, it is estimated that 0.16 m^3 (0.21 cubic yard) is lost annually per 0.3 m (ft.) of beach north of Ocracoke Inlet and 0.26 m^3 (0.34 cubic yard) south of the inlet. Over the distances of 28,000 m (93,000 ft.) and 66,000 m (215,000 ft.), respectively, the loss would be 71,000 m^3 (93,000 cubic yards).

Summary of quantitative changes

A summation of all the volumetric changes that were evaluated shows that 796,000 m^3 (1,041,000 cubic yards) of material must be added to this stretch of the coast annually (Table I, Fig.4).

SOURCE OF SEDIMENT

Six potential sources exist for sediment: (*1*) biogenous deposition; (*2*) reworking of older materials and local erosion; (*3*) from the mainland; (*4*) alongshore from the southwest; (*5*) alongshore from the northeast; and (*6*) from offshore.

Biogenous deposition

Shell fragments are not major constituents of the beach sands of this area. For Portsmouth and Core Banks, south of Ocracoke Inlet (Fig.2), fragments constitute about 5% of the total; north of Ocracoke Inlet, approximately 6.5% of the sediment consists of shells or shell fragments. Because Core Banks and Portsmouth Banks together are slightly more than twice the length of Ocracoke Island (Fig.2), 5.5% is an average value. Biogenous deposits thus can contribute approx-

TABLE I

SUMMARY OF SEDIMENT BUDGET SHOWING AREAS AND AMOUNTS OF LOSS AND GAIN

	Amount	
	(10^3 m^3)	*(10^3 cubic yards)*
Loss areas		
Banks	70	91
Shoals	198	259
Inlets	382	500
Across barriers:		
overwash	75	98
wind	71	93
Total	796	1,041
Sources of gain		
Biogenous	44	57
Alongshore:		
from southwest	89	116
from northeast	326	427
Offshore	337	441
Total	796	1,041

imately 44,000 m^3 (57,000 cubic yards) of the volume of sediment that must be supplied to this stretch of the coast.

Reworking of pre-existing deposits

This segment of the coast has had a net addition of sediment. Obviously then, there must be a net addition from outside of this segment since reworking of pre-existing deposits cannot supply any of the necessary deficit.

From the mainland

Pamlico Sound has a width of about 37 km. Currents appear to be insufficient in the sound to transport sand-size material this distance (ROELOFS and BUMPUS, 1953) and, therefore, it seems unlikely that there is a direct contribution to this stretch of the coast from the mainland.

Alongshore from the southwest

Part of the required sediment input could come from the southwest by means of longshore drift. Because of the predominant longshore drift to the southwest north of Cape Lookout, it is doubtful that any of the required volume north of the Cape could be supplied from the southwest.

The west side of Cape Lookout has been undergoing errosion, concomitant with the accretion on the east side. The average rate of erosion of the cape could supply 29,000 m^3 (38,000 cubic yards) annually. Erosion of Shackelford Banks (Fig.1) could supply 38,000 m^3 (48,000 cubic yards) annually, as indicated by shoreline changes between 1853 and 1913.

JOHNSON (1956), from historical records, estimated that the littoral drift, south of Beaufort Inlet (Fig.1) was 22,600 m^3 (29,500 cubic yards) annually toward the northwest. With this low rate, it is doubtful that much sediment could be added to the area under study although, for the purpose here, it is assumed that the entire amount bypasses Beaufort Inlet and accretes on Cape Lookout or Lookout Shoals.

A summation of the volumes of sediment, available from the three sources south and west of Cape Lookout, shows that 89,000 m^3 (116,000 cubic yards) is potentially available from this direction.

Alongshore from the northeast

The point of input for sediment coming alongshore from the northeast is the north shore of Hatteras Inlet. After subtracting the amount being contributed from the southwest from the total loss, 664,000 m^3 (869,000 cubic yards) would have to bypass this inlet if all the required amount of sediment came from the northeast.

This amount is well within the amount that can be moved in a beach-longshore system. Measured net drift rates along the East Coast of the United States are 306,000 m^3 (400,000 cubic yards) or less annually (JOHNSON, 1956). BRUUN and GERRITSEN (1960) estimate that 765,000 m^3 (1 million cubic yards) bypasses Oregon Inlet (Fig.1) annually.

Of this 765,000 m^3, coastal surveys indicate that 121,000 m^3 (158,000 cubic yards) are added to the coast between Oregon Inlet and Cape Hatteras; Diamond Shoals (Fig.1), south of Cape Hatteras has a minimal annual accretion rate of 382,000 m^3 (500,000 cubic yards) although the rate may be more than twice this amount (U.S. ARMY CORPS OF ENGINEERS, 1964). This leaves a net amount of 261,000 m^3 (342,000 cubic yards) which could move into the study area from north of Cape Hatteras. This rate assumes no loss to deep water at the tip of Diamond Shoals where the Gulf Stream impinges on the end of the shoals.

Cape Hatteras has undergone, over a period of 65 years, an annual loss of 65,000 m^3 (85,000 cubic yards). This volume, coupled with the amount from north of Hatteras, indicates that 326,000 m^3 (427,000 cubic yards) could be the amount of input to the system across Hatteras Inlet.

Offshore source

Unaccounted for by the normal sources of sediment are 337,000 m^3 (441,000 cubic yards) that are necessary to fulfill the annual deficit between Hatteras Inlet and Cape Lookout. The only remaining source area are the sediments and sedimentary rocks on the continental shelf.

An offshore source has been suggested by GILES and PILKEY (1965) in a study of the heavy mineral distributions along the coast in the southeastern United States. SAVILLE (1960, p.789) suggests an offshore source for some of this material along Long Island.

The actual source of the necessary volume of material is not known. The sediment being transported inshore and filling the deficit could be material winnowed from the "relict" sediments on the shelf or, alternatively, could be from erosion and reworking of poorly consolidated Tertiary formations, cropping out through a thin veneer of unconsolidated sediment.

On most of the stable continental shelves of the world, there is a band of "relict" sediments, generally coarser than those nearer shore (NIINO and EMERY, 1961; EMERY and NIINO, 1963; GORSLINE, 1963). On the continental shelf of the southeastern United States, the nearshore sediment has a mean diameter about 0.25 mm while there is a band of "relict" sediment offshore with mean diameter near 0.5 mm. These coarse deposits could be lag deposits from which some of the finer inshore sand has been and is now being winnowed. Only extremely long surface waves would stir up the bottom sediments. Such waves do touch bottom at sufficient depths so as to have an affect nearly to the outer edge of the continental shelf (CURRAY, 1960, p.233; HADLEY, 1964). SHEPARD (1963, pp.176–177) cites several instances where it appears that the continental shelf sediments act as reservoirs of sediment. When this sediment has moved far enough inshore, it is affected by the more common waves and nearshore currents so as to complete the transition from relict to recent.

Tertiary formations are thought to crop out at several localities on the continental shelf (STETSON, 1938). ROBERTS and PIERCE (1967) suggest the existence of outcrops of Miocene rocks under a thin, discontinuous veneer of sediments south of Cape Lookout. Breakdown of these poorly consolidated sedimentary rocks could also provide the necessary material.

CONCLUSIONS

Physiographic changes along a coast can be used to define the sediment budget of a stretch of the coast, if reliable historical records are available over a sufficient time span to smooth out short-term fluctuations.

Application of the method to a stretch of the North Carolina coast indicates that accretion of sediment has occurred at an average annual rate of 796,000 m^3 (1,041,000 cubic yards).

This rate requires an input greater than can normally be supplied by the commonly assumed sources such as longshore drift or transportation from the mainland. To make up the deficit, it is believed that the continental shelf acts as a reservoir of the sediment that is presently accreting in the nearshore area.

The coarse older deposits on many of the outer continental shelves could be lag deposits from which much of the finer sand sizes have been and now are being winnowed out and part is moved onshore. It undoubtedly requires a considerable time period to move the material into relatively shallow water. Eventually, the fine sand reaches water of such depth as to be affected by the more commonly occurring surface waves.

Alternatively, possible outcrops of Tertiary formations, generally masked by a thin veneer of recent sediment, are eroded and reworked to provide a source of sediment.

ACKNOWLEDGEMENTS

Part of this work was done under a grant from the U.S. National Science Foundation, grant G-16362, Dr. R. H. Benson principal investigator. The manuscript was critically read by Dr. David Duane, Coastal Engineering Research Center; Dr. Orrin Pilkey, Duke University; and Mr. Norman Taney, National Engineering and Science Company. Appreciation is also expressed to the Wilmington District, U.S. Army Corps of Engineers for use of their data.

REFERENCES

Athearn, W. D. and Ronne, C., 1963. Shoreline changes at Cape Hatteras. *Naval Res. Review*, 16:17–24.

Bagnold, R. A., 1954. *The Physics of Blown Sand and Desert Dunes.* Methuen, London, 265 pp.

Belly, P., 1964. *Sand Movement by Wind.* U.S. Army, Coastal Eng. Res. Center, Washington, D.C., 38 pp.

Bruun, P. and Gerritsen, F., 1960. *Stability of Coastal Inlets.* North-Holland, Amsterdam, 123 pp.

Caldwell, J. M., 1966. Coastal processes and beach erosion. *J. Soc. Civil Eng.*, 53:142–157.

Curray, J. R., 1960. Sediments and history of Holocene transgression, continental shelf, northwest Gulf of Mexico. In: F. P. Shepard, F. B. Phleger and Tj. H. van Andel (Editors), *Recent Sediments, northwest Gulf of Mexico.* Am. Assoc. Petrol. Geologists, Tulsa, Okla., pp.221–266.

Emery, K. O. and Niino, H., 1963. Sediments of the Gulf of Thailand and adjacent continental shelf. *Geol. Soc. Am. Bull.*, 74:541–554.

Giles, R. T. and Pilkey, O. H., 1965. Atlantic beach and dune sediments of the southern United States. *J. Sediment. Petrol.*, 35:900–910.

Gorsline, D., 1963. Bottom sediments of the Atlantic shelf and slope off the southern United States. *J. Geol.*, 71:422–440.

Hadley, M. L., 1964. Wave-induced currents in the Celtic Sea. *Marine Geol.*, 2:164–167.

Johnson, J. W., 1956. Dynamics of nearshore sediment movement. *Bull. Am. Assoc. Petrol. Geologists*, 40:2211–2232.

Luternauer, J. L. and Pilkey, O. H., 1967. Phosphorite grains: their application to the interpretation of North Carolina shelf sedimentation. *Marine Geol.*, 5:315–320.

Niino, H. and Emery, K. O., 1961. Sediments of the shallow portions of East China Sea and South China Sea. *Bull. Geol. Soc. Am.*, 72:731–762.

Richards, H. G., 1950. Geology of the coastal plain of North Carolina. *Trans. Am. Phil. Soc., New Ser.*, 40:1–80.

Roberts, W. P. and Pierce, J. W., 1967. Outcrop of the Yorktown Formation (Upper Miocene) in Onslow Bay, North Carolina. *Southeastern Geol.*, 8:131–138.

Roelofs, E. W. and Bumpus, D. F., 1953. The hydrography of Pamlico Sound. *Bull. Marine Sci. Gulf Caribbean*, 3:181–205.

SAVAGE, R. P., 1963. Experimental study of dune building with sand fences. *Conf. Coastal Eng., 8th*, pp.380–396.

SAVILLE, T., 1960. Sand transfer, beach control, and inlet improvements, Fire Island Inlet to Jones Beach, New York. *Conf. Coastal Eng., 7th*, pp.785–807.

SHEPARD, F. P., 1963. *Submarine Geology*, 2nd ed. Harper Row, New York, N.Y., 557 pp.

STETSON, H. C., 1938. The sediments of the continental shelf off the eastern coast of the United States. *Papers Phys. Oceanog. Meteorol.*, 5(4); 5–48.

U. S. ARMY COASTAL ENGINEERING RESEARCH CENTER, 1966. *Shore Protection, Planning, and Design.* U.S. Army, Coastal Eng. Res. Center, Tech. Rept. 4, Washington, D.C., 401 pp.

U. S. ARMY, CORPS OF ENGINEERS, 1964. *Ocracoke Inlet to Beaufort Inlet North Carolina.* U. S. Army, Corps Engrs, Wilmington, N.C., 28 pp. (mimeographed report).

U. S. CONGRESS, 1948. North Carolina shoreline, beach erosion study. *House Doc., 763, 80th. Congr.*, 33 pp.

Sediment. Geol., 3 (1969) 5–16

Editor's Comments on Paper 25

25 Leontyev: *Flandrean Transgression and the Genesis of Barrier Bars*

Here O. K. Leontiev amplifies upon his two earlier papers to give us further details concerning his concept of barrier island formation.

Leontiev has found in laboratory models that a submarine bar cannot build through sea level, because of erosive wave action over its crest. He maintains, however, that once developed, the bar can emerge as the sea level regresses. This, of course, requires a higher-than-present sea level during the Holocene, a sea-level curve that Leontiev supports. Within this framework, it is easy to visualize maximum development of a submarine bar at the peak of the Flandrean transgression, emergence during the regression which followed, and present-day erosion as sea level again climbs slowly.

While Leontiev concurs with Zenkovich on the mechanics of transverse sediment transport and submarine bar growth accompanying transgression, he differs on the method of bar emergence and the post-Pleistocene sea-level curve. One other area of agreement concerns the alternative of barrier island development through partial submergence of a preexisting coastal ridge.

Reprinted from *Quaternary Geology and Climate,* H. E. Wright, ed., **16,** 146–149 (1969)
(Natl. Acad. Sci. Publ. 1701)

O. K. LEONTYEV
Faculty of Geography, Moscow State University, Moscow, USSR

25

Flandrean Transgression and the Genesis of Barrier Bars

ABSTRACT

Barrier bars and islands are the largest and most widespread accretionary forms on modern coasts. The fact that these formations are everywhere eroded seems to prove their relict character. Through model studies of bar formation at the Laboratory of Experimental Geomorphology, Moscow State University, it was established that bars can be buried and preserved only under certain conditions involving recession of sea level. It is known that the oceans of the world reached their highest level (3–5 m higher than at present) 5,000–6,000 years ago. In the course of this rise, called the Flandrean transgression, submarine barrier bars were formed; then, sea level receded and the bars emerged above the surface. Because their subaerial position is not in balance with present hydrodynamic conditions, which are characterized by a renewed rise of sea level, the forms are now being eroded.

ZUSAMMENFASSUNG

Sandbänke und Inseln sind die umfangreichsten und am weitesten verbreiteten Akkumulation-formen an den heutigen Küsten. Die Tatsache, dass diese Bildungen überall abgetragen sind, scheint ihren Reliktcharakter zu beweisen. Modelluntersuchungen zur Bildung von Sandbänken im Laboratorium für experimentelle Geomorphologie der Universität Moskau, haben bestätigt, dass Sandbänke nur unter bestimmten Bedingungen, zu denen auch die Absenkung des Meeresspiegels gehört, begraben werden und erhalten bleiben können. Es ist bekannt, dass die Weltmeere ihren höchsten Wasserstand (3–5 m höher als gegenwärtig) vor 5,000–6,000 Jahren erreicht hatten. Während dieses Anstiegs, der sog. Flandrischen Transgression, bildeten sich untermeerische Sandbänke; dann sank der Meeresspiegel und die Sandbänke tauchten auf. Da sie nach dem Auftauchen nicht mehr in Einklang mit den heutigen hydrodynamischen Bedingungen stehen, die durch einen erneuten Anstieg des Meeresspiegels gekennzeichnet sind, werden sie jetzt abgetragen.

BARRIER bars are the largest and most common accretionary coastal formations. According to Vinogradov (1963), they total about one tenth of the length of the shorelines of the world's oceans. The causes of such extensive development should be established.

The origin of offshore bars has been considered mainly in connection with that of lagoon shores, of which the bars are an essential element. Zenkovich (1962) suggested that the bars are "long narrow strips of beach drifts uplifted above sea level and extending at some distance from the mainland parallel to the general direction of sea coast." A lagoon is generally located behind a barrier bar, which is built of material stirred up from the sea bottom. The barrier bar therefore differs essentially from other coastal accretionary forms, many of which are composed of materials from adjacent shores.

Three stages of offshore-bar development are distinguishable: the underwater bar, the insular bar, and the barrier bar proper. Principal conditions for the formation of barrier bars and islands are a considerable reserve of loose material on the sea bottom in front of a forming bar and a gently sloping sea bottom. The approach of dominant waves at right angles to the shore is not essential because the existence of a gently sloping bottom, along with a smooth profile, enables bottom material to be stirred up in the shoreward direction. Such bottom material may be either oolitic or biogenic (as in the bars of the Caspian Sea, Florida, and the Bahamas) or it may be terrigenous. The presence of terrigenous material on submerged coastal slopes indicates that mainland plains were flooded during the postglacial transgression of the sea that was caused by melting of the ice sheet. According to Fairbridge, sea level was 100 m lower than at present 16,000 years ago (by the end of the Wisconsin), and 11,000 years ago (Alleröd) it had risen to a level 30 m lower than at present. Sea level became stabilized at approximately its present level about 6,000 years ago and subsequently has varied within ±3 m.

Coastal plains flooded during the postglacial transgression were mostly accretionary formations, composed of alluvial, fluvioglacial, and colluvial deposits. On the bottoms of such coastal plains, huge masses of deposits began to migrate toward shore during the formation of the continental slope. Accumulation of these deposits near the shore resulted in the formation of initial bars; according to Zenkovich (1962) and others, this initial formation took place during transgression and then shifted in the direction of the mainland as the shoreline of the transgressing sea shifted in this direction.

Many known offshore bars were built of subaerial deposits, including gravel. The Meechken gravel bar near the coast of the Chuckchee Peninsula, in the Arctic Ocean, is supplied by rock debris from a piedmont plain flooded during postglacial transgression. A gravel bar on the Sea of Okhotsk apparently had the same origin, as did the Bohai Bay (Yellow Sea) sand bar, which is built of redeposited alluvial quartz sand (Leontyev, 1961).

According to Zenkovich (1962) and Leontyev (1961), the formation of sand bars is possible only where bottom slopes are very slight—from 0.005 to 0.003. For gravel bars, because of their rare occurrence, bottom-slope requirements have not yet been ascertained.

Zenkovich (1962) suggested that subaerial barrier bars are the final phase of development of submarine bars. However, other works (for example, by Egorov, 1951) have shown that the only link between these two types of coastal feature is that both are formed as a result of transverse shifting of settled sediments. Leontyev noted (1961) that submarine bars and barrier bars and islands differ in size. Barrier bars apparently develop from the very beginning as large features and are not related to the critical depth $2H$ (twice the wave height), as are the submarine bars. The initial stage of development of the submarine barrier bars takes place at depths greater than $2H$ and is probably related to the rebuilding of the submarine profiles of flooded coastal plains (Figure 1).

In the Laboratory of Experimental Geomorphology, Moscow State University, an attempt has been made to simulate the process of submerged-bar formation (Leontyev and Nikiforov, 1965). In the course of this experiment, the bar never rose above the surface of the water when the "sea level" was held constant or was raised.

This finding, however, is not new; Johnson (1919), in contrast to Zenkovich, had already considered the submerged bars as evidence of shore uplift, and he believed that formation of bars as emergent forms is associated with absolute or relative recession of sea level. Egorov (1951), from long study of offshore bars, later concluded also that a sea-level recession, even of short duration, is needed to raise the bar crest to the surface.

On the other hand, much evidence indicates that barrier beach complexes are located mainly on shores of submergence. Thus, an apparent contradiction arises: barrier bars, which should develop only with a recession of sea level, occur typically on shores of submergence.

First, I shall try to explain why the recession of sea level is absolutely necessary for bar emergence. Because of hydrodynamic forces, the bar crest may grow upward only to a depth below the sea surface equal to the height of the waves above it (Figure 2). With such a depth over the crest, the waves above it would break completely. When breaking occurs, maximum rates of water movement are related precisely to the zone of breaking. If the wave breaks in front of a beach, elastic material is shifted and deposited over the beach profile at a decreasing rate of swash upward along the beach slope. Thus, the beach will generally show a distinct

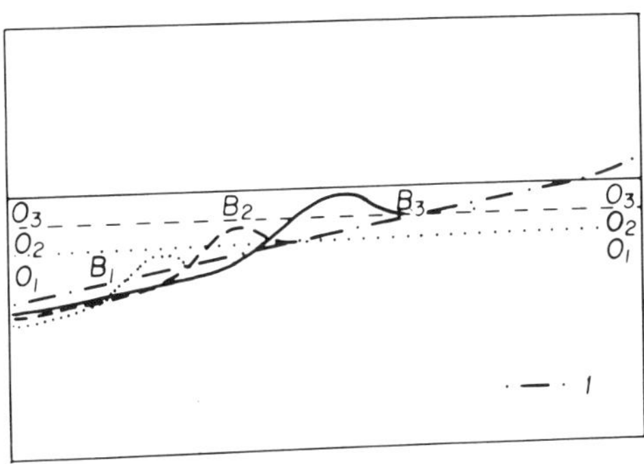

FIGURE 1 Formation of the submarine barrier bar during transgression. O_1, O_2, O_3 are successive stages of sea-level rise; B_1, B_2, B_3 are corresponding phases of bar formation; 1 is profile of flooded coastal plain.

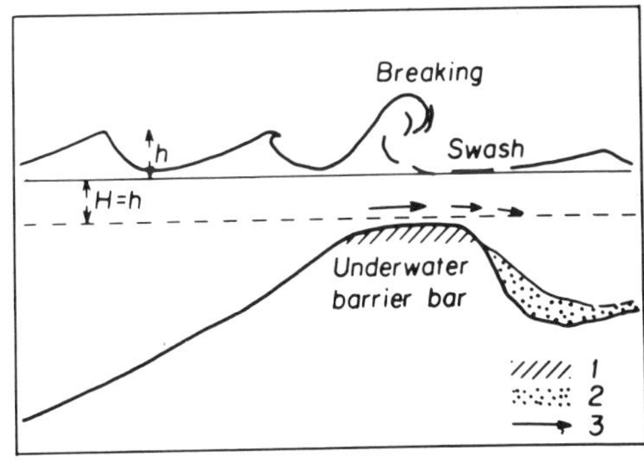

FIGURE 2 Illustration of impossibility of bar growth when the depth over the crest is equal to the wave height (H=h). 1, erosion; 2, accumulation; 3, rates of swash.

demarcation between the area of removal of material from the breaking zone and the accumulation area in the zone of decaying swash rates. When waves break over the underwater bar, material is removed from the crest and deposited on the shoreward slope or at the foot of this slope (Figure 2). The bar is therefore capable of shifting shoreward, but upward growth through accretion at its crest ceases.

If sea level recedes (for example, because of offshore winds), the breaking zone shifts to the seaward slope, and the crest may become an accumulation zone.

Thus, the bar crest cannot emerge above the sea surface if sea level rises or is stable. Yet most bars are associated with shores of submergence. As shown by Zenkovich (1962) and Kaplin (1964), lagoon coasts can be formed in ways other than through the formation of an underwater bar. During a transgression of the sea, an onshore accumulative terrace may be slowly flooded. The seaward margin of the terrace is then higher than the rest of the terrace, either because of differential shore submergence or because of eolian processes (Figure 3), and the lower part of the terrace becomes flooded while the higher part is transformed into a barrier beach or a barrier island.

However, this explanation applied only to certain individual cases. In my opinion, most offshore bars are related to shores of submergence mainly because these bars are mostly relic accumulative formations that developed when the world sea level temporarily receded during general postglacial transgression.

Numerous data on the contemporary erosion of bars show that many of them are relics. Barrier bars along the Atlantic and Gulf coasts of the United States, the Netherland coast, and certain areas of the Black Sea coast, among other regions, are subject to erosion, and appropriate measures are being taken to strengthen these shores. These examples indicate that barrier bars do not correspond with contemporary hydrodynamic conditions, and that many of them are relic formations.

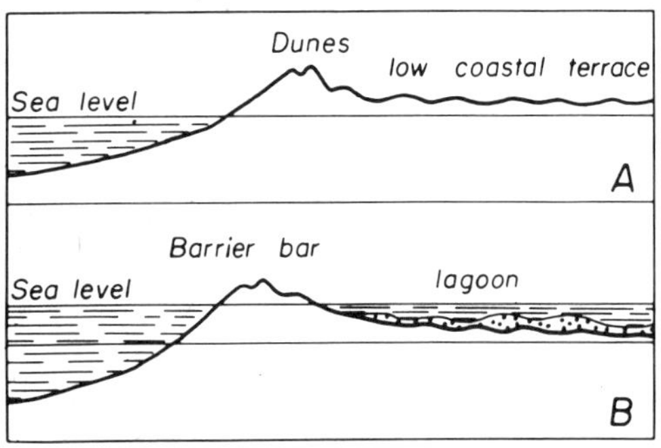

FIGURE 3 Formation of barrier bar by flooding of a low coastal terrace with a higher seaward margin.

Many investigators (e.g., Fairbridge, 1959; Zenkovich *et al.*, 1960) emphasize the irregular development of postglacial transgression. The maximum ocean level of the Holocene was reached under optimum climatic conditions a few thousand years ago. In many works this stage of development of postglacial transgression has been called the "Flandrean transgression," which left terraces in many areas 3-7 m above present sea level (Table 1). The global formation of barrier bars bordering coastal plains apparently took place during the brief recession of sea level that followed the Flandrean transgression.

However, sea level now unquestionably is rising. According to Valentin (1952) and other contemporary investigators, the mean rise of sea level during about the last century has been somewhat higher than that during the upper Holocene. As shown in Table 1, most authors consider the age of Flandrean transgression to be 4,500-5,000 years. Some authors (e.g., Bird, 1963) have stated that the Flanders terrace is represented morphologically by barrier bars.

The postglacial history of a submerged bar—its formation, emergence through sea-level recession, and subsequent submergence—is particularly well described in Fisk's (1959) analysis of geological sections of Padre Island on the Texas coast (Figure 4). The formation of the offshore bar began

TABLE 1 Data on the Occurrence, Age, and Height of Flandrean Terraces

Region	Author	Height[a] (m)	Age (thousands of years)
Netherlands	de Jong, 1960	2-3	Subatlantic
Acadia	Harrison and Lyon, 1963	3	4.5
Crete	Boeckshoten, 1963	7	Flanders
New Zealand	Schofield, 1960	3	Holocene
India	Chatterjee, 1961	5	Flanders
Western Australia	Comm. for research of shoreline change	3	7.5
Gippsland, Australia	Bird, 1963	3-4	6.5, 5
Kent	Oldfield, 1960	6	5.2
Morocco	Markov, 1961	5	Flanders
La Rochelle	Verger, 1960	8	5.5
Eastern Australia	Ruszczyńska, 1961	2	5
Ghana	Ruszczyńska, 1961	3-6	5
Valparaiso, Chile	Paskoff, 1963	5-7	Holocene
Primorie, USSR	Solovyev, 1959	4	4-5
Tanabe, Japan	Mii, 1960	6	Holocene
McMurdo, Antarctica	Speden, 1960	10	6
NE Coast of the Sea of Okhotsk	Zabelin, 1951	7-10	Holocene
Matanzas, Cuba	Ducloz, 1963	4-9	Flanders
Black Sea	Fedorov, 1963	3-4	5-6
Gibraltar	Giermann, 1962	3	4.5
Irish Sea	Marshall, 1962	6-7	Flanders

[a]Above present sea level.

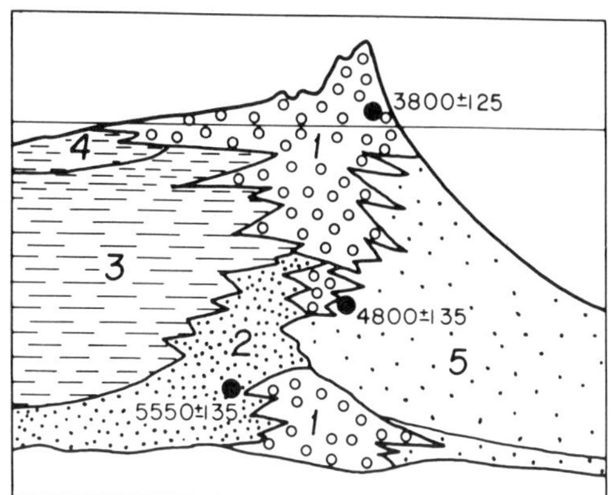

FIGURE 4 Schematic cross section of Padre Island bar (from Fisk, 1959). Deposits of: 1, barrier bar; 2, bay or sound; 3, open lagoon; 4, closed lagoon; 5, shoreface.

5,500-5,700 years ago; its development was interrupted about 5,000 years ago, apparently by the rapid fall of sea level by approximately 3 m, which resulted in emergence of the bar and a sharp increase in salinity of the lagoon behind it. An analogous dating is given by Fedorov (1963), for Black Sea fluctuations in the Holocene; he points out that according to New Black Sea time (which corresponds to Flandrean time), sea level receded during the Phanagerian regression to a level 3 m lower than the present level.

It may be concluded that the wide occurrence of barrier bars is a result of the irregular course of postglacial transgression. The bars were formed during Flandrean time and emerged from the sea because of the subsequent short recession of the sea level. Erosion of the seaward sides of the bars, the result of a renewed increase in the world sea level, shows that the bars are relic forms.

REFERENCES

Bird, E. C. F., 1963, The physiography of the Gippsland Lakes, Australia: *Zeitschr. f. Geomorph.*, v. 7, p. 232-245

Boeckshoten, G. J., 1963, Some geological observations on the coasts of Crete: *Geol. en Mijnb.*, v. 42, p. 241-247

Chatterjee, S. P., 1961, Fluctuations of sea level around the coasts of India during the Quaternary period: *Zeitschr. f. Geomorph.*, suppl. v. 3, p. 48-56

de Jong, I. D., 1960, The morphological evolution of the Dutch coast: *Geol. en Mijnb.*, v. 39, p. 638-643

Ducloz, C., 1963, Etude géomorphologique de la région de Matanzas: Cuba, *Arch. Sci.*, v. 16, p. 351-402

Egorov, E. N., 1951, Observations of the dynamics of offshore bars: *Acad. Sci. U.S.S.R., Proc. Inst. Oceanol.*, v. 6, p. 21-29

Fairbridge, R. W., 1959, Periodicity of eustatic oscillation, p. 97-99, *in* Sears, Mary, ed., *Internat. Oceanographic Cong., Preprints: Amer. Assoc. Advancement Science*, 1022 p.

Fedorov, P. V., 1963, Stratigraphy of the Quaternary deposits in the Crimean-Caucasian coast and some aspects of the geological history of the Black Sea: *Acad. Sci. U.S.S.R., Proc. Geol. Inst.*, v. 88, p. 139-150

Fisk, H. N., 1959, Padre Island and Laguna Madre mud flats, south coastal Texas: Baton Rouge, Louisiana State Univ., Coastal Studies Inst., *II Coastal Geog. Conf.*, p. 103-151

Giermann, G., 1962, Meeresterrassen am Nordufer der Strasse von Gibraltar Bericht: *Naturforschung Gesellsch. Freiburg*, v. 52, p. 111-118

Harrison, W., and Lyon, C. J., 1963, Sea level and crustal movements along the New England-Acadian shore: *Geol. Soc. Amer. Spec. Paper 73*, p. 8-9

Johnson, D. W., 1919, Shore processes and shoreline development: New York, John Wiley & Sons, 584 p.

Kaplin, P. A., 1964, Some regularities in lagoon formation: *Oceanology*, no. 2, p. 290-294

Leontyev, O. K., 1961, The fundamentals of geomorphology of coasts: Moscow Univ. Publ. House, 418 p.

Leontyev, O. K., and Nikiforov, L. G., 1965, On the causes of the global occurrence of barrier bars: *Oceanology*, no. 4, p. 653-661

Markov, K. K., 1961, Problems of paleogeography of the Anthropogen in Morocco: *Bull. Comm. Quatern. Studies*, no. 26, p. 101-109

Marshall, J. R., 1962, The morphology of the Upper Solway salt marshes: *Scottish Geogr. Mag.*, v. 78, p. 81-99

Mii, Hideo, 1960, Coastal geology of Tanabe Bay: *Sci. Reports Tohoku Univ.*, ser. 2, no. 1, v. 34, p. 1-93

Oldfield, F., 1960, Late Quaternary changes in climate, vegetation and sea level in Lowland Lonsdale: *Publ. Brit. Geogr.*, no. 28, p. 99-117

Paskoff, R., 1963, Indices morphologiques d'un stationnement de l'océan Pacifique á 5-7 m au-dessus de son niveau moyen actuel sur le littoral du Chili central: *Soc. Geol. France Comp. Rend.*, no. 6, p. 191-192

Ruszczyńska, A., 1961, Z baden nad czwartorzedowymi zmianami poziomu oceanow: *Przegl. geol.*, no. 2, v. 8, p. 111-112

Schofield, J. C., 1960, Sea level fluctuations during the last 4000 years as recorded by a Chenier Plain, Firth of Thames, New Zealand: *New Zealand Jour. Geol. Geophys.*, v. 3, p. 467-487

Solovyev, V. V., 1959, On the Holocene ingression in the Southern Coastal Area: *Proc., All-Union Res. Geol. Inst.*, pt. 2, p. 193-197

Speden, I. G., 1960, Postglacial terraces near Cape Chocolate, McMurdo Sound, Antarctica, *New Zealand Jour. Geol. Geophys.*, no. 2, v. 3, p. 203-217

Valentin, H., 1952, Die Küste der Erde: *Petermanns Mitt., Erg.*, v. 246, 230 p.

Verger, F., 1960, La transgression flandrienne sur la littoral Atlantique de l'embouchure de la Vilaine á la Rochelle: *Norois*, no. 25, v. 7, p. 48-50

Vinogradov, O. N., 1963, *Representation of the features of morphology, dynamics, and origin of coasts on general geographic maps:* Moscow, *Acad. Sci. U.S.S.R.*, 22 p.

Zabelin, I. V., 1951, On the newest elevation of the north-western coast of the Okhotsk Sea: *Nature*, no. 8, p. 72

Zenkovich, V. P., 1962, The fundamentals of the theory of the development of coasts: Moscow, "Nauka," 712 p.

Zenkovich, V. P., Leontyev, O. K., and Nevesski, E. N., 1960, The influence of the eustatic Postglacial transgression on the development of the coastal zone in the U.S.S.R.: *Coll. Soviet Geologists' Papers at the XXI Internat. Geol. Cong., Acad. Sci. U.S.S.R.*, p. 154-163

Editor's Comments on Paper 26

26 Hoyt: *Chenier Versus Barrier, Genetic and Stratigraphic Distinction*

In his reply to Fisher's discussion paper, Hoyt made a point of not confusing cheniers with barriers, and cited this paper, which was then in press. The issue is well worth considering at this time.

As described by Hoyt, there are three stages in the development of a chenier: (1) progradation of a mud flat; (2) erosion of the seaward edge of the mudflat, reworking of the mudflat deposits, and formation of the ridge; and (3) further progradation of the mudflat, leaving the ridge as a chenier. Therefore, the chenier is bounded on both sides by, and overlies, mudflat sediments. Comparison with any of the foregoing hypotheses of barrier island formation reveals a difference in genesis to begin with. Although the two features may appear similar in fossil form, barrier islands are generally larger in all dimensions than cheniers and, of course, are bounded by a marine environment on the seaward side.

Reprinted from the *Amer. Assn. Petroleum Geologists Bull.*, **53**, 299–306 (1969)

Chenier *Versus* Barrier, Genetic and Stratigraphic Distinction[1]

26

JOHN H. HOYT[2]

Sapelo Island, Georgia 31327

Abstract Barrier islands and cheniers are elongate narrow sand bodies which may appear similar where preserved in the sedimentary record. However, their modes of origin and sequence of development are distinctive. Differentiation of these features is important in the interpretation of the depositional environments, paleogeography, and geologic history of coastal areas.

Chenier development begins with progradation by deposition of clay, silt, and sand sediments. Rapid sedimentation precludes removal of fines. Progradation is followed by a period of reworking, shore retreat, and formation of a ridge along the landward side of the beach. Fines are transported seaward and along the shore. Sand is concentrated on the upper beach and on top of the adjacent marsh and is transported along the shore; it may accumulate in areas not being actively eroded. The contact of the chenier with underlying marsh and mudflat deposits is disconformable in areas of reworking and shore retreat, but the chenier may intertongue with finer sediments where the sand has been transported laterally along the shore. A return of conditions favoring rapid sedimentation reinitiates mudflat progradation and the sand ridge is left as a chenier. Holocene cheniers are commonly less than 15 ft thick.

Barriers originate from a topographic ridge along the landward side of a beach which subsequently is partly submerged. Lagoonal-marsh sediments are deposited behind the barrier; however, continued submergence accompanied by transgression may result in complex intertonguing of barrier and lagoonal-marsh sediments. Barriers also form from spits, but this is not believed to be the general mechanism of barrier formation. Barriers, like cheniers, may be eroded, reworked, and moved landward over the adjacent marsh. Generally, barriers predate the lagoonal-marsh sediments, although, with continued submergence, synchronous deposition may occur. The sand ridge of the chenier develops on, and seaward of, existing marsh and mudflat deposits.

Introduction

In coastal areas there is a relation between sediment supply, wave energy, land-sea stability, and the resulting shoreline features. Because of the erosion, sorting, and reworking that occurs along shorelines, deposits of diverse origins may appear similar in the geologic record. Two similar forms are barriers and cheniers. Both are elongate sand bodies that develop along the coast; however, their origins are distinctly different, resulting from different sedimentologic conditions. Barriers initially are separated from the mainland by a lagoon, estuary, or bay, whereas cheniers form along the prograding shoreline seaward of marsh and mudflat deposits. An additional requirement for cheniers is that they are enclosed on the seaward side by finer grained sediments.

The preserved stratigraphic remains of barriers and cheniers may appear similar making distinction difficult. If it is possible to distinguish between these features, important information is gained concerning the depositional history and sedimentologic factors of the coastal areas.

The purpose of this paper is to review the origin and modification of cheniers and barriers and to consider the characteristics of the resulting deposits. The recognition and distinction of cheniers and barriers in the geologic record are helpful in the search for potential reservoir rocks. There are many problems in the identification of these features; however, a first step is proper interpretation of modern examples. This paper is intended to lead to more accurate use of the terms and a better understanding of the depositional environments.

Cheniers have been described by several investigators, including Russell and Howe (1935), Fisk (1948), Price (1955), Byrne *et al.* (1959), and Gould and McFarlan (1959). The literature on barriers is much more extensive. Classical works were published by Élie de Beaumont (1845), Gilbert (1885), and Johnson (1919). More recent studies include works by Fisk (1959), LeBlanc and Hodgson (1959), Rusnak (1960), and Shepard (1960). Studies by the writer include Hoyt *et al.* (1964), Hoyt and Henry (1967), Hoyt and Hails (1967), and Hoyt (1967b).

Cheniers

The term "chenier" comes from the French word *chêne,* meaning oak, in reference to the stands of large oak trees characteristically growing on higher areas. The classic area of chenier development is along the Louisiana coast, west of the Mississippi delta, where a

[1] Manuscript received, September 15, 1967; accepted, February 10, 1968. Contribution No. 165 from the Marine Institute, University of Georgia, Sapelo Island, Georgia.

[2] Marine Institute and Department of Geology, University of Georgia.

The support of National Science Foundation by grants NSF-GP 1380 and NSF-GA 704 is gratefully acknowledged.

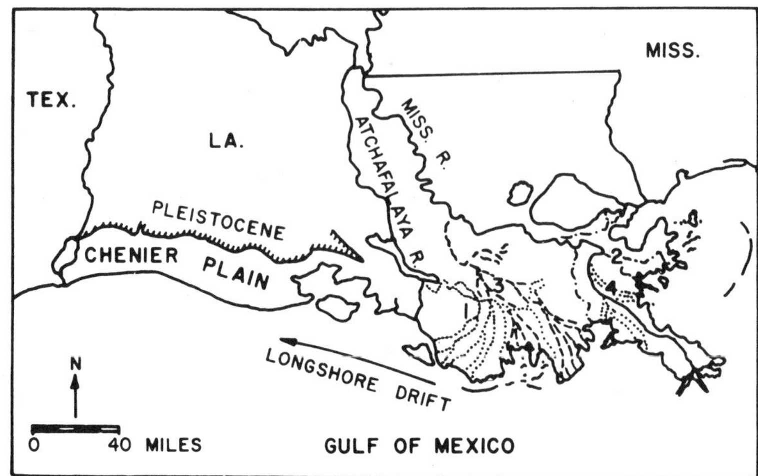

FIG. 1.—Map of Louisiana coast showing chenier plain and Mississippi deltas. **1** = Teche delta, 3,800–2,800 years B. P.; **2** = St. Bernard delta, 2,800–1,700 years B. P.; **3** = Lafourche delta, 1,800–700 years B. P.; **4** = Plaquemines-Modern delta, 1,200–400 years B. P.; and modern Mississippi delta, 450-present years B. P. (after Gould and McFarlan, 1959; Kolb and Van Lopik, 1966).

broad plain, termed a "chenier plain" (Price, 1955), developed during Holocene time and occupies an area seaward of the outcrop of Pleistocene sedimentary rocks (Fig. 1). The plain is more than 20 mi wide and extends 110 mi along the coast. On the east, the chenier plain is bordered by the Mississippi delta and on the west it merges with the barrier islands of the Texas coast. The main features of the chenier plain are a series of coastwise ridges, the cheniers, separated by marshy areas (Figs. 2, 3).

Typical cheniers are 150 to 1500 ft wide, and several tens of miles long. In studied examples (Byrne *et al.* 1959, p. 238; Gould and McFarlan, 1959, p. 268) the sand of the chenier does not extend more than 5–6 ft below sea level, the height of the dunes is commonly only

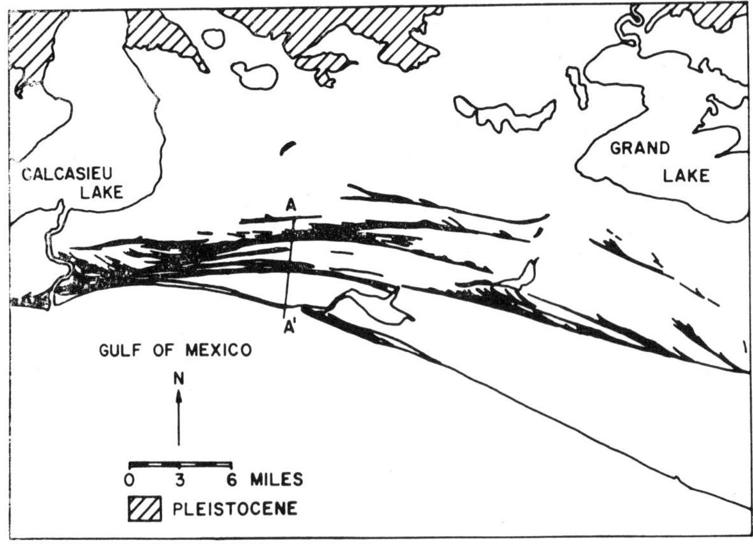

FIG. 2.—Map showing chenier development in central part of chenier plain, Louisiana (after Byrne *et al.*, 1959). Cross section A–A′ shown in Figure 3.

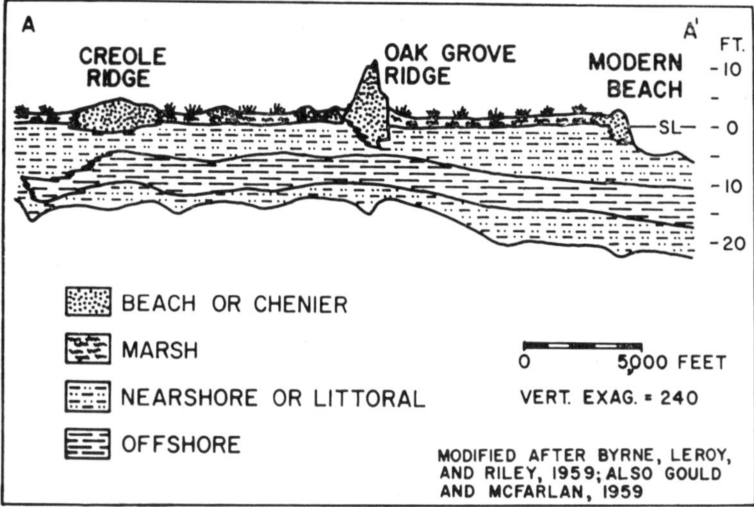

FIG. 3.—North-south cross section A–A′ of part of chenier plain showing facies relations (after Byrne *et al.*, 1959; Gould and McFarlan, 1959). Location of cross section shown in Figure 2. *SL* = sea level.

10–20 ft and the maximum thickness of the chenier is only 18–20 ft. Sediment grain size is fine to very fine sand; shells are common.

The conditions necessary for the development of cheniers are abundant sediment supply, low to moderate wave energy, and land-sea stability. Some periodic variation in sediment supply also is important. These conditions are best met near the mouths of large rivers in areas with broad continental shelves. The chenier plain forms downdrift from the river source area. If wave energy is high, silt- and clay-size sediments are carried offshore rather than being added to the shoreline deposits. If the sediment supply is not large, even moderate wave energy can rework the sediment transported to the coast.

A preliminary phase in chenier development is a supply of sediment from the source river in sufficient quantity to preclude the removal of the silt and clay fraction. This results in a mudflat of relatively unsorted sediment consisting of sand, silt, and clay (Fig. 4). Under these conditions sediment generally is sufficient to prograde the shoreline.

The distributary pattern of the delta changes occasionally altering the area where sediment is introduced into the marine environment; this shift affects the rate of deposition in the adjacent chenier plain. If the river outlet is near the plain, the rate of progradation is increased; if the shift in discharge is away from the plain, progradation may cease and there may be a stabilization or a retreat of the shoreline. During these periods the sand ridges develop and may be driven landward over the marsh. The reworking that accompanies shore stabilization and retreat separates out silt and clay, and transports them seaward, leaving the sand and shell to accumulate on the shore face and to be piled up as beach or dune ridges. Additional

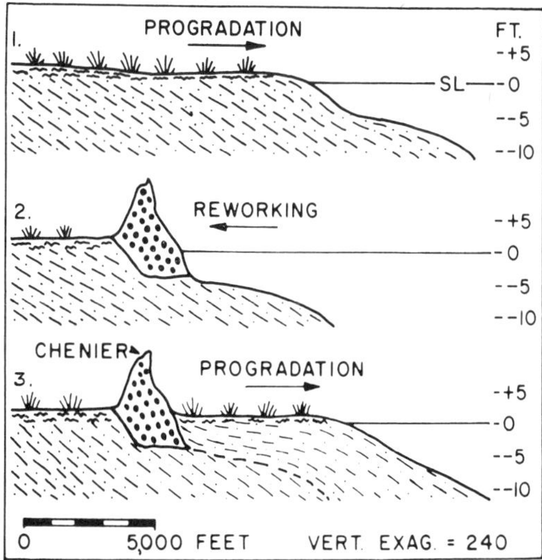

FIG. 4.—Idealized cross sections showing development of chenier. **1.** Mudflat progradation. **2.** Erosion and reworking of mudflat deposits and formation of ridge along shoreline. **3.** Mudflat progradation, ridge becomes chenier. *SL* = sea level.

sand may be introduced at this time by littoral and longshore currents. The size of the ridge produced by these conditions depends on the amount of reworking and the amount of sand introduced by the marine currents. In the course of delta development, additional shifts in the distribution pattern result in a new cycle of shore progradation of mixed sand-, silt-, and clay-size sediments. The sand ridge, or series of ridges, is enclosed landward and seaward in fine-grained sediment and the ridge becomes a chenier.

The areas separating the cheniers are occupied by salt marsh if near the coast or freshwater marsh. The surface of the marsh builds up to high spring tide level, but in places river floodplain sediments raise the level of the marsh areas and may bury the cheniers. Compaction and isostatic movement caused by the weight of the sediment deposited in the coastal area may result in formation of lakes or lagoons by depression of the chenier plain below water level. The width of the marsh areas between the cheniers ranges from a few tens of yards to several hundred yards. The sediment contains appreciable organic matter, silt, and clay. The sediments of the shallow neritic sea floor adjacent to the chenier plain are interbedded sand and silty clay. Farther seaward there is less sand because of the diminishing influence of wave action. The cheniers overlie the shallow neritic deposits and, in places, marsh deposits.

The immediate source of the sand for chenier development (Price, 1955) is commonly an area of erosion upcurrent along the coast. Silt and clay bypass the area of chenier development or move offshore. Where retreat of the shoreline precedes development of the sand ridge, the contact of the sand with underlying mudflat and marsh sediment is erosional; however, sand washed into the marsh by storms, in turn, may be covered by marsh sediment with resultant interfingering of the two facies.

Although the cheniers are the common sand deposits of the chenier plain, in some areas spits also develop and produce a mixture of depositional features. Spits form where a beach intersects a river mouth, inlet, or other irregularity in the shoreline.

The Louisiana chenier plain is more than 20 mi in maximum width, but the cheniers are concentrated in the seaward half of the plain. Shells in Little Chenier, the farthest landward well-developed chenier, have a C^{14} age of 2,775 years (Gould and McFarlan, 1959, p. 264). More than 15 major and minor cheniers have formed in the area seaward of Little Chenier while sea level has been at approximately its present altitude (Byrne *et al.* 1959, pl. 1; Coleman and Smith, 1964; Scholl, 1964). In contrast, Galveston Island, Texas, a barrier island, has prograded nearly 4 mi during a similar period of time (Bernard and LeBlanc, 1965, p. 158).

Fossil cheniers should have dimensions similar to those of modern examples. Transgression or regression should result in some redistribution of surficial sediments, but should not produce sheet sands. It seems unlikely that longer periods of sea-level stability, which probably were common in pre-Quaternary time, would increase significantly the width and thickness of individual cheniers, although the width of the chenier plain might be greater. Composite cheniers rarely exceed 1 mi in width or 25 ft in thickness.

Chenier internal stratification is mainly gently dipping laminations (less than 10°); individual sets are traceable for several yards. Dip may be directed toward the ocean if formed on the foreshore or toward the land if formed by washover. Throughout the deposit shells are abundant in layers up to several inches thick although leaching may reduce shell content in the upper few feet, depending on climate and length of exposure. Grain size characteristically increases from base to top (Byrne *et al.*, 1959, p. 253). Stratification is typical of low coastal dunes. The strata have a large variety of dip directions, and some beds with a high angle dip (greater than 20°) are extensively truncated.

Barriers

Barriers are the commonest type of shoreline feature along the gently shelving coastal plains of the world. They are of many sizes and shapes, but are generally elongate bodies from a few to more than 100 mi long and only a few miles wide. Barriers commonly consist of sand, but also may contain large quantities of gravel or shells. They are bordered on the landward side by lagoons or salt marshes whose waters vary from fresh to hypersaline. This discussion does not include barrier reefs which are developed *in situ* by living organisms.

Although barriers have been studied for many years (Élie de Beaumont, 1845; Gilbert, 1885; Johnson, 1919), their origin has not been explained adequately; classical theories of development from offshore bars must be reeval-

uated in light of modern studies. Élie de Beaumont believed that barrier islands developed from sediment eroded from the sea floor and deposited on an offshore bar, which eventually was constructed above sea level. Gilbert's studies along the shores of Pleistocene Lake Bonneville in Utah indicated that barriers developed from spits which were constructed downdrift from promontories along the shoreline. Later breaching of the spits during storms resulted in barrier islands. Both ideas were reviewed by Johnson who concluded, from a study of coastal profiles, that Élie de Beaumont's hypothesis best fit modern barriers. However, Johnson did not include in his analysis the effect of submergence on barrier development. Modern studies of sea level movements (Jelgersma, 1961, 1966; Scholl, 1964; Shepard, 1963) suggest that sea level in many areas of the world has been rising in relation to the land during the past several thousand years. A change in sea level affects the shape of profiles associated with the barrier and limits the usefulness of profiles in suggesting the mode of origin of the barriers.

Several lines of evidence indicate that barriers have not developed from offshore bars (Hoyt, 1967b). If barriers developed from offshore bars, the area landward of the bar-barrier would have been originally a marine environment with littoral and shallow neritic sediments and organisms. Detailed studies of many barrier coasts have not found the sediments and faunas required for this offshore-bar hypothesis (Fisk, 1959; Hoyt *et al.*, 1964, p. 172; Hoyt and Hails, 1967; Rusnak, 1960, p. 191; Shepard, 1960, p. 212). Observations of bars (Price, 1962) and wave tank studies (Leontyev and Nikiforov, 1966, p. 222; McKee and Sterrett, 1961, p. 17–20) indicate that bars build up to near sea level, but the surf and wave swash maintain the bar below or only slightly above still water level. If barriers developed from bars, then various stages of such development should be available for study along the coasts of the world. Because they are not, it must be concluded that development from offshore bars is not the general mechanism of barrier formation.

The proposed hypothesis of barrier formation incorporates the effects of submergence which accompanied the melting of the continental glaciers of North America and Eurasia and the return of sea level to its present position. Along relatively stable coasts a eustatic rise in the ocean level has flooded the continental margins.

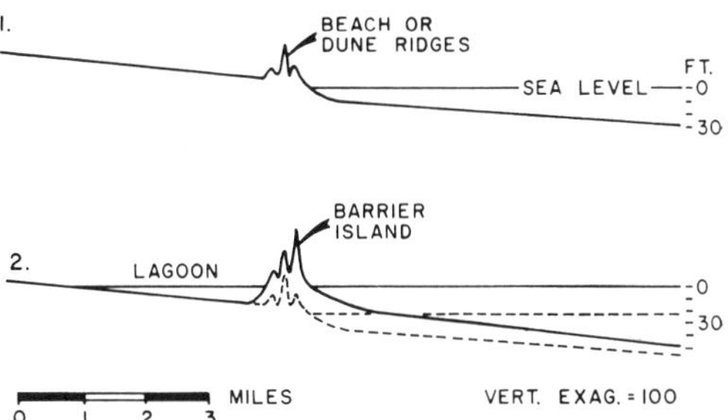

FIG. 5.—Idealized cross sections showing formation of barrier island. **1.** Ridge forms along shoreline. **2.** Submergence partially inundates ridge forming barrier and lagoon (after Hoyt, 1967b).

A ubiquitous feature of shorelines along gently sloping coastal areas is the development of a ridge, of either wind- or wave-deposited sediment, just landward of the shoreline. Sand forms a dune ridge several tens of feet high and coarse-grained sand and gravel may be thrown by wave swash into a beach ridge several yards high. If sufficient sediment is available, a series of dune and/or beach ridges is developed along a prograding shoreline. As submergence continues the ridge is partly covered, the low area on the landward side is inundated and forms a lagoon, and the ridge becomes a barrier (Fig. 5). The barrier may continue to prograde, may remain in place, or may be eroded, depending on the sediment supply, hydrodynamic conditions, and rate of sea-level change. During periods of rapid sea-level rise the barrier may be quickly destroyed in the transgressing surf zone; however, during the late Holocene, when the rate of submergence slowed, conditions apparently were favorable for preservation and growth of incipient barriers. The width of the lagoon is related to the slope of the land surface, the effect of subsidence, and is modified by erosion and sedimentation.

Price (1962) described minor barriers formed from offshore bars during periods of temporary high water associated with storms. The bars built vertically to slightly below the elevated water level and were left as barriers when the water receded. The observed features were small and short lived, and major barriers apparently are not formed in this way.

Barriers may form from spits by breaching and segmentation; however, as in the case of barrier formation from offshore bars, open marine sediments and faunas should be present in the lagoonal area landward of the spit, but are not commonly found. Thus, barrier island formation from spits does not appear to be a general mechanism, although segmented spits may be a factor in island formation in places.

After formation of the barrier, the lagoon begins to fill with sediment. The quiet water is favorable for the accumulation of fine-grained sediment; however, tidal currents also bring sand and coarser sediment and rivers emptying into the lagoon contribute a variety of material, depending on local conditions such as river gradient and discharge, available sediments, and climate. As the lagoon fills with sediments and is converted to a salt marsh, the water is concentrated in tidal channels where strong currents develop. Deep channels are formed in which currents rework and incorporate sediments from the subjacent deposits. In general, the lower energy of the lagoonal environment results in a higher percentage of silt- and clay-size sediments than accumulates in the barrier deposits.

Barrier and lagoon sediments commonly interfinger. The fine sand, silt, and clay of the lagoon are deposited on the landward side of the better sorted dune and littoral sands of the barrier, but may be followed by washover sands from the barrier, which in turn may be covered by lagoonal deposits. The continued repetition of this sequence depends on several factors such as submergence, height and width of the barrier, and the morphology of lagoonal features. Meandering tidal channels, for example, may impinge on the lagoonal side of the barrier, reworking and modifying the contact of lagoon and barrier sediments.

The thickness of barrier sediments is considerably varied and depends on local conditions. In many areas of the Gulf and Atlantic coasts where the barrier islands are well developed, the sand of the barrier and associated deposits are several tens of feet thick, but along some stretches of the same coasts only a vestige of the barrier remains. There is also considerable variation in the height of dunes which develop along the barrier shoreline; some are more than 100 ft high.

Recognition of fossil barriers presents many problems and there may be significant contrasts with modern examples. It is likely that sea-level movements were much slower during previous periods, except for the Pleistocene, and considerably more time was available for sediment accumulation. Fossil barriers may be several times thicker than their modern counterparts; in addition, the transgression, regression, and subsidence which accompanied deposition could result in blanket deposits. With either transgression or regression an upper part of the barrier could be eroded; in the former case, in the surf zone, and, in the latter, by meandering channels of the lagoon salt marsh. Further, migration of the inlet channels, which separate the barrier islands, effectively reworks the barrier sediments and leaves a deposit with characteristics of the inlet environment (Hoyt and Henry, 1965; 1967). Modifications by the inlet include steepening of stratification inclinations, reorientation of stratification inclination direction to reflect the channel flow normal to the coast, and concentration of coarse sediments, including shell and shell fragments, as channel fill. The channel deposits are poorly sorted in comparison to the littoral and sublittoral sediments, and the greater lithologic variety reflects the shifts in current energy in the channel. Primary structures are those typical of channel deposits; however, the bidirectional flow produces stratification which is inclined in places toward the ocean and elsewhere toward the land. Ripple forms in studied areas along the Georgia coast (Hoyt, 1967a) range from small, less than ½-in. amplitude, to sand waves with amplitude of more than 12 ft. Megaripples with amplitudes of 1–3 ft are particularly abundant. Stratification inclination may be as much as the angle of repose, which is commonly about 30°. Littoral sand deposits along the inlet are similar to those along the front of the island. Gentle dips (less than 10°) are characteristic with stratification inclined toward the channel, and individual laminations are traceable for several yards. Dune ridges parallel the inlet but few probably are preserved.

There are similarities in the construction of the ridges associated with barriers and cheniers because both are shaped by similar hydrogenic and eolian processes acting on similar particles. The basic differences are in quantity and types of sediment supplied to the depositional area and in the ability of the marine processes to transport and sort the material. The abundance of fines and the low coastal energy along the chenier plain contribute to the deposition of mudflat and marsh along the shoreline. Along barrier shorelines fines are trapped in the lagoon or are transported offshore and do not accumulate along the shoreline.

The area of merger of the chenier plain with the barrier islands of the East Texas coast has not been studied in detail; however, the transition results from the diminished supply of silt and clay to the Texas barrier coast and the lack of mudflat and marsh deposition along the shoreline. Under these conditions sand is added to the barriers without mudflat deposits between successive ridges. The change from chenier coast to barrier coast is the common transition downdrift from a major source of sediment supply. This sequence should be important in predicting the location of major sand bodies which may be associated with barrier-island deposits.

Conclusions

There is a major difference between cheniers and barriers in sequence of development. In areas of chenier development mudflat deposits initially accumulate along the shoreline and cause rapid progradation. This is followed by a period of reworking, removal of fines, and accumulation of sand and shell ridges. Another period of progradation of muddy sediments leaves the ridges surrounded by finer sediments as a chenier. The mudflat-marsh deposit precedes the formation of the sand ridge. In the development of barriers the elongate sand body is formed first and protects the lagoon where finer grained lagoon and marsh sediments accumulate.

A basic difference in the depositional conditions of cheniers and barriers is the proximity of a major sediment source capable of overloading the distributive agencies along the adjacent shoreline. Near such a source there may be an alternation of progradation and erosion and the development of cheniers. No change in sea level is required. With a smaller influx of sediment, sorting is accomplished more readily and commonly sand accumulates in a subaerial ridge along the shoreline which, with slow submergence, may develop into a barrier. The low area landward of the barrier becomes a lagoon.

Cheniers and barriers can be distinguished by the shape and extent of the deposits. The studied examples of cheniers along the Louisiana coast, for example, rarely extend more than 6 ft below the associated sea level and have a total thickness of less than 20 ft and width of less than 1,500 ft for an individual ridge, or less than 1 mi for composite ridges. In contrast, barriers commonly extend several tens of feet below sea level and, including dunes, may be 100–150 ft thick. Barriers may be several miles wide. In addition, transgression, regression, and subsidence of barrier coasts may result in widespread blanket deposits, whereas along chenier coasts the dimensions of individual ridges are not changed greatly by strandline movements unless the ridges are eroded. Preserved primary structures of barrier deposits may be those mainly formed in migrating inlets. Such deposits have abundant medium- to high-angle stratification (10–30°) dipping toward the land in places and toward the ocean elsewhere. The sediments are poorly sorted in comparison to littoral and sublittoral deposits. The cheniers consist mainly of gently dipping littoral, sublittoral, and washover deposits and some overlying dune accumulations.

References Cited

Bernard, H. A., and R. J. Le Blanc, 1965, Résumé of the Quaternary geology of the northwestern Gulf of Mexico province, *in* The Quaternary of the United States: Princeton, New Jersey, Princeton Univ. Press, p. 137–185.

Byrne, J. V., D. O. LeRoy, and C. M. Riley, 1959, The chenier plain and its stratigraphy, southwestern Louisiana: Gulf Coast Assoc. Geol. Socs. Trans., v. 9, p. 237–259.

Coleman, J. M., and W. G. Smith, 1964, Late recent rise of sea level: Geol. Soc. America Bull., v. 75, p. 833–840.

Élie de Beaumont, L., 1845, Leçons de géologic pratique: Paris, p. 223–252.

Fisk, H. N., 1948, Geological investigations of the lower Mermentau River basin and adjacent areas in coastal Louisiana: Vicksburg, Miss., U.S. Army Corps Engineers, Mississippi River Comm., 78 p.

——— 1959, Padre Island and the Laguna Madre flats, coastal South Texas, *in* R. J. Russell, chm., 2d Coastal Geography Conf.: Louisiana State Univ., p. 103–151.

Gilbert, G. K., 1885, The topographic features of lake shores: U.S. Geol. Survey 5th Ann. Rept., v. 9, p. 207–217.

Gould, H. R., and E. McFarlan, Jr., 1959, Geologic history of the chenier plain, southwestern Louisiana: Gulf Coast Assoc. Geol. Socs. Trans., v. 9, p. 261–270.

Hoyt, J. H., 1967a, Occurrence of high-angle stratification in littoral and shallow neritic environments, central Georgia coast, U.S.A.: Sedimentology, v. 8, p. 229–238.

——— 1967b, Barrier island formation: Geol. Soc. America Bull., v. 78 p. 1125–1136.

——— and J. R. Hails, 1967, Pleistocene shoreline sediments in coastal Georgia; deposition and modification: Science, v. 155, p. 1541–1543.

——— and V. J. Henry, 1965, Significance of inlet sedimentation in the recognition of ancient barrier islands: Wyoming Geol. Assoc. 19th Fld. Conf., p. 190–194.

——— and ——— 1967, Influence of island migration on barrier-island sedimentation: Geol. Soc. America Bull., v. 78, p. 77–86.

——— R. J. Weimer, and V. J. Henry, Jr., 1964, Late Pleistocene and recent sedimentation, central Georgia coast, U.S.A., *in* Deltaic and shallow marine deposits: Amsterdam, Elsevier, p. 170–176.

Jelgersma, S., 1961, Holocene sea level changes in the Netherlands: Meded. Geol. Stinchting, Ser. C-VI-7, 100 p.

——— 1966, Sea-level changes during the last 10,000

years: Royal Meterol. Soc. Internat. Symposium World Climate Proc., p. 54–71.
Johnson, D. W., 1919, Shore processes and shore line development: New York, John Wiley & Sons, 584 p.
Kolb, C. R., and J. R. Van Lopik, 1966, Depositional environments of the Mississippi River deltaic plain —southeastern Louisiana, *in* Deltas: Houston Geol. Soc., p. 17–61.
LeBlanc, R. F., and W. D. Hodgson, 1959, Origin and development of the Texas shoreline: Gulf Coast Assoc. Geol. Socs. Trans., v. 9, p. 197–220.
Leontyev, O. K., and L. G. Nikiforov, 1966, An approach to the problem of the origin of barrier bars (abs.): 2d Internat. Oceanogr. Cong., p. 221–222.
McKee, E. D., and T. S. Sterrett, 1961, Laboratory experiments on form and structure of longshore bars and beaches, *in* Geometry of sandstone bodies: Am. Assoc. Petroleum Geologists, p. 13–28.
Price, W. A., 1955, Environment and formation of the chenier plain: Quaternaria, v. 2, p. 75–86.
——— 1962, Origin of barrier chain and beach ridge (abs.): Geol. Soc. America Ann. Mtg. Program, p. 119.
Rusnak, G. A., 1960, Sediments of Laguna Madre, Texas, *in* Recent sediments, northwest Gulf of Mexico: Am. Assoc. Petroleum Geologists, p. 153–196.
Russell, R. J., and H. V. Howe, 1935, Cheniers of southwestern Louisiana: Geog. Rev., v. 25, p. 449–461.
Scholl, D. W., 1964, Recent sedimentary record in mangrove swamps and rise in sea level over the southwestern coast of Florida, pt. 1: Marine Geology, v. 1, p. 344–366.
Shepard, F. P., 1960, Gulf Coast barriers, *in* Recent sediments, northwest Gulf of Mexico: Am. Assoc. Petroleum Geologists, p. 338–344.
——— 1963, Thirty-five thousand years of sea level, *in* Essays in marine geology in honor of K. O. Emery: Los Angeles, California, Univ. Southern California Press, p. 1–10.

Editor's Comments on Paper 27

Tidal inlets and washover fans have been discussed in a number of preceding papers, but their development has not been adequately dealt with. Jack Pierce has taken this as his task in this paper.

Pierce sees frontal wave attack as unlikely to cut an inlet through a barrier island. Rather, overtopping, from the seaward side of a narrow barrier or from the lagoonal side of a wide barrier, is more conducive to producing the erosive power necessary for inlet cutting. Storm surge across barrier flats, concentrated in tidal channels and creeks, provides enough intensity to surmount the barrier from the lagoonal side. Overtopping of a wide barrier from the seaward side generally results in the development of washover fans. These coarse sands dip landward and interfinger with the finer lagoonal muds.

Reprinted from the *J. Geol.*, **78**, 230–234 (1970)

TIDAL INLETS AND WASHOVER FANS[1]

27

J. W. PIERCE
Division of Sedimentology, Smithsonian Institution, Washington, D.C. 20560

ABSTRACT

Tidal inlets and washover fans are genetically related. The resulting feature is dependent upon barrier configuration, depths in the lagoon adjacent to the barrier, and the direction from which the storm surge came, either from the sea or the lagoon.

Attack on barrier islands from the seaward side by waves overtopping the barrier will result in washover fans on wide barriers where extensive adjoining tidal flats are present. Inlets can be cut by this type attack on narrow barriers where no tidal flats are present. Storm surge from the lagoonal side, if channeled along tidal creeks, can easily cut inlets through a barrier.

INTRODUCTION

The formation of tidal inlets and washover fans are genetically related. Both are formed when a tidal surge associated with a storm causes water to flow across a barrier.

The resulting end product, either a tidal inlet or washover fan, is dependent upon several factors applicable at time of formation. These factors include topography of the barrier, the lagoon floor, and the nearshore sea bottom; the sea-state conditions; the type of storm generating the surge; and the direction, either from the lagoon or the sea, from which the surge comes.

This paper discusses the conditions under which washover fans or tidal inlets will form and which conditions will permit inlet formation from the seaward side or from the lagoon side. Evidence, as shown on aerial photographs, is used to substantiate the theoretical considerations.

IDEAS ON FORMATION OF INLETS

Shaler (1895) believed that tidal inlets were formed by the breaking out through a barrier of dammed-up water. This excess water was land water derived through surface runoff.

Johnson (1919, p. 307) unequivocally rejected this concept. He stated that inlets could be opened only by wave attack on the seaward side of the barrier during high water associated with a storm surge. His usage of wave attack, rather than overtopping, could be construed to imply direct frontal erosion by waves.

Hite (1924), based on personal observations during storms, found that inlets could be opened either by wave attack on the seaward side or by the breaking out of dammed-up waters from the lagoon. The method of formation depended on the type of storm and the configuration of the lagoon.

Despite Hite's short paper, the most commonly cited theory on the method of formation is that attributed to Johnson, or modifications thereof (Brown 1928, p. 512; Lucke 1934, p. 51; Marshall 1951, p. 54; Strahler 1960, p. 423; Shepard 1963, p. 200).

ATTACK FROM THE SEAWARD SIDE

Direct attack by waves on a barrier is generally frontal, thus effectively dissipating the tremendous energy present in the waves. Erosion along the entire beach face occurs, and there is a general regression of the barrier. To be effective in cutting a pass through the barrier, wave energy must be concentrated at some point or along a very narrow reach of the shore. Usually, the wave attack, if concentrated at all, is directed toward projecting points, which often are the widest parts of the barrier and the most improbable places for an inlet to occur. Thus, direct wave attack on the front of the barrier would seem to be ineffective in forming inlets.

Storm waves can, and often do, overtop the storm beach ridge of a barrier. In addi-

[1] Manuscript received July 30, 1969; revised October 16, 1969.

[JOURNAL OF GEOLOGY, 1970, Vol. 78, p. 230–234]

tion to some remaining kinetic energy, the water associated with these waves has a potential energy by reason of its height above still-water level in the lagoon. This water must flow toward the lagoon rather than return directly to the ocean because of its location back of the storm beach ridge. If the flow can be channeled and if frictional losses are minimal, the available energy may be sufficient to cut a channel. Both of these conditions are dependent upon the configuration of the rear part of the barrier and the adjacent floor of the lagoon. Most important are the slope of the barrier surface, the width, and the depth of the lagoon adjacent to the barrier.

Overwash fans are the most likely result of overtopping when the barrier is wide and when extensive adjacent barrier flats are present. The slope of the combined backshore-barrier flat will be low on a wide barrier. As a result, the water never attains a very high velocity and hence does not erode deeply. Because of the distance of travel necessary to cross a wide barrier, a considerable part of the available energy is dissipated by friction. The velocity of the water decreases as the rear of the barrier is approached because of energy loss due to friction and the lower slope present along the rear of the barrier. Deposition of the entrained sediment occurs on the tidal flats and on the rear of the barrier. This deposition builds up the rear of the barrier, further reducing the slope. A series of distributaries is formed. Water associated with later waves is not channeled but flows over a wider area through the distributaries, lessening the probability of erosion to a depth sufficient to cut below still water.

Tidal inlets are the most likely result from overtopping of the storm beach ridge when the barrier is narrow and no extensive tidal flats are present. The slope of the combined backshore-barrier flat is greater because of the narrow width, and frictional losses are not as great because of the lesser distance of travel. Higher water velocities ensue, and more sediment can be moved. The bulk of this sediment is not deposited until the still-water level in the lagoon is reached. Since greater depths adjoin the rear of the barrier, the sediment is deposited in greater depths of water, thus not contributing to an increase in the height of the rear of the barrier nor to the formation of a series of distributaries. The slope is increased because of erosion, and water associated with succeeding waves attains a higher velocity.

ATTACK FROM THE LAGOON SIDE

Circulation in many storms and their rate of advance are such that the wind direction may suddenly reverse during passage of the storm. This is especially true of the tropical storms that periodically strike the east coast of the United States.

The initial winds of such a storm are onshore. Abnormally high water levels occur along the front of a barrier and on the mainland side of the lagoon. Low water levels occur on the lagoon, or back side, of the barrier. This is caused by movement of the lagoonal waters downwind (away from the barrier) and the lag time necessary for the existing inlets to allow the excess water along the front of the barrier to pass into the lagoon.

After passage of the center of the storm, an offshore wind may result. If the forward movement of the storm is relatively rapid, reversal of direction is abrupt, giving rise to low waters on the ocean side of the barrier. A mass of water, previously banked against the mainland by the onshore wind, is unsupported and is pushed toward the barrier by the offshore wind, resulting in an abnormally high storm surge against the rear of the barrier. Such a surge can contain a large mass of water if the lagoon is large.

Little damage will be done by the rise in water level on the lagoon side of the barrier if no way exists to channel the water of this surge. Tidal channels and creeks are nearly universally present in the marshes and reach onto the barrier flats. These creeks are often connected to low spots in the dune field, if such is present. Thus, these creeks give the storm surge access to the backshore across the barrier flat and the dune field. Rather

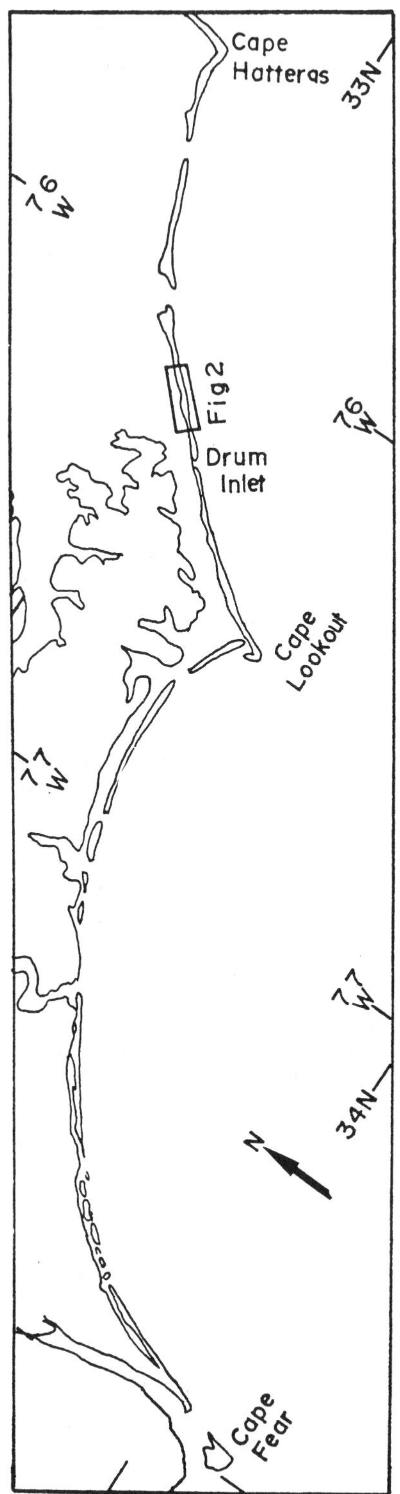

FIG. 1.—Map of a portion of the North Carolina coast showing location of Drum Inlet and that part of the coastline covered in plate 1.

than a general rise in water level along the whole rear of the barrier, the water may be channeled through a few select places after following the straighter tidal channels. Access to the backshore is permitted at only a few places. This concentration of the storm surge onto the backshore at only a few places intensifies the high water effect.

Once the water reaches the top of the storm beach ridge, the slope of the foreshore is relatively steep and is effectively increased by the low water levels on the ocean side of the barrier. High velocities will be attained by the water running down the front of the beach. These velocities will be very effective in eroding a channel below the normal water level. Sediment removed from this channel and deposited at the end of the channel is easily reworked and redistributed by ocean waves and currents.

Shaler's (1895) concept of inlet formation by the breaking out of dammed-up land-derived waters would be a special case of this situation. In his concept, an excessively high water level in the lagoon would be built up by surface runoff. The flow through existing inlets or percolation through the barrier could not keep up with the excess runoff, allowing the water level back of a barrier to reach such a height so as to break out of the lagoon. This would most likely occur in a small barred estuary which would be subjected, at times, to high stream discharge.

EVIDENCE FROM AERIAL PHOTOGRAPHS

Both extratropical and tropical storms affect the coast of the southeastern United States. Extratropical storms follow a path so that an offshore wind precedes the onshore wind. During tropical storms the wind pattern is reversed and an offshore wind follows an onshore one. Extratropical storms, as a rule, are not as severe as tropical storms.

Larger inlets generally have existed throughout historical times and the mode of their formation is obscure. Some may have existed from the time of formation of the

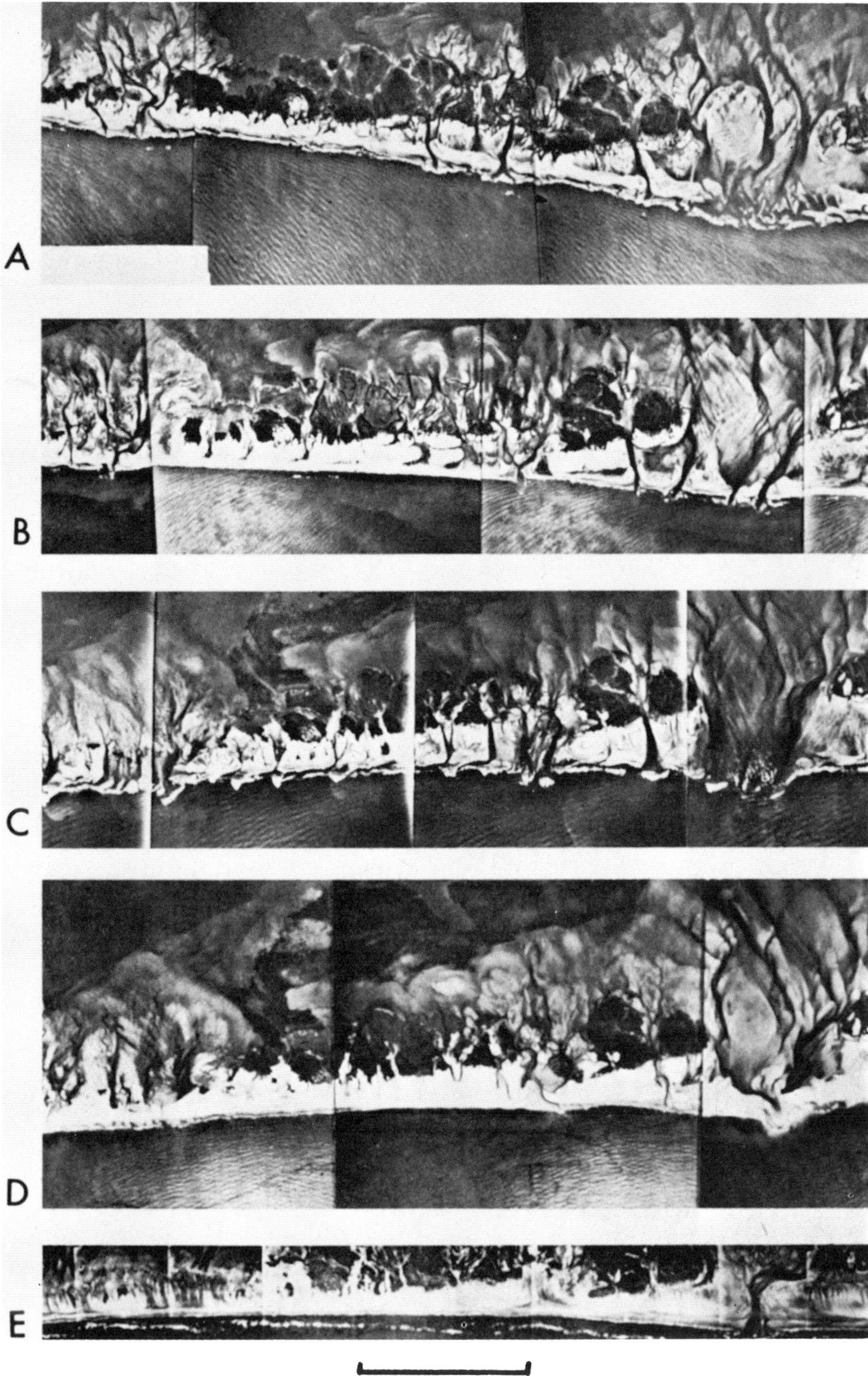
A
B
C
D
E

barrier while the position of others may be inherited from previous transgressive-regressive cycles.

Channels can be cut through barriers by attack from the seaward side as pointed out by El-Ashry and Wanless (1965) and Strahler (1960, p. 423). In both of these cited cases, the barrier appears to be relatively narrow and the lagoon relatively deep adjacent to the barrier.

Several series of aerial photographs, from 1945 to 1962, show the results of tropical and extratropical storms along the coast of North Carolina (fig. 1 and pl. 1). Tropical storms occurred in August 1944, September and October 1954, and September 1958. An extratropical storm affected the coast in March 1962 and was one of the most damaging storms ever to hit the area.

In all cases, open inlets or ones opened during tropical storms and subsequently closed occur as extensions of tidal creeks or channels in the marsh (pl. 1). In most cases, the inlets have been reopened, or rejuvenated, at the same place along the barrier as inlets previously existed. In a few cases, bulges occur in front of the inlets (pl. 1*C*). This may be the result of deposition of sediment during the time of channel cutting.

Drum Inlet (fig. 1) was opened during a hurricane in September 1933. A high storm-surge occurred on the lagoon side of the barrier after passage of the center of the tropical storm. As a consequence of this, water flowing across the barrier toward the ocean cut a channel which is now Drum Inlet (U.S. Army Corps of Engineers 1964).

Few inlets, if any, were opened during the storm of March 1962, with the exception of the one cited by El-Ashry and Wanless (1965). Instead, washover fans were largely the result (pl. 1*E*). Although the storm was one of the most destructive ever to hit this coast, its effects were entirely unlike those brought about by tropical storms.

The amount of sediment deposited on the lagoon floor during a tropical storm seems far less than would be expected if the entire volume of material removed during channel cutting had been deposited on the lagoon floor. This deficiency is quite striking when compared to the volume added to the lagoon floor during the extratropical storm of 1962 (pl. 1).

It is probable that a combination of these contributes to the formation of inlets. Direct frontal attack and overtopping by waves will erode the front of the barrier and lower the height of the storm beach ridge. The ability of the surge from the lagoon to cut through the barrier would be facilitated by lowering of the beach ridge and prior erosion of the front of the barrier.

CONCLUSIONS

Tidal inlets in a wide barrier with extensive tidal flats are generally opened by attack from the lagoon side of the barrier. Washover fans will be the result of an attack of this barrier from the seaward side.

With a narrow barrier and reasonably large depths in the lagoon adjacent to the barrier, channels can be cut from the seaward side by water associated with waves overtopping the storm beach ridge. General wave attack on the front of the barrier will seldom, if ever, result in the cutting of a channel through the barrier.

Acknowledgments.—Financial support for this work was through the National Science Foundation grant G-16362 (R. H. Benson, principal investigator) and Smithsonian Research Foundation RA 3307. Dr. Robert Dolan critically read the manuscript.

Plate 1.—Aerial photographs of a small portion of the North Carolina coast showing the effects of storms. *A*, taken in January 1945; *B*, March 1955; *C*, October 1958; *D*, October 1959; and *E*, April 1962. Tropical storms hit the coast in August 1944; August, September, and October 1954; and September 1958. An extratropical storm hit in March 1962. One inch equals 1 mile.

REFERENCES CITED

BROWN, EARL I., 1928, Inlets on sandy coasts: Am. Soc. Civil Eng. Proc., v. 54, p. 505–553.

EL-ASHRY, M. T., and WANLESS, H. R., 1965, The birth and early growth of a tidal delta: Jour. Geol., v. 73, p. 404–406.

HITE, M. P., 1924, Some observations of storm effects on ocean inlets: Am. Jour. Sci., ser. 5, v. 7, p. 319–326.

JOHNSON, D. W., 1919, Shore processes and shoreline development: New York, John Wiley & Sons, 584 p.

LUCKE, J. B., 1934, A Study of Barnegat Inlet New Jersey and related shoreline phenomena: Shore and Beach, v. 2, p. 45–93.

MARSHALL, NELSON, 1951, Hydrography of North Carolina marine waters, *in* Survey of marine fisheries of North Carolina: Chapel Hill, Univ. of North Carolina Press, p. 1–76.

SHALER, N. S., 1895, Beaches and tidal marshes of the Atlantic Coast: Natl. Geog. Mon. 1, p. 137–168.

SHEPARD, F. P., 1963, Submarine geology (2d ed.): New York, Harper & Row, 557 p.

STRAHLER, A. N., 1960, Physical geography: New York, John Wiley & Sons, 534 p.

U.S. ARMY CORPS OF ENGINEERS, Wilmington (N.C.) District, 1964, Combined hurricane survey interim report, Ocracoke Inlet to Beaufort Inlet, North Carolina: Wilmington, N.C., 25 p. plus appendices.

Editor's Comments on Paper 28

Otvos takes exception with Hoyt's views on two points: the mode of barrier island formation and the distinction between barriers and cheniers.

Invoking a stillstand, Otvos maintains that most of the present Gulf Coast barrier islands began to form about 5,000 to 3,500 years ago. He believes that the islands evolved through the emergence of submarine shoals, nourished essentially by littoral drift and currents. Once established, the barriers migrated parallel, perpendicular, or at oblique angles to the coastline. It is during this migration, Otvos states, that all traces of the origin of such features may be lost. As for the necessity of open marine characteristics on the mainland shore if bars do emerge, he points out that ocean-facing coasts may have marshy shores and mainland lagoonal shores may be sandy. He feels, therefore, that stratigraphy along the mainland shore is not a sufficient argument. Citing barrier island drillhole data and observations in historical times, Otvos claims that there is proof to substantiate the bar emergence concept.

Concerning the matter of barriers and cheniers, Otvos sees no genetic difference between the two. Whether coastal ridges are prograded seaward, multiple bars emerge, or nested spits develop, he claims that the intervening lagoons of the latter two fill in, and they all end up looking quite similar. Furthermore, Otvos finds no significant deviation in the dimensions of barriers and cheniers.

Ervin G. Otvos, Jr., was born in Budapest, Hungary, in 1935. He received a Diploma in geology in 1958 at the Lóránd Eötvös University of Science in Budapest. an M.S. in 1962 at Yale University, and a Ph.D. in 1964 at the University of Massachusetts. Between 1958 and 1968 Otvos worked as a geologist in Hungary, Canada, and the United States. In 1970 he became Head, Geology Division, Gulf Coast Research Laboratory, Ocean Springs, Mississippi, a position he still holds. He is also a lecturer at the University of Southern Mississippi. Otvos' research interests include sedimentology, coastal geomorphology, and Late Tertiary and Quaternary stratigraphy of the northeastern Gulf Coast.

Reprinted from the *Geol. Soc. Amer. Bull.*, **81**, 241–246 (1970)

ERVIN G. OTVOS, JR. *Department of Earth Sciences, Louisiana State University in New Orleans, New Orleans, Louisiana 70122*

Development and Migration of Barrier Islands, Northern Gulf of Mexico

28

ABSTRACT

Historical evidence and drilling results from published sources and U.S. Coast Survey charts[1] indicate that barrier islands form by upward aggradation of submerged shoal areas. Subsequent extensive barrier island migration may completely obscure conditions of formation of the original barrier island. Migration may take place parallel, perpendicular, or at oblique angles to the mainland shoreline and appears to take place much faster when parallel with the shoreline. No evidence indicates barrier island formation from engulfed beach and dune ridges during the early stages of transgression. Many strand plain and chenier ridges form the same way as barrier ridges.

INTRODUCTION

During the past 4000 to 5000 years, extensive barrier island formation and alteration took place along the northern shores of the Gulf of Mexico. Study of barrier island migration and alteration and the review of subsurface drilling information from Padre, Galveston, Horn, and the Pine Island chain helps to answer the problem of barrier island genesis.

In several recent publications, Hoyt (1967, 1968, 1969) stated that barrier island formation from submerged offshore bars and from the breaching of spits is much less frequent and important than formation by engulfment of coastal beach and dune ridges during transgression. Submergence of coastal zones would turn landward areas of coastal beach and dune ridges into lagoons and would turn ridges into barrier islands. Hoyt's main proof is the absence of barrier islands of beach and shallow neritic deposits which would have formed during the submarine growth stages of offshore bars that were developing into barrier islands (Hoyt, 1967, p. 1127).

Island formation by beach-ridge engulfment would be most likely to occur when a stable shoreline with well-developed ridges is engulfed by a relatively sudden transgression which does not erode or push the ridges landward. This should be followed by a somewhat slower sealevel rise during which the islands maintain their upward growth. Such conditions may have existed at the onset of the Holocene transgression in the Gulf area, at about 11,000 B.P. (Bruun, 1962). Most present Gulf barrier islands started to form about 5,000 to 3,500 years ago when this transgression had slowed down or stopped altogether. Except for the Mississippi delta zone, shore stability was established at this time. During the last period of stabilized sealevel, shore and barrier island progradation started at several locations.

Failure to recognize beach and shallow marine sediments landward of barrier islands (Hoyt, 1967, 1969) can be attributed to several factors other than island formation from engulfed beach ridges. Even if we assume the absence of submerged bars that are responsible for bay-sound or estuarine-type reduced salinity environments landward of them, such environments can also prevail along open shores like bays and estuaries. Presence or absence of beach sediments is not sufficient proof either. Open marine shores often have salt marshes instead of beaches. On the other hand, sandy shores without salt marshes may be present on the mainland shores of lagoons, as in Texas. When transgression reached the site of the future lagoons, the first sediments often formed from reworked Pleistocene deposits and are

[1] U.S. Coast Survey (later: "U.S. Coast and Geodetic Survey") Charts H-194 (Cat and Ship Island, 1848); H-363 (Northern Chandeleur Island, 1852–53); H-999 (Breton Sound, 1869); H-1000 (Chandeleur Sound, 1869); H-430 (Horn Island and offshore, 1854); H-190 (Horn Island, 1846); H-362 (Petit Bois Island, 1851).

often indistinguishable from them (Rusnak, 1960). A bay-mouth bar or spit may develop fast enough so as not to allow time for the deposition of noticeable volumes of open marine sediments before sedimentation becomes lagoonal. An example of the speed of spit formation is Nauset Beach spit at Cape Cod, Massachusetts. It has grown as the result of the available abundant sediment supply, 9 km in 120 years (U. S. Army, Chief of Engineers, 1957).

SUBSURFACE EVIDENCE OF BARRIER DEVELOPMENT

The presence of barrier island sands over pretransgression surfaces is not necessarily sufficient proof for barrier island formation through beach and dune ridge engulfment. It must also be proven that the total section between the pretransgression surface and the surface of the islands was formed in the supratidal environments. In the northern Gulf area, drilling information from three recent and one subrecent barrier island provides sufficient data to indicate formation of these islands from submarine shoals and not from engulfed beach ridges:

Padre Island, Texas

Two cross sections across central Laguna Madre and Padre Island indicate (Fisk, 1959, Figs. 12–13) that the present Padre Island barrier sands were deposited over earlier Holocene open lagoon, bay-sound, and shoreface sediments. All these sediments represent prior stages of the same transgressive sequences and are marine and brackish water deposits (Fig. 1a).

Galveston Island, Texas

The barrier island sands exposed on the present island surface (eolian, beach and upper shoreface sands) were found to be underlain by burrowed, laminated, thin-bedded, well-sorted, very fine middle shoreface sands with thin layers of silty clay. The vertical sequence of sedimentary features within the barrier island sand sequence from bottom to top represents the offshore-to-shore facies succession (Bernard and others, 1962, p. 202), clearly indicating the development of the island by simultaneous upward aggradation and seaward progradation of Holocene sediments over the Pleistocene land surface. Galveston Island began as a small offshore bar in 1.5 to 2.5 m of water, and during its longshore growth it has covered the Holocene Brazos River prodelta silty clays toward the southwest. The first sediments on the Pleistocene land surface were deposited in water depths ranging between the low tide level and 14-m depth. During subsequent aggradation the submerged bar slowly

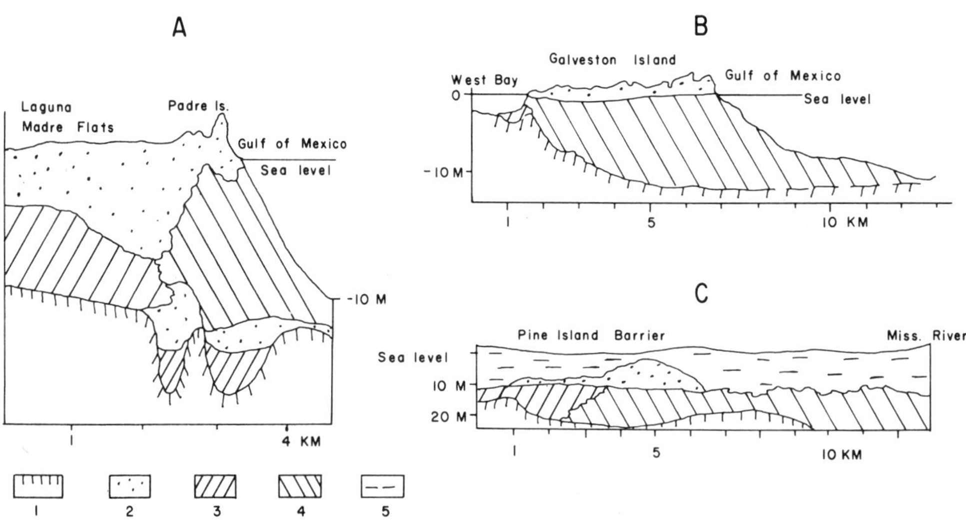

Figure 1. Cross sections across three Gulf Coast barrier islands parallel with the depositional dip. (A) Padre Island (Fisk, 1959, Fig. 13; (B) Galveston Island (Bernard and others, 1962, Fig. 12 a, b, c); (C) Pine Island (Saucier, 1963, Fig. 18). Symbols: 1: Pleistocene; 2: Holocene beach foreshore and eolian complex; 3: Holocene brackish lagoonal, bay-sound and estuarine sediments; 4: Holocene open marine subtidal shoreface sediments; 5: Holocene alluvium.

reached sealevel, and beach and dune sediments started to develop on the newly emerged island (Fig. 1b).

Horn Island, Mississippi

The Pleistocene surface in the Mississippi Sound and adjoining open Gulf area is marked by a yellowish brown, violet-colored, stiff, hard clay, representing an old soil profile. Under Horn Island and offshore under the Gulf, a 3-m thick earlier Holocene grey clay lies over the soil profile, in turn overlain in Horn Island by 12 m of clean quartz barrier sand (Ludwick, 1957, p. 37). This indicates that Horn Island and probably also the similarly developed other Mississippi Sound barrier islands did not form from coastal ridges on the Pleistocene surface.

Pine Island Chain, Louisiana

This subrecent, 80-km-long barrier chain, located under New Orleans, started its formation when sealevel was about 7 m below the present level. It has grown westward from the direction of the present Mississippi shores (Saucier, 1963, p. 52–54, Fig. 18; Weidie, 1968). The Pine Island barriers formed over Holocene bay-sound and nearshore Gulf deposits (Fig. 1c) during an advanced stage of the Holocene transgression. No indication exists that they developed from engulfed coastal ridges on the pretransgression land surface.

CONTEMPORARY ISLAND AND SHOAL FORMATIONS

Hoyt (1967, p. 1126–1127) maintained that only a few small, short-lived barrier islands, located close to the shoreline, formed from offshore bars. Records from the Louisiana and Mississippi coastal areas do not support this idea. Several minor and major examples of barrier island development from underwater shoals took place during historic times in the Chandeleur, Mississippi Sound, and Timbalier Islands' groups, partially in connection with barrier island migration. Barrier islands (for example Horn, Petit Bois, and Timbalier Islands) grew westward by building shoals up and into island surfaces in the direction of littoral drift, while the opposite island ends were reduced by erosion to submarine shoals. In the southern Chandeleur Island group, the 7.5-km-long Grand Gosier Island has developed since 1869, through stages of development and merger of small islands, from the shallow seafloor. During the past 130 years, erosion has also turned whole islands into shoals, among them the Dog Keys between Horn and Petit Bois Islands, Massacre Island (northwest of Petit Bois Island), and Myth Island in the southern Chandeleurs.

EXAMPLES OF BARRIER ISLAND MIGRATION

A few examples of island migration in the northern Gulf illustrate the possibility that barrier islands can be completely detached from their sites of formation and moved several kilometers away. Thus, islands which might have formed by dune and beach ridge engulfment could have been easily lodged, in some instances, over bay-sound, marsh, deltaic, or open marine sediments, depending on the direction of migration. Subsequent migration would completely obscure the conditions of genesis.

Frontal-Lateral Progradation

Galveston Island, Texas. The greatest progradation distance is found in the direction of the littoral drift and currents, but significant progradation also takes place seaward, perpendicular to it. Minor amounts of landward (lagoonward) progradation is provided by eolian and washover deposits that are derived from the barrier islands and are slowly filling the lagoon.

Lateral Progradation

Ship, Horn, Petit Bois, and Dauphin Islands, Mississippi Sound. Historic evidence indicates that these barrier islands migrated dominantly toward the west, in the direction of the predominant littoral drift and currents. Only minor landward and seaward shifts of the island shorelines occurred. In addition to the changes in the positions of the passes between the islands, due to island migration two new inlets are known to have been created since the end of the 18th century. The latest one which separated Ship Island into a western and an eastern half formed as the result of a 1946 hurricane (Fig. 2). The greatest westward advance (Horn and Petit Bois Islands) exceeded 5 km during the past century, while the maximum amount of retreat on ends of the eastern island (*in* Petit Bois Island) was over 12 km. Among the minor islands in the chain, Pelican Island grew southeastward 1.6 km between 1860 and 1909, and Sand Island (at Dauphin Island) since 1860 has been driven northward by almost 1 km

Figure 2. Migration of Mississippi Sound barrier islands between the mid-19th century and 1966 (based *on* U.S. Coast Survey charts and Glezen, 1951). Arrows indicate dates of inlet and pass openings. Shaded area, approximate location of barrier islands in 1848-1856.

(Glezen, 1951, p. 4). A well-documented example of the seaward shift of the Sound shores of one of the major islands, Ship Island, is found at Old Fort Massachusetts. In a century the shore has retreated here toward the Gulf by about 0.3 km (Foxworth and others, 1962, p. 28). Given a much more substantial littoral-drift sediment supply, the Mississippi Sound islands probably could have merged and prograded Gulfward in the same manner as the Texas barrier islands.

A significant amount of shoreward component can be detected in the lateral migration of Timbalier Island, south of Timbalier Bay, Louisiana (U.S. Army, Chief of Engineers, 1962). As indicated on Figure 3, in 69 years the island has shifted northwestward parallel with the direction of the dominant littoral drift. During this migration the island has moved 6 km westward and 1.6 km northward in the direction of the mainland. In this process the island has built itself over bay and marsh sediments deposited in Timbalier Bay.

Predominantly Frontal Barrier Island Retreat

Chandeleur Islands, Louisiana. The Chandeleurs, south of Ship Island, Mississippi, form an approximately 90-km-long arcuate island and shoal chain, trending in a northeast-southwesterly direction toward the presently active Mississippi delta. The width of the islands at sea level is generally less than 1.5 km, but a 6- to 25-km wide, 1- to 2-m thick, discontinuous sand apron adjoins it on the east, including sand bars of 5- to 7-m thickness (Frazier, 1967; Ludwick, 1957). A few borings suggest that the sand thickness below sealevel under the islands is from 5 to 12 m (Ludwick, 1967; Shepard, 1960). The islands and shoals developed from the destruction of two Mississippi River subdeltas which were active from 3000 to 1800 B.P. No evidence exists that the arcuate island pattern reflects the shape or location of the ancient delta fronts. The reworked delta sediments form not only the island and shoal chain but also the seaward sand apron and submerged bars. Under the impact of easterly and southeasterly waves and currents the island-shoal chain is slowly shifting westward. Storm washovers and eolian sediment transportation are the main means of this westward movement. According to Treadwell (1955) between 1852–1875 and 1917, the maximum amount of shift was 0.7 km, as measured along the eastern shorelines of the islands. Ludwick (1957) stated that the amount of migration was 0.4 to 1.6 km in a century. Russell (1936) cited an 0.8-km shoreline shift within 5 years from northern Chandeleur Island. Comparison of 1853 and 1966 charts indicates a total of 3-km westward shore shift in the same area. In the south around Breton Island no significant shoreline shifts can be detected, except for the disappearance and emergence of certain island sections. The rate of island migration in the case of the Chandeleur complex is much slower than migration rates with lateral progradation and retreat, due to the much wider shore front which the coastal erosion must attack. The retreating barrier islands in this case are driven over relict deltaic and Recent bay-sound-type sediments.

East Timbalier Island, Louisiana. A frontal retreat of East Timbalier Island between Timbalier Island and the mainland between 1890 and 1934 amounted to 2 km (U.S. Army,

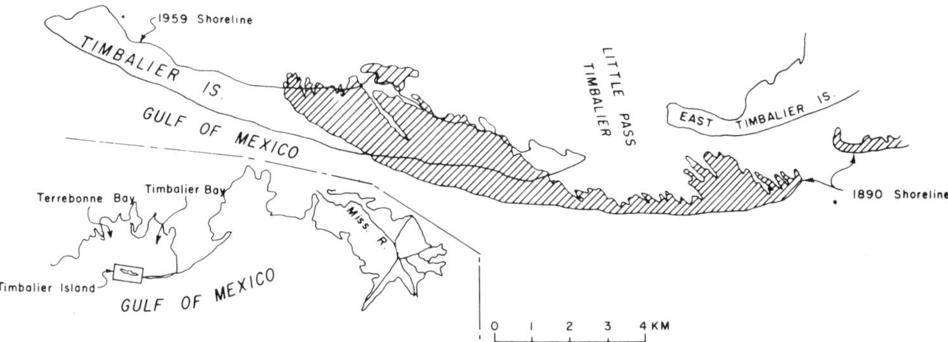

Figure 3. Migration of Timbalier Island between 1890 and 1959. (Adapted *from* U.S. Army, Chief of Engineers, 1962).

Chief of Engineers, 1962, p. 14). During this retreat, fragmentation and large scale area reduction of the island also took place. The coastal retreat was the direct result of the approximately same amount of recession suffered by the mainland shore lying immediately to the east in alignment with the island shoreline. This was and is the area from which sediment through littoral drift reaches the island. Since 1934, the recession of the island and mainland shores has almost stopped; the island was re-integrated from the fragmented islets, has widened itself bayward, and has grown westward 6.6 km.

Isles Dernieres, Louisiana. This barrier island chain is the westward continuation of the Timbalier group. Sparse littoral drift sediment reaches the islands which suffered very extensive erosion and fragmentation during historic times. The thickness of beach and shoreface sands under the sealevel is only about 1 m (U.S. Army, Chief of Engineers, 1962). These sands overlie a few meters of marsh deposits over which they were driven during island retreat. The islands do not retreat in the same manner as the Chandeleurs do; extensive shore erosion occurs on both the seaward and the bayward shores. Shore retreat on the Gulf side amounted to 0.2 to 0.7 km between 1890 and 1959, and it was accompanied by only insignificant lateral drift progradation.

Barrier Island with No Significant Progradation: Grand Isle, Louisiana

The 10- to 12-m thick barrier sand unit overlies delta-front silty clays and silty sands of the *Bayou Blue* Mississippi subdelta (Frazier, 1967). Since the inception of the island about 2000 B.P., no significant seaward or landward progradation or retreat occurred, indicating a dynamic balance between the erosion and accumulation by littoral drift. Only minor amounts of lateral progradation took place on the island.

BARRIERS AND CHENIERS—A GENETIC OR TOPOGRAPHIC DISTINCTION?

Beach ridge progradation of coastal plain, barrier spit, and barrier island shores is essentially an identical process. New ridges added to the land form not only by seaward beach expansion but also by spit formation and the development of nearshore submarine bars aggrading up to sealevel. In the latter case the lagoon between the emerged bar and the land is subsequently filled in, and the bars (barriers) become part of the shore. Strand plains are underlain by a continuous sheet of beach and shoreface sand, while in chenier plains clusters of sandy ridges alternate with tidal flat zones. New ridges along the seaward margins of strand and chenier plains also form as beach ridges and as offshore bars, tied later to the land. Ridge configuration on these mainland plains is very strongly influenced by the location of the coastal streams, the position of which is a lesser factor in the distribution of barrier and also of barrier spit ridges (for example, on Mobile Point, Bolivar, and Matagorda Peninsulas). Seaward accretion by chenier plain growth happened occasionally not only on mainland coasts but also on lower energy former barrier island shores (Tanner, 1966, p. 92).

The dimensions of medium-sized and major

strand plain ridges, cheniers, and chenier clusters are comparable with those of the smaller and medium-sized barriers and barrier islands. Sand thicknesses of 7 to 13 m are found in both types. The largest known chenier ridge complexes equal in size the largest Gulf of Mexico barrier islands, although the western Louisiana cheniers are much smaller than these barriers. Composite chenier ridges 1–3.5 km wide occur in great numbers in the Rhone delta area (Kruit, 1955) and in Surinam, northeastern South America (Brouwer, 1953). The thickness of certain chenier sand ridges was found to exceed 10 m in southeastern Louisiana (Otvos, 1969).

ACKNOWLEDGMENTS

Sincere appreciation is expressed for the courtesy of Gulf Research and Development Company in making several unpublished reports available. The author has greatly profited from his discussions of the subject matter with A. E. Weidie and W. Glezen. Dr. Weidie kindly read the manuscript.

REFERENCES CITED

Bernard, H. A., LeBlanc, R. J., and Major, C. F., 1962, Recent and Pleistocene geology of southeast Texas, *in* Rainwater, E. H., and Zingula, R. P., *Editors*, Geology of the Gulf Coast and Central Texas: Guidebook of excursions: Houston Geological Society, p. 191–206.

Brouwer, A., 1953, Rhythmic depositional features of the East-Surinam Coastal Plain: Geol. en Mijnbouw, Nw. Ser., v. 15, p. 226–236.

Bruun, P., 1962, Sea level rise as a cause of shore erosion: Jour. Waterways and Harbors Div., Am. Soc. Civ. Engineers Proc., v. 88, p. 117–130.

Fisk, H. N., 1959, Padre Island and the Laguna Madre flats, coastal south Texas: 2nd Coastal Geography Conf., Louisiana State Univ., Baton Rouge, Louisiana, p. 101–151.

Foxworth, R. D., Priddy, R. R., Johnson, W. B., and Moore, W. S., 1962, Heavy minerals of sand from recent beaches of the Gulf Coast of Mississippi and associated islands: Mississippi Geol. Survey Bull. 93, 92 p.

Frazier, D. E., 1967, Recent deltaic deposits of the Mississippi River: their chronology and development: Gulf Coast Assoc. Geol. Soc. Trans., v. 17, p. 287–315.

Glezen, W. H., 1951, Changes in the barrier bars of Mississippi Sound recorded on maps and charts from 1710 to 1948, Unpub. memorandum: Gulf Research and Development Co., 8 p. (available on interlibrary loan).

Hoyt, J. H., 1967, Barrier island formation: Geol. Soc. Amer. Bull., v. 78, p. 1125–1136.

——1968, Barrier island formation: Reply: Geol. Soc. Amer. Bull., v. 79, p. 1427–1432.

—— 1969, Chenier *versus* barrier, genetic and stratigraphic distinction: Am. Assoc. Petroleum Geologists Bull., v. 53, no. 2, p. 299–306.

Kruit, C., 1955, Sediments of the Rhone delta: 1) Grain size and microfauna: Rijksuniv. te Groningen, The Hague, Mouton & Co., p. 357–514.

Ludwick, J. C., 1957, Lithofacies patterns on the continental shelf in the northeastern Gulf of Mexico: Gulf Research and Development Co., Rept. no. 58, 77 p.

Otvos, E. G., 1969, A subrecent beach ridge complex in southeastern Louisiana: Geol. Soc. Amer. Bull., v. 80, no. 11, p. 2353–2358.

Rusnak, G. A., 1960, Sediments of Laguna Madre, Texas, p. 153–196 *in* Shepard, F. P., and others, *Editors*, Recent Sediments, Northwest Gulf of Mexico: Tulsa, Oklahoma, Am. Assoc. Petroleum Geologists, 394 p.

Russell, R. J., 1936, Physiography of the Lower Mississippi River Delta, *in* Reports on the Geology of Plaquemines and St. Bernard Parishes: Louisiana Dept. Conserv. Bull., no. 8, p. 3–199.

Saucier, R. T., 1963, Recent geomorphic history of the Pontchartrain Basin: Louisiana State Univ. Studies Coastal Studies, ser. no. 9, 114 p.

Shepard, F. P., 1960, Gulf coast barriers, p. 197–220 *in* Shepard, F. P. and others, *Editors*, Recent Sediments, Northwest Gulf of Mexico: Tulsa, Oklahoma, Am. Assoc. Petroleum Geologists, 394 p.

Tanner, W. F., 1966, Late Cenozoic history and coastal morphology of the Apalachicola River region, western Florida; deltas in their geologic framework: Houston Geol. Soc., p. 83–105.

Treadwell, R. C., 1955, Sedimentology and ecology of southeast coastal Louisiana: Coastal Studies Inst., Louisiana State Univ. Tech. rept. no. 6, 78 p.

U. S. Army, Chief of Engineers, 1957, Chatham, Mass. beach erosion control study: House Document no. 167, Washington, 37 p.

—— 1962, Belle Pass to Raccoon Point, Louisiana, beach erosion control study: House Document no. 338, Washington, 31 p.

Weidie, A. E., 1968, Bar and barrier island sands: Gulf Coast Assoc. Geol. Soc. Trans., v. 18, p. 405–415.

MANUSCRIPT RECEIVED BY THE SOCIETY MAY 26, 1969

PRINTED IN U.S.A.

Editor's Comments on Paper 29

29 Pierce and Colquhoun: *Holocene Evolution of a Portion of the North Carolina Coast*

Pierce and Colquhoun cooperated again in 1970 to reiterate the ideas proposed in their earlier report. The reader will recall that they identified a primary barrier built over a terrestrial surface and a secondary barrier overlying marine deposits.

Based on stratigraphic evidence collected in extensive field work, they suggest that the North Carolina barrier system originated as a primary barrier by the partial submergence of a preexisting ridge, quite possibly a barrier, which occupied the site of shoals and small islands now situated in the lagoon, formed during the Wisconsin regression. They find nothing to indicate that the present landward margin of Pamlico Sound was ever an open mainland shoreline. Longshore drifting of sediment, provided by both headland erosion and transverse transport, prograded the barrier seaward and developed extensive, secondary barrier spits. With continued migration, the primary and secondary type barriers coalesced to form the present-day system. Further reworking will totally destroy all evidence of the origins of the respective segments.

The last sentence of their conclusions is particularly worth noting: "Barriers are not formed by mutually exclusive methods, but can result from submergence of a topographic ridge or spit extension during a transgression, or both." The dominance of one type of an origin over another, so stoutly defended by some earlier workers, would appear, here, to be on the wane.

Reprinted from the *Geol. Soc. Amer. Bull.*, **81**, 3697–3714 (1970)

J. W. PIERCE *Smithsonian Institution, Washington D.C. 20560*
D. J. COLQUHOUN *Department of Geology, University of South Carolina, Columbia, South Carolina 29208*

Holocene Evolution of a Portion of the North Carolina Coast

29

ABSTRACT

The presence of a soil zone under a portion of the large barrier chain, which makes up much of the North Carolina coast, and the absence of this zone under the remainder of the barrier has led to a re-evaluation of the mode of formation and the evolutionary development of this stretch of coastline during Holocene time.

A primary barrier, formed during a rising sea, became detached by flooding of the area behind a mainland beach. This primary barrier apparently reoccupied the position of a barrier formed during a stillstand within the general Wisconsin regression.

The present coastline has evolved from the primary barrier by retreat and migration of headlands as well as formation of secondary barriers, by spit elongation, over the adjacent continental shelf. Approximately 61 percent of the present barrier chain is secondary in nature; whereas the remaining 39 percent is modified primary.

Continued retrogression of the barrier islands eventually will superimpose the secondary barriers over the primary. When this happens, evidence that a secondary barrier existed will be obliterated. The extreme case results when this compound barrier is driven landward until it eventually reaches the mainland, hence a mainland beach. At this time, no evidence remains that any barrier previously existed. The presence of a secondary barrier would suggest the landward presence of a primary barrier or a mainland beach.

INTRODUCTION

Barrier islands are complex features on many of the stable and less stable coasts of the world. Scientific interest is high on the mode of formation of barrier islands, their evolution, and their response to marine processes. This interest is generated by potential loss of valuable recreational land and potential petroleum reservoirs that buried barrier islands may have.

The methodical study of barrier chains was given impetus by the results of the American Petroleum Institute Project 51 on the northwest Gulf of Mexico (Shepard and others, 1960). The tendency has been to use the Gulf of Mexico as a model that is relevant to all barrier islands. Since the publication of this work, many studies have appeared on other barrier islands or chains.

Our paper developed from a study of the large barrier island chain off the coast of North Carolina (Fig. 1). Results include an interpretation of data obtained from the samples of 45 bore holes on or near the barrier chain, interpretation of changes along the barriers from U.S. Coast and Geodetic Survey charts, and a comparison of aerial photographs from different flights. Interpretation deals primarily with the latest part of the Holocene history of the area, but is extended into the Pleistocene when evidence permits.

FORMATION OF BARRIER ISLANDS

Until recently, it was argued that barrier islands formed in the ways proposed by de Beaumont (1845) or by Gilbert (1885). De Beaumont suggested that barriers formed from the erosion products of the sea floor as a result of wave action: first, a submarine bar and trough form; the bar is eventually built above sea level and becomes a barrier. This method was unequivocally accepted by Johnson (1919) who made topographic sections across many barriers. Johnson did not have data from borings through barriers and apparently did not consider the effect of upward growth of barriers during the Holocene rise in sea level.

Gilbert (1885) believed that barriers were formed primarily through longshore drift and

Geological Society of America Bulletin, v. 81, p. 3697–3714, 9 figs., December 1970

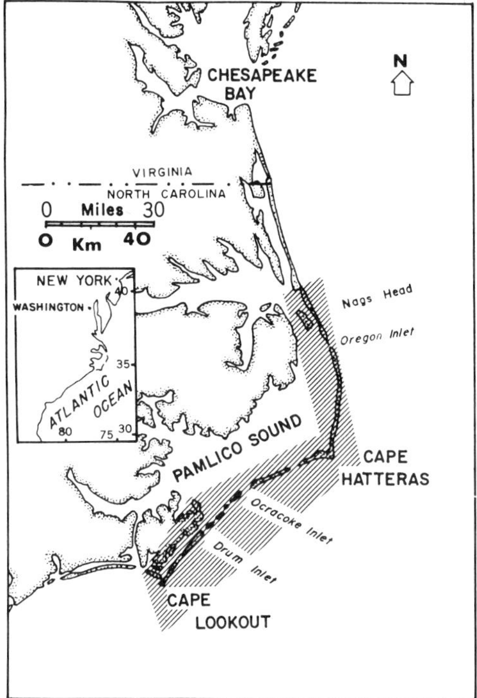

Figure 1. Area of study along the coast of North Carolina.

were spits that had been extended for long distances by digital progradation from headlands or deltas.

Studies of the Gulf Coast barriers did not resolve the question of the relative merits of these conflicting viewpoints, but concluded that each process contributed some material to present-day barriers (Shepard, 1960, p. 214). It has been brought out that barriers along the Texas portion of the Gulf Coast lie, for the most part, on a substrate of weathered material, although some progradation has apparently occurred (Shepard and Moore, 1955; Shepard, 1960).

Colquhoun (1965, 1969a, 1969b), in a study of borings through Pleistocene barriers in South Carolina, found that these older features could be separated into two types, essentially based on the type of substratum on which they rest. Primary barriers were formed at the juncture of the mainland and ocean and rested on subaerially weathered and eroded surfaces. Secondary barriers, on the other hand, rested on continental shelf facies with no evidence of a subjacent subaerially weathered surface. It was postulated that these secondary barriers formed as spits, extending from deltas or headlands, across a nearby continental shelf. More than one secondary barrier could be present. The secondary barrier (or barriers) is located seaward of the primary barrier.

Hoyt (1967) proposed a modified version of a concept, originally proposed by McGee in 1890, for the formation of a barrier chain during a period of rising sea level. As originally proposed by McGee, a mainland beach (Colquhoun's primary barrier) became stabilized along a pre-existing topographic high and maintained its position by upward growth during the rise in sea level. Flooding of the lowland, behind the high, gave rise to lagoons and barrier detachment. As modified by Hoyt, the mainland beach necessarily did not exist along a topographic high, but beach or dune ridges provided the necessary height to permit flooding behind the beach and detachment from the mainland.

Proponents of the various methods of formation of barrier islands suggest that only one method is needed to account for nearly all barrier islands. Except in a minor way, a combination of methods is excluded, although Shepard (1960) stresses that both offshore and longshore sources are important.

Barrier chains are complex features. The end result of the evolution and modification of a primary form, an intermediate stage between the initial shape and a theoretical end result, or a combination of stages are available for study. The barrier-island chain off North Carolina is a classical example of a complex feature. It is far removed, in places, from the mainland; it forms a salient onto the continental shelf; it is exposed to a great deal more wave energy from the Atlantic Ocean than areas to the north or south are (Tanner, 1960). A study of the stratigraphy of these barrier islands should yield evidence on the origin and evolution theories of barrier islands in general.

PHYSIOGRAPHY OF REGION

The barrier islands of North Carolina comprise a portion of the large barrier chain that borders most of the east coast of the United States and extends, with few interruptions, from New Jersey to southern Georgia. The lagoons separating the barrier from the mainland range from narrow to wide and, in some places, are absent.

In the North Carolina portion only, or nearby, are found several of the different types of lagoon–barrier island complexes. A large part of the coastal portion of North Carolina is composed of barrier islands separated from the mainland by wide lagoons; in some instances lagoons are over 25 mi wide (Fig. 1). Near the North Carolina-Virginia line, and in the southern part of the study area, the lagoons are relatively narrow. The barrier chain becomes attached to the mainland just south of Chesapeake Bay and creates a mainland beach.

Man's interference with natural processes is limited along this portion of the North Carolina coast. Except for isolated areas that are privately owned, most of the barrier islands are part of two National Seashores, which have precluded their extensive development and have preserved much of them in a seminatural state. There are no inlets protected by jetties along this stretch of coast and little is done in the way of inlet maintenance by dredging.

Barrier Islands

The islands making up the barrier chain are relatively narrow. The maximum width is about 2.5 mi at Cape Hatteras. Much of the barrier has widths of less than 0.5 mi and long stretches are less than 0.25 mi wide.

Elevations are generally less than 10 ft, except for the rare dune fields. Natural dunes are confined to a relatively few places on the islands; the highest dunes are on Cape Hatteras and Cape Lookout where they rise to elevations of more than 40 ft above sea level. South of Ocracoke Inlet for about 8.5 mi, the west side of the barrier is marked by older, stabilized dunes of higher elevations, which are covered with vegetation. In addition to the natural dunes, a line of artificial dunes has been built by the National Park Service along the barrier from Ocracoke Inlet to Nags Head. Dune restoration also has begun on that portion of the islands immediately south of Ocracoke Inlet.

Prominent beach or dune ridges are present only on Cape Hatteras and Cape Lookout. The beach-dune ridges on Cape Hatteras are truncated mostly on the east, suggesting that Cape Hatteras was located farther east at one time. Those ridges on Cape Lookout, on the other hand, are truncated on the southwest side, indicating that the primary erosion of Cape Lookout is occurring on the south and west sides. Fisher (1967) has pointed out that recurved beach ridges are present at intervals along the barriers, but it is not clear whether these are evidence of recurved spits, inlet migration scars, or possibly the result of wind action. In general, these latter ridges are small and indistinct.

Washover fans are prominent features along much of the west side of the barrier. The processes involved in the formation and nourishment of these features must move large volumes of sediment into the lagoons (Pierce, 1969) and may cause a widening of the barrier by addition of material on the lagoon side.

Five permanent inlets break the continuity of the barrier chain. A large tidal delta has formed inside the barrier line at each inlet. The magnitude of these deltas suggests the loss of large volumes of sand from the longshore system. The seaward tidal deltas, in general, are poorly developed. In most cases, no more than a bar exists. The reworking of seaward deltas is probably due to the relatively high wave energy at this section of the coast.

The beach undergoes seasonal changes, alternating between the low, gentle profile of the summer beach and the steep profile of the winter beach (Leith, 1970). Historically, the beach front has been retrogressing over much of its length. Outcrops of peaty sand are present along a considerable portion of the barrier chain: the stretch of coast from Oregon Inlet (Fig. 1) to about 12 mi north of Cape Hatteras has intermittent peat outcrops, as does the stretch from Drum Inlet (Fig. 1) to about 5 mi north of Cape Lookout. This peaty sand is a barrier flat-lagoon sediment that is exposed and undergoing erosion because of the retrogression of the barrier. In many instances, the outcrop of peaty sand is continuous from just below low-tide level on the front of the beach into the present-day marsh on the west side of the barrier. Continuous outcrops have been traced on the downdrift side of two of the inlets from the beach front to the marsh. These exposures are expected because similar outcrops are present along much of the east coast of the United States (Kraft, 1968; Oaks, 1965; Hoyt, 1970).

Lagoons

The barriers are separated from the mainland by a series of lagoons of various sizes. At the north and south ends of the study area, the lagoons are relatively narrow and the barrier lies from 1 to 7 mi from the mainland. Pamlico Sound (Fig. 1) lies between the barrier and the mainland in most of the area. This lagoon is 25 mi wide by about 55 mi long. Depths in the lagoons are less than 25 ft, except in isolated deep holes where depths may reach 35 ft.

Linear shoals are common in Pamlico Sound (Pickett and Ingram, 1969) and apparently have varied origins (Welby, 1970). A persistent line of shoals and small islands lies in Pamlico Sound about 3.5 mi west of the present barrier, between Ocracoke Inlet and 11 mi north of Cape Hatteras.

The shorelines of the lagoons are very irregular, except for the southern shore of Pamlico Sound. This shore has been straightened by waves which are generated by the prevailing northeasterly winds blowing over the longest available stretch of water inside the lagoon. Irregular shorelines elsewhere suggest that the lagoonal shores have not been exposed to waves from the open ocean.

METHODS

Field Methods

Thirty-three bore holes were drilled on the barrier islands (Fig. 2). An additional 11 holes were drilled on Roanoke Island (Fig. 2) and 7 on the mainland north of Cape Lookout (Fig. 4). All 51 holes were drilled with a truck-mounted auger unit.

Drilling locations were limited to locations of parking areas or other paved areas. Because of of these limitations, the section along the barrier chain is essentially two dimensional. Few cross sections could be drilled across the barrier islands because of inability to get around in the marsh or on the beaches. Most of the holes are located on the barrier flat at the west of the barrier, except in the vicinity of Cape Hatteras.

What may appear to be a serious shortcoming in data acquisition and a hindrance to interpretation is more apparent than real, because of the extremely narrow width of the barrier in this area. In most cases, the holes are within 500 yds of low-tide line and, in some cases, within 100 yds. Thus, any frontal progradation of the barrier, not found in the borings, would amount to less than 500 yds and generally less than 200 yds. The persistent outcrops of peat on the foreshore near low-tide level would indicate also that frontal progradation is not important along much of this barrier chain. Progradation is important in certain places, as suggested by historical changes.

Drilling was accomplished primarily by power augering using a truck-mounted rotary table to turn the auger rods into the ground and a hydraulic lift system to pull them out; otherwise continuous coring procedures were used. No wash boring was done.

The rods were continuous flight augers in

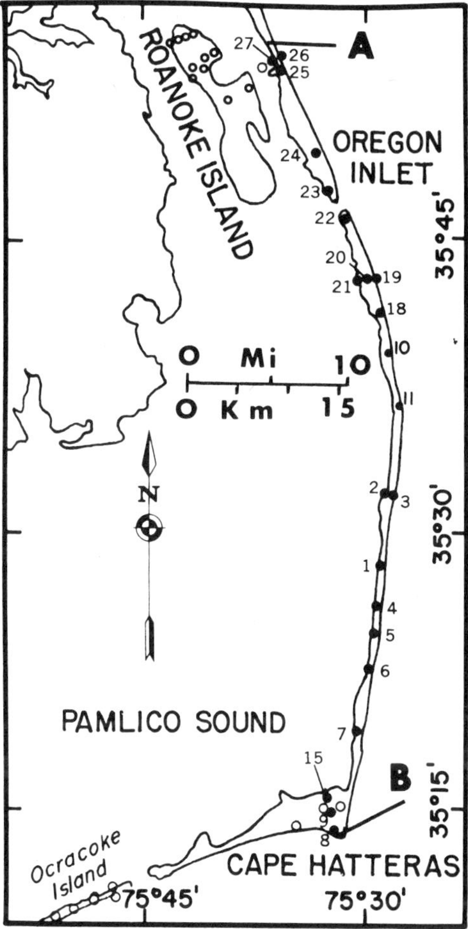

Figure 2. Location of bore holes on the barrier chain. Black dots, holes shown on the cross section, Figure 3; open circles, additional holes drilled, but not included on any cross section.

5-ft lengths. Continuous flight augers permit drilling to depths of several tens of feet without interruption other than stopping to add rods.

The general practice in this work was to auger to approximately 40 ft and then pull the rods. Pulling was accomplished by the hydraulic lift system, if at all possible, rather than using the rotary table to back the rods out of the ground. In a few cases, it was necessary to back part way out of the hole before a straight lift could be accomplished. Reversing of the rotary motion in order to back out of the hole has a tendency to remove the sediment on the lower part of the auger string. If straight pulling is used, sediments on the rods are within about 1 ft of true depth.

Samples were taken at 5-ft intervals or immediately above and below each visible lithologic change. Samples were taken from near the stem of the rods rather than the outer part where a considerable amount of smearing occurs.

If additional penetration was desired, the hole was re-entered and augering continued to a new total depth. The rods were pulled, and samples were taken from the rods which extended into the new hole. No samples were taken from the older portion of the hole.

Elevations of the drill sites range from 2 to 10 ft, as determined by transit surveys where elevations were greatest or by topographic maps in the area of least relief.

Shell Development Company provided descriptions, size analysis, and colored pictures of core samples of three holes drilled on Cape Lookout. These holes were cored continuously from the surface to depths of about 100 ft. Field descriptions of two wash borings drilled by Shell Development Company about 10 mi north of Cape Lookout were also provided.

Gross descriptions of samples recovered from water wells drilled by the U.S. Geological Survey also were studied to determine lithologic change at depths greater than our penetration.

Splits of samples from various environments within the lagoons were obtained from R. L. Ingram and splits from the continental shelf were obtained from Orrin H. Pilkey.

Laboratory Methods

Splits of samples were analyzed for size parameters, by a Rapid Sediment Analyzer, and for percent of carbonate. The fauna in the samples from sixteen holes was examined in detail and compared to the distribution of the forms living in the area today in order to define the environment of deposition of each sample. In some cases, mixed faunas or absence of fossils precluded a biologic determination of environments. In these cases, the samples were examined under a binocular microscope. The combination of visual observations, size parameters, and percent of carbonate permitted a matching of the sample with characteristics distinctive of an environment existing along the barrier chain today as determined by Rochna (1961), Duane (1962, 1964), and Pickett and Ingram (1969). This procedure, plus a rapid examination of the fauna, was used for determination of the environment of deposition of the samples from the borings that were not studied in detail.

DEFINITION OF ENVIRONMENTS

The physical attributes of the sediments from different environments are covered in Rochna (1961), Duane (1962, 1964) and Pickett and Ingram (1969), and are similar to those used by Shepard and Moore (1955) for differentiation of environments along the Gulf Coast of the United States.

Most environmental interpretations rely heavily upon the faunal remains present in the samples. In this study, the entire population of the sample was considered so that no undue reliance was placed on the presence or absence of one form. We were careful to distinguish forms transported from their life habitat and those that have been reworked from Pleistocene strata. At times, such discriminations are exceedingly difficult or impossible, although extremely critical to the interpretation.

STRATIGRAPHY

Barrier Islands, Cape Hatteras Area and North

Twenty-six holes were drilled along the barrier chain from the area of Cape Hatteras northward (Fig. 2). The sequence of interpreted environments in 22 of the 26 holes is shown on Figure 3. The sedimentary sections penetrated in the upper part of 20 of the 26 holes display a sequence of barrier sands overlying lagoonal sediments (Fig. 3). This sequence is expected beneath a barrier undergoing retrogression, as is indicated by historical studies and the outcrops of peaty sand exposed near low-tide level on the front of the barrier. Lagoonal sediments undoubtedly form the base upon which much of the barrier rests.

The remaining six holes did not penetrate the normal section. Two of these were in areas where inlets have existed within historical times and the sediments from these two holes exhibited the characteristics of inlet fill, a "dumped" deposit with mixed faunal assemblages. The other four holes, without lagoonal material under barrier-type sediments, were on Cape Hatteras where marine sediments were found directly beneath the barrier sediments.

Three of the holes drilled in the Cape Hatteras area and one from the north end of the the study area are not shown on the cross section (Fig. 3).

Sediments encountered immediately below

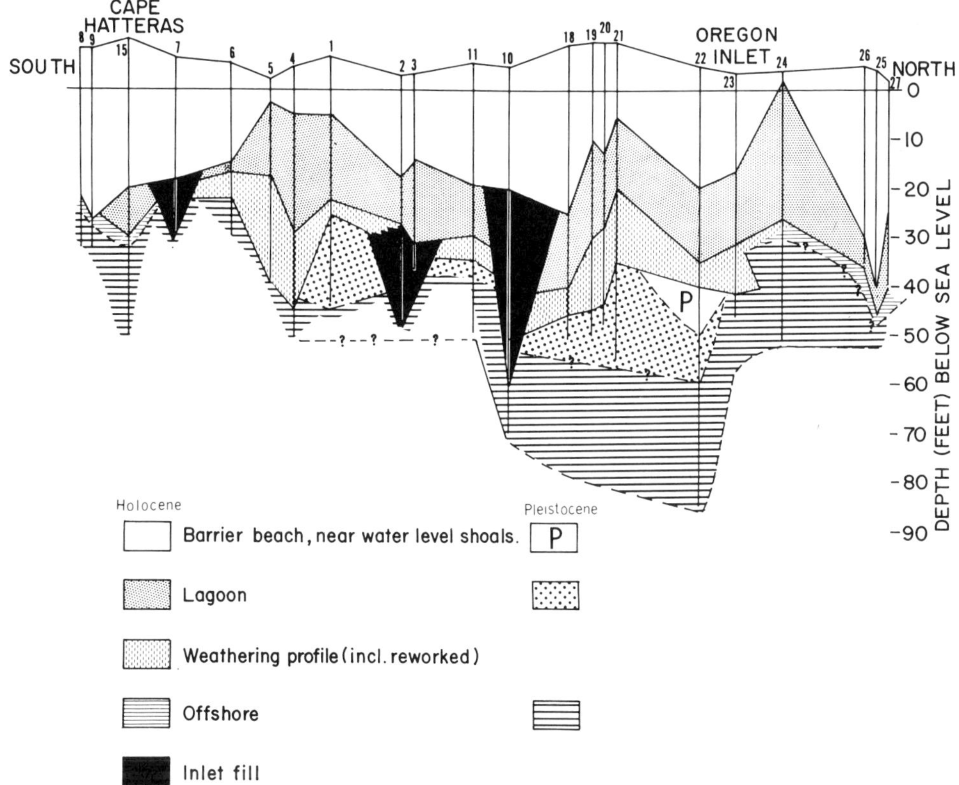

Figure 3. Cross section along the barrier chain. Hole locations shown on Figure 2.

the barrier-lagoon sequence were varied in texture, composition, and color, which resulted from deposition in different energy environments as well as postdepositional changes. Of the twenty holes which penetrated the barrier-lagoon sequence, eleven had reddish-brown, tan or yellow sediments, suggestive of a soil zone; two had reddish sediment, apparently reworked soil; two had marine sediments; one had inlet fill; and four did not encounter sediment other than that deposited in a low-energy environment (Fig. 3).

The most widespread unit below the lagoonal material is highly oxidized, reddish-brown, tan or yellow sediment and is found in eleven holes. The bright colors contrast markedly with the dull gray of the superjacent lagoonal sediment. Texture of the weathered material is variable ranging from a coarse, clean quartz sand to a slightly sandy, micaceous clay-silt. Elevations, at which this zone was encountered, ranged from −16 to −41 ft. These elevation differences, coupled with the large textural differences of the material, suggest that the color change is not the result of circulating groundwater or water-table effects.

If the color is due to circulating ground waters, the effect should be either limited to the coarser, more permeable sediments or the color should be more intense. Such is not the case. It is difficult to visualize how ground water could circulate through some of the oxidized finer sediment, which has little to no visible porosity and little apparent permeability. Water-table effects are not believed the cause of the color change because, logically, the effect should be at approximately the same elevation or, if perched, consistently above a nearly impervious layer, and this does not appear to be the case.

Therefore, the highly oxidized zone is interpreted as a soil zone. This soil zone is believed to have formed during the last low stand of sea level between the time of the regression of the sea from the area of the present coast (possibly about 40,000 yrs B.P.) and the time of its return to near the present level.

The environment of deposition of the sediments constituting the soil zone is difficult to determine because of the weathering effects. Fossils and most of the heavy minerals are absent. Therefore, interpretation of the environment of deposition is based on texture, lateral distribution of the sediment types, and is influenced greatly by the environment of deposition of the underlying sediments.

The oxidized, weathered sediment encountered in five of the eleven holes (4, 5, 6, 22, and 23) was of a much coarser texture than that of the other six. One group of these holes (4, 5, and 6) lies about 11 to 15 mi north of Cape Hatteras while the other two holes, separated from the first group by a distance of 21 mi, lie near Oregon Inlet (Fig. 2).

Hole 22 (Figs. 2 and 3) encountered weathered, medium to coarse sand overlying a humate-stained sand of similar texture and underlain by a marine, silty fine sand. Humate is depositing in the sands on the west side of the present-day barrier. Humic acids are mobile until they contact saline water, where precipitation is nearly instantaneous (Swanson and Palacas, 1965). Hole 23 penetrated dark-brown, iron-stained, gravelly, coarse sand below the Holocene barrier-lagoon sequence. This dark-brown coarse sand, lying over marine sands, constitutes the soil zone in this hole.

The soil zone in holes 5 and 6 consists of reddish-brown, medium to coarse, gravelly sand underlain by marine sand in hole 6; hole 5 bottoms in the iron-stained sand.

The weathered zone in hole 4 consists of two distinct lithologies. Immediately below the Holocene lagoonal sequence lies a reddish-brown, medium-grained sand which is underlain, in turn, by reddish, sandy, semicompact silty clay. Light-gray, fine-grained, marine sands are present under the reddish, silty clay.

The environment of deposition of the weathered sands in each of these five holes probably is part of a barrier island which was formed during a previous transgression-regression cycle. This barrier underwent retrogression as shown by humate-coated sands under clean, weathered barrier sands in hole 23 and silty clay in hole 4. The humate suggests proximity to the lagoon-side of a barrier while the silty clay is interpreted as lagoonal in origin. The presence of marine sands below the weathered zone in these five holes would indicate a progradation of this barrier prior to retrogression.

Six holes (1, 11, 18, 19, 20, 21) penetrate what we interpret as weathered lagoonal sediments (Fig. 2). Fine, silty clay to clayey silt constitute the soil zone in all of these holes and, in all cases, unquestionable lagoonal sediments are encountered below the soil zone.

Two holes (2 and 24) had sediments below the Holocene lagoonal section that were not as highly oxidized as that in the soil zone of the previous eleven holes. Hole 2 had a thin brownish to buff zone, less than 1 ft thick, at 21 ft (−18), containing pebbles up to 1.5 cm in diameter. We found an additional zone of pebbles in this hole at 35 ft (−32). The thin, brownish zone and pebbles are interpreted as soil, which was reworked by inlet cut and fill. The bulk of the sedimentary section in hole 2 has a mixed fauna and many characteristics of inlet deposits. This agrees with the material found in hole 3 nearby, in which no weathered material was found. A thin brown to buff sand at 35 ft (−31) in hole 24 lay below a medium-gray lagoonal, silty sand (Holocene) and overlay a light-gray marine sand. The reworking at hole 24 is believed to have been accomplished by marine agencies rather than inlet cutting as suggested by the absence of sediment characteristic of inlet fill.

The holes at the north end of the cross section, after starting in barrier sediment, penetrated deposits of a low-energy environment to total depth (Figs. 2 and 3). The soil zone is believed to be absent although the possibility exists that drilling depths were insufficient to reach it. In all cases where the soil zone was penetrated, the depth to the top was between 13 and 35 ft. The deepest penetration in the holes at the north end of the area was 50 ft, sufficient to reach the soil zone if it were present.

No weathered profile was found in the Cape Hatteras area to depths of 60 ft. Examination of the published data of water wells at two campgrounds near Cape Hatteras indicated no evidence that this type of soil zone is present to a depth of 153 ft (Kimrey, 1960). In the holes around Cape Hatteras, barrier sediments, or the barrier-lagoon sequence, are underlain by marine sediment.

Roanoke Island

Eleven holes were drilled on Roanoke Island (Fig. 2). In those holes near the east side of the island, beach and backshore sediments were encountered while those borings near the west side of the island penetrated deposits of a barrier-flat or a shallow lagoon. Nearly all of the sediments were highly oxidized. Roanoke Island is a portion of a barrier chain formed during a previous transgression-regression cycle, which is in agreement with a previously published interpretation (U.S. Congress, 1948).

Ocracoke Island

Seven holes were drilled on Ocracoke Island (Fig. 2). All seven started in barrier sediments followed at depth by material deposited in a low-energy environment. Sediment from no other environment was penetrated nor was a weathering profile present to depths drilled (80 ft).

A water-test well, near the center of Ocracoke Island, encountered, to a depth of 60 ft, yellow to brown, fine-grained quartz sand and yellow to brown, coarse- to medium-grained sand. From 60 to 102 ft, the sands are dark gray to brown, medium- to fine-grained, interbedded with dark-gray, fine-grained silty sand and streaks of clay (Harris and Wilder, 1964).

From such a description, it is nearly impossible to interpret the environment of deposition of the sediment at this location. Correspondence from the U.S. Geological Survey indicates that the samples are no longer available for study. Under these circumstances, it is assumed that to depths of 80 ft (the deepest of our holes) the interpretation of a low-energy environment is correct.

The possibility of a soil zone below 80 ft, as might be suggested by the brown coloration in the sample descriptions from the water well, cannot be discounted. It seems highly unlikely that this zone would be equivalent to the weathered zone that was encountered between 13 and 35 ft north of Cape Hatteras, although such a possibility must be considered. A more acceptable interpretation is part of a buried stream channel, formed at a time of lowered sea level (Welby, 1970).

Cape Lookout and Mainland

Aerial photographs show the existence of at least three features that have the characteristics of abandoned barriers, with an east-west trend, on the mainland north of Cape Lookout (Fig. 4). A continuation of the southernmost one, Harkers Island, can be traced a short distance southwest from the area of study. Seven holes were drilled on the three features between Cedar Island and Harkers Island. Again, lack of roads and hard surfaces prevented the best selection of locations for drilling. In addition to these seven holes, Shell Development Company furnished descriptions, pictures, textural data, and carbon 14 dates of samples from three holes drilled on Cape Lookout and made available

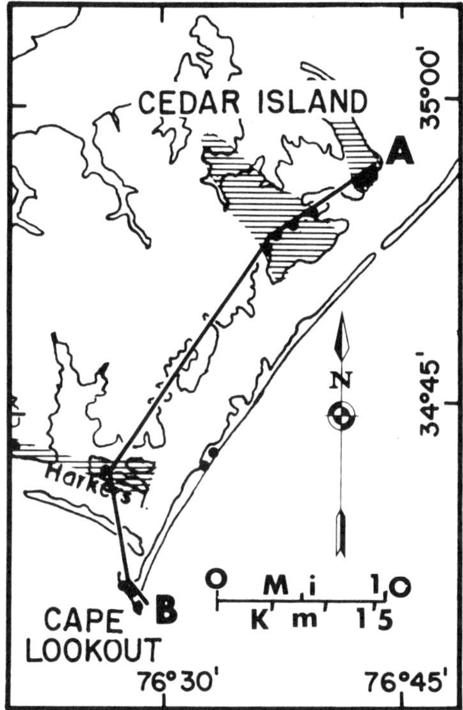

Figure 4. Location of bore holes on Cape Lookout and the mainland north of the Cape. Line of cross section, Figure 5, shown by line A-B. Hachured areas are where topographic features with appearance of abandoned barriers are located.

field descriptions of two wash borings on the barrier north of Cape Lookout.

Sediment types were not as varied as those found under the barrier chain to the north. We found no evidence of inlets cutting out the normal expected sequence nor was a soil zone present.

The three borings made by Shell Development Company on Cape Lookout went from barrier sediments directly into marine sediments (Fig. 5); sediments of Pleistocene age occurred at 42.5 ft (−37.5). Two wash borings drilled by Shell Development Company about 10 mi north of Cape Lookout indicated no soil zone present.

No borings have been made on the present-day barrier west of Cape Lookout, except for some shallow cores to depths of about 8 ft. Barrier sand was present to total depth in these short cores.

The hole on the southernmost of the supposed abandoned barriers, Harkers Island, started in sand that was interpreted as a dune sand because its textural characteristics are

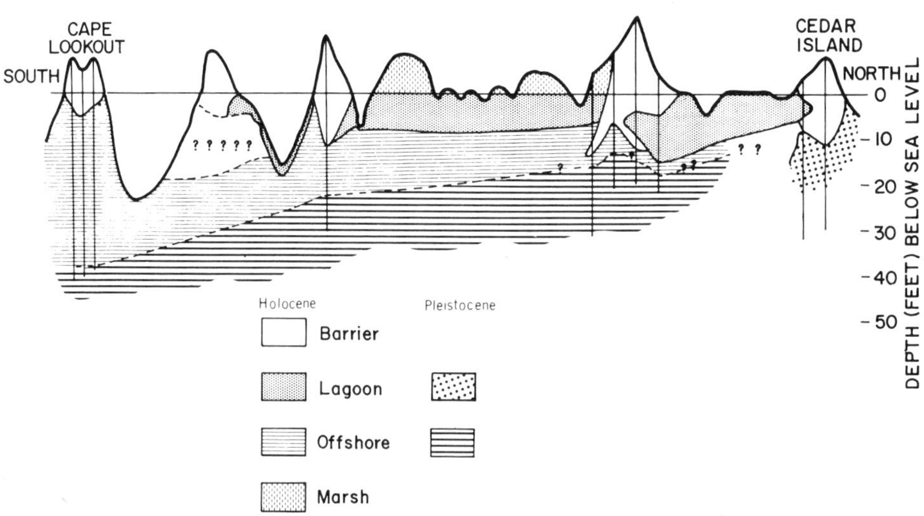

Figure 5. Cross section, Cape Lookout to Cedar Island. Hole locations shown on Figure 4.

similar to the dune sand of the barriers and the topographic expression of the deposits are similar to that of dunes (Figs. 4 and 5). Transition to marine sediments occurs between 22 and 25 ft (−10 to −13) on the basis of samples that are approximately 50 percent broken and worn shell fragments with a strong marine influence. Marine sediments were present to a total depth of 77 ft (−65). A carbon 14 date at 40 ft (−28) gave an age greater than 37,000 yrs B.P., obviously Pleistocene. This age, although indefinite, places the top of the Pleistocene at less than −28 ft and suggests a rise in the Pleistocene surface northward from the −37.5 ft depth at Cape Lookout.

Four holes were drilled on the middle abandoned barrier (Figs. 4 and 5). The section of lagoonal sediments, penetrated in the boring at the south edge of the barrier, probably is not continuous or connected with that found in the hole at the north edge (Fig. 5). The lagoonal sediments in the northern boring would be associated, in part, with this barrier and contemporaneous with the upper part of the marine section in the southern hole. The lagoonal section in the southern hole is believed to have been deposited when the barrier at Harkers Island formed the barrier shore.

Both borings on the northernmost barrier went from beach sediments into lagoonal material. None of the samples from either of these holes had characteristics indicative of a true marine environment. A carbon 14 date at a depth of 32 ft (−29) gave an age of greater than 43,000 yrs B.P.

INTREPRETATION

The samples from all of the borings exhibit characteristics that suggest similarity to those of sediments found in the environments along the present barrier chain, in associated lagoons, and offshore. This would indicate that the environments, in which the older sediments were deposited, were similar to the environments found in the area today and that the coastline and adjacent shelf were similar to the present one.

The different types of sediment encountered below the barrier sediments or barrier-lagoon sequence indicates a migration of environments concurrently with the modification of the shape of the barrier. Sediments deposited in low-energy environments, or marine units, or affected by subaerial weathering occur beneath the Holocene deposits that constitute the present-day barrier.

Formation of the Primary Barrier

The extreme irregularity of the mainland shores of the lagoons strongly suggest that they have not been exposed to oceanic waves. In part, the irregularity may be due to marsh growth, but it seems probable that Pamlico Sound has not been exposed to the ocean waves since the return of the sea after the Wisconsin regression. If such is the case, then the present shoreline could not have been formed by spit extension from the north as proposed by Fisher (1967). If ocean waves had reached Pamlico Sound, there should have been considerable smoothing of the mainland shores during the time necessary for the spit to grow and seal off the lagoons. Also, if the lagoon area had been open, mainland beaches should have been formed by the waves impinging upon the mainland.

It is postulated, therefore, that a primary barrier formed somewhere near the present barrier-island shoreline, effectively closing the lagoons when sea level was slightly less than at present. This primary barrier later became detached by inundation of low areas behind the beach during the slow sea level rise of the past 5000 yrs.

Obviously, there must be some evidence for the existence of the primary barrier. The Holocene shoreline in this area may have stabilized along the trend of a Pleistocene barrier, formed when the sea stood near the present level during a minor stillstand within the general Wisconsin regression. This Pleistocene barrier formed a topographic high along which the initial shoreline stabilized. A continued rise in sea level flooded the lower lying areas behind this shore, detaching the barrier from the mainland. Concurrently, there was an upward growth in the Holocene barrier island. We do not have sufficient data to determine whether this Pleistocene barrier was primary or secondary at the time of formation, although available data suggest that it overlies marine sediments and, hence, would be secondary.

Configuration of the Pleistocene Barrier

Work by Oaks and Coch (1963) and Oaks (1965) show that several barriers coalesced in southern Virginia to form one barrier when the sea last stood near its present level. They extend this barrier into northern North Carolina (Fig. 6).

Samples from the drill holes on Roanoke Island indicate that it is a former barrier island. An extension of the barrier trend, projected into North Carolina by Oaks (1965), will intersect Roanoke Island. Thus, it seems logical that the coalesced barrier in southern Virginia continues southward through Roanoke Island (Fig. 6).

Weathered sediments, associated with the soil zone and probably deposited as part of a barrier island, are present at depth in two widely separated places under the present-day barrier chain. The northernmost localities are near Oregon Inlet (holes 22 and 23); the second set of localities are about 11 mi north

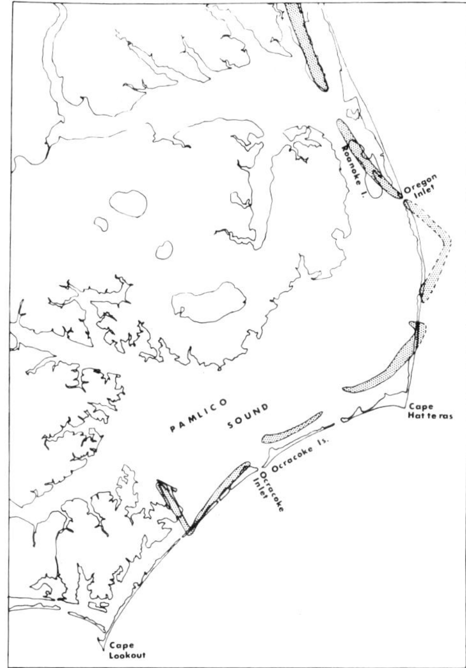

Figure 6. Postulated configuration of the Pleistocene barrier and the Holocene primary barrier which reoccupied the same position. Pleistocene-Holocene primary barrier trend is stippled; present barrier chain, unstippled.

of Cape Hatteras (holes 4, 5, and 6). The area between these localities may have been the site of a lagoon at the time of the existence of the Pleistocene barrier because of the texture of the sediments and the lower elevations of the weathered zone.

If these interpretations are correct, the Pleistocene barrier crossed the present-day coastline near Oregon Inlet and continued eastward for some unknown distance (Fig. 6), before turning southwest and again crossing the present coastline about 11 mi north of Cape Hatteras.

No estimate can be made of the distance out to sea that the Pleistocene barrier, and, consequently, the Holocene primary barrier extended. The distance and shape of the configuration is conjecture; the location of the south leg of the barrier has been influenced by the location of a series of northeast-southwest-trending nearshore shoals.

Nearshore shoals have been noted on the continental shelf off the eastern United States from Long Island to North Carolina (Sanders,

1963; Uchupi, 1968; Kraft, 1969; and Moody, 1964) and off Florida (Hyne and Goodell, 1967). They have been correlated to beach ridges with similar orientation on land (Sanders, 1963; Kraft, 1969) and ascribed to contemporary processes (Moody, 1964; Uchupi, 1968), but their origin is in doubt. It is possible that some may be relict ridges, others may be due to contemporary processes, and others may be relict ridges that are achieving equilibrium by present-day processes. The juxtaposition of the south leg of the Pleistocene barrier north of Cape Hatteras with nearshore shoals is not proven and is merely one of convenience.

The presence of a soil zone under a portion of the barrier chain indicates the approximate location of the primary barrier and, in this case, the Pleistocene barrier. Lack of a soil zone under Ocracoke Island and the area around Cape Hatteras suggests that marine processes have been operative.

In order for waves to pass over this area, the primary barrier must lay west of the present coastline, in what is now Pamlico Sound, or else the barrier did not extend this far south. In the latter case, wave action could impinge upon the mainland shores of what is now Pamlico Sound, but seems to be precluded by the extreme irregularity of these shores.

We believe that the primary barrier and its Pleistocene predecessor followed the line of shoals and small islands, now lying approximately 3 mi west of the present shoreline in the lagoon. These shoals and small islands appear to be much too large to be the result of wave action in Pamlico Sound, where waves generally are small. The trend is roughly parallel to the present barrier trend. The shoals stop approximately where the Pleistocene barrier presumably crossed the coast about 11 mi north of Cape Hatteras.

We must consider that the Pleistocene barrier returned to the position of the present coast near the center of Ocracoke Island as suggested by the presence of brown sand in the water well drilled by the U.S. Geological Survey at depths greater than 80 ft (−65); see Harris and Wilder (1964). This consideration requires that the soil zone lies at a much lower elevation than to the north where minimum elevation was approximately −40 ft. Although this is a possibility, the trend of the Pleistocene barrier probably follows the line of shoals and islands as far as the south end of Ocracoke Inlet. It is here that the line of shoals disappears (Fig. 6). To draw the Pleistocene barrier under the southern half of Ocracoke Island would require explanation for a continuation of the remainder of the shoals.

An extension of this shoal trend southward intersects the rear of the present-day barrier south of Ocracoke Inlet. Along this part of the barrier chain, a series of older appearing dunes may indicate the position of the Pleistocene barrier. These dunes have a trend that diverges approximately 5° from that of the present shoreline, causing the dune line to intersect the present coast about 8.5 mi south of Ocracoke Inlet (Fig. 6). It is believed that the Pleistocene barrier followed this dune line south to its intersection with the coast. A short distance farther south the barrier turned west and joined with the northernmost barrier at Cedar Island. Thus, the Pleistocene "Cape Lookout" lay approximately 25 mi northeast of its present position (Fig. 6).

It must be recognized that our dating in this area is indefinite and that the Holocene "ancestral" Cape Lookout might have been associated with any of the three east-west-trending barriers, which are present on the mainland (Fig. 4). If the Holocene sea reoccupied a barrier shoreline other than that of the northernmost barrier, then the time frame for development of this section of the coast is changed. A change of the time frame, however, does not change the method by which these barriers formed.

Evolution of the Present Coastline

The present barrier coast has evolved from the Holocene primary barrier which reoccupied the site of a Pleistocene barrier shoreline. The configuration of the present barrier shoreline is the result of modification of the shape of the primary barrier as well as progradation over the adjoining shelf.

The projecting headland, north of Cape Hatteras, has been driven landward from its original position to that of the present-day barrier (Fig. 7). The material eroded during the retrogression of this headland, as well as the mass of sand brought in ahead of the rising sea, has been redistributed by longshore drift to form spits both north and south of this headland, which probably acted as a nodal point for longshore drift.

Absence of the soil zone and the juxtaposition of Holocene sediments over marine sediments suggests strongly that parts of the barrier chain have grown by progradation over the sea floor. The sections of the barrier chain to which this applies are from Cape Lookout to at least 10 mi north, from Ocracoke Island to 10 mi north of Cape Hatteras, and from just north of Oregon Inlet to the northern edge of the study area.

Encroachment over the nearshore continental shelf can occur by frontal progradation or by spit extension. Frontal progradation must be considered because it is a common, well-documented method of barrier encroachment over the shelf (Curray, 1969; Bernard and others, 1962). In frontal progradation, successive accretion on the seaward side results in the formation of beach ridges. Curray (1969, p. JC-II-13) suggests that a typical spacing for beach ridges is about 165 ft. This figure is low for the beach-dune ridges of Cape Hatteras and Cape Lookout where a typical spacing is about 310 ft and 260 ft, respectively.

Frontal progradation can be shown to have occurred on the southwest side of Cape Hatteras (U.S. Congress, 1948). Frontal progradation may also be valid just south of Ocracoke Inlet, where the beach may have prograded directly from the line of dunes on the west of the present barrier.

According to Johnson (1919, p. 407), when a longshore current diverges from an existing shoreline, either frontal progradation or spit development may occur. The resulting form depends upon the amount of divergence: with a small amount of divergence, frontal progradation with beach ridges will form; spits will form when the divergence is great. Some features may be compound. The cuspate forelands along this section of the coast result in large changes of trend of the barrier coastline, and undoubtedly result in large divergences of the longshore current from the coast. This should lead to spit elongation if Johnson's theory is correct.

It is believed that most of the barrier shoreline has encroached upon the sea floor by the extension of spits from existing headlands. These spits would originally form as submarine shoals analogous to the shoals presently existing off the ends of Cape Hatteras and Cape Lookout. Continued contribution of material by longshore drift eventually would build the shoals above sea level and thus extend the barrier chain. The spits probably do not parallel the trend of the pre-existing barrier shoreline. Extension of the spits would tend to straighten the coast and lessen the concentra-

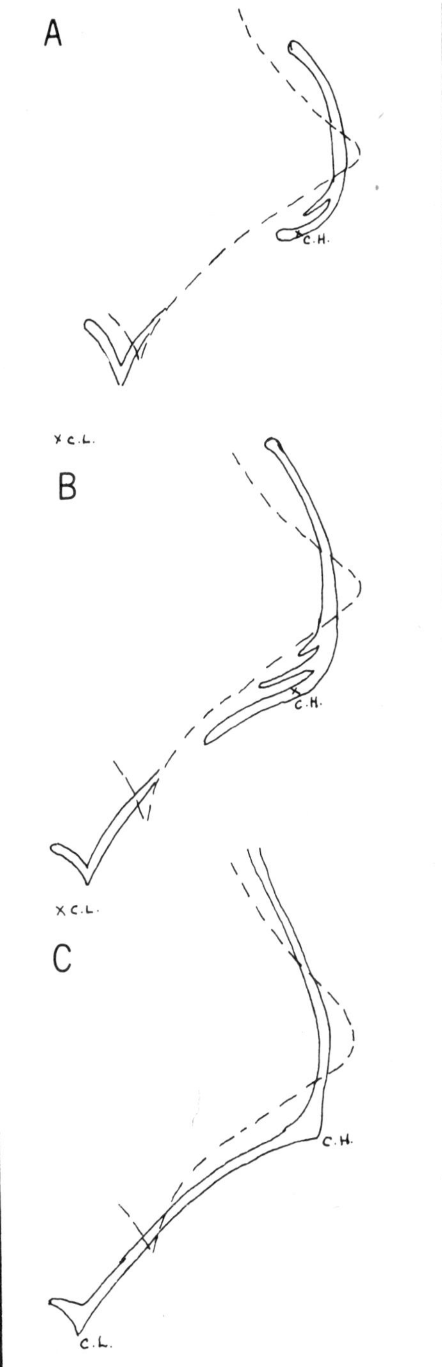

Figure 7. Schematic of evolutionary development of the North Carolina coast. Dashed line, location of primary barrier; solid lines, hypothetical development of spits and shoals. A. Early stage with very little erosion of headlands; B. Intermediate stage; C. Present coast.

tion of wave energy on the salients. These spits and shoals create a low-energy environment on their landward side. Continued wave attack, in the absence of a large supply of detritus, will tend to drive the spit landward (Guilcher, 1958, p. 109 and 189). Examples along the North Carolina coast are attested by peat outcrops along the front of the barriers.

It is impossible for that section of the present barrier chain from 8.5 mi south of Ocracoke Inlet to Cape Lookout to encroach on the sea floor by frontal progradation. No abandoned barrier, nearly parallel and landward of the present barrier, exists from which frontal progradation could have started. This section of the barrier chain appears as a spit that has been extended south by accretion of material and resembles the growth of the present Cape Lookout which accreted at the southern end during historical times (Pierce, 1969).

The barrier island west of Cape Lookout (Fig. 4) also has the appearance of a spit. It is separated from the southernmost abandoned barrier by about 2 mi of lagoon. Such a spacing is considerably more than suggested as typical spacing by Curray (1969) and also more than the spacing now present on Cape Lookout. The middle abandoned barrier north of Cape Lookout is separated from the next one to the south by about 13 mi of low marsh, a rather large spacing for beach-dune ridges.

We also believe that the rest of the barrier chain, where no soil zone is present, has grown by elongation of spits (Fig. 7). This would include the section north of Oregon Inlet and from just north of Cape Hatteras to Ocracoke Island. The horizontal separation from the location of the postulated primary barrier is about 3.5 mi. There is no topographic evidence of additional beach ridges between the present barrier shoreline and the line of shoals which mark the position of the primary barrier.

The re-entrant in the primary barrier opposite Roanoke Island was closed off by spit extension from the south and possibly also from the north (Fig. 7). The present barrier is now about 3.5 mi east of Roanoke Island. Depths of 60 to 70 ft exist 3.5 mi east of the present barrier so that the spits probably enclosed a bay with depths at least as great as 60 ft and, because the barriers have been retrogressing, it may have been as much as 75 ft, which does not imply that the barrier attained a thickness of 75 ft at any time. Throughout much of its existence, the barrier may have been in the form of a submarine shoal such as now exists south of Cape Hatteras and Cape Lookout. Both of these shoals act as obstacles to wave refraction and create wave shadows on their lee side (Pierce and others, 1970).

A similar situation probably existed in front of the present Ocracoke Island. Such circumstances would account for the abnormally thick section of low-energy deposits under Ocracoke Island and under the barrier opposite Roanoke Island.

Longshore drift from the north extended the barrier chain from "ancestral" Cape Lookout to its present position. East-west-trending barriers were formed periodically and as each succeeding east-west barrier was formed, the previous shoreline was abandoned.

RELATIVE IMPORTANCE OF BARRIER TYPES

Along present shorelines, no primary barrier exists in an unmodified form, nor do secondary barriers exist as originally formed. Modification occurs too rapidly on a beach to preserve forms exposed to the force of marine processes for any length of time. This observation may appear to be at variance with reports of topographic forms that resemble barriers on the shelf (Emery, 1968, p. 448D; Curray, 1960) and the previously mentioned nearshore shoals. Many of the relict forms that are drowned barriers appear to be restricted to areas of the continental shelves that were inundated during the early, more rapid rise of sea level. There is some question on the origin of the nearshore features and their response to present-day processes. Some data indicate that these features, if relict in origin, are being reworked by modern processes (Swift, 1969, p. 4–29).

The two theories on the method of barrier island formation, acceptable today, involve spit extension (Gilbert, 1885) or drowning of a low area behind a beach (McGee, 1890; Hoyt, 1967). The barrier chain of northeastern North Carolina can serve as a model by which the relative importance of the two methods can be judged. Continued regression of the secondary barriers that is occurring today will superpose them on the site of primary barriers and juxtapose both barriers on a base of terrestrial sediment or subaerially weathered soil (Fig. 8). Further evolution will lead to destruction of all barriers by driving them onto the mainland, so that the evidence necessary for reconstruction is removed. This destruction has occurred in Delaware (Kraft, 1968) and may be prevalent along much of the coast of the

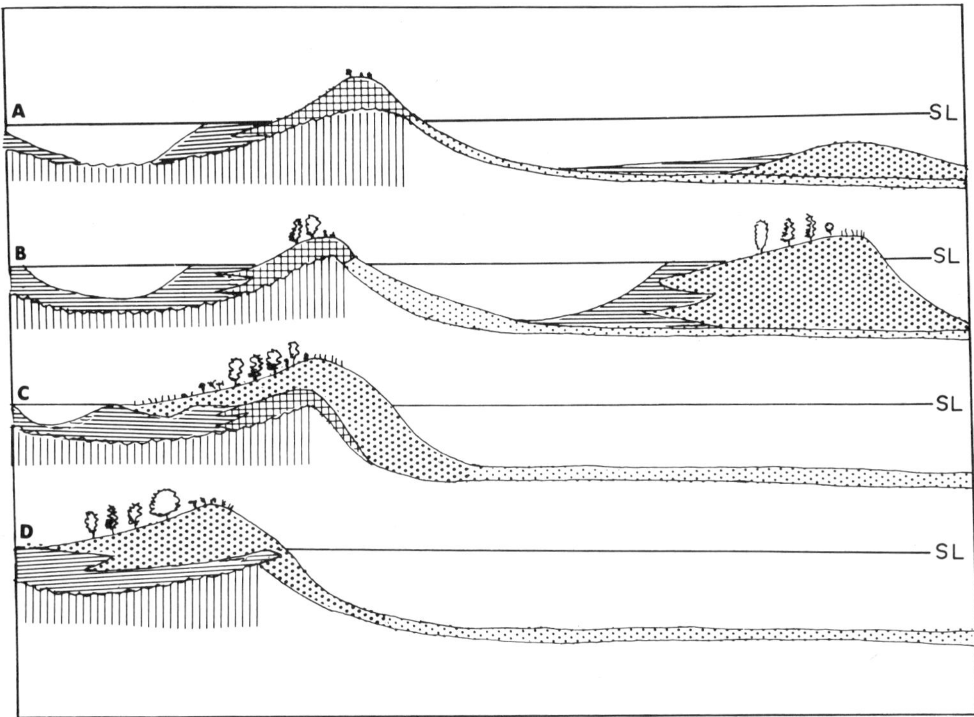

Figure 8. Cross section showing evolutionary development of primary and secondary barriers. A. development of submarine bar offshore giving rise to low-energy area behind it; B. growth of the bar above sea level through contribution of material by longshore drift; C. retrogression of the secondary barrier to point of coincidence with the primary barrier; D. more retrogression such that all evidence of secondary *versus* primary origin destroyed.

southeastern United States, so that it could account for the absence of marine sediments landward of the barrier, as found by Hoyt (1967) in Georgia.

Primary barriers cover 47 mi along the North Carolina coast for which we have data (Fig. 9). This length includes that part of the coast where the secondary barrier is attached to the front of the primary barrier, south of Ocracoke Island. This barrier is considered to be a primary because little evidence would be present at depth to distinguish it as a compound feature.

Secondary barriers make up 73 mi of the North Carolina coast and include the present shoreline west and north of Cape Lookout, Ocracoke Island, to approximately 11 mi north of Cape Hatteras, and from Oregon Inlet north approximately 9 mi.

Thus, of 120 mi of coast, primary barriers account for 39 percent and secondary barriers, for 61 percent.

COMPARISON TO OTHER BARRIERS

Barriers along the Gulf Coast of the United States are one of the most studied barrier chains in the world. According to Shepard (1960), the barrier islands in the western Gulf have been relatively stable within historical times. The chain along the Texas Coast is long, straight, or gently curving, nearly continuous islands.

Matagorda and St. Joseph Islands have prograded little, if any, over "nearshore Gulf" sediments and these islands have existed since their formation with only a slight movement landward, as suggested by lack of nearshore Gulf sediments and presence of bay deposits under the present beaches (Shepard and Moore, 1955, p. 1578). Some progradation may have occurred as indicated by barrier sediments under a large overwash fan on St. Joseph Island (Shepard, 1960, p. 212). Shepard and Moore (1955, p. 1575, Fig. 68) show Matagorda Island resting on a topographic high of weathered bay deposit. This high has about 40 ft of relief, and shows notable similarity to that found under parts of the North Carolina primary barrier chain and, as reported by Colquhoun (1965), under Pleistocene primary barriers in South Carolina.

Galveston Island rests on bay sediments at

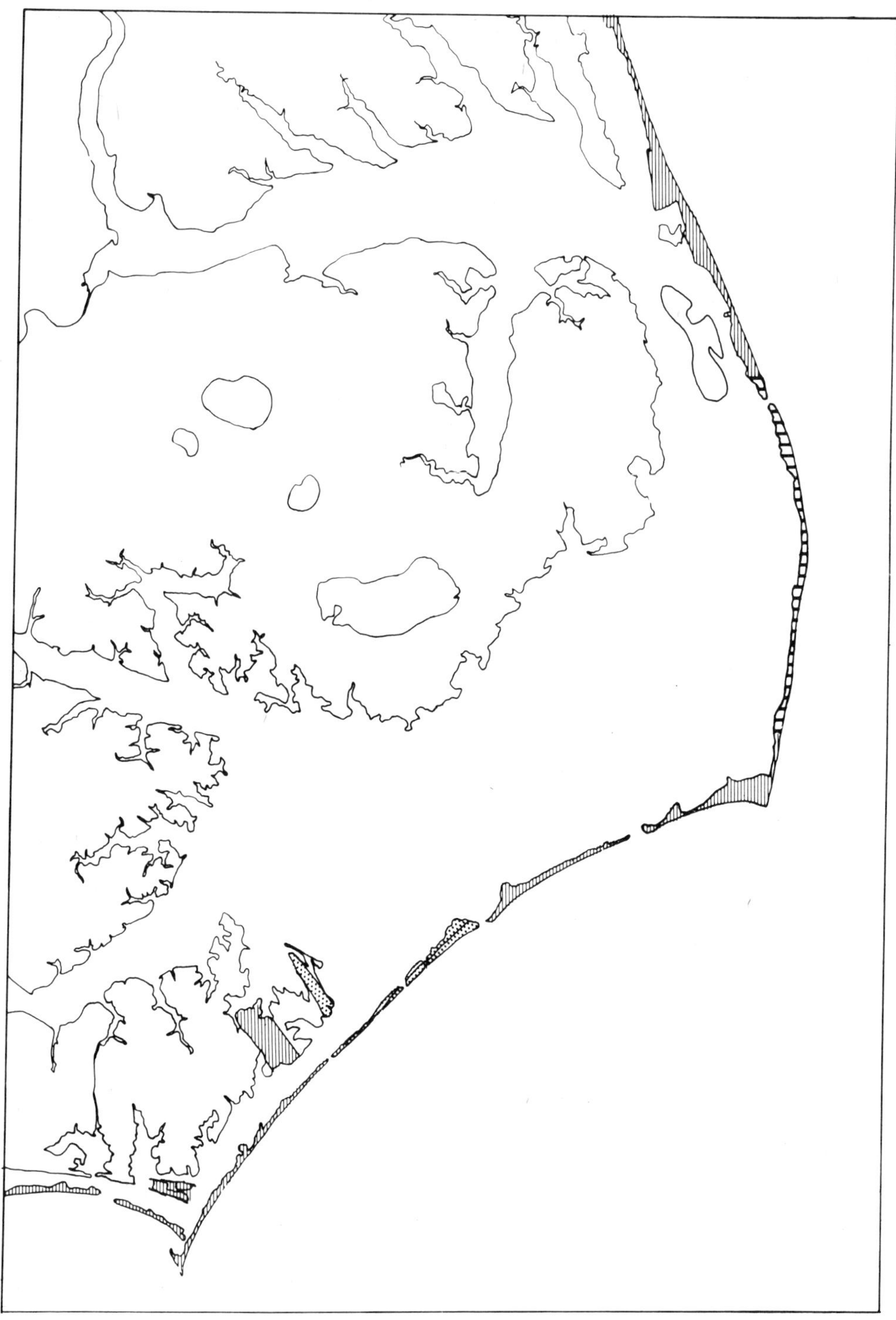

Figure 9. Types of barriers making up the present coast. Vertical hatching, secondary barrier; horizontal lines, eroded and modified primary; dotted area, primary barrier with frontal progradation.

the southwest end and apparently has prograded across nearshore Gulf sediments to the northeast (Shepard, 1960, p. 213). According to Le Blanc and Hodgson (1959), Galveston Island developed from a small island that was extended by spit development and frontal progradation toward the northeast.

At the north end of Padre Island, a weathered surface was encountered at a depth of 19 ft; Pleistocene clay crops out near the shore at one place at the southern end of the island (Shepard, 1960). Shepard interprets samples from a boring near the center of Padre Island to indicate that barrier sands overlie bay sediments. On the other hand, Fisk (1959) shows that the present beach sediments near the center of Padre Island, overlie shoreface sediments and that this barrier has grown upward and seaward. An older beach not connected to the present beach section, is shown near the base of the Holocene section (Fisk, 1959, Figs. 2 and 13). We interpret this older beach as possibly a primary feature with the present barrier beach forming later. Padre Island may have been formed by a coalescence of two spits from two small areas at what are now the north and south ends of the present island.

The barrier islands of the northern Gulf are much more discontinuous (Alabama-Mississippi) or are cuspate spits (Florida) according to Shepard (1960). The islands along the coasts of Mississippi and Alabama have a history of migration (Otvos, 1970). Many of these islands lie on a base of lagoon (sound) deposits, although some overlie marine clay (Shepard, 1960).

The western Florida area, with the cuspate spits, appear to be closer in form to the North Carolina barriers than those of the Texas coast. Cape San Blas has prograded over nearshore sediments and other islands in the same area also have had similar histories (Schnable and Goodell, 1968). These progradations must have taken place as spit extensions.

CONCLUSIONS

Geomorphic features known as barrier chains are complex features that may be primary or secondary, according to the mode of formation.

Primary barriers are the initial form and probably develop by drowning of low-lying land behind a mainland beach during a slight rise in sea level. These primary barriers, where they form the shoreline along the North Carolina coast, have been extensively modified. Preservation has also occurred where submarine shoals and spits have been extended in front of primary barriers.

Shoals have eventually grown above sea level in the form of spits to form secondary barriers which have prograded across the sea floor, not necessarily parallel to the pre-existing barrier shoreline. More than one secondary barrier can be present as shown by the abandoned barriers north of Cape Lookout.

Differentiation of primary barriers from secondary barriers can be made from the substrata upon which they are built. Primary barriers are built upon material that is terrestrial in origin or that has been exposed to weathering. Secondary barriers lie upon marine sediments, but are not to be confused with beach areas formed by frontal progradation.

The coastline of the study area consists predominantly of secondary barriers (61 percent); primary barriers make up 39 percent.

Both primary and secondary barriers occur together along the barrier coastline, although further evolution will juxtapose the two types of barriers. The hypothetical result would be to drive both barriers onto the mainland; consequently, the evidence for a secondary barrier would be destroyed.

Barriers are not formed by mutually exclusive methods, but can result from submergence of a topographic ridge or spit extension during a transgression, or both.

ACKNOWLEDGMENTS

We wish to thank Douglas Nelson, Drew Comer, John Brynes, Roger Hughes and Allan DeWall for analyzing most of the samples. Joel Zipp did much of the work on fauna from the samples with the assistance of Thomas R. Waller and Joseph Rosewater of the Smithsonian Institution and Druid Wilson of the U.S. Geological Survey.

Partial financial support for the project was through Smithsonian grants RA 3307 and RA 3369. Some of the ideas were developed under National Science Foundation grant G16362 to R. H. Benson. Partial support was obtained also through National Science Foundation Grants G4559 and GP1817 to Colquhoun. Acknowledgment is made to the donors of the Petroleum Research Fund administered by the American Chemical Society for partial support of this project through Grant PRF 3223A2 to Colquhoun.

We acknowledge the invaluable data supplied by Shell Development Company on the primitive portion of the barriers where lack of roads made it impossible for us to drill. Sharing of samples from the lagoon by R. L. Ingram and from offshore by Orrin Pilkey was of great assistance.

The manuscript benefited greatly from critical reading by D. B. Duane and J. C. Kraft.

REFERENCES CITED

Bernard, H. A., LeBlanc, R. J., and Major, C. F., 1962, Recent and Pleistocene geology of southeast Texas, *in* Rainwater, E. H., and Zingula, R. P., *Editors,* Geology of the Gulf Coast and Central Texas: Guidebook of excursion for annual meeting, Geological Society of America: Houston, Houston Geological Society, p. 175–224.

Colquhoun, D. J., 1965, Terrace sediment complexes in central South Carolina: Atlantic Coastal Plain Geol. Assoc. Field Conf. Guidebook, 62 p.

—— 1969a, Coastal plain terraces in Carolinas and Georgia, p. 150–162 *in* Wright, H. E., Jr., *Editor,* Quaternary Geology and Climate: Natl. Acad. Sci. Pub. 1701, 162 p.

—— 1969b, Geomorphology of the lower coastal plain of South Carolina: MS-15; South Carolina Development Board, Div. Geol., 36 p.

Curray, J. R., 1960, Sediments and history of Holocene transgression, continental shelf, northwest Gulf of Mexico, *in* Shepard, F. P., Phleger, F. B., and van Andel, Tj. H., *Editors,* Recent Sediments Northwest Gulf of Mexico: Tulsa, Oklahoma, Am. Assoc. Petroleum Geologists, p. 221–266.

—— 1969, Shore zone sand bodies: Barriers, cheniers and beach ridges, *in* Stanley, D. J., *Editor,* The New Concepts of Continental Margin Sedimentation: Washington, D.C., Am. Geol. Inst., p. II1–II18.

de Beaumont, E., 1845, Leçon de geologie pratique: Paris, P. Bertrand, p. 223–252.

Duane, D. B., 1962, Petrology and Recent bottom sediments of the western Pamlico Sound region, North Carolina: Ph.D. dissert., Kansas Univ., Lawrence, Kansas, 115 p.*

—— 1964, Significance of skewness in Recent sediments, western Pamlico Sound, North Carolina: Jour. Sed. Petrology, v. 34, p. 864–874.

Emery, K. O., 1968, Relict sediments on the continental shelves of the world: Am. Assoc. Petroleum Geologists Bull., v. 52, p. 445–464.

Fisher, J. J., 1967, Development patterns of relict beach ridges, Outer Banks barrier chain: Ph.D. dissert., North Carolina Univ., Chapel Hill, North Carolina, 225 p.*

Fisk, H. N., 1959, Padre Island and the Laguna Madre flats, coastal Texas: 2d Coastal Geography Conf., Louisiana State Univ., Baton Rouge, p. 101–151.

Gilbert, G. K., 1885, The topographic features of lake shores: U.S. Geol. Survey 5th Ann. Rept., p. 69–123.

Guilcher, A., 1958, Coastal and Submarine Morphology: Methuen, London, 274 p.

Harris, W. H., and Wilder, H. B., 1964, Groundwater supply of Cape Hatteras National Seashore recreational area, North Carolina: North Carolina Dept. Water Resources, Rept. Inv., no. 3, 22 p.

Hoyt, J. H., 1967, Barrier Island formation: Geol. Soc. America Bull., v. 78, p. 1125–1136.

—— 1970, Littoral surge channels, Cabretta Island, Georgia: Abstracts with Programs, v. 2, no. 3, Geol. Soc. America, p. 218.

Hyne, N. J., and Goodell, H. G., 1967, Origin of sediments and submarine geology of the inner continental shelf off Choctawatchee Bay, Florida: Marine Geology, v. 5, p. 299–313.

Johnson, D. W., 1919, Shore Processes and Shoreline Development: New York, John Wiley & Sons, Inc., 584 p.

Kimrey, J. O., 1960, Ground water supply of Cape Hatteras National Seashore recreational area: North Carolina Dept. Water Resources Rept. Inv., no. 2, 28 p.

Kraft, J. C., 1968, Coastal sedimentary environments, Lewes-Rehoboth Beach Delaware coastal area: Guidebook, 1968 Field Trip, Eastern Sect., Soc. Econ. Paleontologists and Mineralogists, 30 p.

—— 1969, Pre-Holocene paleogeography and paleogeology in the Delaware coastal areas, *in* Abstracts with Programs, Part 1: Geol. Soc. America, p. 34.

LeBlanc, R. J., and Hodgson, W. D., 1959, Origin and development of the Texas shoreline: 2nd Coastal Geography Conf., Louisiana State Univ., Baton Rouge, p. 57–101.

Leith, C. J., 1970, Environmental aspects of beach processes, sediment, and erosion in the coastal region of North Carolina: Abstracts with Programs, v. 2, no. 3, Geol. Soc. America, p. 226.

McGee, W. J., 1890, Encroachment of the sea: The Forum, v. 9, p. 437–449.

Moody, D., 1964, Coastal morphology and processes in relation to the development of submarine sand ridges off Bethany Beach, Delaware: Ph.D. dissert., Johns Hopkins Univ., Baltimore, Maryland, 167 p.*

Oaks, R. Q., Jr., 1965, Post-Miocene stratigraphy and morphology, outer Coastal Plain, southeastern Virginia: Ph.D. dissert., Yale Univ., New Haven, Connecticut, 423 p.*

Oaks, R. Q., Jr., and Coch, N. K., 1963, Pleistocene sea levels, southeastern Virginia: Science, v. 140, p. 979–983.

Otvos, E. G., Jr., 1970, Development and migration of barrier islands, northern Gulf of Mexico: Geol. Soc. America Bull., v. 81, p. 241–246.

Pickett, T. E., and Ingram, R. L., 1969, The modern sediments of Pamlico Sound, North

Carolina: Southeastern Geology, v. 11, p. 53–83.

Pierce, J. W., 1969, Sediment budget along a barrier island chain: Sed. Geology, v. 3, p. 5–16.

Pierce, J. W., So, C. L., Roth, H. D., and Colquhoun, D. J., 1970, Wave refraction and coastal erosion, southern Virginia and northern North Carolina: Abstracts with Programs, v. 2, no. 3, Geol. Soc. America, p. 237.

Rochna, D. A., 1961, Physiography and sedimentology of the Outer Banks between Cape Lookout and Ocracoke Inlet, North Carolina: MS thesis, Kansas Univ., Lawrence, Kansas, 91 p.*

Sanders, J. E., 1963, North-south trending submarine ridge composed of coarse sand off False Cape, Virginia (abs.): Am. Assoc. Petroleum Geologists Bull., v. 46, p. 278.

Schnable, J. E., and Goodell, H. G., 1968, Pleistocene-Recent stratigraphy, evolution, and development of the Apalachecola coast, Florida: Geol. Soc. America Spec. Paper 112, 72 p.

Shepard, F. P., 1960, Gulf Coast barriers, p. 197–220, *in* Shepard, F. P. and others, Recent Sediments; Northwest Gulf of Mexico: Tulsa, Oklahoma, Am. Assoc. Petroleum Geol., 394 p.

Shepard, F. P., and Moore, D. G., 1955, Central Texas coast sedimentation: Characteristics of sedimentary environments, recent history and diagenesis: Am. Assoc. Petroleum Geologists Bull., v. 39, p. 1463–1593.

Shepard, F. P., Phleger, F. B., and van Andel, Tj. H., 1960, Recent Sediments Northwest Gulf of Mexico: Tulsa, Oklahoma, Am. Assoc. Petroleum Geologists, 394 p.

Swanson, V. E., and Palacas, J. G., 1965, Humate in coastal sands of northwest Florida: U.S. Geol. Survey Bull. 1214-B, 29 p.

Swift, D. J. P., 1969, Inner Shelf sedimentation: Processes and products, *in* Stanley, D. J., *Editor*, The New Concepts of Continental Margin Sedimentation: Washington, D. C., Am. Geol. Inst., p. 4-1–4-46.

Tanner, W. F., 1960, Florida coastal classification: Gulf Coast Geol. Assoc. Trans., v. 10, p. 259–286.

Uchupi, E., 1968, The Atlantic continental shelf and slope of the United States: physiography: U.S. Geol. Survey Prof. Paper 529-C, p. 1–20.

U.S. Congress, 1948, North Carolina shoreline, beach erosion study: House Document 763, 80th Cong., 2d sess., 33 p.

Welby, C. W., 1970, Observation on the origin of the North Carolina Outer Banks—results from a geophysical study: Abstracts with Programs, v. 2, no. 3, Geol. Soc. America, p. 248.

Manuscript Received by The Society June 1, 1970

Revised Manuscript Received July 13, 1970

* Available on microfilm or by interlibrary loan.

Editor's Comments on Paper 30

30 Hoyt: *Development and Migration of Barrier Islands, Northern Gulf of Mexico: Discussion*

The debate continued, as Hoyt differed with Otvos, in this paper published shortly after Hoyt's death.

Hoyt takes exception to Otvos's claim to having proved the existence of any barrier island formed from an offshore bar or shoal. He feels that, at most, all that was shown was that Grand Gosier Island was formed by the merger of a few small islands. Hoyt returns to his original argument that evidence of marine fauna, deposits, or shore morphology is entirely lacking along the present mainland coast. These would be found, to some degree, he insists, had the barrier islands been derived from spits or bars; he refutes the claim that any significant salt-marsh growth takes place on open ocean coasts. Since the original stratigraphy is destroyed during migration, Hoyt feels that little can be proved by drilling through displaced barriers. He believes that many of the barriers cited by Otvos are examples of development by the submergence of beach or dune ridges.

Reprinted from the *Geol. Soc. Amer. Bull.*, **81,** 3779–3782 (1970)

JOHN H. HOYT *Marine Institute and Department of Geology, University of Georgia, Sapelo Island, Georgia 31327*

Development and Migration of Barrier Islands, Northern Gulf of Mexico: Discussion 30

Editor's Note: It has come to the attention of The Society that Mr. Hoyt died September 6, 1970, while his manuscript was in press.

In a recent paper, Otvos (1970) discusses several of my publications concerning the formation of barrier islands (Hoyt, 1967, 1968, 1969). Although, in general, agreement is expressed on many points, there are sections of his paper that require clarification.

A major tenet in Otvos' discussion is the affirmation that barrier islands form from offshore bars or shoal areas. In my 1967 paper, I relegated this method of formation to a possible, but minor role. In support of his thesis, Otvos states, "Several minor and major examples of barrier island development from underwater shoals took place during historic times in the Chandeleur, Mississippi Sound, Timbalier Islands' groups. . ." (1970, p. 243). Careful reading of the paper reveals only one example: "In the southern Chandeleur Island group, the 7.5-km-long Grand Gosier Island has developed since 1869, through stages of development and merger of small islands, from the shallow sea floor." Unfortunately, even this one example is not valid because the development of Grand Gosier Island from "small islands" does not establish its formation from an offshore bar. The question, left unanswered in his discussion, is the means of formation of the small islands. All the other examples which Otvos mentions, such as Dog Keys, Massacre Island, and Myth Island, were islands which were partially converted to shoals, a well-understood process. The point, however, is not whether one or two examples of islands formed from offshore bars can be found, but rather that there is a lack of evidence that any major system formed in this way.

There is no evidence given by Otvos to show that the Chandeleur Islands or other island groups in the Mississippi delta area formed from offshore bars. On the other hand, the lack of open marine beach and nearshore deposits and fauna (Kolb and Van Lopik, 1958, 1966) and the lack of open-marine shoreline morphology landward of the barrier islands indicate that these island groups did not form from offshore bars or from submerged shoals. Mollusk shells, which are particularly abundant along Gulf Coast beaches, and other faunal elements should make the identification of open-marine sediments a rather simple matter. To dismiss the absence of open-marine features landward of barrier islands as a result of rapid transition from offshore bar to barrier island is not reasonable without evidence that the development of the offshore bar would precede deposition landward of the bar. If sufficient sediments are available for bar formation, it is likely that recognizable deposits also would accumulate along the shoreline. Although salinities may be low in estuaries, bays, and sounds in areas where large volumes of fresh water are discharged, it should not be assumed that similar conditions prevailed along the shore prior to barrier development. If brackish or marine waters wash the shore, it is reasonable to expect there were similarly saline waters along the shoreline at the time of barrier formation.

As an alternate method of formation, I suggest that the Chandeleur Island group, including the small islands which later became Grand Gosier Island, and other barrier islands surrounding the Mississippi River delta appear to be outstanding examples of the formation of barrier islands by the submergence of beach or dune ridges, or both. Several studies which were incorporated in the papers by Kolb and Van Lopik (1958, 1966) indicate that these islands formed near the distal ends of former deltas. Waves reworked the deltaic deposits and the sandy sediment was piled into ridges. Submergence of the area, augmented by the weight and compaction of the deltaic mass, flooded the area landward of the ridges, forming barrier islands (Chandeleur Islands) and lagoons (Chandeleur Sound). Once the islands formed, they were subjected to a complex history of modification which is common to barrier islands (Hoyt and Hails, 1967; Hoyt and Henry, 1967). In the case of the Chandeleur group, this history included several kilometers of landward (westward) migration which is still continuing. During this migration it is normal that parts of the barrier chain would be destroyed and reformed several times by major storms. The thin sand sheet seaward of the barrier chain is part of the delta deposits which have been reworked by the waves dictating the landward migration of the barriers.

There is no question about the rapidity of spit growth in favorable areas and the 9-km extensions of Nauset Beach at Cape Cod, Massachusetts, in the past 120 years is a fine example (Otvos, 1970, p. 242). However, open marine sediments also accumulated in the area landward of the spit during the several hundreds of years which preceded the formation of the spit. Deposition of open marine sediments and development of open-shore morphology along mainland shores in areas of spit formation is evident from many areas of the world. This situation was illustrated and explained in a recent paper (Hoyt, 1968, Fig. 1), and the process of formation of barrier islands from spits was outlined earlier (Hoyt, 1967, Fig. 4). In areas of active sedimentation, such as deltas, spits may develop rapidly and be quickly breached to form a barrier island. Survey intervals may be inadequate to show the spit phase of barrier development and result in an erroneous interpretation of barrier island formation.

In rejecting the formation of barrier islands by the submergence of the area landward of dune and beach ridges, Otvos (1970, p. 242) requests that, "It must also be proven that the total section between the pretransgression surface and the surface of the islands was formed in the supratidal environments." Unfortunately, when the history of barrier islands is considered, such idealistic preservation is improbable. Otvos' discussion of Padre and Galveston Islands suggests a lack of appreciation of the magnitude of the modification which accompanied the development of barrier systems. His hypothesis that "Galveston Island began as a small offshore bar in 1.5 to 2.5 m of water, . . ." is entirely unsupported by any evidence and is considered unlikely, based on the studies of Fisk (1959), Rusnak (1969), and as mentioned in my previous papers (Hoyt, 1967, 1968). The lagoonal sediments which accumulated in West Bay landward of Galveston Island and Laguna Madre landward of Padre Island indicate that these barrier islands could not have developed from offshore bars as suggested by Otvos (Fisk, 1959; LeBlanc and Hodgson, 1969; Bernard and others, 1962). On the other hand, the beach and dune sediments of these islands offer abundant opportunity for the formation of the islands from submerged beach and dune ridges.

Otvos (1970, p. 242) appears to interpret erroneously a vertical sequence of barrier island sedimentary features which, "From bottom to top represents the offshore-to-shore facies succession" as an indication of barrier island formation from offshore bars. Unfortunately, such a sequence is quite normal for barrier islands and shows the accumulation and progradation of the islands once they have formed, although providing no information about their origin. Barrier islands may show this sequence no matter how they have formed. Also, the indication that "barrier sands were deposited over earlier Holocene open lagoon, bay-sound, and shoreface sediments" (Otvos, 1970, p. 242) tells nothing of the original formation of the barrier islands, but merely shows the degree of modification of the islands prior to a period of stabilization and growth. The examples of Horn Island, Mississippi, and the Pine Island chain, Louisiana, fail to give any indication that they formed as offshore bars, whereas the lack of open-marine sediments and fauna landward of the islands indicates that they did not form from bars or shoals. The presence of barrier sands over bay-sound, lagoonal, and estuarine sediments illustrates the erosion and shoreward shifting of the barrier during some period in its history. Although this erosion would probably destroy sedimentary features which would establish the formation of the islands from partially submerged ridges, it would not disturb all the open-marine deposits which accumulated landward of developing offshore bars, if the islands originated in this way. The fact that the required open-marine sediments are not commonly found argues strongly that the islands did not form from bars or shoals. The lack of these deposits and the sedimentary features and stratigraphic relationships is completely compatible with an origin from submerged beach or dune ridges.

It is idealistic to maintain that coastal ridges must overlie Pleistocene land surfaces to indicate barrier development by submergence of beach and dune ridges. Initially, this may be the case, but, almost universally, erosion, submergence, and transgression will result in the barriers resting on Holocene lagoonal-salt marsh-tidal flat deposits. This is the case at Horn, and Pine Islands, and at many barriers in other areas of the world. This relationship is particularly well shown on a series of diagrams by van Straaten (1965, Fig. 26) for an area near The Hague along the Dutch coast. These diagrams show the development of the initial barrier seaward of the present coast, landward retreat of this barrier with the barrier resting on Holocene tidal flat deposits, and, finally, progradation of the Holocene barrier. The easternmost barrier ridge is shown as much as 14 km landward of the position of initial barrier formation. In these cases the initial and subsequent barriers have provided the protected environment necessary for accumulation of the fine-grained lagoonal and tidal flat sediments.

Otvos' statement (1970, p. 242) that "Open marine shores often (commonly) have salt marshes instead of beaches" is incorrect. A survey of several areas of the world, including most of eastern and Gulf coasts of the United States, the North Sea and English Channel coasts of England, the English Channel and Atlantic coasts of France, coasts of Belgium, the Netherlands, and Germany, parts of southern African coasts, and small sections of the Mediterranean coast reveals only a very few insignificant areas of salt marsh along open-marine shores. A review of pertinent references such as King (1959), Chapman (1960), and Zenkovitch (1967) failed to indicate significant areas of salt marsh along open-marine shores. Marshes are present along the shores of embayments such as The Wash of eastern England, Helgoland Bay, Germany, and near the head of Mont-Saint-Michel Bay, France; however, these protected localities cannot be considered open-marine shores. The only areas of salt marsh which formed exposed to the open-marine environment appear to be a few anomolous areas such as the zero-energy coast in northwest Florida (Tanner, 1960) and areas in the vicinity of some large deltas, such as the MeKong and Irrawady deltas, and along sections of the Louisiana coast, where high turbidity or gentle sea-floor gradients may effectively reduce wave energy reaching the shore, or high sedimentation rates may preclude the development of sand beaches (Hoyt, 1969). Although beaches may develop along lagoonal shores, the associated fauna will generally distinguish the lagoonal environment from the open marine.

In summary, much of the evidence mentioned by Otvos has no bearing on the original formation of barrier islands, but merely repeats information concerning their erosion, progradation, and migration; all aspects that are well understood. Unfortunately, he ignores criteria such as a general lack of open-marine sediment and fauna landward of barrier systems and the absence of marine morphology along mainland shores that would be helpful in rejecting some of the possible ways in which barriers have been considered to form. Although the search for barrier islands which may have formed from offshore bars is losing significance, the continued lack of examples has some implications.

REFERENCES CITED

Bernard, H. A., LeBlanc, R. J., and Major, C. F., 1962, Recent and Pleistocene geology of southeast Texas, *in* Rainwater, E. H., and Zingula, R. P., *Editors*, Geology of the Gulf Coast and Central Texas: Guidebook of excursions: Houston Geological Society, p. 191–206.

Chapman, V. J., 1960, Salt Marshes and Salt Deserts of the World: New York, Interscience Pub. Co., 392 p.

Fisk, H. N., 1959, Padre Island and the Laguna Madre flats, coastal south Texas: 2d Coastal Geography Conf. Proc., Baton Rouge, Louisiana, Louisiana State Univ., p. 101–151.

Hoyt, J. H., 1967, Barrier island formation: Geol. Soc. America Bull., v. 78, p. 1125–1136.

—— 1968, Barrier island formation: Reply: Geol. Soc. America Bull., v. 79, p. 1427–1432.

——1969, Chenier *versus* barrier, genetic and stratigraphic distinction: Am. Assoc. Petroleum Geologists Bull., v. 53, no. 2, p. 299–306.

Hoyt, J. H., and Hails, J. R., 1967, Pleistocene shoreline sediments in coastal Georgia: Deposition and modification: Science, v. 155, p. 1541–1543.

Hoyt, J. H., and Henry, V. J., Jr., 1967, Influence of island migration on barrier-island sedimentation: Geol. Soc. America Bull., v. 78, p. 77–86.

King, C. A. M., 1959, Beaches and Coasts: London, Edward Arnold Ltd., 403 p.

Kolb, C. R., and Van Lopik, J. R., 1958, Geology of the Mississippi River deltaic plain, southeastern Louisiana: U.S. Army Corps Engineers Waterways Expt. Sta. Tech. Rept. no. 3–483, 120 p.

——1966, Depositional environment of the Mississippi River deltaic plain—southeastern Louisiana, *in* Shirley, M. L., *Editor*, Deltas: Houston, Houston Geological Society, p. 17–61.

LeBlanc, R. F., and Hodgson, W. D., 1959, Origin and development of the Texas shoreline: Gulf Coast Assoc. Geol. Soc. Trans., v. 9, p. 197–220.

Otvos, E. G., Jr., 1970, Development and migration of barrier islands, northern Gulf of Mexico: Geol. Soc. America Bull., v. 81, p. 241–246.

Rusnak, G. A., 1960, Sediments of Laguna Madre, Texas, *in* Shepard, F. P., and others, *Editors*, Recent Sediments, Northwest Gulf of Mexico: Tulsa, Am. Assoc. Petroleum Geologists, p. 153–196.

Tanner, W. F., 1960, Florida coastal classification: Gulf Coast Assoc. Geol. Societies Trans., v. 10, p. 259–266.

van Straaten, L. M. J. U., 1965, Coastal barrier deposits in South and North Holland: Meded. Geol. Stichting, new ser., no. 17, p. 41–87.

Zenkovitch, V. P., 1967, Processes of Coastal Development: New York, Interscience Pub. Co., 738 p.

Manuscript Received by The Society April 16, 1970

Marine Institute of the University of Georgia Contr. no. 202

PRINTED IN U.S.A.

Editor's Comments on Paper 31

In reply to Hoyt, Otvos maintains that salt marshes and tidal mud flats on open-marine shores are much more extensive than Hoyt is willing to admit. Beyond that, Otvos points out that: (1) marine sediments are not found in lower salinity basins (bays, estuaries, sounds) and (2) marine sedimentation rates may be too slow off an open coast to leave a significant record before the region becomes a lagoon. Basically, he does not believe that lack of "marine morphology" and its attendant characteristics are sufficient proof of ridge engulfment.

Otvos cites charts and observations showing that Grand Gosier Island, among others, evolved from islands developed upon shoals. Furthermore, the stratigraphy of Horn Island, Ship Island, and the Pine Island Chain, he feels, reveal barrier island growth upon subtidal clayey-silty deposits, not the sort of base expected under a dune or beach ridge.

Agreement with Hoyt is reached on the point that migration of a barrier destroys evidence of its foundation at the time of origin, and Otvos concedes that some barrier development through ridge engulfment may occasionally occur.

Reprinted from the *Geol. Soc. Amer. Bull.*, **81**, 3783–3788 (1970)

ERVIN G. OTVOS, JR. *Department of Earth Sciences, Louisiana State University in New Orleans, Louisiana 70122 and Geology Division, Gulf Coast Research Laboratory, Ocean Springs, Mississippi 39564*

31

Development and Migration of Barrier Islands, Northern Gulf of Mexico: Reply

One of the aims of my paper (Otvos, 1970) was the illustration of processes which may obscure or destroy sedimentary proofs of barrier island genesis. In certain of the presented cases, however, the genetic conditions are clear. Clear-cut examples also should have been cited in defense of the "barriers-through-engulfment" (Hoyt) theory. Examples of late Holocene barrier island formation (for example, Texel and Vlieland) through storm breaching and partial erosion of extensive mainland coastal dune-beach fields and the inundation of the hinterland are well known from the medieval history of the Friesian shores of northern Holland (Putzger, 1969, p. 34, 42, 46, and 51). Perhaps an identical process also was responsible for the formation of the Flandrian island chain along the present northwestern French and the Belgian shores, rejoined to the mainland in the Middle Ages by reclamation. These North Sea shores, *in contrast to our shorelines*, were recently affected by a local (Flandrian) transgression. In Holland the sea level has risen by about 3.5 m during the last 4000 yrs (van Straaten, 1965, p. 71; Zenkovich, 1967, p. 532).

The example which Hoyt *did* mention from the Netherlands, based on the work of van Straaten (1965), illustrates not barrier island formation from beach/dune ridges "immediately landward of the shoreline" by engulfment, but development of a beach ridge on the seaward flank of intertidal flats by aggradation. The easternmost, first permanent beach ridge of the Dutch "coastal barrier" complex by the present city of The Hague, therefore, was separated from the mainland at that time by a tidal flat zone (van Straaten, 1970, written communs.). This is the reason why open-marine sediments were not found landward of these Subboreal-Subatlantic barrier islands. Afterward, this barrier ridge was linked with the mainland by the infilling of the tidal channels and flats; extensive beach progradation seaward over deeper open-sea sediments and simultaneous aggradation preceded the development of the coastal beach-dune plain and the present shore configuration of the Dutch mainland (van Straaten, 1965).

Although not rejecting the probability that in certain instances engulfed dune and beach ridges may turn into barrier islands (Fig. 1A; Hoyt, 1967), the absence of marine deposits landward of barrier islands can be accepted only as a supplementary evidence of this process. A number of other facts may also prevent identification of open-marine shore and nearshore sediments located stratigraphically immediately *below* and *landward* of barrier island sediments (Otvos, 1970, p. 241–242).

No marine sediments can be expected landward of the barrier islands which formed adjoining or within lower salinity basin areas (estuaries, bays, sounds). Such conditions appear to have existed along the northern Gulf shores immediately before and during the end phase of the Holocene transgression (Fig. 1B). Extensive brackish embayments still prevailed at the time when barrier construction started seaward. In addition, it is obvious that barrier islands which formed after this stage between areas of brackish and full-marine sedimentation, also would be bordered landward by brackish deposits. This is the condition into which newly emerged islands, for instance in the Chandeleur and Mississippi Sound offshore island chain, are born. Some marine influence would prevail in the lagoon/sound sides, especially around inlet and pass entrances. Such an effect is evident in the Pontchartrain Embayment (mainland) side of the Pine Island barrier chain where an open Gulf-type molluscan assemblage was found (Rowett, 1957).

Assuming that full salinity conditions prevailed landward of future barrier islands following the transgression, low sedimentation

Geological Society of America Bulletin, v. 81, p. 3783–3788, 1 fig., December 1970

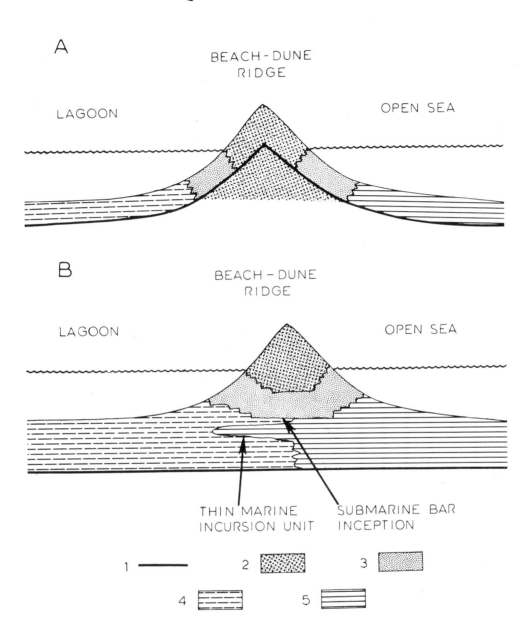

Figure 1. Schematic cross sections comparing two proposed ways of barrier island formation. (A) By marine engulfment and upward growth of mainland beach-dune ridges (Hoyt, 1967). (B) By shoal aggradation (Otvos, 1970). Brief marine incursions may not have taken place. Bar growth started near facies boundary or within lagoonal facies. Symbols: 1: Pre-transgression land surface; 2: Beach foreshore, backshore and dune units; 3: Lagoonal and open marine subtidal shoreface and shoal units; 4: Brackish (estuarine, lagoonal, bay-sound) units; 5: open marine sediments.

rates might have produced only a thin veneer of open-marine sediments, impossible to detect, before brackish deposits started to form over them. At one location in the Mississippi Sound, such sedimentation rates (2.4 cm/100 yrs) for the past 5000 yrs, were interpreted by Ludwick (1957, p. 33).

Beach sands are often found devoid of any fauna. Possible connections between barrier island and barrier spit development were noted also by Ludwick (1957, p. 58) and Saucier (1963, p. 51).

Individual barrier island groups are discussed below with consideration of Hoyt's objections.

CHANDELEUR ISLANDS AND SHOALS

This chain presently occupies a position at least from 5 to 15 km west of the ancient coast of two St. Bernard, Mississippi River, subdeltas which became inactive from 2300 to 1800 B.P. (Frazier, 1967, p. 305–308). Seaward of the chain a wide apron of thin sand sheet and underwater sand bars represent old subaerial and subtidal delta margin deposits, which presumably include dune and beach sediments of the original delta shores which have been reworked by the encroaching Gulf of Mexico. Most of the present barrier island sediments were derived from reworked *inland* delta plain deposits. The arcuate outline of the island chain reflects the shaping effects of the wave approach directions, not positions and configurations of the ancient delta shores. In the absence of accurate map documentation, except for the past 100 yrs, it could be debated perhaps that parts of some larger islands (for example, northern or southern Breton Island) may have shifted as a single unit from the original delta shore positions and may not have turned into shoals during their existence. Even in this unprovable and highly unlikely case, the erosional-accumulational changes which affected such islands during their migration history appear to have been much more thorough than the modifications suffered by the Atlantic coastal barrier islands (Hoyt and Henry, 1967; Hoyt and Hails, 1967). It is hard to accept such a hypothetical genesis as "outstanding example of barrier island formation by submergence of beach or dune ridges, or both" (Hoyt, 1970).

The site of Grand Gosier Island, between Breton Sound and the Gulf of Mexico was covered by 1 to 4 m of water in 1869 (U.S. Coast Survey Chart H-999). By 1917 a number of small islands developed from the shoal and merged later into the large single island. Treadwell (1955, p. 15) stated that "older charts indicate that the process of island construction and degradation has been repeated several times since the first survey. During severe storms the island is destroyed, but it is built again during periods of calm weather." Grand Gosier, along with Curlew, Stake, Boot and several minor islands disappeared again as the result of extreme Hurricane Camille in 1969. At the same time the northern, 32-km-long Chandeleur Island was fragmented into about fifty islets (U.S. Coast and Geodetic Survey, 1970, Chartlet, 1270). The development of numerous small islands from shoals is also documented in the eastern Timbalier Island group (U.S. Army Chief of Engineers, 1962, Pls. 2 and 3). Glezen (1951, p. 6) believed that the Dog Keys off the Alabama shore also might have turned into shoals and later back again to islands during their history. It is clear that under favorable conditions, shoals readily aggrade into barrier islands, *not* only in the course of barrier island migration, but also without the presence of such pre-existing land strips which act as nuclei for further land accretion. Outside the discussed region, another good example is Sakalin Island in the Black Sea, south of the St. George distributary mouth of the Danube River. In less than 65 yrs a shallow-water area, 4.5 to 6 km from the mainland, developed in a 22-km-long and up to 1.5-km-wide sand island (Zenkovich, 1967, p. 576, Fig. 276). Aggradational island formation was documented also by A. O. Lind in Cat Island, the Bahamas (1970, written commun.).

GALVESTON ISLAND

Bernard and others (1962) were quoted by me (Otvos, 1970) for the interpretation of water depths at which the barrier island started its growth (*see also* Bernard and others, 1959, p. 223; Bernard and LeBlanc, 1965, p. 157). These conclusions were based on absolute ages and facies relationships and are supported by the results of another extensive subsurface study of the Galveston Bay–Galveston Island Holocene deposits (Rehkemper, 1969). This paleogeographical reconstruction indicated that at the time of the barrier inception the present island location was covered by the sea (Rehkemper, 1969, p. 41–45, Fig. 2–20). No proof was found for the existence of engulfed mainland beach/dune ridges east of the present island that were capable of serving as nuclei for barrier island progradation and aggradation, in conjunction with the westward littoral drift.

PADRE ISLAND

The subsurface of the island records the final Holocene transgression phase (open lagoon and subtidal shoreface bar sediments), followed by local regression (closed lagoon, tidal flat, intertidal and supratidal barrier sediments), see Fisk (1959). Plentiful local sediment supply aggraded the bars and even large lagoonal areas to and above the sea level. The absence of documented pretransgressional mainland beach/dune ridges and the intensive aggradation suggest that shoaling and aggradation without pre-existing, later engulfed, beach/dune ridges, could have been sufficient in turning underwater shoals into barrier islands.

PINE ISLAND AND MISSISSIPPI SOUND BARRIERS

The available subsurface information from the Pine Island Chain (Saucier, 1963); Horn Island, areas of Horn Island toward the Gulf of Mexico, areas toward the Mississippi Sound of Horn Island (Ludwick, 1957), and from Ship Island (Hahn, 1962) indicates no traces of pretransgressional mainland coastal ridges on the clayey late Pleistocene land surface (Ludwick, 1957). Subtidal Holocene clayey-silty deposits underlie the barrier sands (Ludwick, 1957; Saucier, 1963; Hahn, 1962). The applicability of Hoyt's barrier engulfment theory is made even less likely by the fact that these barriers do not exhibit the sediments of earlier Holocene transgression stages either. At earlier times coastal ridge engulfment by fast-rising sea level was perhaps more feasible. Exactly the opposite conditions existed during barrier island formation: this was the time when the transgression stopped, and shoreline stabilization started (Otvos, 1970, p. 241).

Effects of Hurricane Camille (August, 1969) offered new opportunities for the study of erosional, accumulational, and migration processes affecting not only the Chandeleur Islands, but also the Mississippi Sound barrier islands. They illustrate not only the magnitude of storm erosion, but also the ease and speed with which shoaling and island formation (reconstruction) may take place afterward. Aerial photos, charts, and reports of the U.S. Coast and Geodetic Survey (U.S. Coast and Geodetic Survey, 1969, 1970; *see also* Petersen, 1969, 1970) revealed the opening of a 3.5-km-wide inlet and a narrower channel which cut Ship Island into three parts. The wide inlet overlapped the site of the much narrower 1947 hurricane channel. (An incorrect *1946* date in my previous paper originated from a source reference.) Pelican Island was wiped out completely and Petit Bois Island disappeared entirely for a short period following the hurricane. Large parts of the island reappeared by the time of the second post-hurricane survey, 2 months later. Other islands (Sand, Horn, and Dauphin) lost as much as 1.4 km of their lengths.

SALT MARSHES AND TIDAL MUD FLATS ON OPEN-MARINE SHORES

Salt marsh and mud flat shores which face open seas are not as unique and insignificant as Hoyt stated. Gentle offshore bottom gradients and ample sediment supply account for their persistent and world-wide presence. The Gulf of Papua, Mekong, and Irrawady deltas have these shores (Fisher and others, 1969, Figs. 47–49). Among the more extensive mud flat–salt marsh shores the following few come readily to mind: (1) long stretches in Louisiana between a point 6 km west of Sabine Pass, Texas, and Grand Bay, northeastern shore of the Balize Mississippi subdelta, cumulative length over 400 km (Fisher and others, 1969, Fig. 18; Orton, 1959, p. 10); (2) Heligoland Bight shores on the North Sea between the Weser and Elbe Rivers, West Germany, total length 100 km (Zenkovich, 1967, p. 633); (3) The Wash shores on England's North Sea coast (Chapman, 1960, p. 139; Kestner, 1962); (4) Normandy-Brittany Bay "tangues" and tidal marshes at Mont St. Michel, France (Termier and Termier, 1963, p. 180 and 190); (5) Gulf of Po Hai shores, Yellow Sea, total length over 250 km (Zenkovich, 1967, Fig. 317, p. 652). According to van Straaten (1970, written commun.) in the case of (2), (3), and (4), tidal ranges and the convergence of current channels played a significant role in preventing barrier island formation.

ABSENCE OF "MARINE MORPHOLOGY" ALONG MAINLAND SHORES AS A GENETIC PROOF

Estuarine and lagoonal shores commonly are dominated by tidal flat, marsh and swamp facies as are some sounds and protected bays. The present mainland morphology is not always a conclusive indicator of the shoreline type which was present before or at the time of barrier development offshore. Local regressive and transgressive changes, influx of deltaic sediments, or increase in erosion might have taken place since that time. "Marine morphology" (Hoyt), as indicated above, cannot be equated solely with high-energy coastal environments, nor can lagoonal or sound coastal morphology be associated with low-energy environments alone. Well-developed beach and dune ridges, spits, and even abraded bedrock cliffs are occasionally found along water bodies of reduced salinity, such as sounds and larger protected bays. (*See*, for example, abraded bedrock cliffs and dunes along the Long Island Sound, large dunes along the mainland shores of east Florida lagoons.)

ACKNOWLEDGMENTS

I am greatly indebted to Professor L.M.J.U. van Straaten of Groningen University for his critical reading of the manuscript and for views expressed in a correspondance on the subject matters. Thanks are equally due to Commander Richard H. Houlder who supplied me with a large volume of U.S. Coast and Geodetic Survey documentation on the post-hurricane changes of the barrier islands, including charts, aerial photos, and other information. Part of the presented work was done at the Gulf Coast Research Laboratory, Ocean Springs, Mississippi.

REFERENCES CITED

Bernard, H. A., and LeBlanc, R. J., 1965, Résumé of the Quaternary geology of the northwestern Gulf of Mexico, *in* Wright, H. E., Jr., and Frey, D. G., *Editors*, The Quaternary of the United States: Princeton, New Jersey, Princeton Univ. Press, p. 137–185.

Bernard, H. A., LeBlanc, R. J., and Major, C. F., 1962, Recent and Pleistocene geology of southeast Texas, *in* Rainwater, E. H., and Zingula, R. P., *Editors*, Geology of the Gulf Coast and Central Texas: Guidebook of excursions, Houston Geol. Soc., p. 191–206.

Bernard, H. A., Major, C. F., and Parrott, B. S., 1959, The Galveston barrier island and environs: A model for predicting reservoir occurrence and trend (abs.): Gulf Coast Assoc. Geol. Socs. Trans., v. 9. p. 221–224.

Chapman, V. J., 1960, Salt Marshes and Salt Deserts of the World: New York, Interscience Pub. Co., 392 p.

Fisher, W. L., Brown, L. F., Jr., Scott, J. H., and McGowen, J. H., 1969, Delta systems in the exploration for oil and gas: Austin, Texas, Texas Univ. at Austin, Bur. Economic Geology, 157 p.

Fisk, H. N., 1959, Padre Island and the Laguna Madre flats, coastal south Texas: 2nd Coastal Geography Conf., Louisiana State Univ., Baton Rouge, Louisiana, p. 101–151.

Frazier, D. E., 1967, Recent deltaic deposits of the Mississippi River: Their chronology and development: Gulf Coast Assoc. Geol. Socs. Trans., v. 17, p. 287–315.

Glezen, W. H., 1951, Changes in the barrier bars of Mississippi Sound recorded on maps and charts from 1710 to 1948, Unpub. memorandum: Gulf Research and Development Co., 8 p. (available on interlibrary loan).

Hahn, A. D., 1962, Reconnaissance of titanium resources on Ship Island, Harrison County, Miss.: U.S. Bur. Mines Rept. Inv., no. 6024, 24 p.

Hoyt, J. H., 1967, Barrier island formation: Geol. Soc. America Bull., v. 78, p. 1125–1136.

———1970, Development of Barrier Islands, Northern Gulf of Mexico: Geol. Soc. America Bull., v. 81, p. 3779–3782.

Hoyt, J. H., and Hails, J. R., 1967, Pleistocene shoreline sediments in coastal Georgia: Deposition and modification: Science, v. 155, p. 1541–1543.

Hoyt, J. H., and Henry, V. J., Jr., 1967, Influence of island migration on barrier island sedimentation: Geol. Soc. America Bull., v. 78, p. 77–86.

Kestner, F.J.T., 1962, The old coastline of The Wash: Geog. Jour., v. 128, pt. 4, p. 457–478.

Ludwick, J. C., 1957, Lithofacies patterns on the continental shelf in the northeastern Gulf of Mexico: Gulf Research and Development Co. Rept. no. 58, 77 p.

Orton, E. W., 1959, A geological study of Marsh Island, Iberia Parish, Louisiana: Louisiana Wildlife and Fisheries Comm. Tech. Rept., 28 p.

Otvos, E. G., Jr., 1970, Development and migration of barrier islands, northern Gulf of Mexico: Geol. Soc. America Bull., v. 81, no. 1, p. 241–246.

Petersen, B., 1969, Camille winds altered coast: The New Orleans Times-Picayune, December 26, p. 1–2.

———1970, In Camille's aftermath—shrinking islands: The New Orleans Times-Picayune Dixie Magazine, February 22, p. 10–11.

Putzger, F. W., 1969, Historischer Weltatlas: Bielefeld-Berlin-Hannover, Velhagen & Klasing, 146 p.

Rehkemper, L. J., 1969, Sedimentology of Holocene estuarine deposits: Galveston Bay, *in* Lankford, R. R. and Rogers, J.J.W., *Compilers*, Holocene Geology of the Galveston Bay Area: Houston, Texas, Houston Geol. Soc., p. 12–134.

Rowett, C. L., 1957, A Quaternary molluscan assemblage from Orleans Parish, Louisiana:

Gulf Coast Assoc. Geol. Socs. Trans., v. 7, p. 153–164.

Saucier, R. T., 1963, Recent geomorphic history of the Pontchartrain Basin: Louisiana State Univ. Studies Coastal Studies ser. no. 9, 114 p.

Termier, H., and Termier, G., 1963, Erosion and sedimentation: London-Princeton, Van Nostrand Co., 433 p.

Treadwell, R. C., 1955, Sedimentology and ecology of southeast coastal Louisiana: Louisiana State Univ. Coastal Studies Inst. Tech. Rprt. no. 6, 78 p.

U. S. Army Chief of Engineers, 1962, Belle Pass to Raccoon Point, Louisiana, Beach erosion control study: House Document no. 338, Washington, 31 p.

U. S. Coast and Geodetic Survey, 1969, 1970, Correction Chartlets: Dauphin Island-Sand Island-Pelican Bay area (Chartlets 872 SC; NM 49 Dec. 6, 1969 and 1266; NM 50 Dec. 13, 1969); Western Dauphin Island-Petit Bois Pass-Petit Bois Island (Chartlet 874 SC; NM 51 Dec. 20, 1969 and 1267; NM 3 Jan. 17, 1970); Western Part Petit Bois Island-Horn Island Pass-Eastern Horn Island (Chartlets 414, NM 50 Dec. 13, 1969 and 874 SC; NM 51 Dec. 20, 1969); Ship Island (Chartlet 876 SC; NM 46 Nov. 15, 1969; 1267; NM 3 Jan. 17, 1970); Chandeleur Island-Grand Gosier Island-Breton Island (Chartlets 1270; NM 2 Jan. 10, 1970; 1267; NM 3 Jan. 17, 1970).

van Straaten, L.M.J.U., 1965, Coastal barrier deposits in South- and North-Holland, in particular in the areas around Scheveningen and IJmuiden: Mededel. van de Geol. Stichting, Nieuwe Serie, no. 17, p. 41–87.

Zenkovich, V. P., 1967, Processes of coastal development: New York, Interscience Pub. Co., 738 p.

Manuscript Received by The Society July 1, 1970

Editor's Comments on Paper 32

32 **Dillon:** *Submergence Effects on a Rhode Island Barrier and Lagoon and Inferences on Migration of Barriers*

Dillon's paper provides a closer look at the landward migration of a barrier. In this case the Charlestown–Green Hill barrier beach, developed upon a base of glacial till, outwash, and glaciofluvial sand.

As interpreted by Dillon, the barrier has remained small in size because of lack of sediment. Additionally, storm waves rework the shallow barrier to its base. Through overtopping, sand eroded on the seaward side of the barrier encroaches upon the lagoon in the form of washover fans. As a result of the gradual Holocene submergence, the barrier has been moved slowly landward in this fashion.

Dillon does not believe, as did D. W. Johnson, that a barrier must migrate shoreward. He holds that a rising sea level can completely drown a barrier, modifying it in the process. However, washover during submergence, as in the case cited here, can cause the landward migration of a barrier by "essentially rolling over itself," as Dillon puts it.

William P. Dillon was born in 1936. He received his professional education at Bates College (B.S., Geology, 1958), Rensselaer Polytechnic Institute (M.S., Geology, 1961), and the University of Rhode Island (Ph.D., Oceanography, 1969). He was employed as a research associate at the University of Rhode Island's Narragansett Marine Laboratory from 1961 to 1964. Dillon was an instructor on the faculty of the University of Rhode Island in 1968–1969. In 1969 Dillon joined the faculty of the San Jose State College. During the summer of 1970, he produced a conceptual model of the development of west Africa for Standard Oil Company of California. Dillon became a Geologist on the staff of the U.S. Geological Survey in 1971 and has carried out research on continental margins in the western Caribbean with this organization. He has experience in the general fields of structural geology, sedimentology, and applied geophysics. His principal research interests include the structure and development of continental margins and coastline evolution.

Reprinted from the *J. Geol.*, **78**, 94–106 (1970)

SUBMERGENCE EFFECTS ON A RHODE ISLAND BARRIER AND LAGOON AND INFERENCES ON MIGRATION OF BARRIERS[1]

32

WILLIAM P. DILLON[2]

Graduate School of Oceanography, University of Rhode Island, Kingston, Rhode Island 02881

ABSTRACT

The Charlestown–Green Hill barrier beach–lagoon complex lies along Rhode Island's south shore. Barrier beach and tidal delta sands, lagoonal fine sediments, and lag deposits have been deposited on a base of glacial till, outwash, and glaciofluvial sand. The beach and lagoonal sediments have been derived mainly from reworking of glacial sediments. Tidal currents have built a sand delta at the inshore end of an inlet, but sand on the barrier beach's lagoonal side has been introduced over the barrier. Fine sediments are deposited in the deeper areas of the lagoon. The barrier beach was formed at a lower sea level, and moved landward as the sea transgressed, with lagoonal deposition migrating with the barrier. The small size of this barrier places its base at a shallow depth, resulting in erosion of the entire seaward side by storm waves and also permitting considerable transport of sand across the barrier to the lagoon side. The barrier has remained small because of lack of sand supply. Thus lack of sand supply seems to be the dominant factor in allowing the landward migration of this barrier.

INTRODUCTION

Charlestown–Green Hill Pond is a lagoon behind a barrier beach on the southern coast of Rhode Island (fig. 1). The lagoon is about 8.3 km long and up to 1.5 km wide. The major part of the lagoon has depths of less than 2.0 m with a maximum of about 2.5 m. A tidal channel, located near the middle of the lagoon affords a very restricted connection to the ocean. Primary objectives of this study were to examine the character, structure, and stratigraphy of sedimentary deposits as revealed by subbottom acoustic reflection profiles, cores, and bottom sediment distribution in order to determine mode of evolution of the lagoon.

METHODS

The structure of the sedimentary deposits was determined by the use of an acoustic reflection profiler, the pinger probe (Edgerton and Payson 1964; Edgerton et al. 1964). Fourteen cores were taken to identify the horizons noted in the profiles. Coring was done from a barge by hammering a 5.1 cm-diameter, 3.66 m-long pipe into the sediment. Also a scoop sampler, a hand-operated corer, taking 5.1 × 78.7 cm cores, and a small corer, taking 2.5 × 20.3 cm plugs of sediment were used for various purposes. Currents were checked by tracking the movement of current drogues. Radiography was used to examine cores for structure (Calvert and Veevers 1962; Hamblin 1962). Standard sieving and hydrometer methods were employed for sediment size analysis.

DESCRIPTION OF THE LAGOON

Sediment types, sources, transportation, and deposition.—The sediments may be divided into marine and estuarine deposits, and glacially derived sediments on which the marine deposits are laid. Distribution of recent marine and estuarine sediment types is summarized in figure 2. The barrier beach and tidal delta sands are medium to fine sand, characteristically unimodal, with good sorting. The fine lagoonal sediment, a sand, silt, and clay mixture, generally lacks bedding; silty fine sands usually show quite high concentrations of polychaete fecal pellets. Also these fine sediments contain fairly high concentrations of carbon, ranging from about 1–8 percent organic carbon by weight (by the dichromate reduction

[1] Manuscript received November 16, 1967; revised July 31, 1969.

[2] Present address: Department of Geology, San Jose State College, San Jose, California 95114.

[JOURNAL OF GEOLOGY, 1970, Vol. 78, p. 94–106]

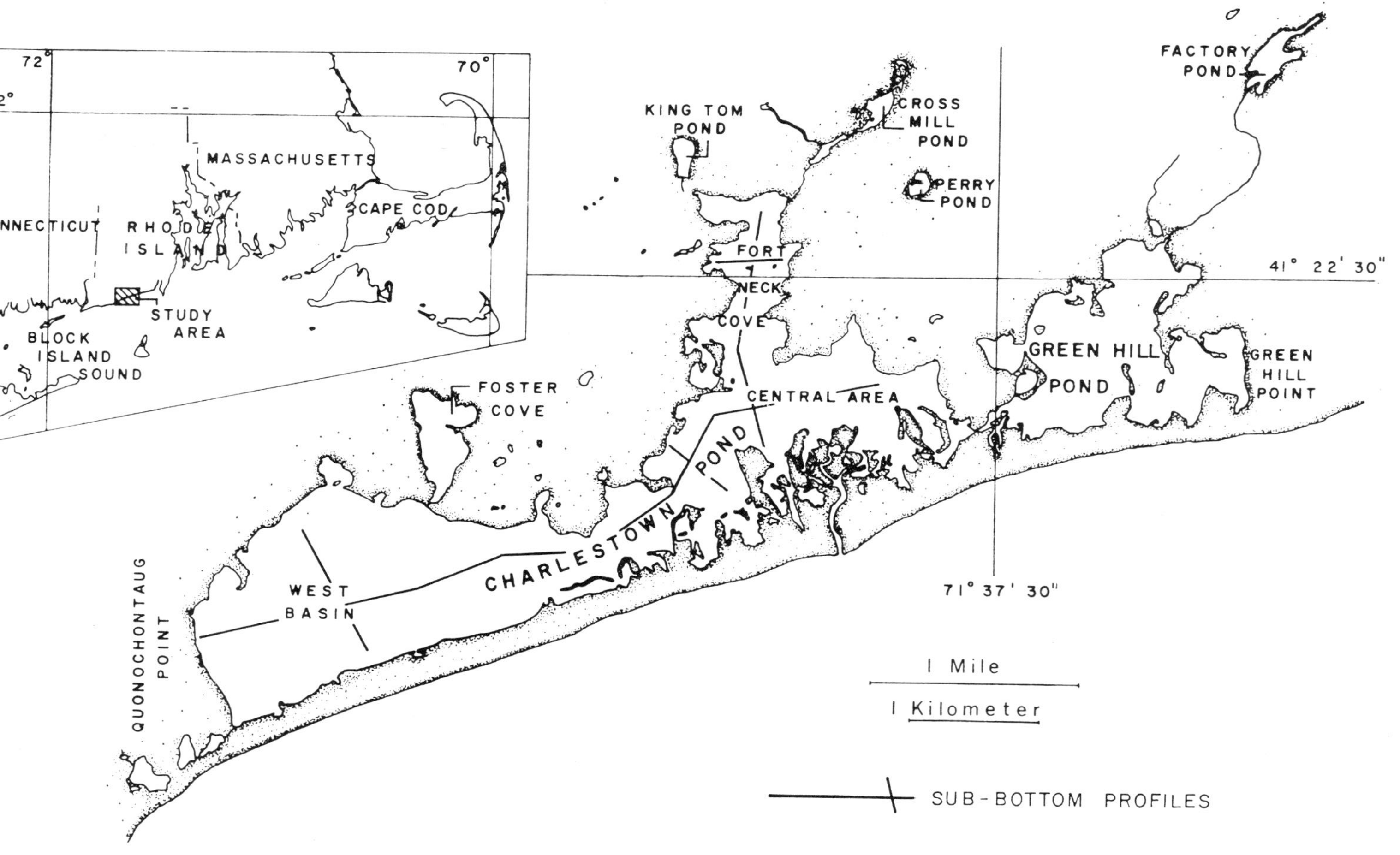

FIG. 1.—Location of study area and positions of all seismic reflection profiles

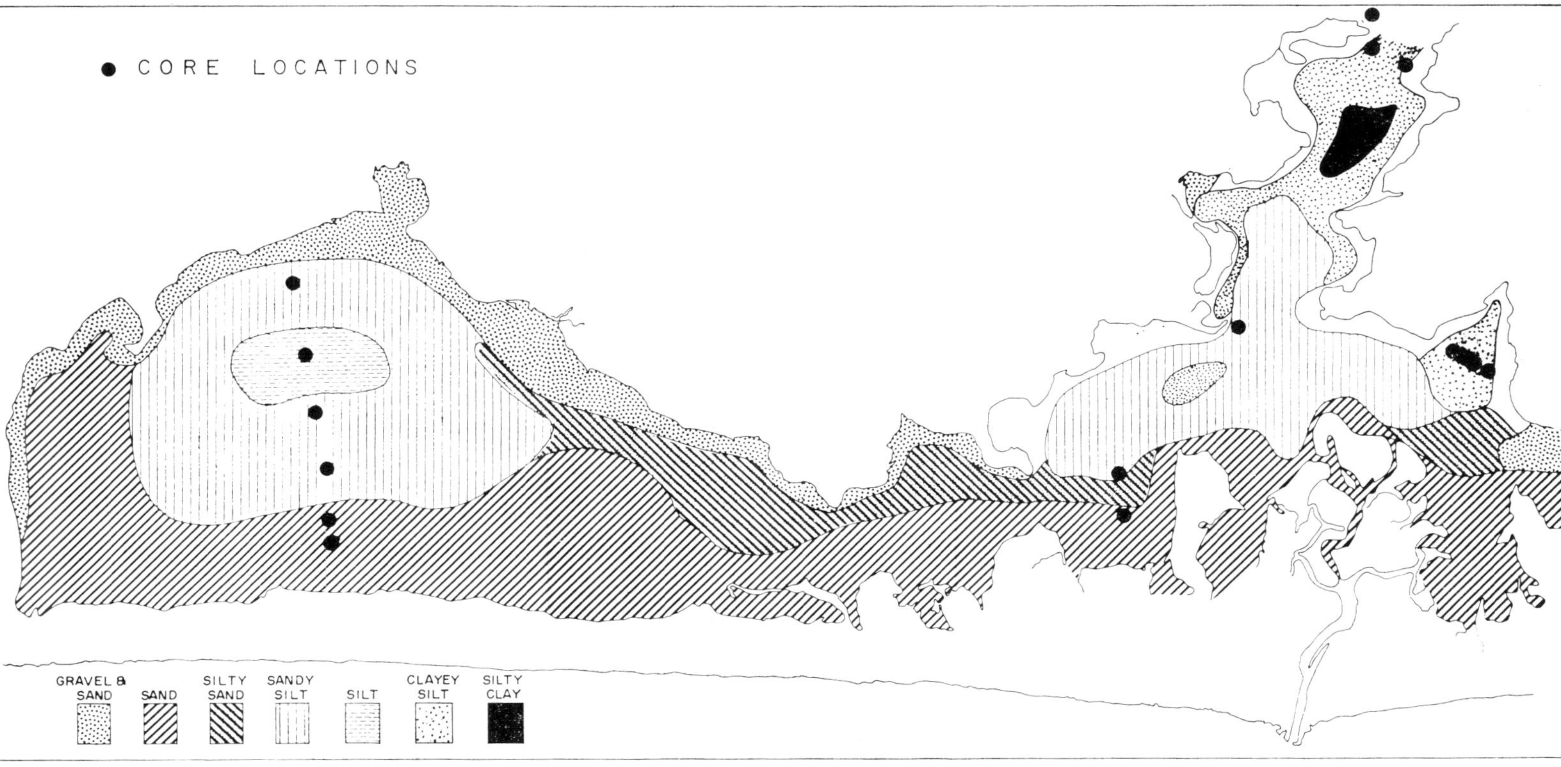

FIG. 2.—Charlestown Pond, generalized sediment distribution based on 113 samples and positions of cores

technique), or possibly twice that amount of plant material. The barrier and tidal delta sands consistently showed less than 1 percent organic carbon, and lagoon sediments displayed a very strong inverse relationship between the percentage of organic carbon and the sand/silt+clay ratio (Dillon 1964, figs. 18, 19, and 20). The lag deposits, which are the third marine-estuarine type, are composed of poorly sorted sand, gravel, and coarser material produced by the reworking and removal of fines from the glacial outwash plain on which the pond developed. Lag deposits are generally bimodal, with very poor sorting. Reworked glacial till and outwash also are present on the sea floor, seaward of the barrier, in water depths greater than about 13 m.

The outwash consists mostly of very poorly sorted, oxidized, roughly stratified outwash "gravel," ranging in size from silt to boulders. Glaciofluvial deposits of very clean, white, well-sorted fine sand to coarse silt (locally called "sugar sand") is characteristically the first sediment found over the outwash in any depression.

The Charlestown Pond Lagoon is basically a dead end for sediment that enters, because inlet velocities are very high compared with currents in the lagoon. However, sedimentation rate is low, as shown by the thickness of lagoonal marine sediment, which is only about 1–1.5 m. Obviously little sediment has been brought into the lagoon. Little fine material is present in the clear ocean water entering the lagoon. Furthermore, almost no sediment is introduced to the lagoon by direct runoff from the land, as the recessional moraine directly to the north blocks stream flow. Also, the outwash plain is very irregular with no runoff channels and is very permeable, tending to absorb rain. Most of the small amount of fine sediment was probably derived from the winnowing of the glacial outwash and till during postglacial submergence, or carried into the lagoon by tidal currents.

Sediment movement and deposition are greatest in the vicinity of the inlet where tidal currents are strongest. A tidal delta has been built at the inside of the inlet by material swept into the lagoon (Meade 1969). The delta is composed of finer sand than that on the ocean side of the barrier. Delta sand has a median diameter of roughly 0.15 mm, while beach sand generally has a median diameter of about 0.2–0.3 mm along the southern coast of Rhode Island (McMaster 1960, fig. 2). Back-barrier sand generally has 0.2–0.3 mm median diameter, and shows no consistent change in size or sorting away from the breachway. Thus most back-barrier sand apparently has been dumped over the barrier from the ocean beach during storms, forming the lobate fans which appear behind blowouts on the barrier. After a hurricane, sand size at one station behind the barrier beach was found to be larger than before the storm, and organisms identified as ocean beach types were found isolated (D. K. Phelps 1963, personal communication).

The clean sand behind the barrier is being reworked by waves, so finer sediments come to rest in the deeper areas (depths of 1–2.5 m) where turbulence and current activity are low. The very finest sediments are deposited in Fort Neck Cove, where tidal currents were found to be weakest of any place in the lagoon system, and where wave activity is greatly reduced because of high banks and generally short fetch available to the wind. The northern shores of the lagoon are characterized entirely by erosional processes; however, the erosion is extremely slow in most places since the shore is armored by the lag deposit.

The seaward side of the barrier shows the usual variations in a beach exposed to the ocean, that is, seasonal changes and variations due to storms. The barrier beach is relatively stable because it is a depositional feature in equilibrium with its environment, supported by the headlands of Quonochontaug and Green Hill Points. The former point is protected by bedrock and a concentration of large boulders, while the latter, although covered by unconsolidated till, is also protected by a lag deposit of boulders. A relative rise of sea level may be the primary cause of recession of this stable shore.

Sand which forms the barrier apparently has come from reworking of the till and outwash as the sea transgressed. No significant amount of sand has been contributed from the land, because the only rivers entering the ocean in the vicinity flow into effective sediment traps (Narragansett Bay and Long Island Sound). Little sand is supplied presently from offshore, since samples taken by Savard (1966) and myself show that sediments seaward of the barrier, all along the southern Rhode Island shore, range from very coarse sand to gravel. These sediments have a mean diameter greater than 0.50 mm and contain very little fine material; they are apparently a product of winnowing of the glacial outwash. Headland erosion is of relatively minor importance (McMaster 1960, p. 404). No other source of beach sand is apparent.

Stratigraphy.—The stratigraphic relationships have been disclosed by seismic reflection profiling and coring. Positions of pinger-probe profiles and of long cores are shown in figures 1 and 2. Detailed drawings of some portions of the subbottom records are shown in figures 3, 6, and 7. The stratigraphic relationships underlying the lagoon may be divided into two general types representative of two different areas; in one type the lagoon has encroached upon the outwash plain, as in West Basin and the channel between West Basin and the central area; in the other an old, sand-floored valley and its alluvial fan have been flooded as in the central area and Fort Neck Cove (fig. 1).

The first type is displayed in a north-south profile across West Basin (fig. 3), which illustrates the irregularity of the surface of the outwash sediments. This surface shows a kettle, scattered boulders, and one large irregular depression, probably a kind of complex kettle. The simpler kettle compares closely in scale and proportions with kettles now exposed above sea level near the lagoon. Radiocarbon dating was performed on wood cored in the large depression (core no. 7, figs. 3 and 4). This wood, found at a corrected (for compaction) depth of about 212–216 cm, or about 360–370 cm below sea level, gave an age of 3,960 ± 80 years before 1950. Actual depth in core was 91.3–93.3 cm, and water depth was 150 cm. Corrections were estimated based on reflection profiling data, a rather unsatisfactory method. Two samples from the base of the peat at a corrected depth of about 380–390 cm below sea level (98.0–103.0 cm uncorrected depth in core) and 390–400 cm (103.0–106.1 cm uncorrected) gave ages of 6,220 ± 80 and 7,300 ± 80 years (Yale Radiocarbon Laboratory nos.: Y1423, Y1424, and Y1642). The age and depth of the wood agree with similar data from Connecticut and Massachusetts, if one assumes an age of a few hundred years for wood at burial. This would indicate that submergence in Rhode Island is comparable to submergence elsewhere in this area, and probably somewhat greater than eustatic (Scholl and Stuiver 1967, p. 447). The peat dates indicate a much slower rate of submergence, if they are accepted. It is probably more reasonable to conclude that some error occurred, such as contamination of the sample with old carbon, perhaps from an old freshwater peat, although botantical work suggested a high-marsh salt peat.

The homogeneous, highly organic, very fine sands and silts, which seem characteristic of lagoonal deposition in water more than 1 or 2 m deep, blanket the outwash surface, thickening southward. The low organic, well-sorted, medium-to-fine-grained barrier sands onlap and interfinger into the lagoonal silt-sand as shown by cores nos. 14 and 15 (fig. 4) and as observed in the pinger record. Interfingering of deposits would be expected, since addition of sand over the barrier bar during storms has produced irregular barrier growth into the lagoon. The simplified structure of this type is basically an onlapping of beds as depicted schematically (fig. 5).

The second stratigraphic type, in which lagoonal sediments onlap glacial stream deposits, is shown in fig. 6. The stream valley apparently evolved from a cluster of kettles which now forms Fort Neck Cove. Figure 7

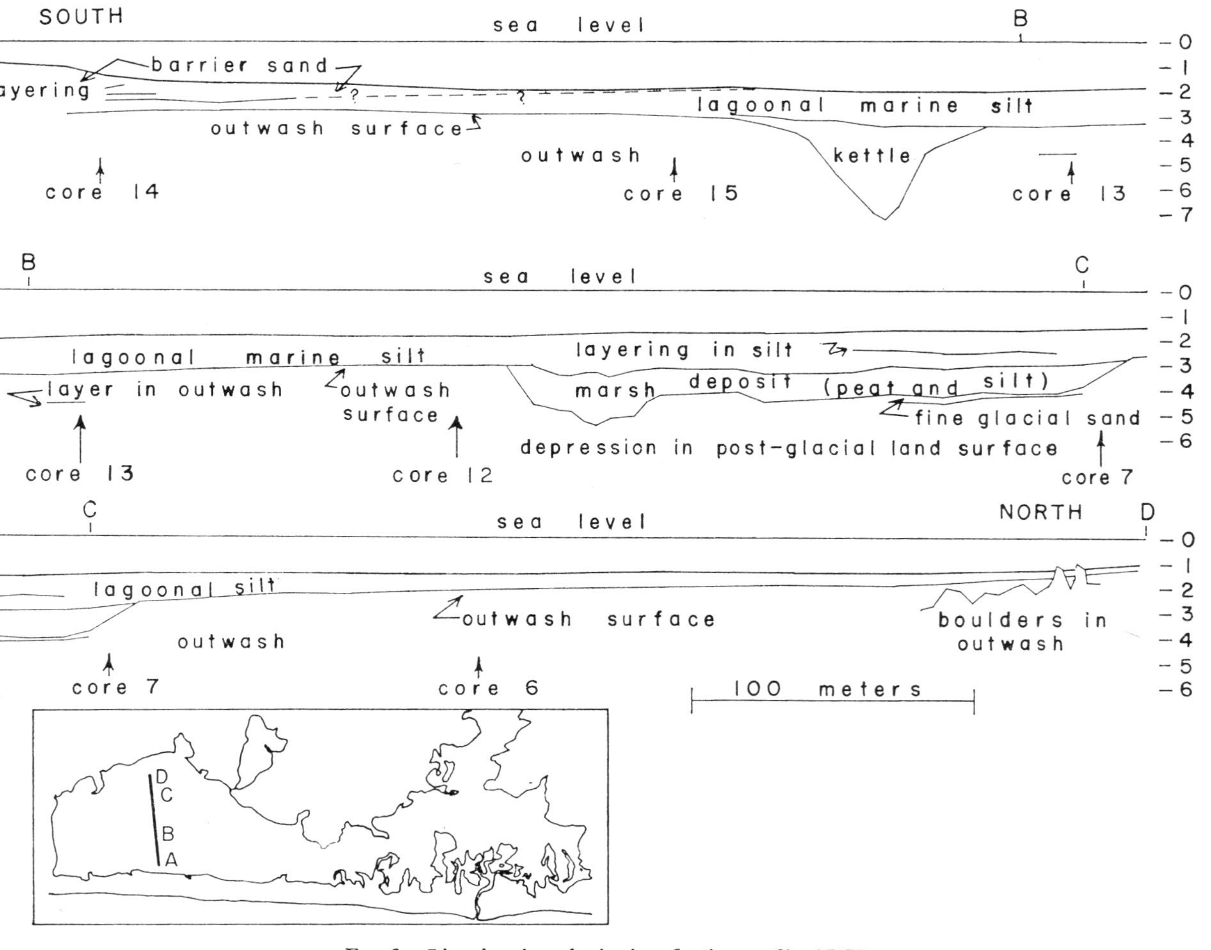

FIG. 3.—Line drawing of seismic reflection profile *ABCD*

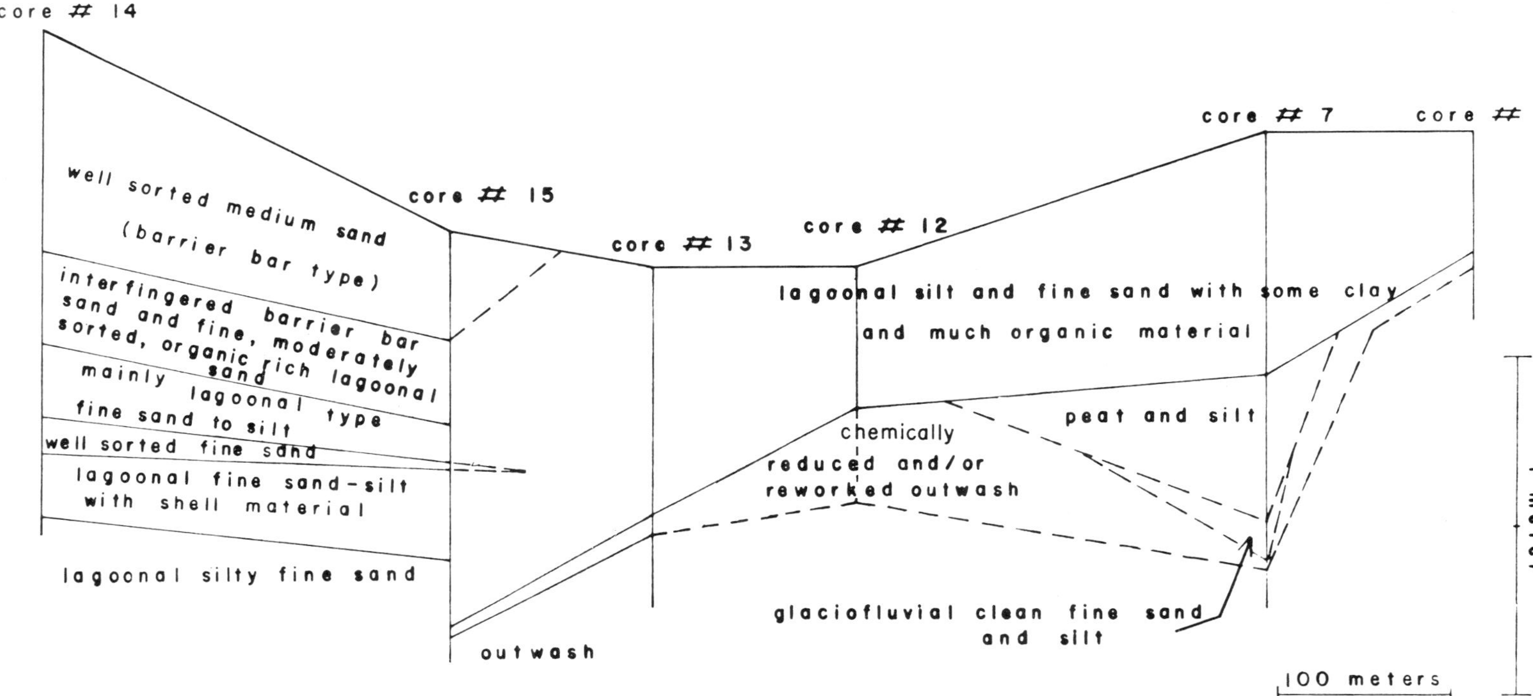

FIG. 4.—North-south cross section of West Basin based on cores

shows a section across part of a former kettle in the cove. The bottoms of the kettles first were filled with a large amount of the very fine glaciofluvial sand, then additional stream sediment formed prograding fans (fig. 6). Two old stream valleys can be traced northward on shore from Fort Neck Cove. Small spring-fed streams still run in these channels, but flow is obviously much less than at the time when the streams were supplied with glacial meltwater. This change is evidenced by the relict, braided stream pattern with gravel bars.

In the southern part of the central area, profiles show that the migration of the tidal delta and its associated channels has created considerable stratigraphic complication. The outwash surface also is more rugged than it is in West Basin, although the presubmergence land surface was somewhat leveled by the deposition of alluvial sand.

FORMATION AND DEVELOPMENT OF THE LAGOON

Three groups of processes, as well as bedrock morphology, must be considered in order to understand the sedimentary relationships and evolution of the lagoon. These processes are the submergence that is responsible for the stratigraphic relationships of sediment types, the glacial activity that produced the foundation on which the lagoon is built and ultimately supplied the sediments, and the coastal processes that have built the barrier beach and caused its migration.

Bedrock contours in the lagoon area and to the north (Bierschenk 1956, pl. 1; La Sala and Hahn 1960; La Sala and Johnson 1960) and well data (W. B. Allen 1962, personal communication) show that the preglacial morphology was characterized by a series of SSE-trending valleys. A "sparker" seismic reflection profile, taken by me about 2 miles seaward of the barrier beach, shows that the valleys turn southward as they continue offshore (McMaster et al. 1968).

The beginning of the retreat of the glaciers in the late Pleistocene resulted in a eustatic rise of sea level. Data from southern

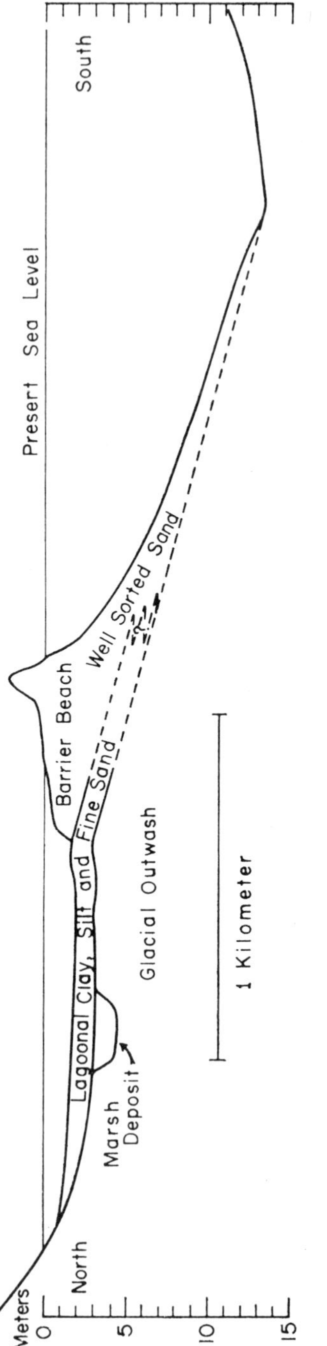

FIG. 5.—Typical stratigraphy of West Basin, the first stratigraphic type. Bathymetric and topographic information is from: U.S. Coast and Geodetic Survey, Hydrographic Survey H 8615 (1961–63); U.S. Army Engineers, U.S. Intra-Coastal Waterways Survey, Narragansett Bay-Fishers Island Sound section, local map sheet no. 6 (1911); U.S. Dept. of the Interior, Geological Survey, Rhode Island (Washington County) Quonochontaug quadrangle (1944); echo sounder surveys by the author. Subbottom trends are generalized from data presented in this paper. Vertical exaggeration is about 50/1.

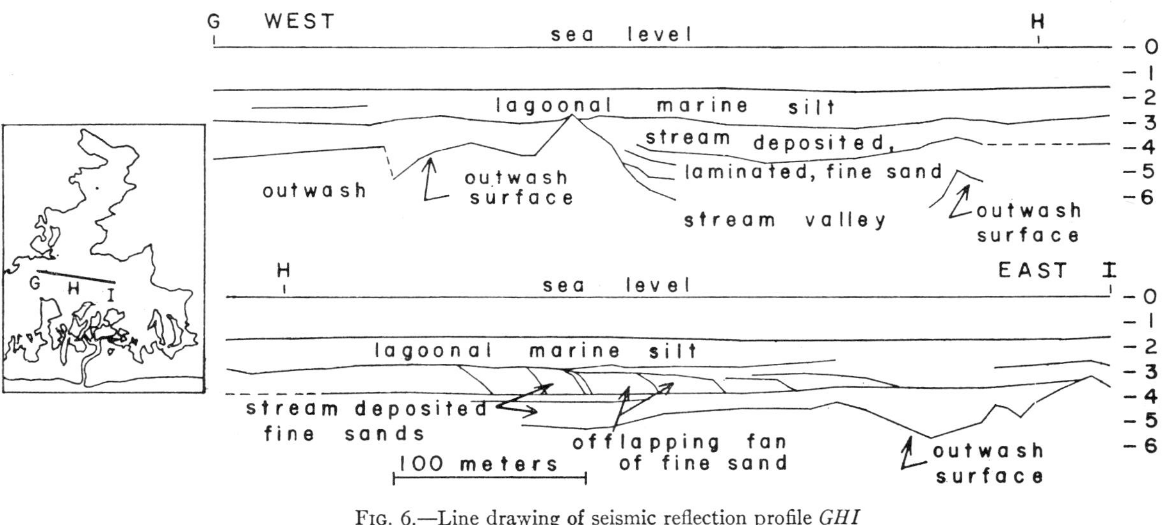

Fig. 6.—Line drawing of seismic reflection profile *GHI*

Massachusetts (Redfield and Rubin 1962), Connecticut (Bloom and Stuiver 1963; Bloom 1964) and New Jersey (Stuiver and Daddario 1963) based on radiocarbon dating of buried plant material, show that from the present to 2,100–3,000 years B.P., rate of submergence in these areas has been about 1 m per 1,000 years. This rate is somewhat greater than the assumed rate of eustatic sea level rise (Scholl and Stuiver 1967, p. 447). Before that time, rates were about 2–3 m per 1,000 years, at least back to 6,000–7,000 years B.P. These values probably indicate crustal downwarping (Redfield and Rubin 1962; Upson et al. 1964; Scholl and Stuiver 1967, p. 448). Similar submergence apparently also has affected the Rhode Island coast, although Kaye and Barghoorn (1964, p. 76) have indicated that no change of sea level has taken place at Boston during the past 2,500–3,000 years.

Kaye (1960, p. 345–347, 350–377) has outlined the glacial history of southern Rhode Island. Approximately 15,000 years ago, a retreat of Wisconsin ice began, and continued until the edge of the ice had passed north of the present lagoon site, roughly 12,000 (?) years ago. Then a minor increase in ice flow resulted in a stillstand of the ice front. When wastage again became relatively more important than ice advance, an ice-cored, sediment-covered ridge remained at the position of that ice front, eventually forming a recessional moraine just north of the present lagoon. Melting to the north supplied water which, to some extent, drained through tunnels in the ice-cored moraine to springs to the south. Most of the area between the ocean and the recessional moraine was covered by a thick layer of outwash, deposited on top of the till during the ice stillstand (La Sala and Hahn 1960; La Sala and Johnson 1960).

In the bedrock valleys, where ice was thicker, blocks of ice were abandoned, subsequently producing lines of kettles, of which Fort Neck Cove is an example. Fine sand to silt remained available to streams, and was deposited extensively in depressions in the outwash plain. A stream which flowed through the Fort Neck Cove glacial valley was supplied with meltwater through the ice tunnels in the recessional moraine. This stream deposited fine sand-silt in the bottoms of the series of kettles through which it flowed, eroded the high spots, and eventually filled the bottom of the valley with glaciofluvial sand. Then additional sediment began to form an alluvial fan downstream, where the valley widened in the present central area of the lagoon (fig. 1). Stream flow through the valley undoubtedly decreased drastically when the ice tunnels collapsed. These appear to have been the only channelways to carry water to the south of the moraine, because no significant southward surface drainageways can be seen. At the time after the glacial ice had melted and before submergence the area which is now West Basin exhibited a sloping, generally flat land surface, irregular in detail. The present Fort Neck Cove was a steep-walled valley with a flat sandy floor, containing a small stream.

At a sea level only 5 m below present

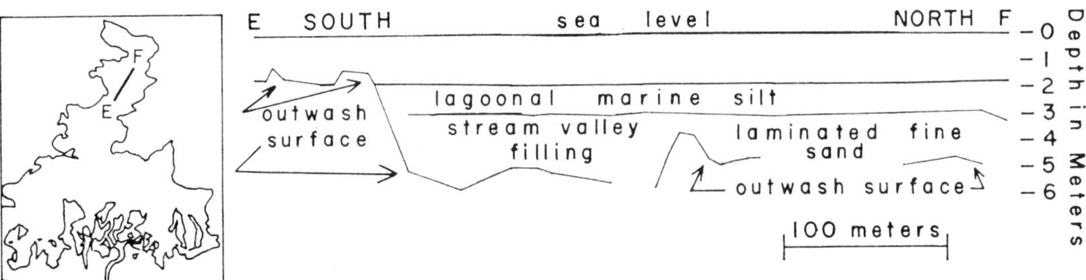

FIG. 7.—Line drawing of seismic reflection profile *EF*

level, perhaps 3,500 years ago, almost none of the present lagoon area would be flooded. The average bottom slope does not steepen abruptly seaward, and the headlands supporting the present barrier beach extend seaward. Therefore a barrier could have existed at this lower (5 m) sea level, because the outwash plain apparently has a flatter slope than the equilibrium gradient for a beach (King 1959, p. 353), given the materials and energy conditions which exist at this location. The fact that no ocean beach sediments were found in cores at any point within Charlestown Pond indicates that ocean waves never beat directly on a shore formed on the outwash, but rather that a barrier did exist during the last few meters of submergence. As the ocean continued its rise, the barrier was not able to build upward and maintain its profile of equilibrium because of a lack of available material. Rather, a shoreward migration was produced which maintained the profile of equilibrium. Thus, the barrier sands overran the lagoonal sediments that had been deposited behind the beach. This migration of the barrier across lagoonal deposits is evidenced by large chunks of lagoonal peat observed on the ocean side of Rhode Island barriers (R. L. McMaster 1967 and A. Ashraf 1967, personal communications). Because the lagoon encroached on glacial sediments, the lagoonal fine sand, silt, and clay also advanced over the outwash and glaciofluvial material. This onlapping of sediments displayed in cores and seismic reflection profiles, accounts for the idealized stratigraphic relationships shown in figure 5. At present, the shoreward movement of the barrier beach is still in progress with the barrier essentially rolling over itself. Of course, the movement is a very slow process since submergence is slow, and the long-term recession of the beach front is hidden in the short-term erosion and deposition.

CONCLUSIONS

Rates of sediment supply to the Charlestown Pond barrier and lagoon are very small in comparison with rates at some other barrier-lagoon complexes, such as those of the Texas coast (Bernard et al. 1959; Shepard 1960; Shepard and Moore 1960; Le Blanc and Hodgson 1961). Sand supply to the Charlestown barrier front is insignificant, and the principal permanent loss from the barrier front is by washover. Sand is supplied to the barrier back almost exclusively by washover and wind transport from the barrier front; practically none is supplied by tidal currents, except in the vicinity of the tidal delta. Virtually the only sediment entering the main parts of the lagoon is the small amount carried in by tidal currents or winnowed from the outwash.

Johnson (1919, p. 376–386) was not correct in assuming that a barrier *must* be pushed shoreward with time. Certainly the evidence shows that a barrier *can* be abandoned and drowned (Curray 1960, pp. 224, 256). If the barrier is large, and if the sand supply is sufficiently great, a barrier can maintain its position and can add to its total volume and its volume above sea level during a period of rising sea level (McKee and Sterrett 1961; Hoyt 1967; Beall 1968; Swift 1968). However, in a sense this is a self-defeating operation, since as a barrier grows in place, it must extend its foot into deeper water (assuming a seaward slope to the bottom). Since the profile of equilibrium remains essentially the same, the volume of sand which must be added for each increment of vertical growth required by each unit rise of sea level will constantly increase. That is to say, the required sand supply will show an accelerating increase with time if rate of sea level rise is constant (fig. 8). As the barrier becomes larger, washover becomes increasingly less important, until it becomes virtually negligible under normal storm conditions (High 1969). Thus, the barrier is forced to grow upward in place and cannot migrate landward. Eventually, barrier growth is not able to keep pace with the rising ocean due to the accelerating requirement for sand. Finally the barrier founders and is somewhat flattened but not completely removed as the surf zone crosses, because most of it is deeper than the interval of highest wave energy. In such a way a barrier might be drowned.

On the other hand, some barriers, including the Charlestown barrier, have migrated landward by washover during submergence. Thus "Brunn's Rule" as proposed by Schwartz (1967, p. 90, 1968) does not apply in such cases because of loss of sand over the barrier. The landward migration of the Charlestown barrier beach seems to have resulted from its small size, which is the consequence of the lack of significant sand supply. The small size results in the barrier being exposed to wave attack down to its "base" at a depth of 13 m. As a result, all of the barrier sand is exposed to wave erosion. Furthermore, because of the small size of the barrier, sand is easily transported over it to the lagoon side. Accordingly, the lack of sediment supply would appear to be the controlling factor in preventing drowning of the barrier and inducing its shoreward migration. This migration has resulted in the preservation of the lagoon during a a period of rising sea level.

ACKNOWLEDGMENTS.—This work was supported by the U.S. Public Health Service grant WP 23. I wish to thank Dr. Robert L. McMaster, who directed much of the size analysis work, for his advice and suggestions, and Dr. J. Towne Conover for his botanical work. Edgerton, Germeshausen, and Grier Inc. supplied the experimental "pinger probe"; this instrument was operated by Capt. Harold Payson USN (ret.) of the Massachusetts Institute of Technology, Geology Department. Thanks are due this organization and Captain Payson for their great help. Dr. Minze Stuiver of Yale Radiocarbon Laboratory supplied C^{14} dates for three samples, and Mr. William B. Allen of the U.S. Geological Survey, Groundwater Division, supplied well logs.

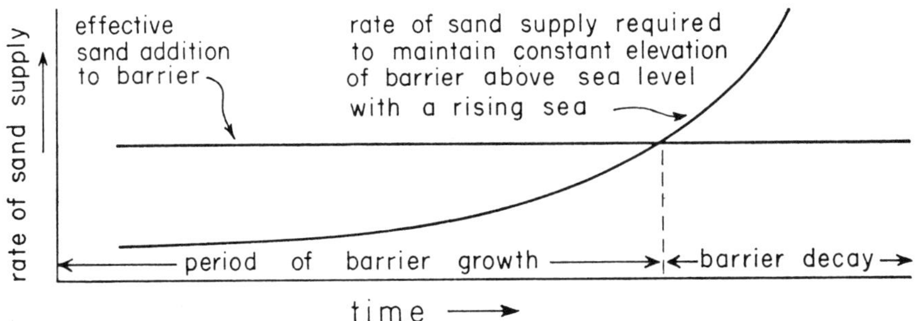

FIG. 8.—Suggested relationship of sand supply and time to barrier growth and decay, when barrier is sufficiently large to impede washover of sand and submergence occurs at a constant rate.

REFERENCES CITED

BEALL, A. O., JR., 1968, Sedimentary processes operative along the western Louisiana shoreline: Jour. Sed. Petrology, v. 38, p. 864–877.

BERNARD, H. A.; MAJOR, C. F., JR.; and PARROT, B. S., 1959, The Galveston barrier island and environs: a model for predicting reservoir occurrence and trend: Gulf Coast Assoc. Geol. Soc., v. 9, p. 221–224.

BIERSCHENK, W. H., 1956, Ground-water resources of the Kingston quadrangle, Rhode Island: U.S. Geol. Survey and Rhode Island Development Council, Geol. Bull. no. 9, 60 p.

BLOOM, A. L., 1964, Peat accumulation and compaction in a Connecticut coastal marsh: Jour. Sed. Petrology, v. 34, p. 599–603.

———, and Stuiver, M., 1963, Submergence of the Connecticut Coast: Science, v. 139, p. 332–334.

CALVERT, S. E., and VEEVERS, J. J., 1962, Minor structures of unconsolidated marine sediments revealed by X-radiography: Sedimentology, v. 1, p. 287–295.

CURRAY, J. R., 1960, Sediments and history of Holocene transgression, continental shelf, northwest Gulf of Mexico, *in* SHEPARD, F. P.; PHLEGER, F. B.; and VAN ANDEL, TJ. H., eds., Recent sediments, northwest Gulf of Mexico: Tulsa, Okla., Am. Assoc. Petroleum Geologists, p. 221–266.

DILLON, W. P., 1964, Geology of the lagoon, Environmental relationships of benthos in salt ponds: Narragansett Marine Lab., Graduate School of Oceanography, Univ. Rhode Island, Ref. 64-3, p. 98–207.

EDGERTON, H. E., and PAYSON, H., 1964, Sediment penetration with a short-pulse sonar: Jour. Sed. Petrology, v. 34, p. 876–880.

———; ———; YULES, J.; and DILLON, W., 1964, Sonar probing in Narragansett Bay: Science, v. 146, p. 1459–1460.

HAMBLIN, W. K., 1962, X-radiography in the study of structures in homogeneous sediments: Jour. Sed. Petrology, v. 32, p. 201–210.

HIGH, L. R., 1969, Storms and sedimentary processes along the northern British Honduras coast: Jour. Sed. Petrology, v. 39, p. 235–245.

HOYT, J. H., 1967, Barrier island formation: Geol. Soc. America Bull., v. 78, p. 1125–1136.

JOHNSON, D. W., 1919, Shore processes and shoreline development: New York, John Wiley & Sons, 584 p.

KAYE, C. A., 1960, Surficial geology of the Kingston quadrangle, Rhode Island: U.S. Geol. Survey Bull. 1071, p. 341–396.

———, and BARGHOORN, E. S., 1964, Late Quaternary sea level change and crustal rise at Boston, Massachusetts with notes on the autocompaction of peat: Geol. Soc. America Bull., v. 75, p. 63–80.

KING, C. A. M., 1959, Beaches and coasts: London, E. J. Arnold, Ltd., 403 p.

LA SALA, A. M., JR., and HAHN, G. W., 1960, Ground-water map of the Carolina quadrangle, Rhode Island: Rhode Island Water Resources Coordinating Board and U.S. Geol. Survey Map GWM 9.

———, and JOHNSON, D. E., 1960, Ground-water map of the Quonchontaug quadrangle, Rhode Island: Rhode Island Water Resources Coordinating Board and U.S. Geol. Survey Map GWM 11.

LE BLANC, R. J., and HODGSON, W. D., 1961, Origin and development of the Texas shoreline, symposium on late Cretaceous rocks, Wyoming and adjacent areas: Wyoming Geol. Assoc., p. 254–375. Reprinted from Gulf Coast Assoc. of Geol. Soc., Trans., v. 9, 1959.

MCKEE, E. D., and STERRETT, T. S., 1961, Laboratory experiments on form and structure of longshore bars and beaches, *in* PETERSON, J. A., and OSMOND, J. C., eds., Geometry of sandstone bodies: Tulsa, Okla., Am. Assoc. Petroleum Geologists, p. 13–28.

MCMASTER, R. L., 1960, Mineralogy as an indicator of beach sand movement along the Rhode Island shore: Jour. Sed. Petrology, v. 30, p. 404–413.

———; LACHANCE, T. P.; and GARRISON, L. E., 1968, Seismic reflection studies in Block Island and Rhode Island Sounds: Am. Assoc. Petroleum Geologists Bull., v. 52, p. 465–474.

MEADE, R. H., 1969, Landward transport of bottom sediments in estuaries of the Atlantic Coastal Plain: Jour. Sed. Petrology, v. 39, p. 222–234.

REDFIELD, A. C., and RUBIN, M., 1962, The age of salt marsh peat and its relation to the recent changes in sea level at Barnstable, Massachusetts: (U.S.) Natl. Acad. Sci. Proc., v. 48, p. 1728–1735.

SAVARD, W. L., 1966, The sediments of Block Island Sound: Unpub. Master's thesis, Univ. Rhode Island, Kingston, 67 p.

SCHOLL, D. W., and STUIVER, M., 1967, Recent submergence of southern Florida: a comparison with adjacent coasts and other eustatic data: Geol. Soc. America Bull., v. 78, p. 437–454.

SCHWARTZ, M. L., 1967, The Brunn theory of sea-level rise as a cause of shore erosion: Jour. Geology, v. 75, p. 76–92.

——— 1968, The scale of shore erosion: *Ibid.*, v. 76, p. 508–517.

SHEPARD, F. P., 1960, Gulf Coast barriers, *in* SHEPARD, F. P.; PHLEGER, F. B.; and VAN ANDEL, TJ. H., eds., Recent sediments, northwest Gulf of Mexico: Tulsa, Okla., Am. Assoc. Petroleum Geologists, p. 197–220.

———, and MOORE, D. G., 1960, Bays of central Texas coast, *in* SHEPARD, F. P.; PHLEGER, F. B.; and VAN ANDEL, TJ. H., eds., Recent sediments, northwest Gulf of Mexico: Tulsa, Okla., Am. Assoc. Petroleum Geologists, p. 117–152.

STUIVER, M., and DADDARIO, J. J., 1963, Submergence of the New Jersey coast: Science, v. 142, p. 951.

SWIFT, D. J. P., 1968, Coastal erosion and transgressive stratigraphy: Jour. Geology, v. 76, p. 444–456.

UPSON, J. E.; LEOPOLD, E. B.; and RUBIN, M., 1964, Postglacial change of sea level in New Haven harbor, Connecticut: Am. Jour. Sci., v. 262, p. 121–132.

Editor's Comments on Paper 33

33 Schwartz: *The Multiple Causality of Barrier Islands*

By 1971 geology textbooks around the world recognized individual, or various combinations of, methods of barrier islands development, as proposed by Hoyt, Fisher, Otvos, and earlier workers. Evidence appeared to be available to support genesis by ridge submergence, breaching of spits, or emergence of submarine bars. The workers in this field agreed, somewhat condescendingly, that means other than that which they championed could produce some barrier islands. The argument seemed to center on the matters of degree and particular locale. Although not meant as a compromise, this paper by M. L. Schwartz attempted to establish recognition of the multiple causality of barrier islands.

Schwartz sees in the work summarized by the earlier Colquhoun, Pierce, and Schwartz abstract, the nucleus of the ideas developed here. The group's primary barrier is the engulfed ridge of Hoyt; the secondary barrier is, alternatively, Fisher's breached spit or the emergent bar advocated by Otvos. The latter formed during a rise (Zenkovich) or fall (Leontiev) in sea level. A simple and tentative classification system is proposed encompassing these types plus a composite, or combination, of the others. Although not as yet universally accepted, this paper presents a unifying synthesis of over 100 years of debate.

Maurice Leo Schwartz was born in Fort Worth, Texas, in 1925. He served in the U.S. Navy during 1944–1946, and, following World War II, was self-employed for 13 years. Schwartz studied at Columbia University, where he received the B.A. degree in 1963, the M.S. in 1964, and the Ph.D. in 1966. He was first a lecturer, then an instructor at Brooklyn College from 1964 to 1968. In 1968 he became an assistant professor at Western Washington State College and in 1971 was promoted to his present position of Associate Professor of Geology and of Education. He was awarded a Fulbright–Hays research scholar grant to study coastal problems in Greece in 1973. Schwartz is interested in, and has published on, earth science education, coastal processes, coastal geology related to archeology, and sea-level-change aspects of marine geology. Besides the present *Barrier Island* collection, he is editor of the previous *Spits and Bars* volume in the Benchmark Papers in Geology series.

Reprinted from the *J. Geol.*, **79,** 91–94 (1971)

GEOLOGICAL NOTES

THE MULTIPLE CAUSALITY OF BARRIER ISLANDS[1]

33

MAURICE L. SCHWARTZ
Department of Geology, Western Washington State College, Bellingham, Washington 98225

ABSTRACT

There is wide disparity in the current literature concerning the origin of barrier islands. The acceptance of multiple causality is advocated here, and a new classification is proposed.

Three basically different explanations of the formation of barrier islands are now proposed: (1) upbuilding of offshore bars; (2) cutting of inlets through spits; and (3) submergence of ridgelike coastal features. Current coastal geology texts and journal papers support: (1) the predominance of barrier islands of one origin; (2) the acceptance of two out of three proposed origins; or (3) the recognition of one origin as the cause for barrier islands in a particular region. The broader question considered here is not a matter of degree, combination, or locale; but a plea for the acceptance of multiple causality, somewhat akin to Chamberlin's (1965) method of multiple working hypotheses. It is deemed feasible that any of the above causes, proceeding independently or in concert, can form barrier islands.

The current texts on coastal geology tend to recognize a combination of origins for barrier islands. Bird (1969, p. 119) maintains that, "The two main modes of barrier origin are by longshore growth of spits . . . and by development of emergent beaches offshore, but many barriers have had a composite origin." In a similar vein Shepard (1963, p. 188) advocates barrier-islands formation through the truncation of spits developed by shore drift and the building up of offshore bars through sea level; while King (1959, p. 348) favors only the latter. In his classic work, Zenkovich (1967, p. 110, 237, 547) comes closest to the essence of this paper by describing barrier islands formed from offshore bars, spits, and engulfed coastal features, and by citing field areas illustrative of all of these modes of origin. His discussion of barrier islands resulting from offshore bars includes those formed by both rising and falling sea levels.

Hoyt (1967, 1968) is the foremost proponent of barrier-island formation through submergence of coastal ridge features. He proposes dune or beach ridges, generated at the mainland shore, becoming barrier islands as sea level rises and forms lagoons in the low region behind the ridges (fig. 1). His examples are taken mainly from the Georgia and Texas coasts. Hoyt does recognize spit and offshore bar origins in some instances, but does not regard them as the general mechanism of barrier-island formation. His main objection is that open-marine sediments and organisms, and open-ocean beaches, are not commonly found landward of barrier islands, which would be expected in the case of spit or offshore bar origin. Then too, he feels that field and laboratory evidence does not support upbuilding of offshore bars through sea level.

Fisher (1968), working along the Middle Atlantic states, takes exception with Hoyt and holds that complex spits are the cause of barrier-island formation. He proposes the breaching of spits to form barrier islands

[1] Manuscript received March 17, 1970; revised June 26, 1970.

[JOURNAL OF GEOLOGY, 1971, Vol. 79, p. 91–94]

and the inlets that separate them from their source and each other (fig. 2). Fisher does agree with Hoyt that the lack of open-marine sediments landward of barrier islands precludes offshore bar origin, but argues that subaerial vegetation, as would be expected behind a ridge, is not found under the lagoons. He differs too on the effects of coastal processes in producing, engulfing, and preserving ridges.

Otvos (1970) advocates an offshore bar origin, and rejects Hoyt's engulfed beach and dune ridges. He cites a number of Gulf Coast barrier islands as proven examples of upward aggradation of submerged shoal areas (fig. 3) and subsequent barrier-island migration, arguing that near-shore low-salinities and coastal marshes could explain the landward absence of open-marine organisms and beach sediments, respectively. Presumably, the barrier island develops through "building shoals up and into island surfaces" (Otvos 1970, p. 243), "constructive waves" (King 1959, p. 352–253), or "the formation

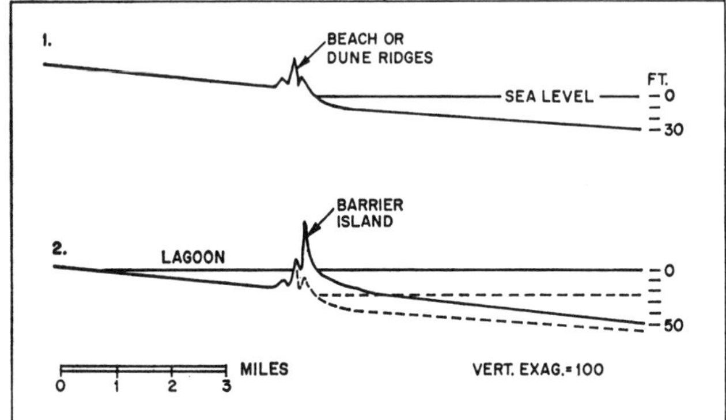

Fig. 1.—Formation of barrier islands by submergence. *1*, Beach or dune ridge forms adjacent to shoreline. *2*, Submergence floods area landward of ridge to form barrier island and lagoon (Hoyt 1967).

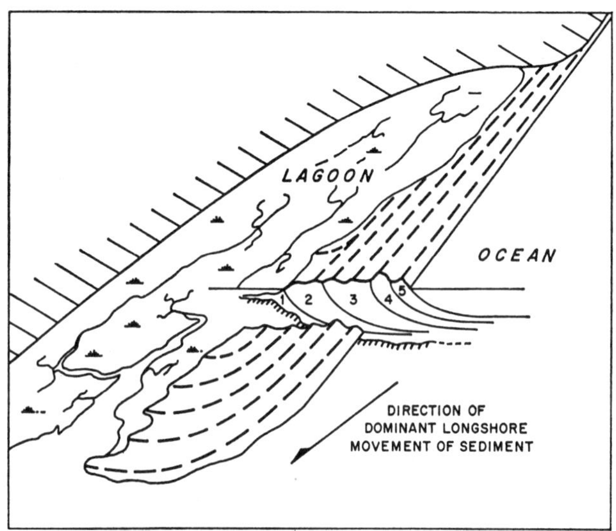

Fig. 2.—Development of barrier islands through breaching of complex spits (Fisher 1968)

of a submarine bar, on which a beach barrier is subsequently formed" (Zenkovich 1967, p. 237–238). Otvos claims that ridge-submergence barrier islands would require special conditions which have not existed in the last 3,500–5,000 years, the age of most present Gulf Coast barrier islands.

A report by Colquhoun et al. (1968) cites field and laboratory investigations indicative of three different origins. Two types of barrier islands were defined: (1) a primary, developed at a higher shoreline, bounded landward by back-barrier sediments overlying a former land surface, and formed during marine transgression and migrating landward with a sea-level rise; and (2) a secondary, developed seaward of and after the primary, underlain by continental-shelf sediments, and formed during a stable sea level or slow fall in sea level. The primary barrier island forms by engulfment of primary strandline ridges, while the secondary forms through breaching of spits or upbuilding of offshore bars. Wave-tank experiments produced the one primary and two secondary types. In this and other reports Colquhoun (1965, 1969) cites extensive stratigraphic evidence from terrace complexes to support a primary barrier island formed by beach ridge engulfment, and a secondary barrier island formed seaward of the primary, through littoral drift at a slightly later time, by either spit growth and breaching or emergence of an offshore bar. Development of either secondary type precludes progradation with a fall in sea level of the primary.

It would seem that the criteria of the principal investigators are embraced by Colquhoun's primary and secondary barrier islands. Is it possible that his primary is Hoyt's submerged beach ridge, and his secondary Fisher's breached spit and Otvos's emergent offshore bar; the latter two developed, respectively, (1) when sea-level rise has virtually ceased (a near stable sea level), or (2) when there is a slow fall in sea level?

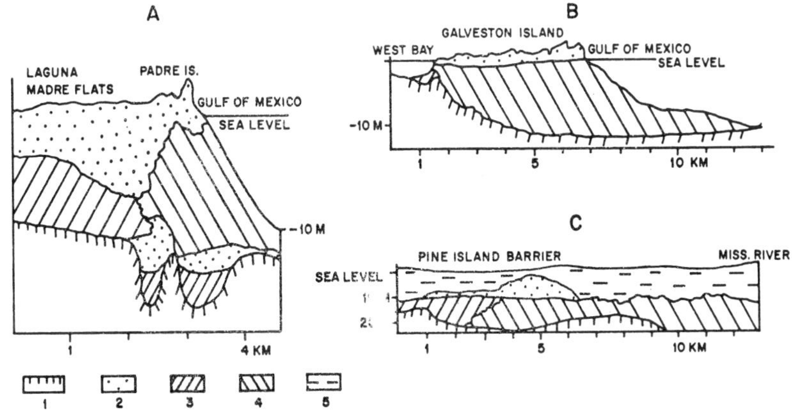

FIG. 3.—Gulf Coast barrier islands cross-sections: *A*, Padre Island; *B*, Galveston Island; *C*, Pine Island. Symbols: *1*, Pleistocene; *2*, Holocene beach foreshore and eolian complex; *3*, Holocene brackish lagoonal, bay-sound and estuarine sediments; *4*, Holocene open-marine subtidal foreshore sediments; *5*, Holocene alluvium (Otvos 1970).

In view of growing evidence that there are various modes of origin for barrier islands, it is proposed that multiple causality be accepted, and the following tentative classification is submitted:

I. Primary
   1. Engulfed beach ridges
II. Secondary
   1. Breached spits
   2. Emergent offshore bars
      a. Sea level rise
      b. Sea level fall
III. Composite
   (Combinations of two or more of the above)

ADDENDUM: After submitting the manuscript for this paper the author found that Leontyev (1969) proposes the genesis of barrier islands through the emergence of submarine barrier bars during a fall in sea level.

ACKNOWLEDGMENTS.—Permission to adopt J. J. Fisher's, J. H. Hoyt's, and E. G. Otvos's figures, valuable discussions with D. J. Colquhoun, and helpful manuscript review by R. W. Fairbridge and D. A. Rahm, are all gratefully acknowledged.

REFERENCES CITED

BIRD, E. F. C., 1969, Coasts: Cambridge, Mass., M.I.T. Press, 246 p.

CHAMBERLIN, T. C., 1965, The method of multiple working hypotheses: Science, v. 148, p. 754–759.

COLQUHOUN, D. J., 1965, Terrace sediment complexes in central South Carolina: Atlantic Coastal Plain Geological Association field conference: Columbia, Univ. South Carolina Press, 62 p.

——— 1969, Coastal Plain terraces in the Carolinas and Georgia, U.S.A., *in* Quaternary geology and climate: Washington, D.C., Natl. Acad. Sci., p. 150–162.

———; PIERCE, J. W.; and SCHWARTZ, M. L., 1968, Field and laboratory observations on the genesis of barrier islands (Abs.): Geol. Soc. America annual meeting, Mexico City, p. 59–60.

FISHER, J. J., 1968, Barrier island formation: discussion: Geol. Soc. America Bull., v. 79, p. 1421–1426.

HOYT, J. H., 1967, Barrier island formation: Geol. Soc. America Bull., v. 78, p. 1125–1136.

——— 1968, Barrier island formation: reply: Ibid., v. 79, p. 1427–1432.

KING, C. A. M., 1959, Beaches and coasts; London, Edward Arnold, Ltd., 403 p.

LEONTYEV, O. K., 1969, Flandrean transgression and the genesis of barrier bars, *in* WRIGHT, H. E., JR., ed., Quaternary geology and climate: INQUA Cong., 7th, Proc., v. 16, Natl. Acad. Sci. Pub. 1701, p. 146–149.

OTVOS, E. G., 1970, Development and migration of barrier islands, northern Gulf of Mexico: Geol. Soc. America Bull., v. 81, p. 241–246.

SHEPARD, F. P., 1963, Submarine geology: New York, Harper & Row, 557 p.

ZENKOVICH, V. P., 1967, Processes of coastal development: Edinburgh, Oliver & Boyd, 738 p.

Editor's Comments on Paper 34

34 Cooke: *Holocene Evolution of a Portion of the North Carolina Coast: Discussion*

C. W. Cooke is represented again by this short article discussing the Pierce and Colquhoun paper. Although agreeing with their observations, he differs with the authors on their conclusions.

Pierce and Colquhoun believed the North Carolina barrier islands to be submerged ridges, located upon a barrier system of an earlier low sea level, which merged with spit segments. As Cooke sees it, the Outer Banks formed offshore as an emergent bar. Pushed landward by wave action, the Banks were finally superimposed upon a relict barrier which had been formed during Silver Bluff time, when sea level was slightly higher. Later regression of the sea provided a period of subaerial weathering, and a soil zone developed over the exposed terrace surface. This is, of course, this same soil zone that Pierce and Colquhoun cite as evidence of ridge engulfment!

Reprinted from the *Geol. Soc. Amer. Bull.*, **82,** 2369–2370 (1971)

C. WYTHE COOKE *The Princess Issena, P. O. Box 5368, Daytona Beach, Florida 32020*

Holocene Evolution of a Portion of the North Carolina Coast: Discussion

34

Pierce and Colquhoun (1970) have brought to light some very interesting facts that may apply not only to the Hatteras Banks (which they do not mention by name) but also to Cape Canaveral and many other barrier islands throughout the world. Although I do not question the accuracy of any of their observations, some conclusions drawn from them may be debatable. The following summary is quoted from the abstract of their report:

> The presence of a soil zone under a portion of the large barrier chain, which makes up much of the North Carolina coast, and the absence of this zone under the remainder of the barrier has led to a re-evaluation of the mode of formation and the evolutionary development of this stretch of coastline during Holocene time.
>
> A primary barrier, formed during a rising sea, became detached by flooding of the area behind a mainland beach. This primary barrier apparently reoccupied the position of a barrier formed during a stillstand within the general Wisconsin regression.
>
> The present coastline has evolved from the primary barrier by retreat and migration of headlands as well as formation of secondary barriers, by spit elongation, over the adjacent continental shelf. . . .

The location of many barrier islands is determined by the configuration of the sea bottom. The Hatteras Banks are no exception. The Outer Banks (the "primary barrier" of Pierce and Colquhoun) stand between the shallow water of Pamlico Sound and the deeper water of the Atlantic Ocean. Presumably they were built of sand gouged from the bottom and piled up by storm waves breaking on the shoals. Continued erosion by storm waves is gradually pushing them shoreward. This progradation may be most active where shallow water extends farthest out to sea, as at Cape Hatteras. The shore of the mainland has been little modified by the much weaker waves in Pamlico Sound.

The topography of the sea bottom off the coast of North Carolina is inherited from a Pleistocene land surface that was drowned by the rise of sea level resulting from the latest deglaciation. The ocean today covers most of the nearly flat Silver Bluff terrace (Cooke, 1945, p. 248; 1966, p. 7), the lowest of a series of Pleistocene coastal terraces.

The ocean and Pamlico Sound were about one fathom deeper during Silver Bluff time. It was probably during this epoch that the barrier, referred to the Pleistocene by Pierce and Colquhoun (1970, p. 3706), was built. Because of the greater depth of water, the Pleistocene (Silver Bluff) barrier stood somewhat closer to the mainland than the original location of the Outer Banks, which have crept landward until they now cover part of the Pleistocene barrier. Though sand eroded from the Pleistocene barrier presumably contributed to the growth of the Outer Banks, the spits so formed, if any, were probably much less extensive than the parts built up in situ by wave action.

After Silver Bluff time, sea level fell a considerable distance below its present location. Pamlico Sound and the adjacent part of the Atlantic became land – the Silver Bluff terrace. The soil zone reported by Pierce and Colquhoun in drill holes doubtless dates from this low-water epoch.

If the Outer Banks ever were a beach ridge on the mainland and became detached from it by rising water, as postulated by Pierce and Colquhoun (1970, p. 3695), the connection must have been temporary, for the flat terrace behind them would have been flooded as soon as the rising water topped its level. The banks now stand far out to sea.

Geological Society of America Bulletin, v. 82, p. 2369-2370, August 1971

REFERENCES CITED

Cooke, C. W., 1945, Geology of Florida: Florida Geol. Survey Geol. Bull. 29, 339 p.

—— 1966, Emerged Quaternary shore lines in the Mississippi Embayment: Smithsonian Misc. Colln., v. 149, no. 10, 41 p.

Pierce, J. W., and Colquhoun, D. J., 1970, Holocene evolution of a portion of the North Carolina coast: Geol. Soc. America Bull., v. 81, p. 3697-3714, 9 figs.

MANUSCRIPT RECEIVED BY THE SOCIETY APRIL 8, 1971

Editor's Comments on Paper 35

Pierce and Colquhoun take exception to two points in Cooke's commentary. While doubting, but allowing for, the emergence of submarine bars, they deny that the barrier system under discussion was developed in this fashion. They still see the North Carolina barrier system as engulfed ridges and breached spits.

They differ less with Cooke on the subaerial weathering of the earlier barrier during the interval between the Wisconsin regression and the Holocene transgression. Only the site of that barrier is debated. Unlike Cooke, who visualizes the barrier as farther landward at the time of greatest transgression, Pierce and Colquhoun believe it to have been formed nearer to its present site during regression.

Reprinted from the *Geol. Soc. Amer. Bull.*, **82**, 2371 (1971)

J. W. PIERCE *Division of Sedimentology, Smithsonian Institution, Washington, D. C. 20560*
D. J. COLQUHOUN *Department of Geology, University of South Carolina, Columbia, South Carolina 29208*

Holocene Evolution of a Portion of the North Carolina Coast: Reply

35

We are in general agreement with most of Cooke's discussion regarding our recent paper. We disagree with only two ideas put forth by Cooke and on one of these only to a minor degree.

Cooke mentions the gouging of sand from the sea floor and the piling up by storm waves on the shoals. He also mentions that "the spits so formed, if any, were much less extensive than the parts built up in situ by wave action." We interpret these statements to mean that Cooke accepts the concept of barrier island formation as presented by de Beaumont (1845). The ability of breaking orbital waves to build a barrier above sea level is questionable and, except in unusual circumstances, submarine bars will not be built above sea level (Kuelegan, 1948; King, 1960, p. 337; McKee and Sterrett, 1960). Submarine bars are built by waves but they will migrate landward and eventually become part of a pre-existing subaerial feature. The sand that makes up the present barrier came from a redistribution of the mass of sand pushed ahead of the transgressing sea, a cannibalization of the Pleistocene substrate (Swift, 1969).

We are in general agreement that the Pleistocene barrier, which acts as base for the present barrier over part of its extent, formed within the regressive part of the transgressive-regressive cycle associated with what Cooke calls the Silver Bluff terrace. This Pleistocene barrier may be only one of several formed within the total cycle since it has been well documented to the north (Oaks, 1965) and to the south (Colquhoun, 1965, 1969) that several barriers may actually be present on one of the "terraces." This barrier was weathered during the time span between the last Wisconsin regression and the Holocene transgression.

Our data does not support the idea presented by Cooke, without substantiation, that the Pleistocene (Silver Bluff) barrier was located landward of the original location of the present barrier. The fact that sea level may have been slightly higher at the time of the maximum transgression associated with the formation of the Silver Bluff terrace does not dictate the level of the sea at the time when a barrier may form during the regressive phase of the cycle. Sea level probably was near or slightly lower than the present level at the time of the formation of the Pleistocene barrier discovered in our drilling.

REFERENCES CITED

Colquhoun, D. J., 1965, Terrace sediment complexes in central South Carolina: Atlantic Coastal Plain Geol. Assoc. Field Conf. Guidebook, 62 p.

—— 1969, Coastal plain terraces in Carolinas and Georgia, *in* Wright, H.E., Jr., ed. Quaternary geology and climate: Natl. Acad. Sci. Pub. 1701, 162 p.

de Beaumont, E., 1845, Lecons de geologie pratique: Paris, P. Bertrand, p. 223-252.

King, C.A.M., 1960, Beaches and coasts: London, E. Arnold Ltd., 403 p.

Kuelegan, G. H., 1948, An experimental study of submarine sand bars: Washington, D.C., U.S. Beach Erosion Board Tech. Rept. 3, 40 p.

McKee, E. D. and Sterrett, T. S., 1960, Laboratory experiments on form and structure of longshore bars and beaches, *in* Peterson, J. A., and Osmond, J. C., eds., Geometry of sandstone bodies: Am. Assoc. Petroleum Geologists, p. 13-28.

Oaks, R. Q., Jr., 1965, Post-Miocene stratigraphy and morphology, outer coastal plain, southeastern Virginia [Ph.D. dissert.]: New Haven, Connecticut, Yale Univ., 423 p.

Pierce, J. W., and Colquhoun, D. J., 1970, Holocene evolution of a portion of the North Carolina coast: Geol. Soc. America Bull. v. 81, p. 3697-3714, 9 figs.

Swift, D.J.P., 1969, Inner shelf sedimentation: Processes and products, *in* Stanley, D. J., ed., The *new* concepts of continental margin sedimentation: Am. Geol. Inst. Rept., p. 4-1— 4-46.

Manuscript Received by the Society February 29, 1971

Editor's Comments on Paper 36

John Hails does not differ basically with Pierce and Colquhoun on the methods by which the North Carolina barrier islands were generated. He takes the authors to task, rather, on the credibility of bore hole data versus natural exposures. Widely spaced bore holes, Hails insists, do not provide substantive evidence of lateral variations, relief, or contacts. He believes that detailed examination of the sediments where exposed in construction excavations, bluffs, and pits or trenches, offers the only opportunity to interpret correctly the evolution of the barriers.

Hails questions in particular, Pierce and Colquhoun's assumption that a reddish sediment is reworked. Furthermore, he wonders if the pebbles at bore hole 2 could be ironstone nodules and is puzzled by the present-day formation of humate in the barrier sands. Thus, as stated above, it is the matter of drawing conclusions from scanty sedimentological evidence that Hails objects to, not the overall thesis that Pierce and Colquhoun have proposed.

John R. Hails is presently Deputy Head of the National Environment Research Council's Unit of Coastal Sedimentation in Somerset, England. Graduated from London University with a B.Sc. (Hons.) degree in Geography, with ancillary Geology, Hails then was awarded a Commonwealth Research Scholarship to the University of Sydney to undertake research in coastal geomorphology and sedimentology for a Ph.D. degree. He has lectured at several North American universities as well as at the University of Sydney and London University. While in the United States, Hails conducted research in Georgia with the support of the National Science Foundation. He was appointed Deputy Head of the Unit of Coastal Sedimentation in 1970.

Reprinted from the *Geol. Soc. Amer. Bull.*, **82,** 3525–3526 (1971)

JOHN R. HAILS *Natural Environment Research Council, Unit of Coastal Sedimentation, Beadon Road, Taunton, Somerset, England*

Holocene Evolution of a Portion of the North Carolina Coast: Discussion

36

In a recent publication Pierce and Colquhoun (1970) reexamined the mode of formation and development of the barrier island chain off the coast of North Carolina. From the evidence of a highly oxidized zone (which probably pertains to a paleosol) under part of this barrier, they concluded that the coastline has evolved from a primary barrier by retreat and migration of headlands, and the formation of secondary barriers, by spit elongation, over the adjacent continental shelf.

Pierce and Colquhoun attempted to reconstruct the latest part of the Pleistocene and Holocene history of this part of the eastern United States coast by examining sediments recovered exclusively from bore holes. The stratigraphic sequence of deposits reported by them is fairly representative of barrier coastlines undergoing retrogression and, in fact, similar sequences have been recorded farther south on the east coast.

The usefulness of soil profiles, in general, and relatively deep oxidized weathered horizons, in particular, as means for distinguishing age difference between deposits has been acknowledged by geomorphologists working in both eastern Australia and the eastern United States (Hails, 1968; Hails and Hoyt, 1969a, 1969b; Hoyt, 1969; Langford-Smith and Thom, 1969; Thom, 1965, 1967). Although there is little disagreement with the general interpretation advanced by Pierce and Colquhoun, one can argue that some of their conclusions are inevitably tenuous since only 45 widely spaced bore holes were drilled along a coastline exceeding 100 miles in length.

Since Colquhoun's exposition on the origin of primary and secondary barrier islands has been reviewed recently by Schwartz (1971), attention is focussed in this very brief discussion on some of the limitations of bore hole data for determining depositional environments.

Despite the voluminous literature explaining the evolution of barriers in various parts of the world during the Pleistocene and Holocene epochs, surprisingly little detailed attention has been paid to post-depositional changes in barrier sands and lagoonal sediments resulting from diagenetic and pedogenic processes. The effects of these processes are particularly evident in deeply weathered Pleistocene sediments where solution, chemical replacement and/or redisposition by percolating ground water have destroyed diagnostic sedimentary structures. Because weathering processes modify deposits it is often difficult to identify sedimentary contrasts between depositional environments. Considerable caution must be exercised, therefore, if geologic or geomorphic interpretation of coastal environments is based solely on bore hole data, with no corroborative evidence. Such data do not necessarily provide adequate information about lithological variations within stratigraphic units, erosional unconformities manifest in such features as truncated soils, variations in local topography, episodic barrier growth or erosion, and sea level at the time of barrier formation.

Recent detailed studies of Pleistocene barrier island sediments exposed in road and rail cuts, river bluffs, trenches, and pits in South Carolina by Thom (1967), in Georgia by Hails and Hoyt (1969a, 1969b), and in Florida by Hoyt (1969) show, quite conclusively, that weathering profiles extend far deeper into some barrier sands than others. Sometimes there is very little textural difference between various soil horizons. The widely distributed podzolic soils, developed mainly in barrier sands on the lower Atlantic coastal plain, display distinct ranges in color, from brownish-yellow to red-brown, as a result of progressive differences in degree of weathering as well as in mineral composition. The illuvial horizon of the podzols varies in thickness from one locality to another, and generally contains a silty-clay matrix, in marked contrast to the loose sands of overlying horizons. Highly oxidized clay lenses are clearly

Geological Society of America Bulletin, v. 82, p. 3525-3526, December 1971

visible in some barrier sands and mottling is often quite prominent. Also a micro-podzolic or bisequel profile can be traced in the leached or eluvial horizon of some of the podzols, and truncated soils occur where barrier dissection has post-dated soil profile development.

The difficulty in reconstructing the sedimentary history of an area as large as the one studied by Pierce and Colquhoun can be appreciated; nevertheless, a few points warrant comment. On page 3702 they report that "Of the twenty holes which penetrated the barrier-lagoon sequence, eleven had reddish-brown, tan or yellow sediments, suggestive of a soil zone; two had reddish sediment, apparently reworked soil. . . . The most widespread unit below the lagoonal material is highly oxidized, reddish-brown, tan or yellow sediment and is found in eleven holes." The description suggests that the oxidized sediments pertain to a buried paleosol, but no reason is cited for why the reddish sediment, in particular, should be considered as a reworked soil. One would like to know on what criteria the distinction was made.

The pebble zone found in bore hole 2 could be part of a paleosol too, but whether or not it has been reworked is debatable. Unfortunately, Pierce and Colquhoun only give details on the size of the pebbles, and do not give their composition. It is feasible that the pebbles are nodular ironstone. Similar deposits have been reported in weathered barrier sands, for example, at the contact between the eluvial and illuvial horizons of podzolic soils (Thom, 1967).

The fact that humate is depositing in the sands on the west side of the modern barrier in North Carolina is of considerable interest since hitherto, this dark brown to black water-soluble organic substance has been associated in the main with Pleistocene sediments. In fact, in eastern Australia its presence in Pleistocene barrier sands is one way in which these older deposits can be distinguished from Holocene sediments. However, in eastern Australia, Pleistocene barrier and dune sands are nearly always impregnated or cemented since, in the past, soluble and colloidal organic substances were carried by surface and subsurface waters to subsurface sand environments where flocculation and precipitation of humate took place. According to Pierce and Colquhoun, the weathered barrier island sands encountered in bore holes 4, 5, 6, 22, and 23 are mainly humate coated, and not cemented, and this evidence seemingly supports their view that humate was precipitated on the lagoon side of a barrier undergoing retrogression.

It is known that humate accumulates in and beneath marsh deposits, near ground water seepages, or as a type of organic sediment in brackish or saline water, but as Swanson and Palacas (1965) correctly point out, many questions about the origin and geochemistry of humate still remain unanswered.

In conclusion, the present writer's view is that environmental interpretations and the study of paleosols must be based on supporting evidence from natural exposures as well as bore hole data. Color variation within sediments must be evaluated critically; it is not necessarily a reliable line of evidence since diagenetic and pedogenic processes can be quite complex.

REFERENCES CITED

Hails, J. R., 1968, The Late-Quaternary history of part of the Mid-North Coast, New South Wales, Australia: Inst. British Geographers Trans., no. 44, p. 133–149.

Hails, J. R., and Hoyt, J. H., 1969a, An appraisal of the evolution of the Lower Atlantic Coastal Plain of Georgia, U.S.A.: Inst. British Geographers Trans., no. 46, p. 53–68.

—— 1969b, The significance and limitations of statistical parameters for distinguishing ancient and modern sedimentary environments of the Lower Georgia Coastal Plain: Jour. Sed. Petrology, v. 39, no. 2, p. 559–580.

Hoyt, J. H., 1969, Late Cenozoic structural movement, Northern Florida: Gulf Coast Assoc. Geol. Socs. Trans., no. 19, p. 1–9.

Langford-Smith, T., and Thom, B. G., 1969, New South Wales coastal morphology, *in* Packham, G, H., ed., The geology of New South Wales: Geol. Soc. Australia Jour., v. 16, p. 572–580.

Pierce, J. W., and Colquhoun, D. J., 1970, Holocene evolution of a portion of the North Carolina Coast: Geol. Soc. America Bull., v. 81, p. 3697–3714.

Schwartz, M. L., 1971, The multiple causality of barrier islands: Jour. Geology, v. 79, no. 1, p. 91–94.

Swanson, V. E., and Palacas, J. G., 1965, Humate in coastal sands of northwest Florida: U.S. Geol. Survey Bull. 1214-B, 29 p.

Thom, B. G., 1965, Late Quaternary coastal morphology of the Port Stephens-Myall Lakes area, New South Wales: Royal Soc. New South Wales Jour. and Proc., v. 98, p. 23–36.

—— 1967, Coastal and fluvial landforms: Horry and Marion Counties, South Carolina: Louisiana State Univ. Studies Coastal Studies Ser. no. 19, Tech. Rept. 44.

MANUSCRIPT RECEIVED BY THE SOCIETY MAY 3, 1971

PRINTED IN U.S.A.

Editor's Comments on Paper 37

Opening with a bit of levity seldom found in the papers of august scientific journals, Pierce and Colquhoun are quick to agree with Hails that there are interpretive problems associated with color changes found in bore holes. In the next three paragraphs of this short paper, they answer the specific points raised in Hails's discussion.

Based on its presence in a section of inlet fill, the reddish zone was judged to be reworked. In the case of the pebbles, that term was used to suggest rounding through abrasion; in addition, they were all quartz. Thus nodular ironstone is ruled out. Finally, they note that Hails's own source on humate reported on its precipitation in Florida.

This exchange of papers between Hails and Pierce and Colquhoun, then, has proved to be a matter of refinement rather than differences on major principles.

Reprinted from the *Geol. Soc. Amer. Bull.*, **82**, 3529 (1971)

J. W. PIERCE *Division of Sedimentology, Smithsonian Institution, Washington, D.C. 20560*
D. J. COLQUHOUN *Department of Geology, University of South Carolina, Columbia, South Carolina 29208*

37

Holocene Evolution of a Portion of the North Carolina Coast: Reply

One of the enjoyable aspects of the discussion section of the Geological Society of America Bulletin is that the authors of the original article always get the last word.

Dr. J. R. Hails, in the preceding discussion has pointed out the problems associated with the interpretation of color changes, found in bore holes, which may result from diagenetic or pedogenic processes. With this, we could not agree more. It is presumed that this part of the discussion is in reference to the different sediment types associated with the soil zone and the environmental interpretations placed on the changes (Pierce and Colquhoun, 1970, p. 3702). As noted in the original article, the interpretation is based not only on texture but on lateral distribution of the different types, and the environment of deposition of the underlying sediments both from bore holes and surface samples. Although not explicitly stated, it was implied that the weathered zone was transitional downward into unweathered material, as is expected with soil zones, although convention requires a solid line somewhere in the illustrations. We feel that our original explanation to rule out circulating ground waters is still sufficient (p. 3702).

Dr. Hails asks what criteria were used to distinguish a reworked soil zone from an in-place soil zone (Pierce and Colquhoun, 1970, p. 3702). On page 3703 an explanation is given, although it is admitted that it is interpretative. The interpretation of one of the bore holes (Hole 2) is based on this reddish zone occurring within a section of inlet fill and on the presence of inlet fill in a bore hole with lithologic similarities less than 200 yards away. The other bore hole with reworked soil zone (Hole 24) is a bit more tenuous. There is a zone at −31 feet that has a very slight reddish cast. It appears to be either less weathered or reworked and we felt that the presence should be mentioned for completeness. A change of this zone from reworked to in-place would not change the over-all interpretation.

Examination of several text books and glossaries indicates that a certain amount of transport and rounding is given, implied, or associated with the term pebble (Pettijohn, 1957; Am. Geol. Inst., 1960; Folk, 1961; Challinor, 1964). We felt that use of the term pebble implied a clastic origin and obviously ruled out diagenetic features, such as nodular ironstone. For the record, the pebbles that we examined were quartz.

The precipitation of humate, at present, should not have been such a great surprise since the article of Swanson and Palacas (1965) who reported on its precipitation in Florida. In the original article by us, it is stated that the sands are humate stained, not cemented, along the west side of the barrier. The origin is far from understood and its association with the Pleistocene may be one of time, water-table depth, and circulating ground waters.

We appreciate the Discussion by Hails and the chance to answer his questions.

REFERENCES CITED

American Geological Institute, 1960, Glossary of geology and related sciences, 2d ed.: Washington, D.C., Am. Geol. Inst., 397 p.

Challinor, J., 1964, A dictionary of geology, 2d ed: New York, Oxford Univ. Press, 289 p.

Folk, R. L., 1961, Petrology of sedimentary rocks: Austin, Texas, Hemphill's, 154 p.

Pierce, J. W., and Colquhoun, D. J., 1970, Holocene evolution of a portion of the North Carolina coast: Geol. Soc. America Bull., v. 81, p. 3697–3714.

Swanson, V. E., and Palacas, J. G., 1965, Humate in coastal sands of northwest Florida: U.S. Geol. Survey Bull. 1214-B, 29 p.

Pettijohn, F. J., 1957, Sedimentary rocks, 2d ed.: New York, Harper and Brothers, 718 p.

MANUSCRIPT RECEIVED BY THE SOCIETY JUNE 7, 1971

Geological Society of America Bulletin, v. 82, p. 3527-3528, December 1971

Editor's Comments on Paper 38

The extent of barrier-lagoon coasts throughout the world cited in the introduction to this volume was taken from the Cromwell abstract included here. Compared with Berryhill's definition-restricted 5,710 km, Cromwell's 32,038 km is more nearly in agreement with the reports of Leontiev and Zenkovich. It would appear, then, that 10 to 13 percent of the world's total coastline consists of barrier coasts. This is a sizable proportion, lending further credence to the importance of this fragile environment to mankind.

John E. Cromwell was born in Philadelphia, Pennsylvania, in 1942. He served in the U.S. Army for three years, and later spent one year with the Sea Floor Studies Branch of the Naval Underseas Research Center. Cromwell received the A.B. degree in geology at Lafayette College in 1969, and is now a doctoral candidate at the Scripps Institution of Oceanography, Geological Research Division. He is presently studying a barrier-lagoon system in southern Mexico.

Reprinted from *Abstr. Vol., Second Natl. Coastal and Shallow Water Research Conf.*, 50 (1971)

38

Barrier Coast Distribution: A World-Wide Survey

JOHN E. CROMWELL

The continent-by-continent world-wide distribution of barrier-lagoon coasts has been determined to compare with the results reported by Berryhill et al. (*A.A.P.G. Bull.*, **53**/3: 706–707, 1969). A standard operational definition was used, which included all linear, detrital, topographic features along present coastlines which were less than 10 meters above sea level and which are impounding, or at one time have impounded, bodies of water between themselves and the mainland. Measurements were made with a hand-operated, direct reading, map measurer, for the most part on Operational Navigation Charts, scale $1{:}10^6$. The results of the measurements are reported in the table, compared with those of Berryhill et al.

The discrepancy with Berryhill et al.'s figures is due to their use of a more restrictive operational definition of barrier coasts. The 13.1 percent of the combined continental coastline (243,775 km) accounted for by barrier coasts compares favorably with the 10 and 13 percent proportions of the world's total coastline as given by O. K. Leontyev (*Coastal Research Notes*, **12**, 5–7, 1965) and V. P. Zenkovitch (*Processes of Coastal Development*, 1967), respectively. For this reason, the magnitudes and distributions of this survey are likely to be more reasonable. Both surveys, however, indicate the very large share of the world's barrier coasts possessed by North America. Research in progress is involved with determining the significance of this distribution.

Continent	Barriers of Berryhill Survey	% of Total	Barriers of This Survey	% of Total	% of Continent's Coastline
North America	3,220	56.5	10,765	33.6	17.6
Europe	800	14.0	2,693	8.4	5.3
South America	560	9.8	3,302	10.3	12.2
Africa	490	8.6	5,984	18.7	17.9
Australia	320	5.6	2,168	6.8	11.4
Asia	320	5.6	7,126	22.2	13.8
Total	5,710 km	100.1	32,038 km	100.0	

Editor's Comments on Paper 39

E. C. F. Bird is, without a doubt, Australia's foremost student of barrier coasts. Since a number of his previous papers on the subject are held by copyright and unavailable at the present time, Bird has most generously written an original article for publication in this volume. Presented here, for the first time, is a synthesis of all of Bird's Australian barrier studies to date.

Bird opens with a description of barriers and barrier islands at Coorong, East Gippsland, New South Wales, and Queensland. On the matter of genesis, he admits the possibility of bar emergence, but is doubtful that any of the Australian barriers have been formed in this manner. He believes, rather, that the barriers in his study area were generated by breaching of spits or partial submergence of coastal ridges. Citing the many factors involved in barrier development, Bird stresses his conviction in the multiple ways by which barriers may be formed.

Eric Charles Frederick Bird was born in 1930. He studied geography, geology, and ecology at the University of London, where he received the B.Sc. in 1953 and the M.Sc. in 1955. He went to Australia in 1957 as a Research Scholar at the Australian National University; his doctoral thesis, on the Gippsland Lakes, was completed there in 1959. Bird was an Assistant Lecturer, Kings College, London, in 1960; Lecturer, University College, London, from 1960 to 1963; and Senior Lecturer, Australian National University, from 1963 to 1965. In 1966 he was appointed to his present position as Reader in Geography at the University of Melbourne. Bird has published a textbook and more than 50 articles on the subject of coastal geomorphology.

39

Australian Coastal Barriers

E. C. F. BIRD

Enormous quantities of sand have been deposited on the coastal fringes of Australia during Quaternary times to form beaches and barriers, with associated dune systems interspersed with rocky headlands and cliffy sectors. From Broome in the northwest, around the west and south coasts of the continent, and as far east as Wilson's Promontory (Fig. 1), these coastal sands are predominantly calcareous, consisting mainly of shell debris; they contain more than 50 percent, and sometimes more than 90 percent, calcium carbonate. From Wilson's Promontory along the east coast to Cape York, around the Gulf of Carpentaria, and along the north coast of the continent, the sands are predominantly quartzose, of terrigenous origin, derived mainly from cliff erosion and fluvial sediment yield; they usually contain less than 10 percent calcium carbonate.

This sedimentological contrast is important geomorphologically (Bird, 1967a; Jennings and Bird, 1967), for calcareous sands become lithified by superficial and internal layered precipitation of cementing carbonates to form a calcarenite (often known as dune limestone or calcareous aeolianite), whereas quartzose dunes remain unconsolidated, except for the formation of sandrock (also known as coffee rock or humate) layers at the water table by accumulation of downwashed organic materials and iron oxides. Pleistocene dunes and beach ridges of lithified calcareous sand are thus preserved in solid form, and may later be attacked by marine erosion to produce rugged cliffs fronted by shore platforms or planed-off reefs, as on the ocean coast south of Melbourne. Similar landforms of equivalent age built of unconsolidated sand are less durable and more likely to have been rearranged or dispersed by subsequent erosion processes. Coastal barriers of lithified calcarenite that formed during Pleistocene times have been penetrated and dissected by the sea during the marine transgression in Holocene times (i.e., the last 20,000 years) to open up coastal embayments or elongated sounds behind barrier spits and islands, as in Shark Bay and Cockburn Sound on the west coast of Australia. In South Australia a distinct outer bar-

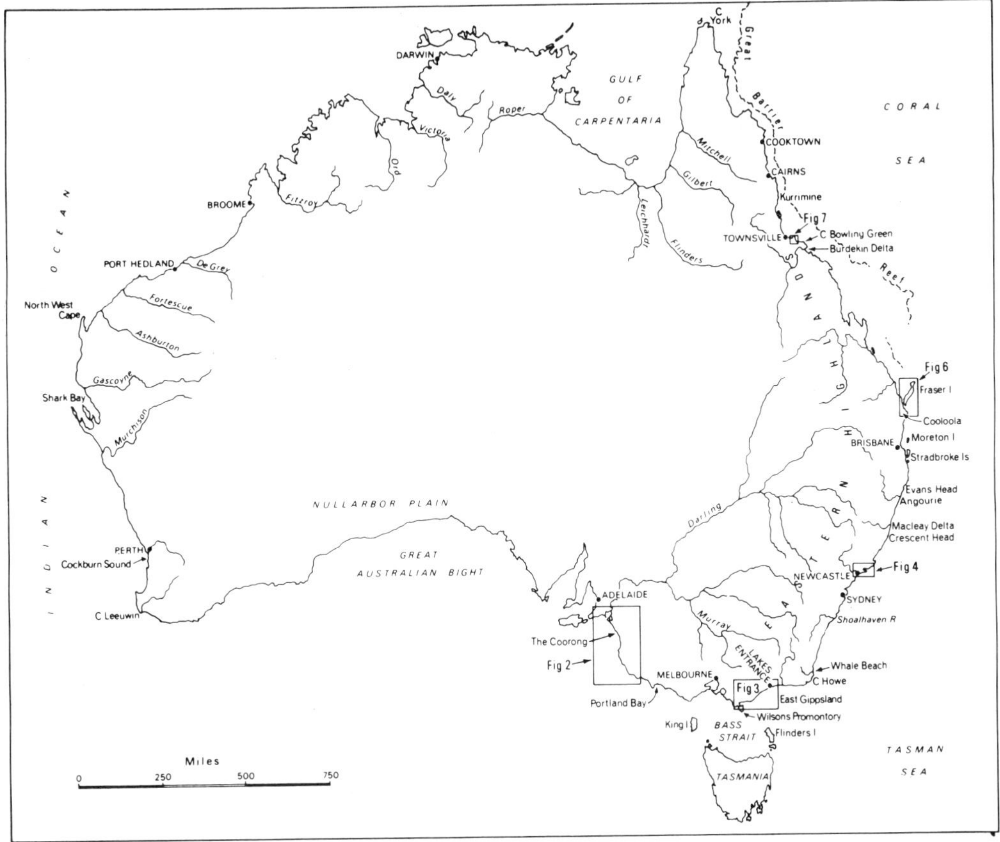

Figure 1. Location map.

rier of calcareous sands exists on the coast in front of the Coorong lagoon, and similar barriers of quartzose sand are found on the East Gippsland coast, as well as on several sectors of the east coast of Australia. The features of these barriers can be illustrated with reference to examples that have been investigated by geomorphologists.

Coorong Barrier

The South Australian coast between Goolwa and Kingston (Fig. 2) presents a fine example of a coastal barrier some 120 miles long, backed for much of its length by a shallow lagoon up to 2 miles wide known as The Coorong. Near Goolwa the barrier is interrupted by an outflow channel through which floodwaters from the River Murray escape to the sea. West of this gap the barrier is known as the Sir Richard Peninsula; to the east it is referred to as Younghusband Peninsula.

Fenner (1931) described this barrier as a spit, implying that it had grown along the coast as the result of deposition from a longshore current, but Sprigg (1952) rejected this explanation on the grounds that there was no evidence for any such cur-

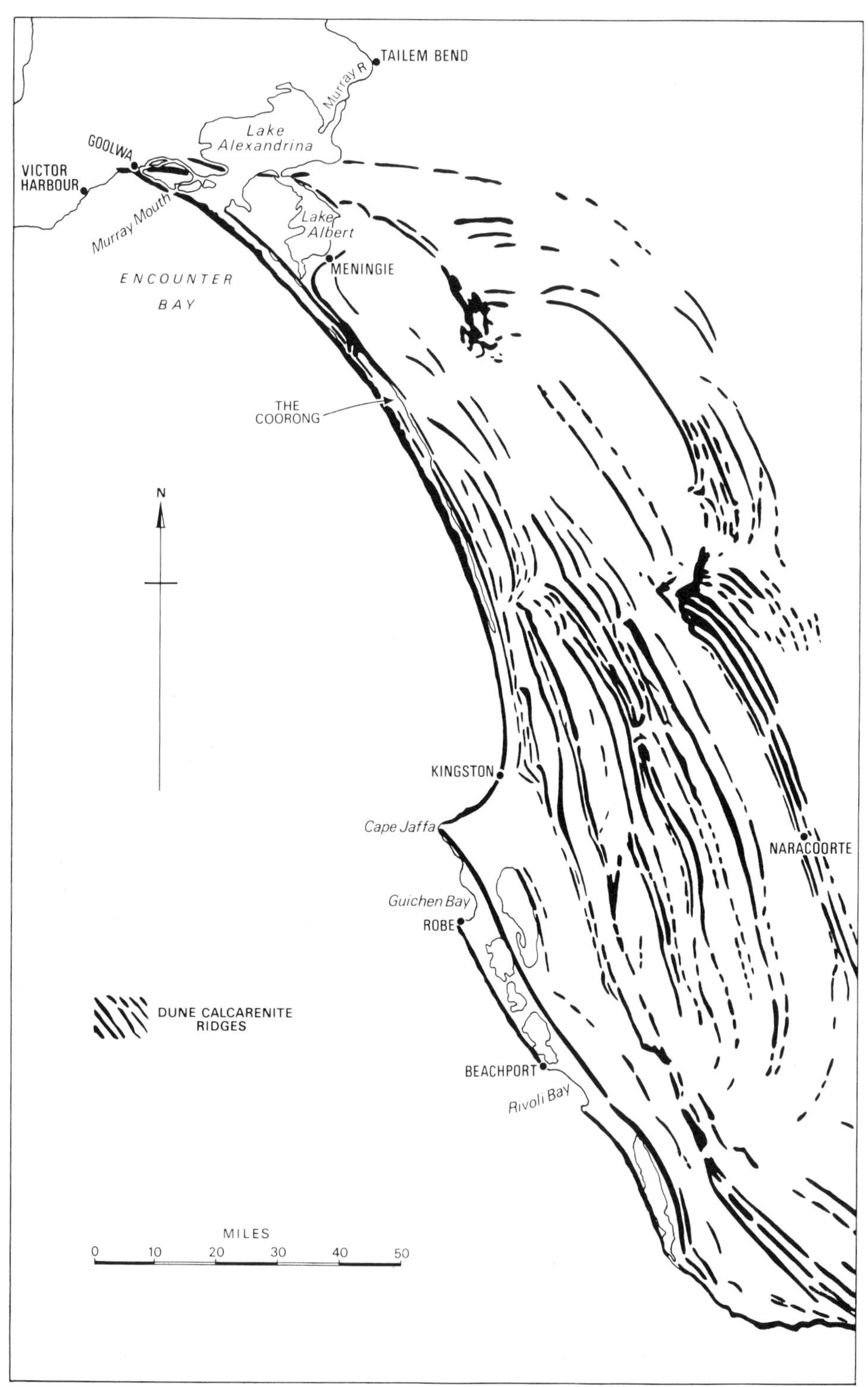

TAILEM BEND
Murray R.
Lake
Alexandrina
GOOLWA
VICTOR
HARBOUR
Murray Mouth
Lake
Albert
MENINGIE
ENCOUNTER
BAY
THE
COORONG
N
KINGSTON
Cape Jaffa
NARACOORTE
Guichen Bay
ROBE
DUNE CALCARENITE
RIDGES
BEACHPORT
Rivoli Bay
MILES
0
10
20
30
40
50

rent, and no sign of the landward recurves that would surely have marked stages in longshore spit growth. Tindale (1947) suggested that the barrier originated when an offshore bar, formed on the sea floor parallel to the earlier coastline, emerged as the result of a fall in sea level, but this hypothesis is too simple for the Coorong barrier, which embodies sectors of older, probably Pleistocene, dune calcarenite topography partly obscured by Holocene dunes (Sprigg, 1952).

The Coorong lagoon is backed by another dune calcarenite ridge and in places contains elongated shoals and islands which are residuals of an intervening much dissected ridge, also of dune calcarenite. There are many such ridges, roughly parallel, at increasing elevations inland, marking a succession of barriers developed during Quaternary times, with intervening swampy plains on the sites of successive emerged lagoons similar to the present Coorong (Crocker and Cotton, 1946). Farther south, the Cape Jaffa ridge, and the dissected coastal ridge cliffed on its seaward side between Robe and Beachport, are also part of this Pleistocene sequence. The ridge at Naracoorte has a shoreline about 190 feet above present sea level, but the contributions of land uplift and sea lowering to this emergence have not been distinguished. Each of the dune calcarenite ridges ascends in level southeastward, indicating that tectonic uplift and transverse warping accompanied the Quaternary eustatic oscillations of sea level in this region. The broadening and deepening of the Coorong lagoon toward the Murray mouth was attributed to accompanying subsidence of the northwestern area by Sprigg (1952), and it seems likely that the truncated Robe and Cape Jaffa ridges had northward continuations across what is now the sea floor off the Coorong barrier, similar to the offshore relics of barriers identified in Portland Bay by Boutakoff (1963).

The origin of the Coorong barrier is thus to be seen in terms of dissection and reworking of earlier dune calcarenite barriers at and seaward of the present coast rather than in longshore spit growth or emergence of an offshore bar. For much of its length this barrier bears irregular transgressive dunes, partly stabilized by vegetation, with little variation in soils (Correll and Lange, 1963), but near Kingston there are sectors in which parallel beach ridges are discernible. These represent intermittent progradation of a sector of shoreline receiving sand accretion during a phase of sea-level stability or coastal emergence (which could here be due to continuing land uplift) following the Holocene marine transgression.

East Gippsland Barriers

The coast east of Wilson's Promontory (Fig. 3) is bordered by barriers composed largely of quartzose sand. In the Gippsland Lakes region there are three distinct barriers enclosing coastal lagoons within a former marine embayment backed by bluffs marking an earlier cliffed coastline. They comprise a prior barrier, the initiation of which preceded the enclosure of the Gippsland Lakes, an inner barrier interrupted by a former outlet to the sea east of Sperm Whale Head, and an outer barrier, the seaward margin of which forms the Ninety Mile Beach (Bird, 1961a; 1965a).

Figure 2. Southeast coast of South Australia (after Blackburn, 1965).

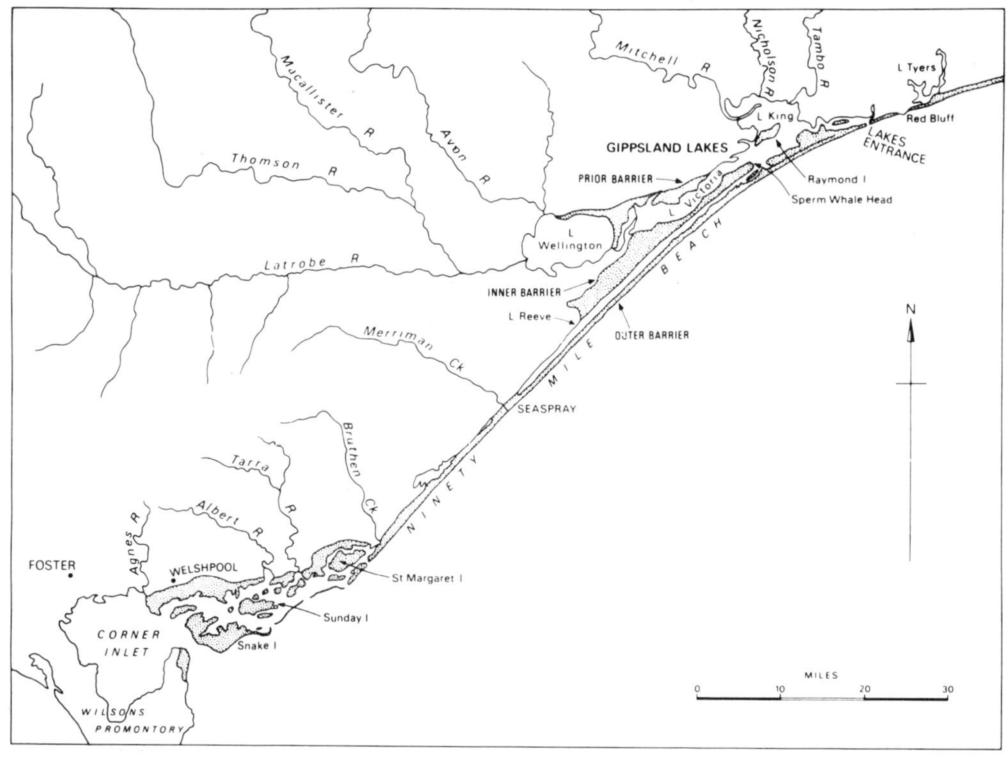

Figure 3. East Gippsland coast.

To the southwest, near Seaspray, there is only a single outer barrier, separated by a tract of lagoons and swamps from degraded sea cliffs to the rear. Toward Wilson's Promontory a second broad embayment is occupied by a complex chain of barrier islands east of Corner Inlet (Jenkin, 1968). The Gippsland Lakes and Corner Inlet both occupy areas of tectonic subsidence, which continued into Quaternary times, while the intervening sector has been uplifted.

East of the Gippsland Lakes the outer barrier extends along the coast, backed by lagoons and swamps and the degraded sea cliff, but interrupted by occasional rocky headlands and by the variable outlets from rivers, the largest of which is the Snowy. Locally, dunes from the outer barrier have spilled inland across the swamp tract and mounted the degraded sea cliff.

As with the Coorong barrier, the original explanation for the East Gippsland barriers was longshore growth in response to a powerful coastal current. This was thought to have come into existence following the submergence of an earlier land isthmus to form Bass Strait (Gregory, 1903). Again, there is no evidence for such a current, and the barriers have been shaped largely by ocean waves that arrive from the southeast, refracted by contact with the continental shelf into patterns that anticipate, and determine, the shape of the Ninety Mile Beach (Bird, 1961a). Nevertheless, relics of recurves are found in the East Gippsland barriers, indicating that longshore spit growth occurred during early stages of their development. In the

Gippsland Lakes sector, this resulted from sand drifting by waves generated by the prevailing southwesterly winds, but toward Wilson's Promontory the available fetch for waves generated by these winds is reduced, and spits have grown southwestward in response to waves generated by easterly winds.

The supply of terrigenous sand to the East Gippsland coast is now limited because the barriers have developed in front of former cliffed sectors (with the minor exceptions of Red Bluff and the rocky promontories farther east) and many of the rivers flow into the Gippsland Lakes, which intercept the modern fluvial yield of sand and gravel. Merriman Creek contributes some sand and gravel to the Ninety Mile Beach near Seaspray during times of floodwater discharge, as does the Snowy River on a larger scale farther east, but under present conditions the only possible source of substantial sand supplies to prograde the Ninety Mile Beach is the sea floor.

The Ninety Mile Beach is not now prograding, except in limited sectors adjacent to the stone jetties that border the artificial outlet from the Gippsland Lakes (Bird, 1960), but the parallel beach and dune ridges present on the outer barrier show that intermittent progradation did occur earlier in Holocene times, when the sand delivered to the shore must have drifted in from the sea floor. It is inferred that unconsolidated quartzose sands were deposited on the emerged sea floor during the last episode of low sea level, and were then collected and swept shoreward by wave action during the Holocene marine transgression. Jennings (1959) detected sea floor ridges near Flinders Island that probably represent relics of submerged dune topography. It is possible that these unconsolidated sands were in the form of barriers that had developed during the preceding phase of falling sea level, and that these were dissected and reworked during the Holocene marine transgression to produce the sand supplied to the present outer barrier.

The inner and prior barriers are present in the embayments, where tectonic subsidence, continuing into Pleistocene times, prepared the way for marine incursion. The prior barrier in the Gippsland Lakes region includes sand and gravel derived in part from the reworking of fluvial deposits, some of which are still preserved in aggraded terrace remnants on the north shore of Lake Victoria. The ridges are roughly parallel to the present ocean shoreline, and owe their alignments to refracted ocean waves, except on Raymond Island, which has a recurved spit structure. Subsequently the inner barrier developed across the mouth of the embayment, growing originally as a spit that was later widened by the addition of beach ridges and dunes built parallel to the present ocean shore. Once the inner barrier had enclosed the embayment to form a lagoon system, the prior barrier began to be reshaped into a series of growing cuspate forelands separated by scoured embayments, and similar irregularities formed on the landward shore of the inner barrier. This pattern was a response to waves generated on the enclosed lagoons, and it led to the separation of Lake Wellington from Lake Victoria by a large cuspate foreland, a process termed segmentation (Zenkovich, 1959). There is some difficulty in drawing a line between original Pleistocene topography on the prior and inner barriers and areas that have attained their present configuration within Holocene times: Bird (1963) and Jenkin (1968) offer differing interpretations of the extent to which the existing barriers can be regarded as Pleistocene features.

The original formation of the prior and inner barriers took place in Pleistocene times because these barriers were dissected by the East Gippsland rivers when they became incised during the last low sea-level phase. An outlet then developed through the inner barrier east of Sperm Whale Head, with a river flowing southward over the emerged sea floor to the lowered shoreline. During this phase, westerly winds developed blowouts on the landward side of the inner barrier, which grew into large parabolic dunes that disrupt the original ground plan of parallel ridges on Sperm Whale Head. These dunes became stabilized either late in Pleistocene or early in Holocene times, and now bear a cover of woodland and scrub, in contrast with the heath vegetation on the more deeply leached sands of the undisturbed ridges characteristic of the Pleistocene terrain on the inner barrier.

The Holocene marine transgression brought the sea back to the East Gippsland embayment, flooding the dissected barrier system, and bringing sands that were deposited to form the outer barrier. East of Sperm Whale Head some of the Holocene sands were deposited in parallel dune ridges that overlap the southern margin of the inner barrier, which in this sector appears to have subsided since Pleistocene times. The outer barrier originated as a chain of barrier islands which became lengthened and united by longshore drifting. The earliest barrier islands formed southwest of the Gippsland Lakes on a sector that may have been uplifted during Holocene times. Outlets from the Gippsland Lakes were deflected eastward by longshore drifting and sealed off, except for the intermittent, migrating, outlet that still existed when explorers arrived in the 1840s. In 1889 this was replaced by the present artificial outlet cut at Lakes Entrance (Bird, 1961b).

The final stage in barrier evolution here has been retrogradation of the Ninety Mile Beach, accompanied by the development of blowouts that are spilling landward, disrupting and burying the parallel beach ridges and dunes that had formed during the preceding phase of outer barrier progradation. These parallel beach ridges resulted from intermittent progradation when the sea level was stable or falling and sand was being supplied to the shoreline. The present retrogradation may result from a rise of sea level, a diminution in sand supply, an increase in storminess in coastal waters, or a combination of these factors.

The barrier islands east of Corner Inlet are of intricate configuration, and intersecting patterns of beach ridges and dunes betoken a complex history (Jenkin, 1968). An inner zone of deeply leached quartzose dunes equivalent to the Pleistocene barriers in the Gippsland Lakes region is bordered seaward by barrier islands, notably Sunday Island and Snake Island, of Holocene origin. The chief contrast with the Gippsland Lakes is in the persistence of five outlets to the sea, probably a consequence of greater tide range. Spring tides of about 8 feet at the entrance to Corner Inlet, compared with 3 feet at Lakes Entrance, generate more powerful tidal currents which have maintained outlets. The configurations of the southwestern end of the Ninety Mile Beach, the outlying barrier beaches, and the spits bordering the tidal channels are subject to much variation from year to year as the result of the contest between tidal currents, tending to maintain transverse channels, and wave action, tending to build up the barrier beaches and extend them laterally by spit growth.

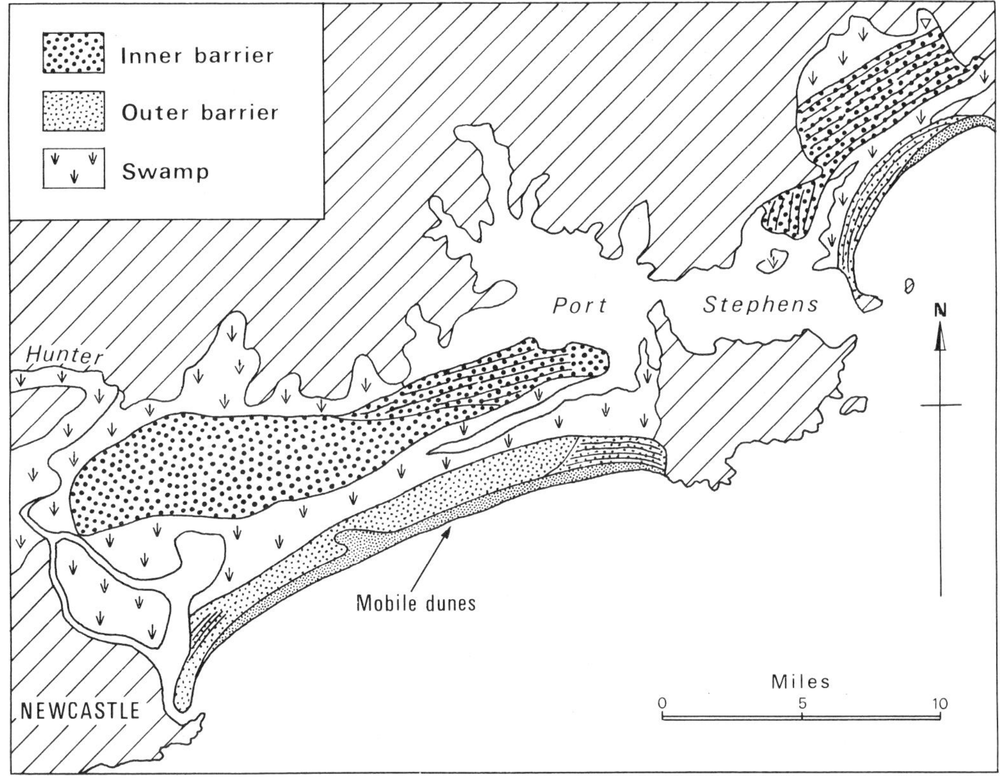

Figure 4. Coastal barriers near Newcastle, New South Wales (after Thom, 1965).

Barriers on the New South Wales Coast

The New South Wales coast has a number of asymmetrical embayments with sandy shores curving between protruding rocky headlands, many of which are former islands tied to the mainland by sand deposition. The sandy shores are often the seaward fringes of barriers that have a ground plan of beach and dune ridges parallel to the present shoreline, indicating past progradation on alignments determined by refracted ocean swell. Parts of these are now being overrun by blowouts and parabolic, or transgressive, dunes spilling inland.

In several sectors there are distinct inner and outer barriers with an intervening lagoon and swamp tract. The inner barrier is as much as six miles wide, but the outer barrier is usually less than half a mile in width (Langford-Smith, 1969). In the Newcastle district (Fig. 4) Thom (1965) showed that the inner barrier, which consists of widely spaced (300–600 feet), subdued, and deeply leached quartzose sand ridges bearing heath and heath-woodland, and underlain by illuvial sandrock, is of Pleistocene age; it was incised by rivers during the last low sea-level phase, and radiocarbon dates from peat, charcoal, and sandrock that postdate the formation of the inner

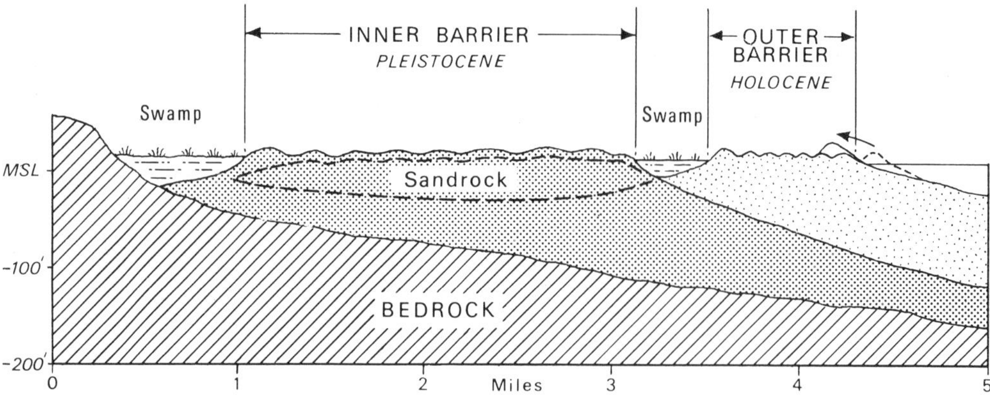

Figure 5. Schematic cross-section through New South Wales coastal barriers (after Thom, 1965).

barrier ranged up to 13,000 years B.P. The outer barrier has been added during and since the Holocene marine submergence, and has more pronounced, relatively closely spaced (60–300 feet) quartzose sand ridge topography bearing dune scrub or woodland, on sands that show much less leaching and no underlying sandrock (Fig. 5).

Similar results were obtained by Hails (1968) from a study of inner and outer barriers farther north, between Crescent Head and the Macleay delta. Hails and Hoyt (1968) noted that barrier development was more extensive in the vicinity of the mouths of large rivers in northern New South Wales, and inferred that these had supplied the coastal environment with quartzose sand derived from granitic outcrops in the hinterland during Quaternary times. Here, as elsewhere, there is a suggestion that the present outer barrier may incorporate sand derived from the dissection and reworking of barrier systems that formerly existed seaward of the present shoreline, presumably formed during the last phase of sea lowering. Shoreward drifting has supplied sand to the prograding barriers, and Hails (1968) attributed the present phase of retrogradation and landward migration of dunes behind the eroding shoreline to a decline in sand supply.

Langford-Smith and Thom (1969) reported further radiocarbon dates, confirming a Pleistocene age for the inner barrier on the New South Wales coast, and noted that the swales between ridges on the inner barrier were typically 15 to 20 feet above present high water mark, indicating that this barrier formed during an episode of higher sea level. In some sectors, the outer barrier is missing, and the present coastline truncates the inner barrier, exposing sandrock on the shore as at Evans Head. Warner (1971) found that inner barrier sandrock on the coast near Angourie had been exposed and trimmed by marine erosion during an episode when the sea stood 3 to 7 feet above its present level some 28,000 to 30,000 years ago, prior to the last low sea-level phase and the subsequent addition of a Holocene outer barrier. The possibility of an interstadial high sea level at about this date has been indicated by other radiocarbon assays on the New South Wales coast (Langford-Smith, 1972) and on emerged Pacific island shores.

South of Sydney, large quantities of sand have moved shoreward during and since the Holocene marine transgression to build beaches and barriers and form

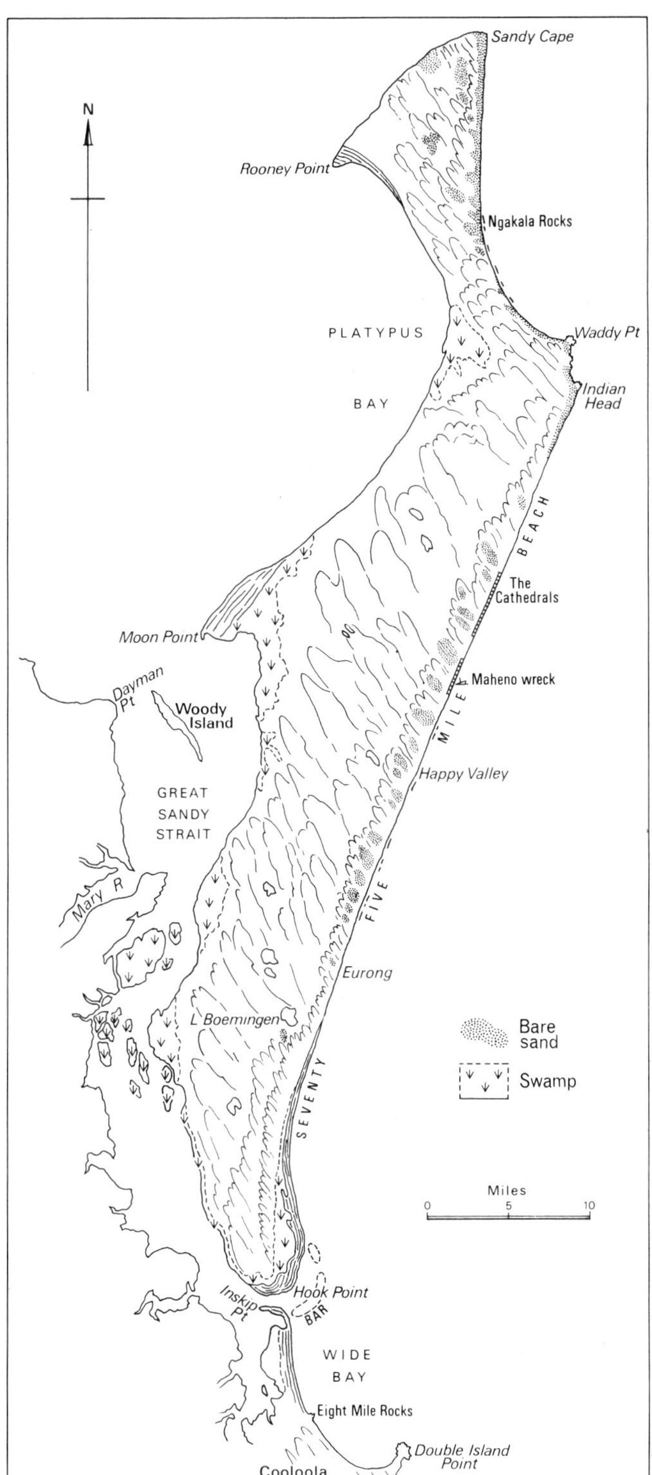
N
Sandy Cape
Rooney Point
Ngakala Rocks
PLATYPUS
BAY
Waddy Pt
Indian Head
SEVENTY FIVE MILE BEACH
The Cathedrals
Maheno wreck
Moon Point
Dayman Pt
Woody Island
GREAT SANDY STRAIT
Happy Valley
Mary R
Eurong
L Boemingen
Bare sand
Swamp
Miles
0
5
10
Hook Point
Inskip Pt
BAR
WIDE BAY
Eight Mile Rocks
Double Island Point
Cooloola

Figure 6.
Fraser Island, Queensland.

thresholds of sand that are moving into estuarine inlets produced by marine submergence of river valley mouths (Bird, 1967b). The modern sediment yield from most of these rivers is trapped within the estuarine inlets, and direct nourishment of beaches and barriers by fluvial sand is exceptional, but at Whale Beach a steep barrier of coarse, poorly sorted sand is being supplied with similar material from the Towamba River during flooding (Hails, 1967); and adjacent to the mouth of the Shoalhaven, barriers have been prograded with sand delivered by floodwater discharge (Wright, 1970).

Barrier Islands of Southeast Queensland

North of the New South Wales border, the pattern of inner and outer barriers behind asymmetrical embayments gives place to a series of massive coastal sand accumulations in southeast Queensland. The Stradbroke Islands, Moreton Island, Cooloola, and Fraser Island are all marked by overlapping waves of transgressive parabolic dunes now largely fixed beneath a cover of forest and scrub vegetation (Coaldrake, 1960; 1962). These represent successive phases of dune migration during Pleistocene times, and as the axes of the parabolic dunes run parallel with modern wind resultants it is deduced that southeasterly winds were as dominant as they are now during the phases of dune advance (Hails, 1964a). Superimposed dune formations, resting in part on a basement of older Quaternary sediments, produce a topography with summits up to 919 feet above sea level. Lakes and swamps are perched amid the dunes, and streams drain many of the interdune hollows.

On North Stradbroke and the southern part of Fraser Island (Fig. 6) a Holocene outer barrier, with ridges parallel to the present shoreline, has been added seaward of the margin of the Pleistocene dune terrain; intervening lagoons and swamps have locally been invaded or overrun by landward-spilling blowouts. On the northern part of Fraser Island, Holocene sands directly abut the Pleistocene dune terrain, and have spilled inland over it in sectors where the foredune vegetation has proved inadequate to retain them (Bird, 1972a). The Seventy-five Mile Beach on Fraser Island shows long sectors where the margin of the Pleistocene dune terrain has been trimmed back to expose reefs and foreshore outcrops of underlying sandrock, as well as gullied cliffs in the basal Pleistocene Coloured Sands formation. The configuration of the sandy ocean shoreline is related to wave patterns produced by the predominant southeasterly swell in sectors bounded by protruding bedrock headlands such as Indian Head on Fraser Island and Cape Moreton and Point Lookout on the northeastern extremities of Moreton Island and North Stradbroke Island. Southeasterly wave action also generates a northward drift of sand, producing accretion at the northern end of the Seventy-five Mile Beach, alongside Indian Head.

These massive barrier islands evidently represent a long-term accumulation of quartzose sand that has been carried northward along the east coast of Australia from deposits delivered to the coast by the rivers of northern New South Wales.

Barrier Formations in Northeast Queensland

Sandy depositional features on the Queensland coast north from Fraser Island are built on a much smaller scale than those to the south, and are more directly related to local fluvial sources of sand supply (Bird and Hopley, 1969; Bird, 1972b). Ocean swell is largely excluded here by the Great Barrier Reef, but waves generated by the prevailing southeasterly winds cause northward drifting of sand supplied to the shore at river mouths. On Burdekin Delta a series of spits runs northward from

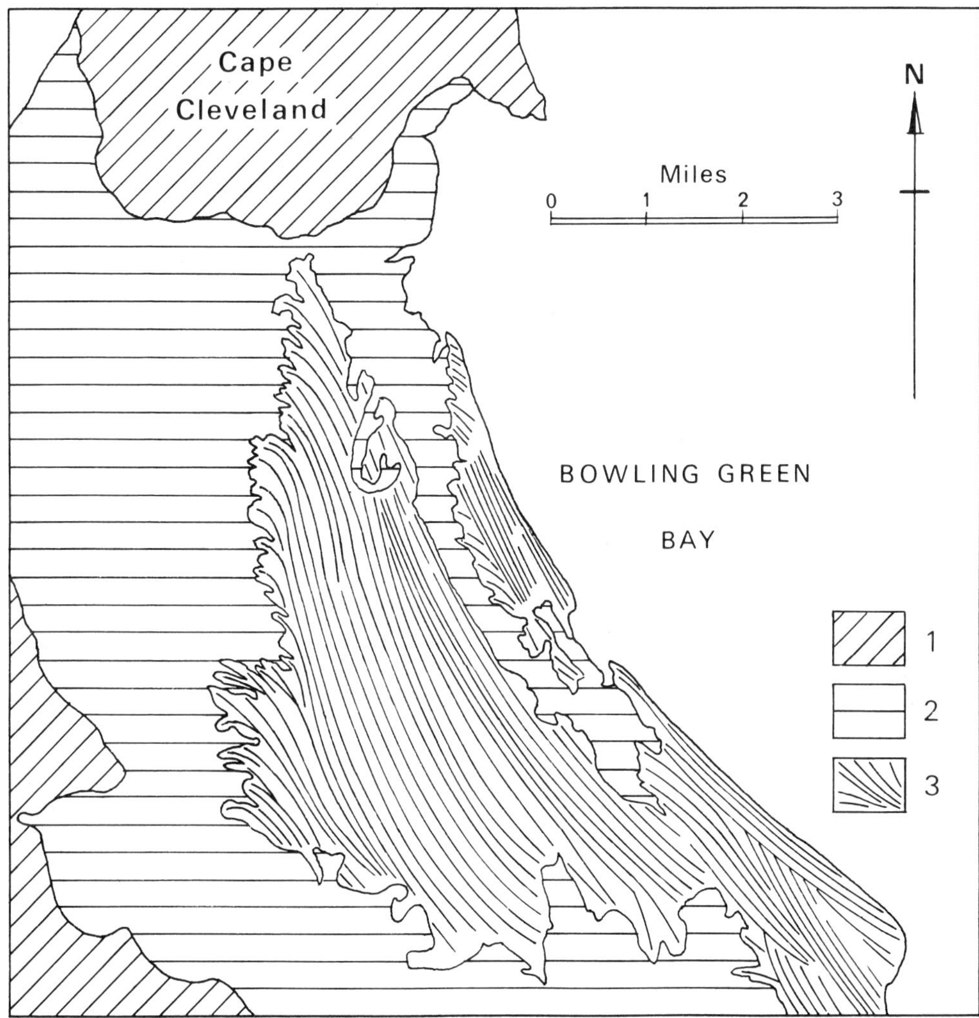

Figure 7. Part of the barrier system near Cape Cleveland, north of the Burdekin Delta, Queensland: 1, bedrock and hillslope deposits; 2, swamp land; 3, beach ridges on Holocene barrier (after Hopley, 1970).

distributary mouths, culminating in Cape Bowling Green. The Holocene barrier formed near Cape Cleveland, on the northern side of this delta, shows landward recurves on its inner margin (Fig. 7), signifying an early history of successive spit growth before parallel ridges were added during subsequent progradation (Hopley, 1970).

Many of the northeast Queensland barriers rest, at least in part, on clay plain foundations, evidently former tidal swamp flats. In some, the sand ridges are spaced out on swampy plains in a manner similar to the "cheniers" of Louisiana, but elsewhere the ridges are roughly parallel with intervening swales, as in the beach-ridge plain on which Cairns is built (Bird, 1970). Dunes are poorly developed, especially in the humid tropical sector between Townsville and Cooktown (Bird and Hopley, 1969), but locally the beach ridges have a dune capping. On the Burdekin delta, higher dunes have formed in sectors where formerly prograded beach ridge plains have been trimmed back by marine erosion (Hopley, 1970).

Inner barriers of the type found in New South Wales are not extensive in northeast Queensland, but near Kurrimine an inner barrier of presumed Pleistocene age exists landward of the Holocene outer barrier (Bird and Hopley, 1969).

Discussion

Several problems have been raised in the course of investigating Australian coastal barriers. The first is the problem of their initiation. Some have been formed by the longshore growth of spits, but others appear to have originated offshore. The latter may have developed when offshore bars (submerged at least at high tide) were built upward by sand accretion to form barriers without any change in the relative levels of land and sea. Such vertical accretion is accompanied by landward migration in the scheme proposed by de Beaumont in 1845 (Johnson, 1919), but actual examples of offshore bars becoming barriers in this manner are rare, and around the Australian coast such bars either remain in a submerged or intertidal position, or are driven onshore by wave action to be incorporated in the beach.

A second possibility is the emergence of an offshore bar as a barrier during an episode of land uplift or sea lowering. This was suggested by Merrill in 1890 (Johnson, 1919) and has recently been discussed by Leontyev and Nikiforov (1966). It is an attractive hypothesis for those who accept the view that the Holocene marine transgression carried the sea to a higher level, 6 to 10 feet above the present (Fairbridge, 1961; Gill, 1961; and others), for outer barriers could then have come into existence as the result of the subsequent emergence as the sea fell back to its present level. However, the concept of a higher Holocene sea level has been rejected, notably by those who have worked on the New South Wales coast (Hails, 1964b; Thom, 1965; Langford-Smith and Thom, 1969), and the question remains controversial in Australia (Thom et al., 1969; Gill and Hopley, 1972; Thom et al., 1972). Tectonic uplift could have led to emergence of offshore bars as barriers, but although Holocene uplift may have influenced barrier evolution locally, it is doubtful that it has occurred in the pattern required for outer barrier initiation on the Australian coast generally.

Another suggestion, based largely on studies of the structure and stratigraphy of North American coastal barriers, is that they originated as the result of partial submergence of a beach or dune ridge previously built up adjacent to a former shoreline, so that a lagoon is formed to landward. This seems to have been the view of McGee in 1890 and Ganong in 1908 (Johnson, 1919), recently revived by Hoyt (1967). A barrier formed in this way may either migrate landward or remain in position, with or without the subsequent addition of prograded beach ridges or foredunes. This explanation requires a particular balance between rate of submergence, rate of sand supply to the shore, and incidence of storm wave activity. The preexisting coastal ridge must be submerged at a rate which permits a lagoon to form to landward but does not allow it to be overrun by landward migration of the barrier engendered by the submergence. A relative rise of sea level may result in shoreline recession and the initiation of dunes which are driven in over the barrier by onshore winds; if the barrier is low-lying, sand may be swept over it by occasional storm washovers. These effects can be offset if there is a continuing sand supply, sufficient to prograde the shoreline despite submergence, or if stormy conditions are infrequent.

The origin of the preexisting coastal sand ridge deserves further attention. If slow submergence produces lagoons and barriers and rapid submergence drowns barriers altogether, a coastal ridge of this kind is evidently formed during an episode of stillstand, or perhaps emergence. This accords with the concept of an oscillating Holocene marine transgression (Fairbridge, 1961) if coastal deposition occurs during phases of stillstand or emergence and lagoons develop behind barriers in phases of slow submergence. However, it is possible that barriers drowned in phases of rapid submergence can be reworked into offshore bars that become new barriers during a later episode of emergence.

Barrier initiation is thus a consequence of several interacting factors, including coastal configuration, transverse coastal profiles, rates of emergence or submergence resulting from changing levels of land and sea, incidence of stillstands, availability of suitable sediment in the coastal zone, and wave regimes related to climatic conditions prevailing in coastal waters. We cannot expect any one explanation of barrier initiation to be universally applicable (Schwartz, 1971).

Hoyt (1967) noted the absence of open sea sediments landward of many existing barriers as evidence against an origin by upward growth, or emergence, of offshore bars. In Australia the landward margins of lagoons behind outer barriers are commonly the modified Pleistocene shorelines at the seaward fringes of inner barriers, and the question is whether these were also exposed to the open sea before the outer barrier came into existence. In some sectors they were, notably on parts of the Gippsland coast, where the outer barrier grew originally as a longshore spit, but in the Coorong region, where the outer barrier incorporates sectors of older dune calcarenite, it is doubtful if the landward margins of the lagoon have been exposed to the open sea in Holocene times. On the New South Wales coast, outer barriers appear to have been formed at or before the phase of maximum marine submergence in Holocene times, either as a coastal sand ridge that became partially submerged (Hails and Hoyt, 1968) or as a transgressive barrier that developed during an oscillation of the rising sea level and then rose and advanced to become anchored on its present alignments as Holocene submergence came to an end.

There is no evidence of new barrier initiation at the present time on the Australian coast, apart from longshore spit growth of the kind seen on the Burdekin delta. Sandy shorelines in Australia are generally retrograding, and outer barriers show blowouts and transgressive dunes spilling inland behind receding shores. This has been correlated with a modern phase of rising sea level. Shoreward drifting of sand, which previously prograded the barriers, seems to have come to an end. Sea floor sand supplies can hardly have run out simultaneously around so much of the Australian coast, and indeed there is evidence of abundant sand offshore (Phipps, 1963), some of which could be pumped in to replenish eroding beaches. Cessation of shoreward drifting is more likely to be a consequence of the sea level rise, leading to a reshaping of the transverse coastal profile, and resulting in shoreline retreat. A sea-level rise would also account for the onset of retrogradation on barrier shorelines still receiving fluvial sand supplies, as at the mouth of the Shoalhaven (Wright, 1970). A recent increase in storminess in Australian coastal waters would yield a similar result, but there is no other evidence to support this hypothesis, and meteorological records are too brief to compare modern weather conditions with earlier norms.

The pattern, height, and spacing of beach ridges and parallel foredunes associated with barriers are also related to the several interacting factors mentioned above, as well as to wind regimes and the effectiveness of dune stabilizing vegetation. Successive formation of beach ridges and parallel foredunes is confirmed by landward sequences showing increasing age of soil and vegetation features (Bird, 1961a). Davies (1957) proposed a "cut and fill" explanation for the formation of parallel ridges on alignments determined by refracted wave patterns, and in Tasmania found evidence for seaward declination of ridge and swale heights in Holocene beach ridges, which he took as evidence for coastal emergence (Davies, 1961). It is certainly easier to envisage shoreline progradation and successive beach ridge formation during a phase of emergence, or during a sea-level stillstand, than during submergence, which tends to promote shoreline retreat and must lead to drowning of earlier swales and ridges as lagoon or water-table levels rise. No satisfactory explanation has yet appeared for the wider spacing of the more subdued parallel ridges on inner barriers, compared with outer barriers, on the east coast of Australia.

The fine sequence of emerged barriers inland from the Coorong is not matched elsewhere around the Australian coast. Separation of distinct ridges is uncommon on the calcareous sector, where coastal sands are often present as transgressive, partly overlapping dune structures. The spaced ridges behind the Coorong could be the outcome of an exceptional rate of Pleistocene land uplift here. On the quartzose sector, where depositional features are less durable, barriers have not been found above the 25-foot contour, but the dunes on the coastal plateau up to 200 feet above sea level inland from the Gippsland Lakes could be the rearranged sands of emerged Pleistocene barriers antedating the prior barrier in that region (Bird, 1965b). On the New South Wales coast, older and higher shorelines are marked by boulder beaches up to 100 feet above present sea level (Langford-Smith and Hails, 1966).

Finally, the continuity of Australian coastal barriers deserves comment. On coasts exposed to ocean swell, where large masses of sand have been deposited, barriers are interrupted only at wide intervals by river mouths or tidal inlets. This is a function partly of the sparseness of rivers in the arid zone, and partly of generally small tidal ranges and weak tidal currents. On the Queensland coast north of Fraser Island and on the north coast of Australia, where ocean swell is attenauted or excluded, sand deposition has been on a smaller scale, river mouths are more frequent, and tide ranges often larger. In these sectors the barriers are fragmented, and it is here that the best Australian examples of barrier islands—as yet, little studied—are to be found.

References

Bird, E. C. F. 1960. The formation of sand beach ridges, *Austr. J. Sci.*, **22**: 349–50.

———. 1961a. The coastal barriers of East Gippsland, Australia, *Geogr. J.*, **127**: 460–8.

———. 1961b. Landform changes at Lakes Entrance, *Vict. Nat.*, **78**: 137–46.

———. 1963. The physiography of the Gippsland Lakes, Australia, *Zeitschr. Geomorph.*, **7**: 233–45.

———. 1965a. The evolution of sandy barrier formations on the East Gippsland coast, *Proc. Roy. Soc. Victoria*, **79**: 75–88.

———. 1965b. *A geomorphological study of the Gippsland Lakes*. Australian National University, Department of Geography Publication G/1.

———. 1967a. Coastal lagoons of southeastern Australia, in J. N. Jennings and J. A. Mabbutt (eds.), *Landform Studies from Australia and New Guinea*: 365–85.

———. 1967b. Depositional features in estuaries and lagoons on the South Coast of New South Wales, *Austr. Geogr. Studies*, **5**: 113–24.

———. 1970. Coastal evolution in the Cairns district, *Austr. Geogr.*, **11**: 327–35.

———. 1972a. Dune stability on Fraser Island, *Queensland Nat.*, in press.

———. 1972b. The origin of beach sediments on the North Queensland coast, *Earth Sci. J.*, in press.

Bird, E. C. F., and Hopley, D. 1969. Geomorphological features on a humid tropical sector of the Australian coast, *Austr. Geogr. Studies*, **7**: 89–108.

Blackburn, G., Bond, R. D., and Clarke, A. R. P. 1965. Soil development associated with stranded beach ridges in southeast South Australia, *C.S.I.R.O. Soil Publ. 22*.

Boutakoff, N. 1963. The geology and geomorphology of the Portland area, *Geol. Surv. Victoria, Memoir 22*.

Coaldrake, J. E. 1960. Quaternary history of the coastal lowlands of southern Queensland, *J. Geol. Soc. Australia*, **7**: 403–8.

———. 1962. The coastal sand dunes of southern Queensland, *Proc. Roy. Soc. Queensland*, **72**: 101–16.

Correll, R. L., and Lange, R. T. 1963. Significant trends of surface lime in coastal dunes on Younghusband Peninsula, *Austr. J. Sci.*, **26**: 59–60.

Crocker, R. L., and Cotton, B. C. 1946. Some raised beaches of the lower southeast of South Australia and their significance, *Trans. Roy. Soc. S. Australia*, **70**: 64–82.

Davies, J. L. 1957. The importance of cut and fill in the development of beach sand ridges, *Austr. J. Sci.*, **20**: 105–11.

———. 1961. Tasmanian beach ridge systems in relation to sea level change, *Proc. Roy. Soc. Tasmania*, **95**: 35–40.

Fairbridge, R. W. 1961. Eustatic changes in sea level, *Phys. Chem. Earth*, **4**: 99–185.

Fenner, C. 1931. *South Australia: a geographical study*. Melbourne.

Gill, E. D. 1961. Changes in the level of the sea relative to the land during the Quaternary era, *Zeitschr. Geomorph.*, Supplementband **3**: 73–79.

Gill, E. D., and Hopley, D. 1972. Holocene sea levels in Eastern Australia—a discussion, *Mar. Geol.*, **12**: 223–33.

Gregory, J. W. 1903. *Geography of Victoria*. Melbourne.

Hails, J. R. 1964a. The coastal depositional features of southeastern Queensland, *Austr. Geogr.*, **9**: 207–17.

———. 1964b. A critical review of sea-level changes in Eastern Australia since the Last Glacial, *Austr. Geogr. Studies*, **3**: 63–78.

———. 1967. Significance of statistical parameters for distinguishing sedimentary environments in New South Wales, Australia, *J. Sed. Petrol.*, **37**: 1059–69.
———. 1968. The Late Quaternary history of part of the midNorth Coast, New South Wales, Australia, *Trans. Inst. Brit. Geogr.*, **44**: 133–49.
Hails, J. R., and Hoyt, J. H. 1968. Barrier development on submerged coasts: problems of sea-level changes from a study of the Atlantic Coastal Plain of Georgia, U.S.A., and parts of the East Australian coast, *Zeitschr. Geomorph.*, Supplementband **7**: 24–55.
Hopley, D. 1970. The geomorphology of the Burdekin Delta, North Queensland, James Cook University, Department of Geography *Monograph 1.*
Hoyt, J. H., 1967. Barrier island formation, *Geol. Soc. Amer. Bull.* **78**: 1125–36.
Jenkin, J. J. 1968. The geomorphology and Upper Cainozoic geology of Southeast Gippsland, Victoria, *Geol. Surv. Victoria, Memoir 27.*
Jennings, J. N. 1959. The submarine topography of Bass Strait, *Proc. Roy. Soc. Victoria,* **71**: 49–72.
Jennings, J. N., and Bird, E. C. F. 1967. Regional geomorphological characteristics of some Australian estuaries, in G. H. Lauff (ed.), *Estuaries:* 121–8.
Johnson, D. W. 1919. *Shore processes and shoreline development.* New York.
Langford-Smith, T. 1969. Coastal sand barrier, *Austr. Geogr.*, **11**: 176–8.
———. 1972. New South Wales report on Quaternary shorelines, *Search,* **3**: 103.
Langford-Smith, T., and Thom, B. G. 1969. New South Wales coastal morphology, *J. Geol. Soc., Australia,* **16**: 572–80.
Leontyev, O. K., and Nikiforov, L. G. 1966. An approach to the problem of the origin of barrier bars, *Intern. Oceanographic Congr.*, Abstracts: 221–2.
Phipps, C. V. G. 1963. Topography and sedimentation of the continental shelf and slope between Sydney and Montague Island, New South Wales, *Austr. Oil Gas. J.*, **10**: 40–46.
Schwartz, M. L., 1971. The multiple causality of barrier islands, *J. Geol.*, **79**: 91–4.
Sprigg, R. C. 1952. The geology of the southeast province of South Australia, with special reference to Quaternary coastline migrations and modern beach development, *Geol. Surv. S. Australia, Bull. 29.*
Thom, B. G. 1965. Late Quaternary coastal morphology of the Port Stephens–Myall Lakes area, New South Wales, *Proc. Roy. Soc. New South Wales,* **98**: 25–36.
Thom, B. G., Hails, J. R., and Martin, A. R. H., 1969. Radiocarbon evidence against higher postglacial sea levels in eastern Australia, *Mar. Geol.*, **7**: 161–8.
Thom, B. G., Hails, J. R., Martin, A. R. H., and Phipps, C. V. G., 1972. Postglacial sea levels in eastern Australia—a reply, *Mar. Geol.*, **12**: 233–42.
Tindale, N. B., 1947. Subdivision of Pleistocene time in South Australia, *Records S. Australian Museum,* **8**: 619–52.
Warner, R. F. 1971. Dating some inner barrier features at Angourie, Northern New South Wales, *Search,* **4**: 140–1.
Wright, L. D. 1970. The influence of sediment availability on patterns of beach ridge development in the vicinity of the Shoalhaven River delta, New South Wales, *Austr. Geogr.*, **11**: 336–48.
Zenkovich, V. P. 1959. On the genesis of cuspate spits along lagoon shores, *J. Geol.*, **67**: 269–77.

Editor's Comments on Paper 40

It is only fitting that in this final paper we examine man's influence on barrier islands. Robert Dolan has been interested in this subject for some time and has worked on the problem with P. G. Godfrey, one of whose publications is cited in Dolan's report.

According to Dolan, the barrier islands of North Carolina were in a state of active equilibrium in the early 1930s. Subsequent attempts at stabilization have been, he believes, detrimental in terms of long-range goals. Dune stabilization upsets the delicate natural balance between barrier and sea, causing further erosion within the present rise in sea level. Dolan calls for a review of the geological and ecological implications of barrier dune stabilization and suggests that other approaches may be more desirable.

For further details on this most vital topic, the interested reader is strongly urged to read the many other fine papers published, or now in press, by Dolan and Godfrey.

Robert Dolan, Associate Professor of Environmental Sciences at the University of Virginia, received his Ph.D. degree in 1965 in coastal geomorphology and geology from Louisiana State University. He was on the staff of the Coastal Studies Institute for several years before joining the faculty at Virginia. His research interest is focused on beach processes, coastal geomorphology, and man's impact on barrier islands of the Atlantic Coast of the United States. Three of Dolan's most recent papers extend his discussion of coastal processes and the role of man as an agent of landscape change: "Man's Impact on the Barrier Islands of North Carolina" (*American Scientist,* in press); "Effects of Hurricane Ginger on the Barrier Islands of North Carolina" (*Bulletin,* Geological Survey of America, in press); and, "Crescentic Coastal Landforms" (*Zeitschrift für Geomorphologie,* in press).

Reprinted from *Science,* **176,** 286–288 (1972)

Barrier Dune System along the Outer Banks of North Carolina: A Reappraisal

40

Abstract. *Barrier dune development has been encouraged by man along the Outer Banks of North Carolina to stabilize the barrier islands. This modification of a delicately balanced natural system is leading to severe adjustments in both geological and ecological processes.*

In the early 1930's the barrier islands of North Carolina were in what might be called a natural or equilibrium state. Changes were rapid, but the system was well adapted to accommodate powerful natural forces. The first steps to stabilize the islands were taken by the Works Progress Administration–Civilian Conservation Corps (WPA–CCC) in the 1930's by encouraging sand accumulations with brush fences. This was followed by extensive dune stabilization by the National Park Service in the 1950's. In building the high coastal dunes along the Outer Banks (Fig. 1), man has created a new state in the beach system that may be detrimental to the long-range stability of the barriers and may become more difficult and costly to manage than the original natural system.

In an article in *Science*, Houston (*1*), a National Park Service biologist, said,

> . . . Criteria for management of a park ecosystem must, of necessity, differ from criteria for other uses of land, since park management involves preventing or compensating for the influence of man. The objectives for natural areas appear to be ecologically feasible if it is recognized that these areas have a finite capacity for absorbing man's consumptive and disruptive influences.

In the case of the barrier dunes, man's disruptive influence is linked, most directly, with geological processes; however, Godfrey (*2*), also a biologist with the National Park Service, has established an important coupling of the geological and ecological implications of barrier dune stabilization.

The "natural condition" for the mid-Atlantic barriers is simply a wide range of sand deposit responses to various wave conditions. Like fluvial systems, in which streams adjust in cross section to accommodate the water flow, beaches adjust in cross section to accommodate wave run-up. When the wave run-up

Fig. 1 (left). The coastal barrier dunes of the Outer Banks of North Carolina. These dunes were originally developed by man in the 1930's and have since been encouraged and expanded with sand fences and grass plantings. (a) General view on Bodie Island (1969). The sharp line between the active beach and the stabilized dune complex is a 3-m erosional scarp. During severe winter northeasters the dunes are commonly eroded back, up to several meters. (b) Some of the original WPA-CCC sand fencing near Cape Hatteras (1936–1937). (c) Sandbag revetment and stabilized dunes near Cape Hatteras. This is the same area as in Fig. 1b, but 25 years later. Fig. 2 (right). (a) Beach system under moderate wave attack. (b) Beach system under severe wave attack. Comparison of Figs. 1a and 2a shows how man's activities have altered the characteristics of these zones.

of the surf zone is high, as during storms, the active beach expands, both landward and seaward. When the run-up is low, as during the summer months, the active beach zone contracts.

Most of the time this process of expansion and contraction is of minor significance, geologically or economically, because it is confined to the central part of the active zone, where little change in the sand deposit is involved. Under these conditions, the cross section required to accommodate the wave run-up is analogous to the stream cross section at low river stage. In the fluvial system the flow is confined within the system's natural levees most of the time, so the bed can easily accommodate the stream discharge. In the beach system, the berm serves as the topographic constraint for wave run-up most of the time (Fig. 2a).

During extreme events, such as hurricanes or winter northeasters, the beach cross section must make major adjustments to lengthen the distance of the run-up and thus dissipate the increased energy. In the offshore region this is manifested as an extension of the active zone beyond the outer bar. At the subaerial end of the profile, if the increased energy level is high enough, the wave run-up extends into the zones normally associated with eolian processes and eolian landforms—namely, the sand dunes and adjacent sand flats (Fig. 2b).

Man's efforts to stabilize the beach system naturally focus on the area that is (i) only occasionally penetrated by storm surge, (ii) suitable terrain for permanent development, and (iii) amenable to simple modification. The active beach does not fit any of these criteria, but the dune area can be controlled. Although intuition suggests that by stabilizing the dunes one succeeds in stabilizing the entire beach system, the dunes are, unfortunately, a response element of the system, not a forcing element.

The rationale behind the construction and stabilization of barrier dunes is that the dunes confine the upper limit of the wave uprush within the swash zone and prevent the undesirable effects of overwash and channel development. For permanent stabilization, this requires a beach system that is stable through time. Erosion of the barrier islands is, however, a well-understood geological process (*3*). Beach stability, for any period of time, is a function of sea level and the amount of sedimentary material supplied to the coast. If the sea level rises or the amount of material required to sustain the system is reduced, the beach system migrates landward, with areas of backshore becoming foreshore and areas of dune becoming backshore. If the dune areas are stabilized, the new system must adjust to any changes that occur in the barrier itself. That is, if the sea level rises and the beach zones shift landward, the dunes must also shift, or the beach sector of the system is reduced in width, the energy dissipation process is changed, and the entire system is forced out of equilibrium.

The most important changes brought about by dune stabilization and beach narrowing are associated with the restriction and consequent steepening of the run-up profile, which forces the process of energy dissipation to take place in a limited cross section of the beach, so that more energy is dissipated per meter of beach cross section. If the energy dissipation ratio is 5 : 1, a certain combination of slope, width, and sand size of the beach will result. If the width of system is reduced by 20 percent and the ratio becomes 4 : 1, a greater amount of energy is being dissipated per meter of beach profile. The material deposited across the upper limit of the swash zone, about 10 m wide along the Outer Banks, is the finest material within the beach system. This is a consistent pattern that has been established by years of beach sampling (*4*). When the system is compressed, it is the finer materials, in the upper part of the swash zone, that are most vulnerable to winnowing processes.

Coarser beaches are steeper, and if they become steep enough to initiate wave reflection a new set of interactions is triggered, which may result in a concentration of the energy dissipation process in narrower zones. The increased stress and turbulence across the narrower beach results in higher attrition and winnowing rates for the sand grains and leads to accelerated losses of fine sand (0.25 mm). This is an important process that accounts for a major loss of beach material (*5*).

In the 13 years since the National Park Service initiated its major effort to stabilize the dunes and beaches along the Outer Banks of North Carolina, a process of beach narrowing, relative to the dune system, has been underway. Originally, the distance from the dune fields to the shoreline was about 100 to 125 m; since then, erosion has reduced this distance by 30 to 40 m, leaving beach widths of 70 to 100 m in most areas.

Figure 3 illustrates a set of measurements taken from several strips of aerial photographs dated between 1945 and 1969, and Table 1 provides estimates of the changes that have occurred along the Outer Banks since the high barrier

Table 1. Changes along the Outer Banks of North Carolina since high barrier dunes were constructed.

Topographic zones	Ratio × 100 (%)	
	January 1945	August 1969
Beach to active sand zone	51.5	54.5
Beach to island width	20.2	10.9
Active sand zone to island width	42.1	22.0
Dune to island width	21.9	11.2

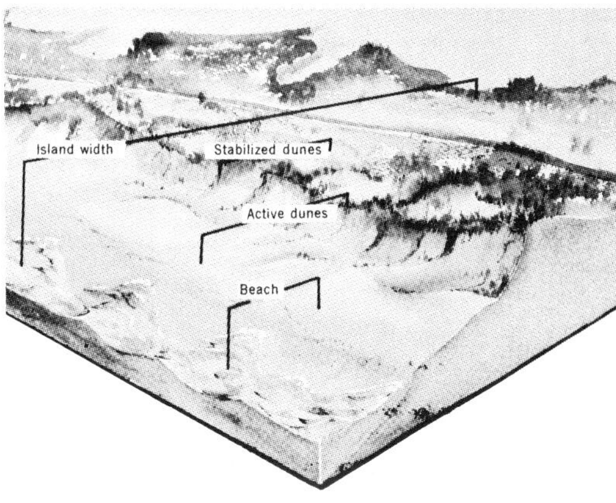

Fig. 3. Measurements taken from aerial photographs to determine the degree of change in the beach and dune systems. The barrier islands can be divided into three distinct biophysical zones, the beach, the zone of active sand dunes, and the stabilized zone, which extended to the sound-side shoreline.

dunes were constructed. The topographic zones in Table 1 include the beach and the active dunes, and these are collectively called the active sand-movement zone. The active sand zone can easily be detected on aerial photographs, since the dense cover of the dune grasses usually marks a sharp transition between the stable and the active areas of the barrier islands.

The results in Table 1 show that in 1945 the beach represented 51.5 percent of the active sand-movement area and by 1969 it represented 54.5 percent. The reason for this increase is that when the dune system is stabilized, the active sand zone, which is critical to the long-term equilibrium of the barrier system, is no longer a dynamic element of the system. When the dunes are reduced to essentially an inactive state, the actual beach becomes the only dynamic element of the system. It can also be seen that during this period the beach width was reduced by 9.3 percent, the active sand zone was reduced by 20.1 percent, and the dune area was reduced by 10.7 percent. The latter figure is highly significant as it indicates that although extensive dune areas have been stabilized, new dune areas are not developing to replace those being eroded on the ocean side of the barrier.

Since it is impossible to move the stabilized system inland with each reduction in the beach profile, which does, in effect, occur in a natural system, the inevitable has taken place. The beaches are narrower, steeper, and coarser, and wave reflection is common. The sediment most vulnerable to these changes is the fine materials (0.20 to 0.40 mm) that are the natural supply for renourishment of the native dune system. This sediment is associated with the uppermost limit of the uprush on an altered beach (*4*).

With little prospect for a reversal of the trend of the sea level to rise, and little hope for a change in the sediment balance, it appears that although the densely vegetated barrier dunes reduce the amount of windblown sand and overwash material, this continued effort to stabilize the beaches and dunes may be detrimental to the long-term equilibrium of the barrier islands themselves. Major amounts of fine sand are being lost by winnowing to the offshore and by increased attrition within the swash zone. Equally important, the National Park Service has recently started using immovable materials to protect some beach areas (Fig. 1c). Sandbags and revetments cause a great deal more swash reflection than even the oversteepened beaches. The loss of sand is a costly price to pay for temporary stabilization.

With the rapid deterioration of the barrier dune systems along the Outer Banks of North Carolina in recent years and the large expenditures of resources necessary to reestablish or maintain them, or both, this research suggests that this is an appropriate time for the National Park Service and the Corps of Army Engineers to review the basic concept of dune construction in light of the geological and ecological implications. This is not to suggest that we should simply return the barrier island to nature and adjust to the whims of the elements, but rather that we should consider very carefully the long-term implications of our present decisions. This is particularly desirable now that the National Park Service has nine large seashore areas within its control (*6*), and several others may be added in the near future.

ROBERT DOLAN

Department of Environmental Sciences, University of Virginia, Charlottesville 22903

References and Notes

1. D. B. Houston, *Science* **172**, 648 (1971).
2. P. J. Godfrey, *Oceanic Overwash and Its Ecological Implications on the Outer Banks of North Carolina* (Office of Natural Science, National Park Service, Washington, D.C., 1970).
3. D. B. Stafford and J. Langfelder, *Photogramm. Eng.* **37**, 565 (1971).
4. R. Dolan, *Prof. Geogr.* **18**, 210 (1966).
5. P. H. Kuenen, *Sedimentology* **3**, 29 (1964).
6. Cape Cod, Fire Island, Assateague Island, Cape Hatteras, Cape Lookout, Cumberland Island, Gulf Islands, and Padre Island.
7. Supported by the Office of Natural Science, National Park Service.

24 August 1971; revised 8 November 1971

Man's impact on the morphology of North Carolina islands is obvious when the cover photo (taken in the early 1930's) of the Cape Hatteras Lighthouse and vicinity is compared with recent photo (below) of the same area. Although the National Park Service has been successful in stabilizing parts of the islands, the process may have serious geologic implications. See page 286. [Cover photo, © National Geographic Society]

Author Citation Index

Subject Index